mk

Collagen

Peter Fratzl

Editor

Collagen

Structure and Mechanics

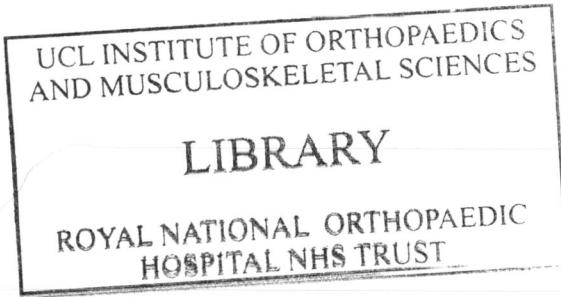

Springer

Editor
Peter Fratzl
Max Planck Institute of Colloids and Interfaces
Department of Biomaterials
14424 Potsdam
Germany

ISBN: 978-0-387-73905-2   e-ISBN: 978-0-387-73906-9

Library of Congress Control Number: 2008921914

Printed on acid-free paper

9 8 7 6 5 4 3 2 1

springer.com

Preface

Collagen type I is the most abundant protein in mammals. It has outstanding mechanical properties and is present in virtually every extracellular tissue with mechanical function. In tendons and ligaments, collagen transmits the force from muscles to bones and stores elastic energy. Smooth walking would not be possible without these properties. Collagen also represents most of the organic matrix of bones and tooth dentin and confers them their fracture resistance. It is a major constituent of skin and blood vessels and is even present in muscles, which could not function without a collagen-rich matrix around the contractile cells. A slightly different type of collagen (type II) is a critical component of a tissue as soft as articular cartilage. The function of collagen is not only mechanical. In the cornea of the eye, for example, the ordering of collagen fibrils confers transparency in addition to mechanical stability.

The versatility of collagen as a building material is mainly due to its complex hierarchical structure. Adaptation is possible at every level, leading to a great variety of properties, to serve a given function. The basic building block of collagen-rich tissues is the collagen fibril, a fiber with 50 to a few hundred nanometer thickness. These fibrils are assembled into composite materials with a variety of more complex structures, which may have anisotropic or nearly isotropic mechanical properties, depending on fiber arrangement. In bone and dentin, collagen is combined with mineral to yield very stiff tissues. In tendon or cornea, collagen is combined with other organic molecules, such as proteoglycans.

Given the size of the collagen fibrils, the important structures are often in the nanometer scale, and recent progress in characterization methods has revealed many details of these structures and of how they relate to the mechanical behavior. Moreover, a large number of molecules have recently been identified as members of the collagen family. This is described in the first part of this book (Chapters 2–4), which is devoted to the structure and biochemistry of collagens. The second part of this book (Chapters 5–8) discusses mechanical properties and the mechanisms at the origin of deformation, fatigue and fracture of collagen-based materials, both from an empirical and from a theoretical viewpoint. The major part of the book (Chapters 9–17) addresses a particular collagen-based tissue per chapter, including tendons and ligaments, artery walls, cornea, bone and dentin, among others. The last two chapters focus on the more special issues of genetic collagen diseases and collagen-based tissue engineering. Chapter 1, finally, is meant as an introduction to the subject in general and to the other chapters in the book. It also gives elementary definitions of mechanical quantities needed throughout.

In this way, this text approaches collagen-based tissues from very different perspectives, highlighting structure and biochemistry, general principles of mechanical behavior, as well as structure, composition and mechanical behavior of a number of important tissues in our body. We hope that by this interdisciplinary approach, the book will be useful as an introduction and as a reference for advanced students, researchers and engineers in very diverse fields, such as materials science and engineering, collagen biochemistry and biophysics, as well as tissue engineering and regenerative medicine. This diversity is also reflected in the different backgrounds of the authors, who are all well-recognized specialists in the fields which they are covering. In addition to providing the best introduction and up-to-date reference, we really hope to transfer to the reader some of our excitement about the beauty of structure and mechanics of collagen-based tissues.

Contents

Contributors

N.C. Avery
Veterinary sciences, Bristol University, Bristol, UK, Nick.Avery@bristol.ac.uk

A.J. Bailey
Collagen Research Group, Division of Molecular and Cellular Biology, University of Bristol, Bristol, UK, a.j.bailey@bristol.ac.uk

V.H. Barocas
Department of Biomedical Engineering, University of Minnesota, Minneapolis, MN, USA, baroc001@umn.edu

A.A. Biewener
Department of Organismic and Evolutionary Biology, Harvard University, Cambridge, MA, USA, abiewener@oeb.harvard.edu

R.D. Blank
Section of Endocrinology, Department of Medicine University of Wisconsin School of Medicine and Public Health and Geriatrics Research, Education and Clinical Center William S. Middleton VAMC, Madison, Wisconsin, New York, NY, USA, blankr@hss.cdu; rdb@medicine.wisc.edu

A.L. Boskcy
Department of Biochemistry and PBSB Weill Medical College, New York, NY, USA, BoskeyA@HSS.EDU

M.J. Buehler
Civil and Environmental Engineering, MIT, Cambridge, MA, USA, mbuehler@MIT.EDU

J. Currey
Department of Biology, University of York, York, UK, jdc1@york.ac.uk

P. Fratzl
Max Planck Institute of Colloids and Interfaces, Potsdam, Germany, fratzl@mpikg.mpg.de

H.S. Gupta
Max Planck Institute of Colloids and Interfaces, Potsdam, Germany,
Himadri.gupta@mpikg.mpg.de

G.A. Holzapfel
Royal Institute of Technology, School of Engineering Sciences, Stockholm,
Sweden, gh@hallf.kth.se, gh@biomech.tu-graz.ac.at

D.J.S. Hulmes
CNRS Lyon, Lyon, France, d.hulmes@ibcp.fr

R.F. Ker
Institute of Integrative and Comparative Biology, The University of Leeds, UK,
r.f.ker@leeds.ac.uk

M. Kjær
Bispebjerg Hospital, Institute of Sports Medicine, Copenhagen, Denmark,
m.kjaer@mfi.ku.dk

W.J. Landis
Northeastern Ohio College of Medicine, Rootstown, OH, USA, wjl@neoucom.edu

S.P. Magnusson
Bispebjerg Hospital, Institute of Sports Medicine, Copenhagen, Denmark,
p.magnusson@mfi.ku.dk

K.M. Meek
University of Cardiff, Cardiff, UK, MeekKM@cardiff.ac.uk

P.P. Purslow
Department of Food Science, University of Guelph, Ontario, Canada,
ppurslow@uoguelph.ca

E.A. Sander
Department of Biomedical Engineering, University of Minnesota, Minnesota, MN,
USA

F.H. Silver
Rutgers University, NJ, USA, silverfr@umdnj.edu, fhsilver@hotmail.com

T.J. Wess
University of Cardiff, Cardiff, UK, WessTJ@cardiff.ac.uk

P. Zaslansky
Max Planck Institute of Colloids and Interfaces, Potsdam, Germany,
paul.zaslansky@mpikg.mpg.de

Chapter 1
Collagen: Structure and Mechanics, an Introduction

P. Fratzl

Abstract Collagen type I is the most abundant protein in mammals. It confers mechanical stability, strength and toughness to a range of tissues from tendons and ligaments, to skin, cornea, bone and dentin. These tissues have quite different mechanical requirements, some need to be elastic or to store mechanical energy and others need to be stiff and tough. This shows the versatility of collagen as a building material. While in some cases (bone and dentin) the stiffness is increased by the inclusion of mineral, the mechanical properties are, in general, adapted by a modification of the hierarchical structure rather than by a different chemical composition. The basic building block of collagen-rich tissues is the collagen fibril, a fiber with 50 to a few hundred nanometer thickness. These fibrils are then assembled to a variety of more complex structures with very different mechanical properties. As a general introduction to the book, the hierarchical structure and the mechanical properties of some collagen-rich tissues are briefly discussed. In addition, this chapter gives elementary definitions of some basic mechanical quantities needed throughout the book, such as stress, strain, stiffness, strength and toughness.

1.1 Collagen-Based Tissues

Collagen is among the most abundant fibrous proteins and fulfills a variety of mechanical functions, particularly in mammals. It constitutes the major part of tendons and ligaments, most of the organic matrix in bone and dentin; it is present in skin, arteries, cartilage and in most of the extracellular matrix in general. Collagen is also used by invertebrates, for example in the byssus threads, by which mussels are attached to rocks (Waite et al. 2003).

More generally, fibrous polymers are the major building blocks in all types of load-bearing tissues from unicellular organisms in water to plants and animals (Jeronimidis 2000). These fibers include polysaccharides, such as cellulose and chitin, extremely abundant in plants and in insect cuticles, respectively, and a variety of proteins. In addition to collagen, structural proteins include keratin, predominant in hair and nails, silk as used by spiders, for example, and actin, present in muscle and the cytoskeleton of every cell. Many more fibrous proteins with more

P. Fratzl (ed.), *Collagen: Structure and Mechanics*,

specialized functions are used by different organisms. A general introduction to this theme can be found in Elices (2000) and Shewry et al. (2003).

From a mechanical viewpoint, fibrous tissues are very special (Vincent 1990, Jeronimidis 2000). Such materials are usually much stronger in fiber direction than perpendicular to it. As a consequence, properties are anisotropic, which can be reduced by special construction principles, such as plywood assemblies to form laminates (Weiner and Wagner 1998). In fact, a large variety of overall properties can easily be achieved by a clever assembly of fibers. We know from our daily life experience how, based on fibers, it is possible to make strong ropes for uniaxial loading or tissues with high or low elasticity, depending on the weaving. In a very similar way, tissues and whole organs are constructed with fibers by hierarchical assembly leading to a large variety of mechanical properties (Fratzl 2003). Understanding the hierarchical structure of biological materials is, therefore, a key to the understanding of their mechanical properties (Tirrell 1994, Fratzl and Weinkamer 2007). Collagen is no exception to this and we find it in very complex hierarchical structures with quite different properties, such as elastic skin, soft cartilage, and stiff bone and tendon.

Hierarchical structuring has the advantage to allow for optimization and mechanical adaptation at every structural level (Fratzl and Weinkamer 2007). For most biological materials (Wainwright 1982), including plants (Niklas 1992, Mattheck 1998), the internal architecture determines the mechanical behavior more than the chemical composition does. To illustrate this fact, let us take two extremely simple examples. The first is a composite of stiff fibers in a soft matrix and the second is a honeycomb structure. Using the Voigt and the Reuss model for a fiber composite as sketched in Fig. 1.1a (Hull and Clyne 1996), where the volume fraction of the matrix phase is called Φ, the elastic young modulus in axial and lateral directions are given as a function of the moduli E_m and E_f of matrix and fibers, respectively, by

$$E_A = \Phi\,E_m + (1 - \Phi)\,E_f, \text{ and } E_L = \left(\Phi\,E_m^{-1} + (1 - \Phi)\,E_f^{-1}\right)^{-1} \qquad (1.1)$$

If the Young's moduli (for a definition see below) of matrix and fiber are very different, these two expressions predict an extremely different mechanical behavior of the

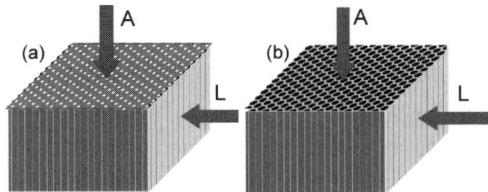

Fig. 1.1 Model for a fiber-reinforced composite (**a**) and a honeycomb structure (**b**). The matrix phase is shown in grey. The fibers in (**a**) are indicated by white circles and the holes in the honeycomb (**b**) by black hexagons. The A and L directions correspond to loading in axial and lateral directions, respectively

composite loaded in axial and lateral directions. Similarly, considering a honeycomb structure as sketched in Fig. 1.1b, the elastic moduli in axial and lateral directions, respectively, can be estimated to be (Gibson and Ashby 1999)

$$E_A/E_m \propto \Phi, \text{ and } E_L/E_m \propto \Phi^3 \qquad (1.2)$$

Again, with thin walls in the honeycomb structure (that is, $\Phi << 1$), the difference between lateral and axial mechanical properties can be orders of magnitude. These two examples show that by simple structuring, mechanical properties can vary enormously, even though the chemical composition is the same. In particular, the local fiber orientation plays a major role for adapting the mechanical properties of most biological materials.

Taking the example of bone (Cowin 2001, Currey 2002), structural optimization means that every bone in our body will – according to its mechanical function – have a slightly different arrangement of the basic building blocks, the mineralized collagen fibrils. The consequence is that bending stiffness, fracture resistance and other mechanical properties will differ from site to site as a consequence of the different local architecture. Hence, it is very difficult to talk about the (mechanical) properties of a tissue, such as bone or tendon, in a general way, since these properties depend on local architecture and on the actual mechanical needs dictated by the environment in the living body. This is the reason why (hierarchical) structure and mechanical properties are addressed in this book simultaneously. Indeed, they cannot be considered separately.

This book focuses on mammalian collagen type I which is the major organic constituent of tendon, bone and dentin, and plays a crucial role in blood vessels, skin and extracellular matrix, in general. The first part (Chapters 2–4) concentrates on general aspects with reviews of the various collagen types and assemblies (Chapter 2), the structure of the collagen fibril, the basic building block of all tissues based on collagen I (Chapter 3), and the chemistry and biology of cross-links between collagen molecules, which are of particular importance for the mechanical behavior (Chapter 4). General aspects of the mechanical behavior of collagen-based tissues are reviewed in the second part of this book. This includes damage and fatigue (Chapter 5), viscoelasticity and energy storage in extracellular matrices, including tendon, skin and cartilage (Chapter 6), and nanoscale deformation mechanisms, both from an experimental (Chapter 7) and from a theoretical materials mechanics viewpoint (Chapter 8). This is complemented by a review of the biological repair mechanisms in extracellular matrix tissues (Chapter 9). A number of special collagen-rich tissues are finally addressed in the third part of this book, including tendon and ligament (Chapter 10), artery walls (Chapter 11), the extracellular matrix of skeletal and cardiac muscle (Chapter 12), cornea and sclera (Chapter 13), bone and calcified cartilage (Chapter 14) and dentin (Chapter 15). The last two chapters address genetic collagen diseases (Chapter 16) and biomimetic collagen tissues, particularly in the context of tissue engineering (Chapter 17).

This book considers a wide selection of collagen-rich tissues in mammals but avoids a particular focus on those tissues, which have been most widely studied in

the context of structure and mechanical properties because of their medical importance, namely bone (see. e.g., Cowin 2001, Currey 2002, Weiner and Wagner 1998,
Fratzl et al. 2004) and cartilage (see, e.g., Bader and Lee 2000), as there are several
relevant text books and review articles.

Chapter 1 introduces some basic mechanical parameters in an elementary way,
which will be needed throughout this book. It also gives general comments on the
hierarchical structure and the mechanical properties of collagen-rich tissues, pointing out where more details can be found in this book.

1.2 Basic Mechanical Parameters

1.2.1 Stress and Strain

When an elementary volume is subjected to tensile forces, for example, the material
will typically elongate in the direction of the applied forces. The relative elongation
is called strain, often denoted ε (see Fig. 1.2). Usually, this elongation leads to a
contraction of the material in the direction perpendicular to the applied stress, by a
relative amount $\nu\varepsilon$, where the coefficient ν is called Poisson ratio. For an isotropic
piece of material, the relative increase of the volume during uniaxial stretching is
$1-2\nu$ which means that the Poisson ratio has an upper bound of $\nu \leq 1/2$, because
the specimen volume is not expected to shrink under the influence of tensile forces.
A typical value for the Poisson ratio is $\nu = 1/3$. The load (force in N) divided by
the surface area A ($A = L^2$, for the sketch in Fig. 1.1) is called stress (units Pascal).

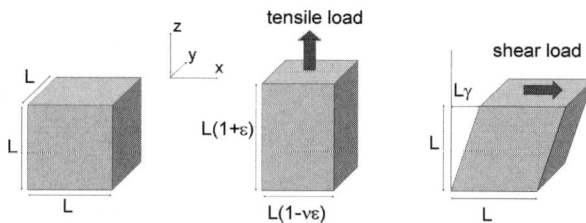

Fig. 1.2 When a cubic piece of material is subjected to tensile load along the vertical (z-direction)
only, its length L is increased by $L\varepsilon$. The relative elongation ε is called (tensile) strain. In most
cases, the dimensions of the cube subjected to tensile load will contract perpendicularly to the
load direction. The ratio ν of the contraction in the x-direction (or in the y-direction) relative
to the elongation in z-direction is called the Poisson ratio. When the load is tangential to the top
surface, (*right picture*), shear deformation occurs. The shear is measured by the parameter γ, which
(for small deformations) corresponds to the tilting angle of the cube edge initially parallel to the
z-direction

Generally, stresses and strains are not just uniaxial and need to be described
by a tensor. More general definitions can be found in text books on mechanics of
materials, see for example, Hull and Clyne (1996).

1.2.2 Elastic and Viscoelastic Behavior

A first inspection of the mechanical behavior of materials is possible by measuring the relation between stress σ and strain ε. Figure 1.3 shows some examples for different types of mechanical behavior, when the material is subjected to a tensile stress σ. Typically the stress increases first linearly, when the strain increases. The ratio between stress and strain (that is, the slope of the $\sigma-\varepsilon$ curve at small ε, see Fig. 1.3) is called Young's modulus E, and $\sigma = E\,\varepsilon$. For larger strains, this linearity is not necessarily conserved. When the same $\sigma-\varepsilon$ curve is followed during a release of the stress and during build up, the mechanical behavior is called elastic. A linear elastic material is the special case where the $\sigma-\varepsilon$ curve is linear. In many cases, materials show a linear elastic behavior at small deformations, but do not return to their original shape when the stress has exceeded a critical value (often called yield stress). Such a material behavior is called plastic and the permanent elongation is the associated plastic deformation. Bone, for example, shows some plastic behavior (see Chapter 14 and Fig. 14.6).

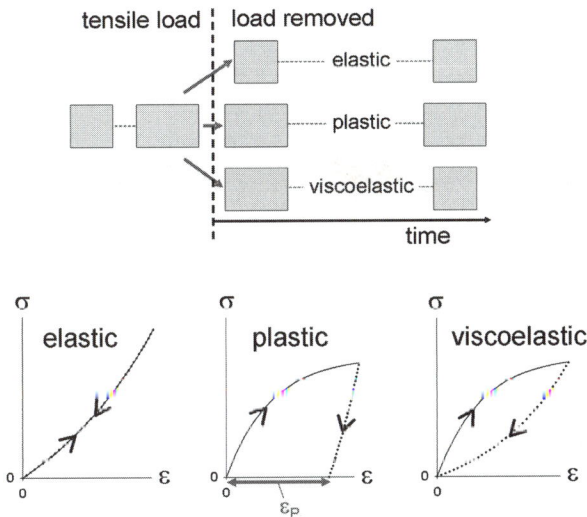

Fig. 1.3 Materials behave differently when the load is removed after deformation (*left*). An elastic material returns immediately to its original shape, a plastic material keeps the deformed shape forever and a viscoelastic material returns slowly to its original shape. Materials often have a combination of those properties: An elastoplastic material relaxes partially and retains only part of the deformation. Similarly, a viscoplastic material loses gradually a part of the deformation but a fraction of it stays forever. In a stress–strain experiment (*right*), the stress σ is measured as a function of strain ε. Full lines in the $\sigma-\varepsilon$ curves show increasing stress and dotted lines decreasing stress. For elastic materials, these two lines coincide. If this line is non-linear (as occurs for many polymeric materials), the elasticity is called non-linear. Close to the origin (near zero stress), nearly all materials are linearly elastic. For plastic materials, a permanent strain ε_P remains after sufficiently large deformation. For viscoelastic materials, there is a hysteresis. The elastic modulus E is defined as the slope of the full line close to the origin

Many biological tissues, including collagen, are viscoelastic. One possible reason is that a viscous fluid (often just water) is flowing during deformation. The simplest model for such behavior is the so-called Kelvin model, where a dashpot (providing fluid friction) is supposed to act in parallel to an elastic spring. The equation of deformation then depends both on the strain ε and its time derivative $\dot{\varepsilon}$, and reads (Puxkandl et al. 2002)

$$\sigma = E\,\varepsilon + \eta\,\dot{\varepsilon} \qquad (1.3)$$

As a consequence, the mechanical behavior becomes time dependent. This is illustrated by a simple example below. A more complete discussion of viscoelastic behavior in biological materials can be found, e.g., in (Vincent 1990).

Now we consider the simple example where the externally applied stress σ is first increased to σ_0 at constant speed during time t_0 and then decreased at the same speed (see Fig. 1.4). To solve Eq. (1.3) in this case, we rewrite it in non-dimensional units

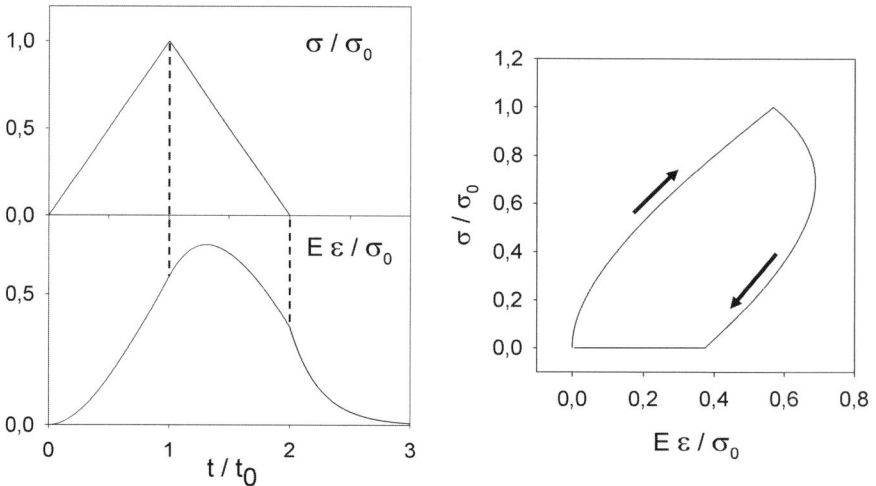

Fig. 1.4 Solution of Eq. (1.4) for $\lambda = 1/2$ and an imposed time variation of the stress as shown on the top left. The applied stress first increases and then decreases linearly, before staying completely at zero. Due to the viscoelastic behavior of the material, the strain is not increasing as fast as the stress, since some of the stress is needed to compensate the forces generated by viscosity rather than by strain in the material. Note that the strain continues to increase even after the stress has reached its peak value (*left broken line*), as a consequence of the relaxation of the frictional forces inside the material. Only after the stress has dropped further, the strain decreases but does not reach zero when all the stress is removed (*second broken line*). It then takes a while before all the strains are removed in a creep process. The consequence for the stress–strain curve (*right*) is that the curves corresponding to increasing and to decreasing stress do not overlap. The area between the two curves corresponds to the energy per unit volume dissipated by viscous friction during the mechanical cycle

$$y = x + \lambda \frac{dx}{d\tau}, \quad \text{where} \quad \begin{cases} y = \sigma/\sigma_0 \\ x = E\,\varepsilon/\sigma_0 \\ \tau = t/t_0 \\ \lambda = \eta/(E\,t_0) \end{cases} \qquad (1.4)$$

Fig. 1.4 shows the solution of this equation for $\lambda = 1/2$ and an imposed time variation of the stress. The stress–strain curve becomes fairly complex and dependent on the strain rate. In particular, the σ–ε curves are not the same during increase and decrease of the applied stress. Moreover, energy is being dissipated by viscous friction in the material. This may be important for damping or dissipating the energy in an impact (see Chapter 5). It may also cause energy loss in a cyclic movement (see Chapter 6).

1.2.3 Stiffness, Strength and Toughness

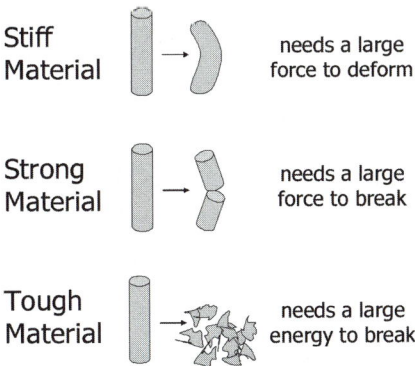

Young's modulus E, introduced in the previous section, describes the stiffness of a material, that is, its resistance against deformation when subjected to a given stress. Stiff materials are needed to transmit forces (for example, in tendons and ligaments) and to resist deformation. Stiffness is especially crucial for transmitting forces when the material is loaded in bending. This may be one of the reasons why bones are reinforced with extremely stiff mineral particles. Stiffness is also needed for resistance against buckling when a bar is loaded in compression along its axis.

But stiffness is by no means the only critical mechanical property (see Fig. 1.5). The strength of a material is defined as the maximum stress it can sustain before breaking. The strength is often denoted by σ_f. High strength is needed to allow

Stiffness, strength and toughness

Stiff Material → needs a large force to deform

Strong Material → needs a large force to break

Tough Material → needs a large energy to break

weak link problem

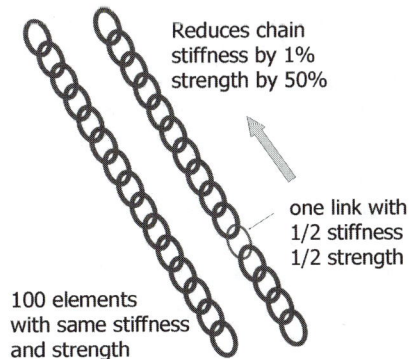

Reduces chain stiffness by 1% strength by 50%

one link with 1/2 stiffness 1/2 strength

100 elements with same stiffness and strength

Fig. 1.5 Some mechanical characteristics of a material, and the weak link problem in a chain

carrying high loads. Tendons, ligaments or bones need to be strong. Technical materials used for construction are usually both stiff and strong. There is, however, a subtle but essential difference in how these properties are affected by defects in a homogeneous material. This may be understood by the "weak link" problem (Fig. 1.5). A simple calculation shows that if one of 100 identical elements in a chain is replaced by a link with half the stiffness and half the strength, the overall stiffness of the chain is reduced just by 1% but the overall strength is only half. This means that the stiffness (as a bulk property) is hardly influenced by small defects while the strength depends heavily on local properties and on defects. As a consequence, the strength of ceramics is almost completely controlled by the size and the amount of defects in the material, to an extent, where the strength becomes a statistical property of the ceramic (depending on the defect distribution) rather than an intrinsic one (Lawn 1993).

This dependence on defects and material inhomogeneities is even worse for yet another crucial material property, the toughness (see Fig. 1.5). The toughness is linked to the energy needed to propagate a crack through the material and to break it. The larger the energy needed, the tougher the material is. Brittle ceramics, for example, have a very low toughness, but they are typically very stiff. Indeed, a major way for a material to dissipate energy in an impact is to deform rather than to fracture. Therefore, many materials which deform easily are also tough, while very stiff materials (such as ceramics) have a higher chance to be brittle. This is a major dilemma in materials design, since both stiffness and toughness are needed for many applications. It is quite interesting to observe (see below) that collagen-based tissues, such as tendon or bone, represent an excellent compromise between stiffness, strength and toughness.

A first estimate for the toughness is given by the area under the stress–strain curve when the material is loaded until failure. This area corresponds to the energy per unit volume needed to break the material, measured in J/m^3. A better definition is given in terms of the energy per unit of crack advance area, needed to elongate a pre-existing crack: This energy is often called work of fracture or energy release rate, G, and measured in J/m^2. The quantity $K = \sqrt{E\,G}$ is usually referred to as fracture toughness (see, e.g., Hull and Clyne 1996). Since, Young's modulus, E, is given in Pa, the fracture toughness is measured in somewhat unusual units of $Pa\sqrt{m}$. The difference between these two estimates is that the fracture toughness describes the energy dissipation by an existing crack which moves, while the area under the stress–strain curve includes the initiation/nucleation of a macroscopic crack. Details of fracture mechanics in the context of collagen-rich tissues can be found in Chapter 5. Usually, the toughness of a material is determined by the ability of its microstructure to dissipate deformation energy without propagation of a critical crack. Polymers are often able to dissipate energy by viscoplastic flow or by the formation of non-connected microcracks, and collagen-rich tissues are no exception (see Chapters 5, 7, 14 for tendon and bone). Well-known toughening mechanisms in ceramics are crack ligament bridging and crack deflection. Most interestingly, all these phenomena are typically identified in tough collagen-rich tissues, such as bone (Peterlik et al. 2006).

1.3 Mechanical Properties of Collagen-Based Tissues

To illustrate some of the mechanical properties introduced above, Fig. 1.6 shows typical stress–strain curves for a tendon from the mouse tail (Misof et al. 1997) and for parallel-fibered bone obtained from a certain region in the cortex of bovine femur (Gupta et al. 2006). Both tissues have in common that they were loaded in tension in the direction of the collagen fibrils. While this is the usual loading case for tendons, both the loading and the structure can be much more complex in bone. Indeed, bone is often loaded in compression and in bending rather than just in tension. Moreover, the bone structure consists very rarely of parallel collagen fibrils. In most cases (see Fig. 1.8), the structure is much more complex with a plywood-like arrangement of fibrils (Weiner and Wagner 1998, Giraud-Guille 1988).

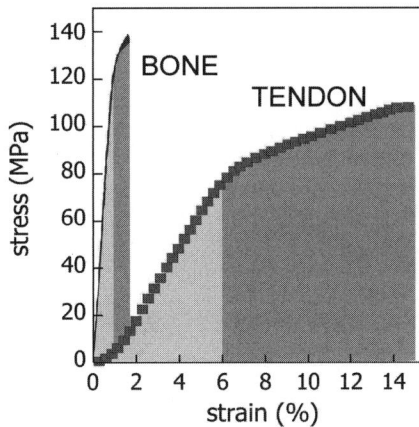

Fig. 1.6 Typical stress–strain curves for mouse tail tendon (Misof et al. 1997) and for parallel-fibered bone (Gupta et al. 2006). The area shaded in darker gray corresponds to the energy per unit volume dissipated in the post-yield deformation. The area shaded in lighter gray indicates the elastic energy per unit volume stored during deformation

The difference in the mechanical behavior between tendon and bone shown in Fig. 1.6 is, therefore, not due to different fiber architectures but due to the fact that bone is mineralized. The inclusion of tiny calcium-phosphate mineral particles leads to a considerable stiffening of the tissue, from roughly 1 GPa for tendon (Chapter 10) to about 20 GPa for bone (Chapter 14), as is visible by the change in the initial slope of the curve in Fig. 1.6. The price to pay for this higher stiffness is a smaller area under the stress–strain curve, which indicates a lower toughness of bone compared to tendon. Generally, the work of fracture decreases with the mineral content (Chapter 14). The light shaded area shows the elastic region of the stress–strain curve and corresponds to the elastic energy per unit volume which may be stored in the tissue during a loading cycle (see Chapter 6). Note that the stress–strain curve of tendon is not quite linear in the elastic region at small strains. The origin of this behavior and the deformation mechanisms of bone and tendon at the fibril and sub-fibril levels are discussed in detail in Chapter 7. Due to the complex hierarchical structure of collagen, its description in terms of theoretical models has just started. The state of the art in the hierarchical modeling of collagen is reviewed in Chapter 8. Other types of models for skin and artery walls are discussed in Chapter 11.

The dark shaded areas in Fig. 1.6 highlight the regions of post-yield deformation where damage and plastic deformation occurs in the tissue. These processes are irreversible but contribute to the dissipation of mechanical energy and, therefore, to the toughness of these tissues. Damage, fracture and fatigue of collagen-rich tissues are reviewed in Chapter 5. Most interestingly, there is a biological response to mechanical stimuli which leads to the adaptation of biological tissues as well as to damage repair. This involves mechanosensitivity of the cells in the tissue and a process called remodeling, whereby damaged tissue is removed and replaced by new tissue, according to mechanical needs (Cowin 2001, Fratzl and Weinkamer 2007, for a discussion of remodeling in bone). Chapter 9 reviews tissue adaptation and remodeling in extracellular matrix and, in particular, tendon.

1.4 Hierarchical Structure of Collagen-Based Tissues

Collagen-rich tissues are typically based on the collagen fibril as an elementary building block. These fibrils have a thickness in the range from 50 to a few hundred nanometers and are assembled in a complex hierarchical way into macroscopic structures. The hierarchical structure of tendon is shown as an example in Fig. 1.7. Several levels of fiber-like structures can be seen in this figure. At the lowest level, there are the collagen molecules, which are triple-helical protein chains with a length of about 300 nm. Details on the synthesis and the assembly of such collagen type I molecules, as well as many others of the collagen family are given in Chapter 2. The collagen molecules are further assembled by a parallel staggering into fibrils. As indicated in Fig. 1.7, neighboring

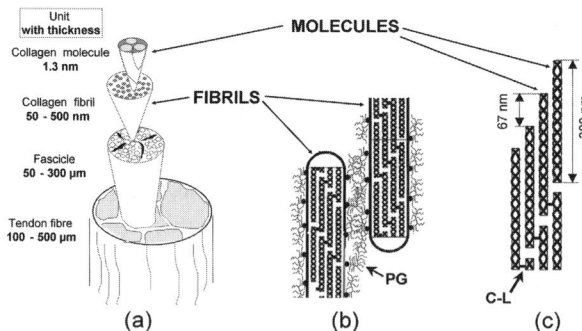

Fig. 1.7 Sketch of the hierarchical structure of tendon (adapted from Fratzl and Weinkamer 2007). (**a**) shows some of the units, fiber, fascicle, fibril and molecule. (**b**) and (**c**) show details of the interaction between fibrils and molecules, respectively. The proteoglycan-rich matrix between fibrils is indicated by PG and cross-links between collagen molecules by C-L

molecules are shifted by about 67 nm inside the fibrils. The three-dimensional structure of collagen fibrils are reviewed in detail in Chapter 3. One aspect, particularly important for the mechanical properties of fibrils, corresponds to covalent cross-linking between molecules which develops only with maturation of the tissues. All aspects related to cross-linking are reviewed in Chapter 4. Once collagen molecules have been synthesized, they assemble spontaneously outside the cell (see Chapters 2 and 3). This can be used for the biomimetic assembly of collagen tissues for tissue engineering and other applications (see Chapter 17).

Moreover, the fibrils are joined by a matrix rich in proteoglycans which also contributes significantly to the mechanical behavior of tendons (Chapter 7). There is a wide range of genetic diseases which affect collagen molecules and, as a consequence, their ability to assemble into the right type of hierarchical structures. The most notable example is osteogenesis imperfecta, the brittle bone disease. Such diseases and their consequences for structure and mechanical properties of collagen-based tissues are reviewed in Chapter 16.

The mechanical functions of tendons and ligaments are discussed in Chapter 10. While tendons and ligaments are mostly needed to transfer forces from the muscles to the bones and to store elastic energy for better locomotion (see Chapter 6), other collagen-rich organic tissues have to fulfill quite different functions. In the case of the cornea, for example, the tissue not only confers mechanical stability to the eye bulb, but also needs to be transparent, which puts certain structural constraints on the tissue. The knowledge in this field is reviewed in Chapter 13. Skin and artery walls need to be flexible but arteries also need to resist blood pressure. This implies yet a quite different arrangement of collagen fibrils in combination with elastomeric molecules, such as elastin, for example. Chapter 11 reviews the structure of artery walls and discusses biomechanical models of this tissue. Even muscles could not function without a scaffold rich in collagen in which the muscle cells are embedded. Chapter 12 reviews the structure and function of extracellular matrix in muscle tissue.

Finally, in bone and dentin, the organic matrix alone is not sufficient to provide the stiffness required for these tissues which have to carry considerable (compressive) loads. The stiffening is provided by the inclusion of mineral particles into the collagen-rich tissue. The mineral phase occupies about half the volume in compact bone. Figure 1.8 shows the example of cortical (compact) bone from a human femur. The basic building block is the mineralized collagen fibril (Fig. 1.8d,e) with a structure very similar to what is found in tendon (Fig. 1.7) but containing plate-shaped mineral particles with 2–4 nm thickness made of carbonated hydroxyapatite. These fibrils are then assembled in a plywood-like fashion to lamellar bone (Weiner and Wagner 1998). Figure 1.8c shows an example of lamellar bone surrounding a blood vessel. The unit comprising the central canal in an osteon and most of the compact bone in a human femur is formed by osteons (Fig. 1.8b). The mechanical properties of bone and dentin depend to a large extent on the degree of mineralization and on the fiber architecture. Chapters 14 and 15 discuss these relations for bone and for dentin, respectively.

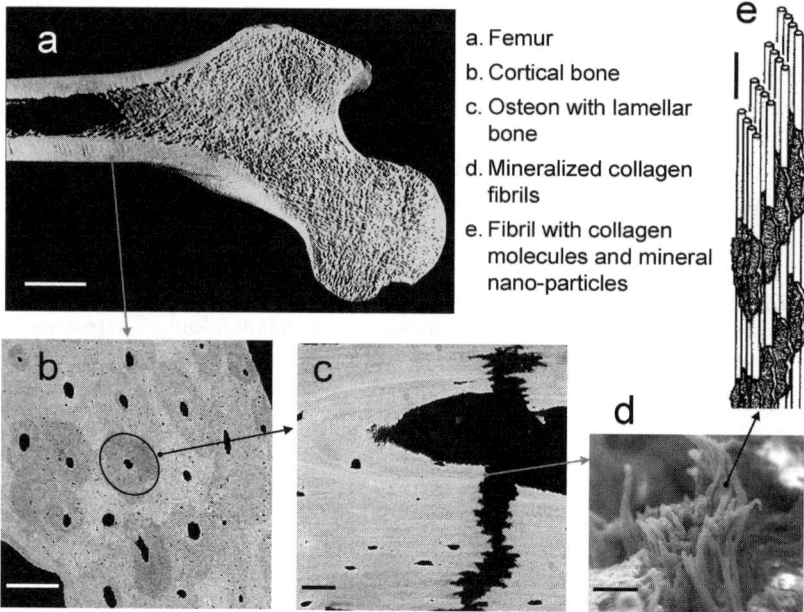

a. Femur

b. Cortical bone

c. Osteon with lamellar bone

d. Mineralized collagen fibrils

e. Fibril with collagen molecules and mineral nano-particles

Fig. 1.8 Hierarchical structure of cortical bone (adapted from Fratzl et al. 2004 and from Peterlik et al. 2006). The length of the bar is 20 mm, 200 μm, 20 μm, 2 μm and 100 nm in (**a**)–(**e**), respectively. The ellipse in (**b**) indicates an osteon of the type shown in (**c**). In addition, (**c**) shows a crack propagating through lamellar bone in an osteon during a fracture experiment (Peterlik et al. 2006)

References

Bader D, Lee D (2000) Structure-properties of soft tissues. Articular Cartilage. Chapter 4, pp. 73–104, in: M. Elices, Ed. Structural Biological Materials: Design and Structure Property Relationships. Pergamon: Amsterdam.

Cowin SC (2001) Bone Mechanics Handbook. CRC Press: Boca Raton.

Currey JD (2002) Bones – Structure and Mechanics. Princeton University Press: Princeton.

Elices M, Ed. (2000) Structural Biological Materials, Design and Structure-Property Relationships. Pergamon: Amsterdam.

Fratzl P (2003) Cellulose and collagen: from fibres to tissues. Curr Opin Coll Interf Sci 8:32–39.

Fratzl P, Gupta HS, Paschalis EP, Roschger P (2004) Structure and mechanical quality of the collagen-mineral nano-composite in bone. J Mater Chem 14:2115–2123.

Fratzl P, Weinkamer R (2007) Nature's hierarchical materials. Prog Mater Sci 52:1263–1334.

Gibson LJ, Ashby MF (1999) Cellular Solids, Structure and Properties. 2nd. edition, Cambridge University Press: Cambridge.

Giraud-Guille MM (1988) Twisted Plywood Architecture of Collagen Fibrils in Human Compact-Bone Osteons. Calcif Tissue Inf 42:167–180.

Gupta HS, Seto J, Wagermaier W, Zaslansky P, Boesecke P, Fratzl P (2006) Cooperative deformation of mineral and collagen in bone at the nanoscale. PNAS 103:17741–17746.

Hull D, Clyne TW (1996) An Introduction to Composite Materials, 2nd edition, Cambridge University Press: Cambridge.

Jeronimidis G (2000) Chapters 1 and 2, pp. 3–29. In: M. Elices, Ed. Structural Biological Materials: Design and Structure Property Relationships. Pergamon: Amsterdam.

Lawn BR (1993) Fracture of Brittle Solids, 2nd edition. Cambridge Solid State Science Series, Cambridge University Press: Cambridge.

Mattheck C (1998) Design in Nature: Learning from Trees. Springer-Verlag: Berlin; New York.

Misof K, Landis WJ, Klaushofer K, Fratzl P (1997) Collagen from the osteogenesis imperfecta mouse model shows reduced resistance against tensile stress. J Clin Invest 100:40–45.

Niklas KJ (1992) Plant biomechanics: an Engineering Approach to Plant form and Function. University of Chicago Press: Chicago.

Peterlik H, Roschger P, Klaushofer K, Fratzl P (2006) From brittle to ductile fracture of bone. Nat Mater 5:52–55.

Puxkandl R, Zizak I, Paris O, Keckes J, Tesch W, Bernstorff S, Purslow P, Fratzl P (2002) Viscoelastic properties of collagen: sunchrofron radiation investigations and structural model. Phil Trans Roy Soc London B 357:191–197.

Tirrell DA, Ed. (1994) Hierarchical Structures in Biology as a Guide for new Materials Technology. National Academy Press: Washington.

Shewry PR, Tatham AS, Bailey AJ, Eds. (2003) Elastomeric Proteins, Structures, Biomechanical Properties, and Biological Roles. Cambridge University Press: Cambridge.

Vincent JFV (1990) Structural Biomaterials, Princeton University Press: Princeton.

Wainwright SA, Biggs WD, Currey JD, Gosline JM (1982) Mechanical Design in Organisms. Princeton University Press: Princeton.

Waite JH, Vaccaro E, Sun C, Lucas J (2003) Chapter 10, pp. 189–212, in Shewry et al. 2003.

Weiner S, Wagner HD (1998) The material bone: Structure mechanical function relations. Ann Rev Mater Sci 28:271–298.

Chapter 2
Collagen Diversity, Synthesis and Assembly

D.J.S. Hulmes

Abstract The vertebrate collagen superfamily now includes over 50 collagens and collagen-like proteins. Here, their different structures are described, as well as their diverse forms of supramolecular assembly. Also presented here are the various steps in collagen biosynthesis, both intracellular and extracellular, and the functions of the collagen-specific post-translational modifications. Assembly of collagen fibrils, both in vitro and in vivo, is reviewed, including the mechanisms that control this process and the interactions involved. Finally, recent developments in the supramolecular assembly of collagen-like peptides are discussed.

2.1 Introduction

Collagens come in all shapes and sizes. The hallmark of a collagen is a molecule that is composed of three polypeptide chains, each of which contains one or more regions characterized by the repeating amino acid motif (Gly-X-Y), where X and Y can be any amino acid. This motif allows the chains to form a right-handed triple-helical structure (Fig. 2.1), with all glycine residues buried within the core of the protein, and residues X and Y exposed on the surface. Depending on the genetic type of collagen, this triple helical motif can be a major or minor part of the molecule, other regions consisting of different non-collagenous domains. While the presence of a triple-helical motif is a necessary condition for being called a collagen, it is not a sufficient one. Some proteins contain these motifs but are not called collagens, mainly because they were named on the basis of their specific biological functions, rather than the structural role traditionally ascribed to collagens. But here also the distinction has become blurred, with new biological functions being discovered for both collagenous and non-collagenous domains. The question of deciding what is and what is not a collagen is best avoided; here I refer simply to the collagen superfamily, being all proteins containing a collagen-like triple-helical motif.

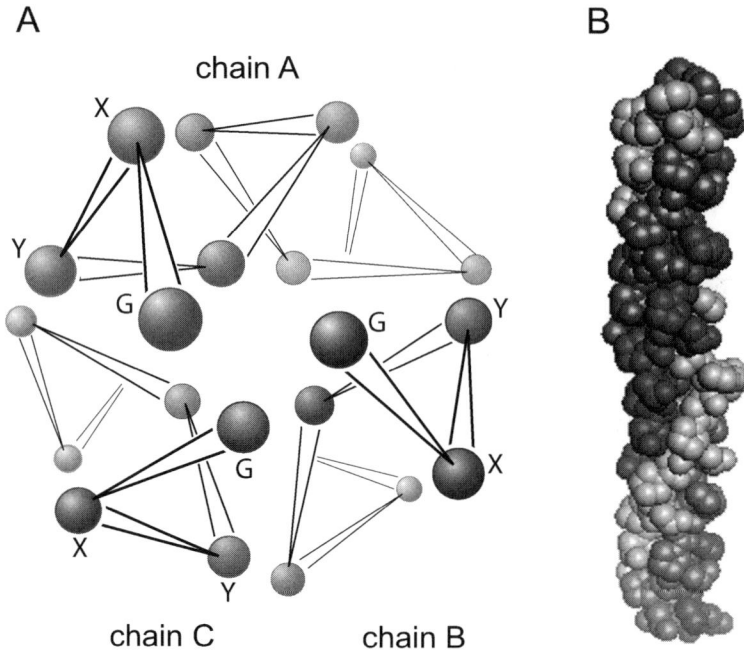

Fig. 2.1 Triple-helical structure of collagen. **A** Viewed along the molecular axis (α-carbons only), showing the paths of the individual polypeptide chains and the locations of residues in the Gly-X-Y triplets (G = Gly; from (Beck and Brodsky, 1998), with permission). **B** Viewed from the side (molecule tilted toward the reader at the top; space-filling representation), showing the right-handed helical twist

2.2 Fibrillar Collagens

In humans, there are currently 28 different proteins known as collagens, as well as about the same number of other members of the collagen superfamily (Kadler et al., 2007; Myllyharju and Kivirikko, 2004; Ricard-Blum et al., 2005). They can be grouped into a number of subfamilies (Table 2.1). From the biomechanical point of view, of most interest are the fibrillar collagens (Hulmes, 2002; Kadler, 1995). These proteins give rise to classical collagen fibrils, as seen by electron microscopy (Fig. 2.2), characterized by a repeating banding pattern with a so-called D period-icity of 64–67 nm, depending on the tissue. Within the fibril, collagen molecules of length 300 nm and width 1.5 nm are staggered with respect to their neighbors by multiples of D. The three-dimensional structure of collagen fibrils is discussed in greater detail in Chapter 3. The classical fibrillar collagens include: type I, the most widely occurring collagen found in skin, tendon, bone, cornea, lung and the vascu-lature; type II, which has a more specific tissue distribution being limited essentially to cartilage; type III, found in relatively elastic tissues such as embryonic skin, lung

Table 2.1 Collagens and collagen-like proteins in vertebrates

Sub-family	Members
Fibrillar collagens	Types I, II, III, V, XI, XXIV and XXVII
Fibril associated and related collagens	Types IX, XII, XIV, XVI, XIX, XX, XXI and XXII
Beaded filament forming collagen	Type VI
Basement membrane and associated collagens	Type IV, VII, XV and XVIII
Short chain collagens and related proteins	Types VIII and X; C1q; hibernation-related proteins HP-20, HP-25 and HP-27; emilins 1 and 2; adiponectin; CTRPs 1-7: inner ear (saccular) collagen
Transmembrane collagens and collagen-like proteins	Types XIII, XVII, XXIII and XXV/CLAC-P; ectodysplasins; macrophage scavenger receptors I-III; MARCO; SRCL; gliomedin; CL-P1
Collectins and ficolins	Mannan binding protein; surfactant proteins A and D; conglutinin; CL-43; CL-46; CL-L1; CL-P1; L-, M- and H-ficolins
Other collagens and collagen-like proteins	Emu1; collagen XXVI/Emu2; collagen XXVIII; acetylcholinesterase tail subunit

and blood vessels; type V, found as a quantitatively minor collagen in association with collagen I, with particularly high amounts in cornea; and type XI, a quantitatively minor component of cartilage in association with collagen II.

As in all collagens, each fibrillar collagen molecule consists of three polypeptide chains, called α chains. Molecules can be homotrimeric, consisting of three identical α chains, as in collagens II and III, or heterotypic, consisting of up to three genetically distinct α chains. Individual α chains are identified by the following nomenclature: $\alpha n(N)$, where N is the Roman numeral indicating collagen type and n is the number of the α chain. Thus, the chain composition of collagen II is $[\alpha 1(II)]_3$, while that of collagen I, a heterotrimer with two identical $\alpha 1$ chains and a third distinct $\alpha 2$ chain, is $[\alpha 1(I)]_2 \alpha 2(I)$. While collagen I has only two different α chains, collagen V has three, leading to different chain stoichiometries such as $[\alpha 1(V)]_2 \alpha 2(V)$ and $\alpha 1(V) \alpha 2(V) \alpha 3(V)$. Collagen XI has three different α chains, where the $\alpha 3$ chain is essentially the same as $\alpha 1(II)$. In addition, there are several variants of the $\alpha 1(XI)$ and $\alpha 2(XI)$ chains due to alternative splicing, corresponding to the variable region (Fig. 2.2), of the corresponding mRNAs. To even further increase diversity of these quantitatively minor fibrillar collagens, hybrid forms of collagen V and XI molecules have also been reported.

The common characteristic of the classical fibrillar collagens is a long central triple-helical region in each α chain, consisting of a continuous $(Gly-X-Y)_n$ repeat, where n is 337–343 (depending on collagen type). In the case of collagens I, II and III, this region is flanked by short non-helical regions called telopeptides, typically

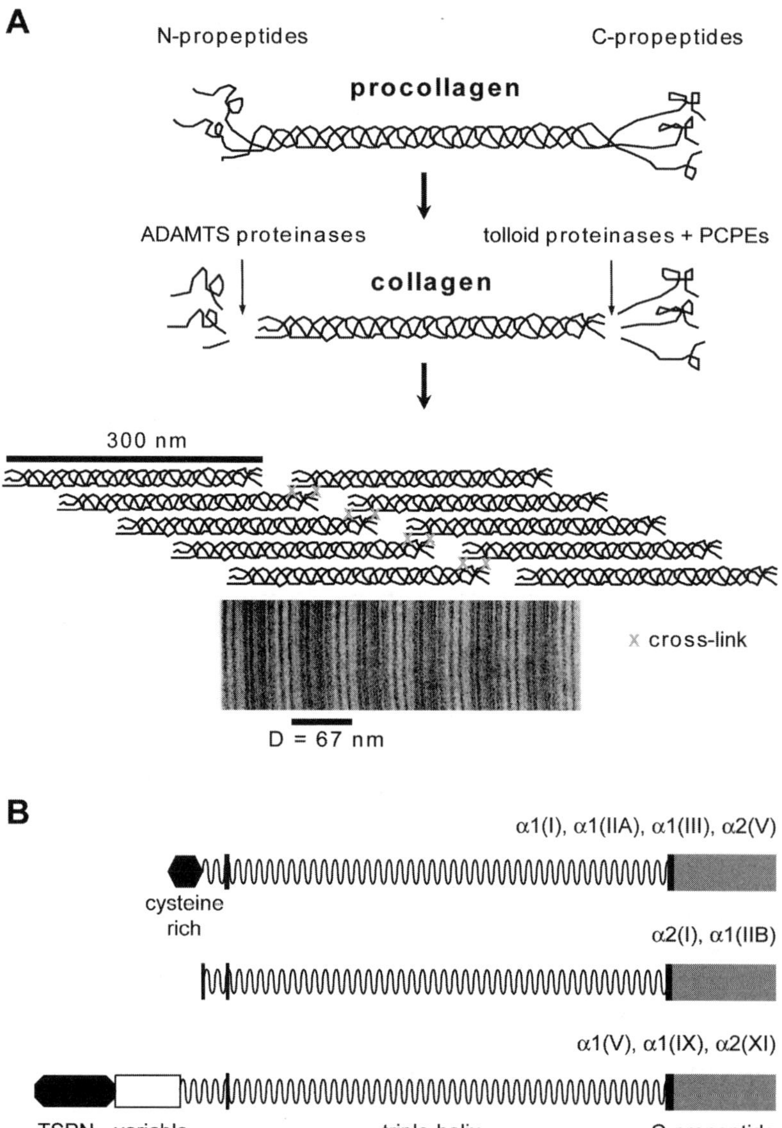

A

N-propeptides   C-propeptides

procollagen

ADAMTS proteinases   tolloid proteinases + PCPEs

collagen

300 nm

X cross-link

D = 67 nm

B

α1(I), α1(IIA), α1(III), α2(V)

cysteine
rich

α2(I), α1(IIB)

α1(V), α1(IX), α2(XI)

TSPN   variable   triple-helix   C-propeptide

Fig. 2.2 Fibrillar procollagens and fibril assembly. **A** N- and C-terminal processing leads to spontaneous assembly of collagen fibrils, which are subsequently stabilized by the formation of covalent cross-links. **B** Domain structures of fibrillar procollagen chains

about 20 residues in length, at both N- and C-termini. All fibrillar collagens are synthesized in the form of soluble precursor molecules, called procollagens, with large N- and C-terminal propeptide domains. The C-propeptides are removed during the later stages of biosynthesis (Section 2.4.4), usually by specific metalloproteinases, leaving the short C-telopeptides. The extent of N-terminal processing depends on collagen type. While for procollagens I, II and III the N-propeptides are completely removed (albeit relatively slowly in the case of procollagen III) leaving short N-telopeptides, N-terminal processing of procollagens V and XI leaves large N-terminal extensions, which help modulate fibril formation (Section 2.5.2).

Little is known about the recently discovered fibrillar-like collagens XXIV and XXVII, though these appear to be associated with types I and II containing tissues, respectively. They differ from the classical fibrillar collagens in that the (Gly-X-Y) region is relatively short (329 triplets), and this region is interrupted by one (collagen XXIV) or two (collagen XVII) short imperfections in the (Gly-X-Y) repeat. Collagen XXVII has recently been shown to form non-striated filamentous structures (Plumb et al., 2007).

2.3 Non-fibrillar Collagens

While fibrillar collagens are the main focus of this chapter, it is of interest to compare these with the diverse forms of assembly of the non-fibrillar collagens (Ricard-Blum et al., 2000).

2.3.1 Basement Membrane and Associated Collagens

The non-fibrillar collagen, about which most is known, is type IV or basement membrane collagen (Hudson et al., 2003). Basement membranes (or basal laminae) are specialized structures found at tissue boundaries, underlying epithelial, endothelial, fat, muscle and nerve cells. Seen by electron microscopy, they are relatively thin sheets (typically 40–50 nm) composed of collagen IV, laminins, heparin sulphate proteoglycans and nidogens. Collagen IV molecules are longer than the fibrillar collagens, and they contain several discontinuities in the (Gly-X-Y) repeat. Also, unlike fibrillar collagens, there is no precursor form of collagen IV and the N- and C-terminal extensions, called respectively the 7S and NC1 domains, are intimately involved in supramolecular assembly. Collagen IV molecules associate via tetramerization of 7S domains, dimerization of NC1 domains, triple-helical interactions and interactions with other basement membrane components to form an open meshwork structure (Fig. 2.3). This structure is stabilized by disulphide bonding in the 7S region, as well as by lysine-derived cross-links initiated by lysyl oxidase (Section 2.4.5; also Chapter 4). There are six different collagen IV α chains, resulting in three types of heterotrimer: $[\alpha 1(IV)]_2 \alpha 2(IV)$, $\alpha 3(IV)\alpha 4(IV)\alpha 5(IV)$ and $[\alpha 5(IV)]_2 \alpha 6(IV)$.

Collagen VII has the longest triple-helical region amongst vertebrate collagens, about 420 nm in length, including interruptions, and is flanked by non-collagenous NC1 and NC2 domains at the N- and C-termini, respectively. This collagen is found underlying the basement membrane at the dermal–epidermal junction, in the form of anchoring filaments (Fig. 2.3). Mutations in collagen VII are the cause of

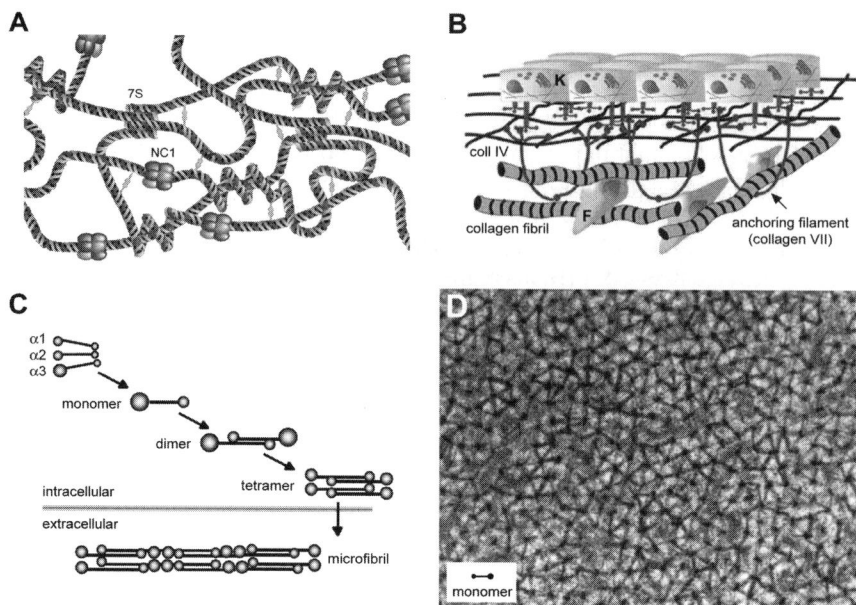

Fig. 2.3 Assembly of non-fibrillar collagens. **A** Basement membrane collagen IV (from (Hudson et al., 2003), with permission). **B** Collagen VII anchoring filaments connecting the epidermal basement membrane to collagen fibrils in the dermis (from (Brittingham et al., 2006), with permission). **C** Intra- and extracellular steps in the assembly of collagen VI microfibrils (from (Baldock et al., 2003), with permission). **D** Hexagonal network formed by collagen VIII in Descemet's membrane of the cornea (courtesy of R. Bruns)

severe blistering disorders called epidermolysis bullosa. Within anchoring filaments, about two molecules in length, bundles of collagen VII molecules aligned in register are arranged tail to tail with a short C-terminal overlap. Antiparallel assembly is triggered by proteolytic processing of the NC2 domain, while the large NC1 domain remains intact and interacts with collagen IV and laminin V (Brittingham et al., 2006).

Collagens XV and XVIII consist of several collagenous domains and together are referred to as multiplexins (for *multiple* triple *he*lices with *in*terruptions). Both are associated with basement membranes, and both carry covalently linked glycosaminoglycan chains, and are therefore also proteoglycans. In addition, both include a C-terminal fragment, endostatin, that is proteolytically cleaved from the molecule and has anti-angiogenic properties (Ricard-Blum et al., 2005).

2.3.2 Collagen VI

Collagen VI is a relatively ubiquitous collagen with important roles in maintaining tissue integrity (Baldock et al., 2003; Knupp and Squire, 2005). The molecules contain relatively short triple-helical regions, about one third the length of fibrillar collagens, with large N- and C-terminal regions made up mostly of von Wille-brand factor A domains. Each molecule is a heterotrimer α1(VI)α2(VI)α3(VI). In the case of the α3(VI) chain, the triple-helical region makes up less than 10% of the entire amino acid sequence. As with several other collagens (e.g., types II, IX, XI, XII and XIII), further diversity in the α chains is introduced at the transcriptional level through the formation of alternatively spliced variants. Collagen VI molecules assemble into the so-called beaded filaments, with a periodicity of 110 nm (Fig. 2.3). Unlike fibrillar collagens, there is no enzymatic processing of the N- and C-terminal non-collagenous regions. In addition, unlike all other collagens, supramolecular assembly begins inside the cell with the formation of antiparallel dimers and tetramers. Tetramers stabilized by disulphide cross-linking then associate in the extracellular matrix to form beaded filaments or microfibrils.

2.3.3 Collagens VIII and X

Collagens VIII and X, the so-called short-chain collagens, have specific tissue locations, underlying endothelial cells and in the hypertrophic zone of cartilage during endochondral ossification, respectively. The triple-helical region, which contains many interruptions, is about half the length of a fibrillar collagen. While collagen X is an α1(X) homotrimer, collagen VII, with both α1(VIII) and α2(VIII) chains, appears to occur as both homotrimers and heterotrimers. Molecules have both N- and C-terminal non-collagenous regions, the latter corresponding to the widely occurring C1q domain (Ghai et al., 2007). There is no proteolytic processing of these regions, rather they help in the formation of hexagonal supramolecular networks in, for example, Descemet's membrane of the cornea (collagen VIII; (Stephan et al., 2004); Fig. 2.3) and calcifying cartilage (collagen X; (Kwan et al., 1991)).

2.3.4 FACITs

The collagens known as FACITs (Fibril Associated Collagens with Interrupted Triple helices) have become a relatively large group (currently eight different types; see Ricard-Blum et al., 2005). The founder member, collagen IX, is an important component of cartilage collagen fibrils, along with collagens II and XI. Unlike the latter, however, collagen IX does not self-assemble into fibrils. The collagen IX molecule is composed of three relatively short collagenous regions (with interruptions) and four non-collagenous regions. The molecule is an α1(IX)α2(IX)α3(IX) heterotrimer, and the α2(IX) chain sometimes carries a chondroitin/dermatan sulphate glycosaminoglycan chain, thereby making collagen IX a part-time proteoglycan. Collagen IX molecules coat the surface of collagen II/XI cartilage fibrils, with the large N-terminal NC4 domain of the α1(IX) chain being available for interactions with other extracellular matrix components. The presence of the NC4 domain in the α1(IX) chain is itself subject to regulation due to the use of alternative transcriptional start sites. Collagen IX is covalently cross-linked to both collagens II and XI in the cartilage fibril, through lysine-derived cross-links; recent studies have shown how flexibility within the non-collagenous domains allows these cross-links to form (Eyre et al., 2004). Strongest homology within the FACITs is in the C-terminal collagenous (COL1) and non-collagenous (NC1) domains, and all have a thrombospondin N-terminal-like domain (also found in the fibrillar collagens V and XI) just before the first collagenous domain (starting from the N-terminus). Collagen IX can be considered as the founder member of a subgroup of FACITs with three or more collagenous domains, which also includes collagens XVI, XIX and XXII (Ricard-Blum et al., 2005). Unlike collagen IX, however, collagens XVI, XIX and XXII do not associate specifically with collagen fibrils and have distinct tissue distributions. Collagen XXII, for example, is specifically localized to tissue junctions (Koch et al., 2004).

Collagen XII has also been shown to be associated with the collagen fibril surface, and has been localized to dense connective tissues rich in collagen I, such as tendons, ligaments, skin and cornea. The molecule is an $[\alpha 1(XII)]_3$ homotrimer, each chain consisting of two collagenous domains (making up only about 10% of the amino acid sequence) and three non-collagenous domains. The non-terminal non-collagenous domain (NC3) is particularly long and includes multiple fibronectin type III and von Willebrand factor A domains. Two alternative spliced forms of collagen XII are known, differing in the length of the NC3 domain, the longest form being a proteoglycan-like collagen IX. Collagen XII expression is upregulated by mechanical stress (Chiquet et al., 2003). Collagen XII can be considered as the founder member of the second subgroup of FACITs, characterized by the presence of just two collagenous domains. Collagens XIV and XX are closely related, while collagen XXI is by far the smallest member of this subgroup.

2.3.5 Other Collagens and Collagen-Like Proteins

The transmembrane collagens and collagen-like proteins have become one of the largest groups and have diverse functions in cell adhesion and signaling (Franzke et al., 2005). Lastly, the collagen-like proteins of the immune system (C1q, collectins and ficolins) have collagen-like stalks and globular heads, the former leading to self-association into bundles containing up to six molecules (Holmskov et al., 2003).

2.4 Collagen Biosynthesis

Collagen biosynthesis has been studied in greatest detail with regard to the fibrillar collagens. The process is complex, involving numerous intracellular and extracellular steps, all of which contribute to the structure and biomechanical properties of the final fibrils. The first event following synthesis of procollagen α chains on the ribosome is their import into the rough endoplasmic reticulum. There they undergo a series of post-translational modifications resulting in the assembly of procollagen molecules (Fig. 2.4) (Myllyharju, 2005). These steps include modification of proline residues to hydroxyprolines, modification of lysines to hydroxylysines, N- and O-linked glycosylation, trimerization, disulphide bonding, prolyl *cis–trans* isomerization and folding of the triple helix. Molecules then transit the Golgi network where they are packaged into secretory vesicles prior to export into the extracellular matrix. Procollagen processing occurs during or shortly after secretion followed by assembly of fibrils. Finally, fibrils are stabilized by the formation of covalent cross-links.

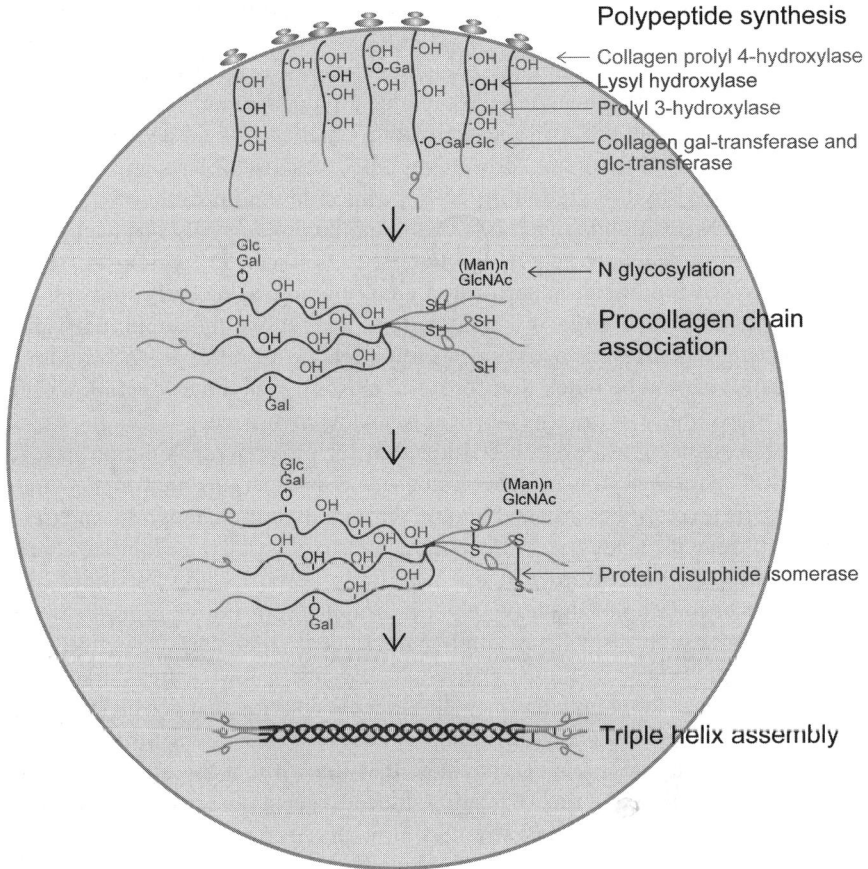

Fig. 2.4 Post-translational modifications and assembly of the procollagen molecule in the endoplasmic reticulum (from (Myllyharju, 2005), with permission)

2.4.1 Post-translational Modifications of Polypeptide Chains

It has long been known that stability of the collagen triple helix is related to the total content of the amino acids proline and hydroxyproline, which together make up about 20% of the total amino acids in human fibrillar collagens. This is a result of the cyclic nature of the side chains that restricts flexibility about the peptide bond. Prolyl hydroxylation provides further stabilization, as shown by the approximately 30°C drop in melting temperature (see below) when this process is prevented. Hydroxylation occurs mainly on prolines in the Y position of the Gly-X-Y triplet, through the action of prolyl 4-hydroxylases (P4H), and also to a much lesser extent on prolines in the X-position, through the action of prolyl 3-hydroxylases (P3H) (Fig. 2.5). The enzymes responsible are 2-oxoglutarate and non-heme-Fe(II)-dependent dioxygenases, which require Fe^{2+}, 2-oxoglutarate, O_2 and ascorbate for activity (Myllyharju, 2005). In humans, three forms of P4H (Myllyharju, 2005) and three forms of P3H (Morello et al., 2006) have been described. P4H is a tetramer consisting of two α subunits and two β subunits. The β subunit is common to all forms of the enzyme and also exists as an isolated protein with both enzyme and chaperone activity, protein disulphide isomerase (PDI). Different forms of P4H therefore differ in the α subunit, which shows tissue specificity, the type II enzyme being the major form in chondrocytes, osteoblasts and endothelial cells. While prolyl hydroxylation in the Y position is the norm, hydroxylation of prolines in the X-position is much less common, there being only one such residue per α chain in collagens I and II, in contrast to collagen IV which has several. Nevertheless, this modification performs a vital function, as recently shown by loss of prolyl 3-hydroxylase activity which leads to a recessive metabolic bone disorder resembling lethal/severe osteogenesis imperfecta (Cabral et al., 2007; Morello et al., 2006). It should be noted that for steric reasons, prolyl hydroxylation occurs only prior to assembly of the triple-helical procollagen molecule.

The mechanism of triple-helix stabilization by prolyl hydroxylation has generated much controversy but a general consensus is emerging that this is due to two effects: (i) increased hydrogen bonding through hydration networks and (ii) the electron-withdrawing (inductive) effect of the hydroxyl group that stabilizes the exo pucker of the (4R) stereoisomer in the Y position (Brodsky and Persikov, 2005). Denaturation or melting of the triple helix is a highly cooperative process, as measured by circular dichroism for example, which leads to a sharp transition with a characteristic melting temperature that is usually just a few degrees above body temperature. This is thought to be an evolutionary adaptation that permits the micro-unfolding that prevents misalignment and is necessary for interactions with other molecules, including degradative enzymes. It is known that the apparent melting temperature depends on the rate of heating. Indeed, recent observations have shown that collagen denaturation is an extremely slow process, and that the equilibrium melting temperature is actually below body temperature (Leikina et al., 2002). This means that collagen molecules only appear to be stable at body temperature, due to kinetic effects, and further stabilization comes from interactions with chaperones and supramolecular assembly (Section 2.4.2).

Lysyl hydroxylation is also an important post-translation modification in collagens. Hydroxylysine plays important roles in collagen cross-linking (see Chapter 4) and is a substrate for O-linked glycosylation (see below), thereby affecting fibril formation and other protein–protein interactions. Three forms of lysyl hydroxylase are known, LH1, LH2 and LH3, which like the prolyl hydroxylases reside

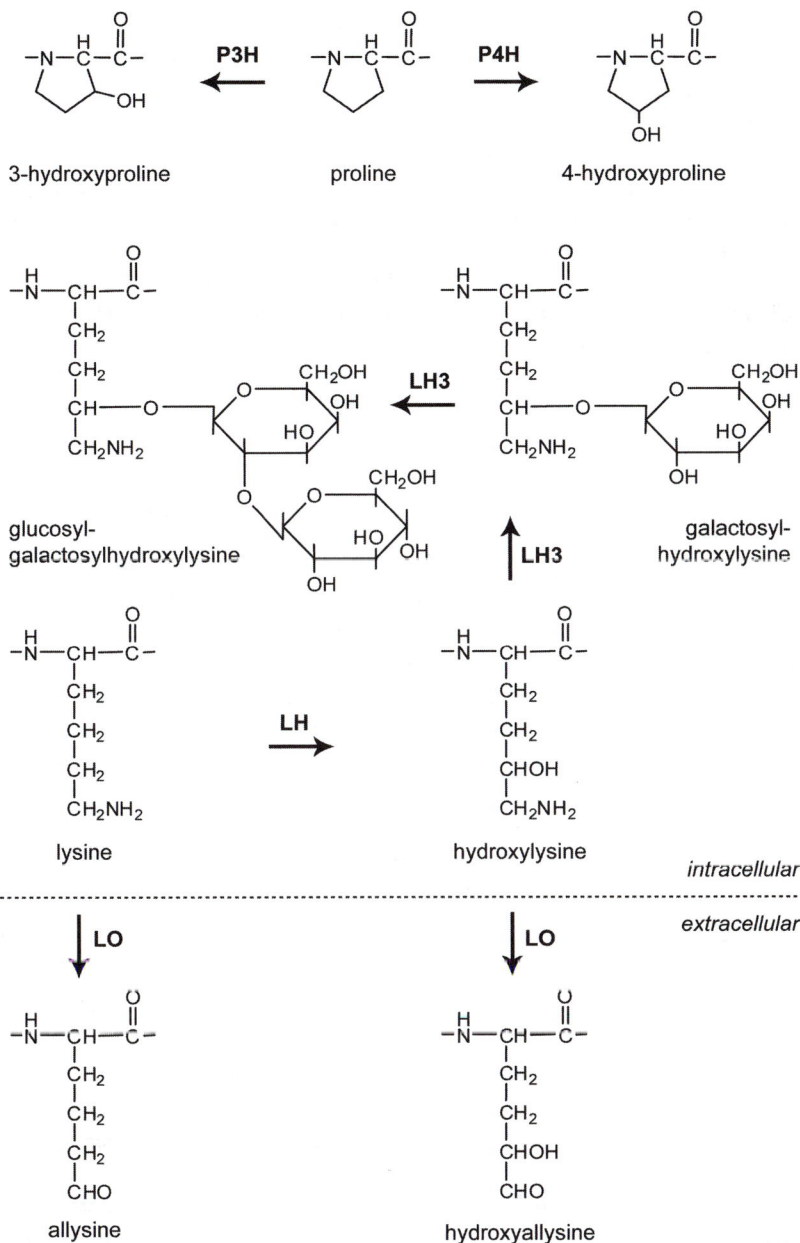

Fig. 2.5 Collagen-specific post-translational modifications of proline and lysine residues, and the enzymes involved. P3H = prolyl 3-hydroxylases; P4H = prolyl 4-hydroxylases; LH = lysyl hydroxylases; LO = lysyl oxidases

in the endoplasmic reticulum. Also like prolyl hydroxylases, these enzymes are 2-oxoglutarate dioxygenases requiring Fe^{2+}, 2-oxoglutarate, O_2 and ascorbate for activity (Myllyla et al., 2007). The extent of lysyl hydroxylation varies between tissues, being relatively high, for example, in collagens IV and VI an also in embryonic tissues. Only lysines in the Y position of the Gly-X-Y triplet can be hydroxylated. LH2 has been shown to specifically hydroxylate lysine residues in the telopeptide regions. Telopeptide lysines and hydroxylysines are substrates for the cross-link initiating enzymes lysyl oxidases (Section 2.4.5), and the nature of the subsequent cross-linking is determined by the state of hydroxylation of the telopeptide lysines (see Chapter 4). This was recently demonstrated by a deficiency in LH2 which leads to Bruck syndrome, a form of osteogenesis imperfecta with joint contractures (van der Slot et al., 2003). Dysfunctional LH1, on the other hand, leads to Ehlers syndrome type VI, characterized by neonatal kyphoscoliosis, generalized joint laxity, skin fragility and severe muscle hypotonia at birth (Yeowell and Walker, 2000).

O-linked glycosylation of hydroxylysines is specific to collagenous proteins and involves the covalent attachment of galactose and then glucose from UDP carriers (Fig. 2.5). The enzyme(s) responsible was for a long time an enigma, but recent work has shown that the activity resides in the multifunctional enzyme LH3 which, in addition to being a lysyl hydroxylase, is also a galactosyltransferase (GT) and a glucosyltransferase (GGT). LH3 is therefore able to catalyze the three consecutive reactions required for the formation of glucosylgalactosyl-hydroxylysine. The molecule comprises a single polypeptide chain, with amino acids involved in the GT and GGT activities near the N-terminus and those involved in LH activity near the C-terminus. Identification of LH3 has allowed the functional significance of the LH and GGT activities to be investigated by genetic inactivation, either of the whole LH3 molecule or just the LH region (Ruotsalainen et al., 2006). The results showed that inactivation of LH had relatively little effect (probably due to partial compensation by other forms of LH), but inactivation of the entire molecule, including the GGT activity, was embryonic lethal, apparently due to disruption of basement membranes. Thus, O-linked glycosylation of collagens is essential for normal development.

2.4.2 Chain Association and Triple-Helix Formation

Following prolyl and lysyl hydroxylation and O-linked glycosylation in the rough endoplasmic reticulum, procollagen α chains then associate to form the procollagen molecule. In recent years, it has become clear that collagenous domains on their own have difficulty in assembling into trimers. Trimerization requires additional domains, adjacent to the $(Gly-X-Y)_n$ domains, to facilitate the assembly process. In the case of the fibrillar collagens, several lines of evidence point to a role for the procollagen C-propeptide domain in the initiation of trimerization. This involves two events: specific chain recognition to assure correct chain stoichiometry and formation of a stable nucleus to favor triple-helical folding. Control of chain stoichiometry is particularly important in cells that simultaneously produce several collagen types, to result in the correct homotrimeric and heterotrimeric chain combinations. The amino acid sequences of the C-propeptide domains of the fibrillar procollagens are highly conserved, with the exception of a discontinuous sequence of 15 amino acids near the middle. It was shown some years ago (Lees et al., 1997), by swapping this region between different C-propeptide domains, that this sequence contains the information required for chain recognition. Regarding the formation of a stable nucleus, a further contribution was the recent identification of α-helical coiled-coil domains at the start of all known procollagen C-propeptide sequences (McAlinden et al., 2003). Alpha-helical coiled coils (not to be confused with collagen α chains!) are widely occurring oligomerization motifs (Parry, 2005), which result from the association of polypeptide chains containing the heptad repeating sequence $(abcdefg)_n$, where residues a and d are typically small hydrophobic amino acids and residues e and g are often charged. Formation of dimers, trimers or higher oligomers allows the α-helical chains to associate together in a rope-like structure (hence "coiled coils") thereby burying surface-exposed hydrophobic residues. In the procollagen molecule, four such heptad repeats are present at the start of the procollagen C-propeptides, separated from the collagen triple-helical region by the C-telopeptide. Similar juxtapositions of collagen triple-helical and coiled-coil domains are found in the collectins (Holmskov et al., 2003), where the coiled-coil domains have been shown to be essential for trimerization.

While triple-helix formation of fibrillar procollagens is initiated in the C-terminal region, with subsequent triple-helical folding in the C- to N-direction, this is not always the case in collagens. In transmembrane collagens and collagen-like proteins, for example, coiled-coils domains are found N-terminal to collagenous domains, where they help direct triple-helical folding in the N- to C-terminal direction (Snellman et al., 2007). Indeed, coiled-coil domains are almost ubiquitous in the collagen superfamily (McAlinden et al., 2003) and can occur C-terminal, N-terminal or between collagenous domains. Collagens without coiled-coil domains include collagen IV, C1q domain containing collagens and collagen-like proteins, and ficolins (which have C-terminal fibrinogen-like domains). Thus while coiled coils are likely to be the most widely used trimerization domains in collagens, alternative trimerization domains (collagen IV NC1, C1q, fibrinogen-like) may be used. One possible advantage of using trimerization domains to initiate triple helix assembly is to avoid misalignment of collagen chains, i.e., slippage of adjacent chains by multiples of the $(Gly-X-Y)_n$ repeat.

Following chain recognition and trimerization, a number of proteins come into play to guide the process of triple-helix formation during assembly of the procollagen molecule. These include prolyl 4-hydroxylase (P4H) which, in addition to its enzymatic function (Section 2.4.1), also acts as a chaperone protein by selectively binding to unfolded procollagen α chains, thereby preventing premature triple-helix formation (Walmsley et al., 1999). As previously mentioned, the β subunit of P4H is the same as another resident protein of the endoplasmic reticulum, protein disulphide isomerase (PDI), whose main function is to catalyze the formation and rearrangement of disulphide bonds (Ellgaard and Ruddock, 2005). In the case of procollagen assembly, PDI is involved in trimerization through the formation of intra- and intermolecular disulphide bonds in the propeptide regions. PDI also binds selectively to unfolded chains thereby also performing a chaperone role (Bottomley et al., 2001). Additional chaperone proteins associated with quality control of intracellular collagen assembly are BiP/Grp78 and Grp94 (Koide and Nagata, 2005).

The peptide bond can exist in two conformations, *cis* or *trans*. For most amino acids, the *trans* conformation is strongly energetically favored. For prolyl- and hydroxyprolyl-containing peptide bonds, however, because of the cyclic nature of the side chains, the energy difference is smaller. Consequently, many of the proline and hydroxyprolines in newly synthesized procollagen α chains are in the *cis* conformation. Because only *trans* peptide bonds can be incorporated into the triple helix, *cis–trans* isomerization has been found to be a rate-limiting step in assembly of the procollagen molecule. This process can be accelerated by the enzymes known as prolyl *cis–trans* isomerases, or immunophilins, which includes the cyclophilins and the FK506 binding proteins (Barik, 2006). Like P4H and PDI, immunophilins also exhibit chaperone functions independent of their enzymatic activities.

In contrast to proteins that interact selectively with unfolded collagen chains, HSP47 is a collagen-specific chaperone that binds preferentially to folded triple helices (Koide and Nagata, 2005). HSP47 is a resident protein of the endoplasmic reticulum and is homologous to the serine protease inhibitors serpins. It is specifically expressed by cells that synthesize large amounts of collagen, where it performs a vital role as shown by the embryonic lethal phenotype of HSP47 knockout mice (Ishida et al., 2006). HSP47 binds to triple-helical collagen, preferentially to $(Gly-X-Y)_n$ repeats containing arginine at the Y position, and is thereby thought to stabilize the triple helix which is inherently unstable at physiological temperature (Section 2.4.1). By binding to the triple helix, it is also thought to prevent premature aggregation of procollagen molecules in the endoplasmic reticulum (Koide and Nagata, 2005). HSP47 dissociates from procollagen molecules as they transit the Golgi compartment (Section 2.4.3), probably as a result of a decrease in local pH.

2.4.3 Intracellular Transport and Secretion

From the endoplasmic reticulum, newly synthesized procollagen molecules proceed to the Golgi complex (or Golgi for short). Exit from the endoplasmic reticulum appears to occur via budding of carriers called ERGICs (Endoplasmic Reticulum to Golgi Intermediate Compartments) or VTCs (Vesicular Tubular Clusters) in a COPII-dependent process. COPII (coat protein complex II) is a multisubunit complex that consists of scaffold proteins, GTPases, GTPase activators and GDP/GTP exchange factors and has the ability to curve membranes, and thereby form vesicles (Fromme and Schekman, 2005). The transport of such large molecules as procollagen (300 nm in length) is difficult to reconcile with the conventional view of ER to Golgi transport involving 60–70 nm COPII-coated vesicles. Recent work (Mironov et al., 2003) has lead to the view that for such large cargos, while COPII is essential for budding, it is not itself incorporated into the carriers.

The state of aggregation of procollagen molecules during transport through the Golgi, the *trans*-Golgi network and subsequent secretion into the extracellular matrix has been a subject of investigation for over 30 years. Early studies (see Canty and Kadler, 2005) revealed bundles of procollagen molecules aligned in register (also known as SLS crystallites) in secretory vesicles, which were subsequently found to be secreted as bundles into the extracellular matrix (Hulmes et al., 1983). More recent studies have shown that procollagen molecules are found in this form throughout their transit through the Golgi, which occurs without leaving the lumen of the cisternae, thus supporting the "cisternal maturation" model (Bonfanti et al., 1998). It is not clear what makes procollagen molecules associate in this way within vacuoles, rather than in D-staggered array as in fibrils. With purified procollagen I, the presence of the propeptides impedes, but does not prevent, D-staggered assembly at the high concentrations likely to be found during intracellular transport (Mould et al., 1990). It is possible that interactions with other secreted components might favor non-staggered aggregation into bundles.

2.4.4 Procollagen Processing

Enzymatic removal of procollagen propeptides, or procollagen processing, is carried out by metalloproteinases belonging to the ADAMTS (Porter et al., 2005) and BMP1/Tolloid-like families (Hopkins et al., 2007), as well as the furin-like proprotein convertases (Seidah and Prat, 2005).

The ADAMTS (for "a disintegrin and a metalloproteinase with thrombospondin repeats") family has 20 members, of which ADAMTS-2, -3 and -14 are implicated in N-terminal processing of fibrillar procollagens (Porter et al., 2005). These enzymes are large multidomain proteins, including a prodomain (cleaved by furin), a reprolysin-like zinc metalloproteinase domain, disintegrin, cysteine-rich and spacer domains and multiple thrombospondin type-I domains. Defective procollagen N-proteinase activity is associated with a hereditary disorder originally found in cattle and sheep, called dermatosparaxis, and later found in humans, called Ehlers–Danlos syndrome type VIIC, characterized by extreme skin fragility and laxity (Section 2.5.2). All three enzymes cleave the amino-propeptides from procollagen I, while ADAMTS-2 cleaves procollagen III, and ADAMTS-3 cleaves procollagen II. They also have somewhat different tissue specificities, such as ADAMTS-2 in skin, lungs and aorta and ADAMTS-3 in cartilage and bone (Le Goff et al., 2006). It is of interest to note that ADAMTS-2 interacts with the FACIT collagen XIV, suggesting a possible role for fibril-associated collagens in procollagen processing (Colige et al., 1995).

BMP1/Tolloid-like proteinases (Hopkins et al., 2007) are largely responsible for the C-terminal processing of fibrillar procollagens. Originally identified as the products of genes involved in the induction of bone formation (BMP1, for bone morphogenetic protein-1) and in dorso-ventral patterning during embryonic development (Tolloid, originally identified in *Drosophila*), this family currently includes four members in mammals: BMP1, mTLD (mammalian Tolloid), mTLL1 (mTLD-like 1) and mTLL2 (mTLD-like 2). These are zinc-dependent metalloproteinases with multiple domains, including a proregion (cleaved by furin), a catalytic astacin domain and a number of CUB (complement-uegf-BMP1) and EGF (epidermal growth factor)-like domains. While mTLD, mTLL1 and mTLL2 have five CUB domains and two EGF domains, BMP1 (a product of alternative splicing, encoded by the same gene as mTLD) lacks the two C-terminal CUB domains and the second EGF domain. All four enzymes can cleave the C-propeptides from the major fibrillar procollagens (types I–III), though BMP1 seems to be the most active in vitro. In the case of the minor fibrillar procollagens V and XI, C-terminal processing is complex, with furin-like cleavage appearing to play a major role in the proα1(V), proα1(XI) and proα2(XI) chains and BMP1/Tolloid-like proteinases cleaving the proα2(V) chain. Furthermore, unlike the major fibrillar procollagens, BMP1/Tolloid proteinases cleave within the large N-terminal regions of the proα1(V), proα1(XI) and proα2(XI) chains, between the TSPN and variable domains (Fig. 2.2).

In addition to the fibrillar collagens, BMP1/Tolloid-like proteinases cleave a large number of extracellular substrates (Hopkins et al., 2007), including the non-fibrillar collagen VII, the transmembrane collagen gliomedin, various proteoglycans, laminin 5 and prolysyl oxidases (Section 2.4.5). They are also involved in the activation of a large number of growth factors, including transforming growth factor-β, a key regulator of extracellular matrix deposition and turnover. The activity of BMP1/Tolloid-like proteinases is regulated by additional proteins of the extracellular matrix, procollagen C-proteinase enhancers (PCPE)-1 and -2 (Hopkins et al., 2007). Like BMP1/Tolloid-like proteinases, PCPEs also contain CUB domains, but are devoid of intrinsic proteolytic activity. Instead, they can enhance the activity of BMP1/Tolloid-like proteinases by up to 20-fold. It has recently been shown that PCPE-1 specifically enhances the action of BMP1 during C-terminal processing of the major fibrillar procollagens (Moali et al., 2005).

For a long time, it was thought that procollagen processing is an extracellular event. Recent studies, however, support the idea that processing and fibril assembly (Section 2.5.5) can begin in Golgi to plasma membrane carriers (Canty et al., 2004). Using pulse-chase and differential extraction procedures, these authors demonstrated the presence of procollagen processing intermediates, without N- or C-propeptides, in intracellular compartments, consistent with the observation that furin activation of proBMP1 and proADAMTS2 can also occur prior to secretion (Leighton and Kadler, 2003; Wang et al., 2003). Procollagen processing probably also occurs in close association with the plasma membrane. For example, targeted deletion of the extracellular matrix protein SPARC results in increased procollagen processing probably as a result of increased retention by cell-surface receptors (Rentz et al., 2007).

2.4.5 Covalent Cross-Linking

The final step in the biosynthesis of collagens is the introduction of covalent cross-links to stabilize the different forms of supramolecular assembly. In the case of the fibrillar collagens, cross-linking is for the most part initiated by members of the lysyl oxidase family of copper-dependent amine oxidases (Lucero and Kagan, 2006; Molnar et al., 2003). This includes five members: lysyl oxidase

(LOX), lysyl oxidase-like (LOXL), LOXL2, LOXL3 and LOXL4. All five are highly homologous in their C-terminal region, which contains the catalytic domain and a cytokine receptor-like domain. In the case of proLOX, the catalytic region is preceded by a proregion that is cleaved by members of the BMP1/Tolloid-like family. Since cleavage of the proLOX propeptide activates LOX, and cleavage of the procollagen C-propeptides trigger fibril formation (Section 2.5.2), BMP1/Tolloid-like proteinases regulate both collagen cross-linking and assembly. The precursor form of LOXL contains a unique proline-rich region and is also processed by BMP1 at different sites, leading to multiple forms of the mature enzyme (Borel et al., 2001). The remaining lysyl oxidase-like enzymes, LOXL2, LOXL3 and LOXL4, form a distinct subgroup that is characterized by the presence of four scavenger receptor domains, N-terminal to the catalytic domain. It is not known whether these enzymes are proteolytically processed.

The effect of lysyl oxidase activity on fibrillar collagens is to convert lysine or hydroxylysine residues in the N- and C-terminal telopeptide regions to corresponding peptidyl aldehydes (Fig. 2.5). Once formed, aldehydes spontaneously condense with other aldehydes or unreacted lysines and hydroxylysines to form a variety of intra- and intermolecular covalent cross-links (see Chapter 4). LOX activity is strongly increased by assembly of collagen into fibrils, presumably as a result of LOX binding to highly conserved sequences that become juxtaposed to the telopeptide regions when adjacent molecules are mutually staggered by 4×67 nm (4D) (Kuhn, 1987). LOXL is also active on a fibrillar collagen substrate (Borel et al., 2001). Non-fibrillar collagens taking part in lysyl oxidase initiated cross-linking include types IV, X and IX (Ricard-Blum et al., 2000). Little is known about the substrate specificity of the other lysyl oxidases. There is evidence for an association of LOXL2 with basement membrane collagen IV, and indeed LOXL2 also seems to act intracellularly through interactions with the transcription factor Snail, a regulator of the epithelial–mesenchymal transition (Peinado et al., 2005). LOX also seems to have an intracellular role, as a tumor suppressor, which appears to be mediated by its propeptide region (see Lucero and Kagan, 2006). Lysyl oxidase activity can be inhibited by lathyritic agents, the mostly widely used being β-aminopropionitrile (BAPN).

Another enzyme involved in collagen cross-linking in vivo is tissue transglutaminase (TG2), a multifunctional enzyme with both cross-linking (extracellular) and GTPase (intracellular) activities (Verderio et al., 2005). Cross-linking involves the formation of an isopeptide bond between the γ-carboxyamide group of specific peptidyl glutamine residues and ε-amino group of peptidyl lysines. Several collagens are cross-linked by TG2, including collagens II, III, V, VII and XI (Grenard et al., 2001).

2.5 Assembly of Fibrillar Collagens

2.5.1 Reconstitution of Fibrils In Vitro

The formation of collagen fibrils in vitro has been a subject of extensive research for over 50 years (see Kadler et al., 1996). Initial observations were on the reconstitution of fibrils from cold solutions of collagen in dilute acid. The major fibrillar collagens are soluble at low pH, for example in dilute acetic acid. When the pH is adjusted to around neutral, and the temperature raised to around physiological, fibril formation occurs spontaneously resulting in banded fibrils. The structure of the resulting fibrils is influenced by several parameters, including buffer composition, intactness of the N- and C-telopeptides, the presence of other types of collagen, the presence of macromolecules other than collagens and the order of the initiating procedure.

The influence of buffer composition on fibril formation has been studied by several authors, notably by Williams et al. (1978) who recommended diluting cold solutions of collagen I at concentrations up to 1 mg/ml in 5 mM acetic acid with an equal volume of optimum buffer consisting of 60 mM TES (N-[tris (hydroxymethyl)methyl-2-amino]ethanesulphonic acid), 60 mM sodium phosphate, 270 mM NaCl, pH 7.3, and then raising the temperature to up to 30°C. In these conditions, fibril morphology is optimized, as judged by the presence of compact fibrils with a clear D-periodic banding pattern. The presence of phosphate is important for producing well-banded fibrils. Another important parameter is the temperature at which fibril formation is carried out, fibrils formed at lower temperatures (20°C) having larger diameters than those formed at higher temperatures (34°C). Finally, the initiation procedure also has an effect, the so-called "warm start" procedure (warming first and then buffer neutralization) giving better formed fibrils (i.e., less non-banded filaments) than with the "cold start" procedure (neutralization and then warming) (Holmes et al., 1986).

Intactness of the N- and C-telopeptides has a marked effect on fibril formation. Commercially available fibrillar collagens are of two types, either acid soluble or pepsin soluble. In acid-soluble collagen, the telopeptides are usually intact, and these help both to initiate fibril formation and to produce long cylindrical fibrils. Pepsin-soluble collagen is produced by preparing tissue extracts with pepsin, which digests most protein structures except for the collagen triple helix. As a consequence, the non-helical telopeptides are mostly removed by this procedure, making self-assembly more difficult. Selective removal of the N-telopeptides results in so-called D-periodic symmetric (DPS) fibrils, in which molecules assemble in an antiparallel manner throughout the fibril length, while partial loss of the C-telopeptides results in relatively short cigar-shaped D-periodic cigar-shaped tactoids (Kadler et al., 1996). The importance of telopeptide interactions has been further demonstrated by Prockop and Fertala (1998) who showed that exogenous peptides corresponding to N- or C-telopeptides could inhibit fibril formation.

Fibril formation is an entropy-driven process, in which self-assembly results in burying surface-exposed hydrophobic residues within the fibril, thereby increasing entropy in the solvent. Assembly proceeds by a nucleation and growth mechanism, as monitored typically by turbidimetry at a wavelength of around 300 nm. Light scattering, or turbidity, is proportional to both the concentration of collagen and the mass of the assembled structures. Turbidity curves classically show three phases: an initial lag phase with no change in turbidity, a rapid growth phase and finally a plateau region. During the lag phase, small numbers of collagen molecules associate to form metastable nuclei, upon which further molecules accrete during the growth phase. The initial interaction seems to be a pair of $4D$ staggered molecules with a short N- and C-terminal overlap, mediated by the N- and C-telopeptides (Ward et al., 1986). Growth in fibril length and width then occurs via longitudinal and lateral interactions.

While fibrils formed in vitro appear at first glance to be approximately cylindrical, detailed observation of unstained fibrils by scanning transmission electron microscopy (STEM), which can measure the mass per unit length of individual fibrils, shows that they have pointed tips (see Kadler et al., 1996). When reconstituted from acid-soluble collagen, fibrils have relatively sharp N-terminal tips and blunt C-terminal tips.

2.5.2 Fibril Formation De Novo from Procollagen

As a more physiological alternative to the reconstitution of collagen fibrils from acid solutions, fibrils can also be formed in vitro from procollagen, in physiological buffers, in the presence of procollagen N-proteinase (ADAMTS-2) and procollagen C-proteinase (BMP-1). The propeptides impede assembly thereby increasing solubility, largely due to the presence of the C-propeptide region (see below). Because of this, it is convenient to begin with a partially processed form of procollagen with the N-propeptides removed, called pC-collagen, and trigger fibril formation by the addition of BMP-1. When this is done, the initially formed (unipolar) fibrils also have pointed tips, with a sharp N-terminus and a blunt C-terminus. Unlike fibrils reconstituted from acid-soluble collagen however, these de novo fibrils then go on to acquire pointed tips at both ends. This is due to the formation of bipolar fibrils in which molecules have their N-termini pointing toward one of the tips, with an abrupt change in molecular orientation somewhere in the middle. It was subsequently shown that such bipolar fibrils are relatively common in vivo, indicating that de novo fibril formation in vitro can reproduce at least some of the features of the in vivo system (Kadler et al., 1996). Also of note is that the shape of the tips is determined by the rate of C-terminal procollagen processing, fast processing generating blunter tips than slow processing (Holmes et al., 1996).

STEM analysis of de novo fibrils formed in vitro (Holmes et al., 1992) shows that the axial mass distributions for both sharp and blunt tips are linear, i.e., the number of molecules added per D period is constant (being greater for blunt tips). This is an important observation, which is at variance with simple conical shapes for the tips which would be expected to give quadratic axial mass distributions. The linear increase shows that the tips are paraboloidal in shape. Theoretical simulations have shown that this is consistent with fibril growth being determined by kinetic mechanisms, such as diffusion-limited aggregation (Parkinson et al., 1994).

Procollagen processing has a dramatic influence on fibril formation. This has been shown by studying procollagen I assembly in vitro, with or without the N- or C-propeptides. When the C-propeptides are removed from pC-collagen, fully processed collagen molecules readily assemble into fibrils at concentrations greater than a so-called critical concentration of approximately 1 μg/ml. When only the C-propeptide is removed from procollagen, the processing intermediate pN-collagen begins to assemble at concentrations approximately 100-fold greater, but unlike fully processed collagen, pure pN-collagen forms D-periodic sheet-like structures. Such sheets can be several microns wide, but their thickness is limited to 8 nm (Hulmes et al., 1989). It seems that persistence of the N-propeptide region prevents lateral aggregation in $1D$, probably by steric hindrance as a result of the N-propeptides folding back against the main triple-helical region. When both the N- and C-propeptides are intact, procollagen molecules have also been found to assemble into similar sheet-like structures, but only at concentrations greater than 1 mg/ml (Mould et al., 1990). Thus, the presence of the C-propeptides increases solubility by approximately 1000-fold.

As mentioned earlier, defective N-proteinase activity is associated with a hereditary disorder originally found in cattle and sheep (dermatosparaxis) and later found in humans (Ehlers–Danlos VIIc) characterized by extreme skin fragility. When examined by electron microscopy, collagen fibrils in the skin are no longer approximately cylindrical as in normal tissue, but instead are thin and branched. By treating mixtures of procollagen and pC-collagen with C-proteinase, resulting in mixtures of pN-collagen and collagen, respectively, it was possible to reproduce exactly the shape of the dermatosporactic fibrils in vitro, with a pN-collagen:collagen ratio of about 2:1 (Hulmes et al., 1989). In contrast, when the pN-collagen:collagen ratio was less than 1:5, fibrils were approximately circular in cross-section. These observations are consistent with the idea that persistence of the N-propeptide distorts fibril shape by increasing the surface area to volume ratio, in order for the N-propeptides to be surface located and not buried within the fibril. Another way of increasing the surface area to volume ratio is by making fibrils with small diameters. Thus, as suggested by Fleischmajer et al. (1990), it is likely that limited persistence of the N-propeptide has a role to play in fibril diameter regulation.

2.5.3 Heterotypic Fibril Assembly

Another important factor that influences fibril assembly is the formation of copolymers consisting of different collagen types. Such heterotypic fibrils are the norm in most connective tissues, for example, collagens I, III and V in skin, collagens I and V in cornea and collagens II and XI in cartilage. N-terminal processing of procollagen III is relatively slow, thus heterotypic collagen I/III fibrils in skin contain significant amounts of pN-collagen III, which have been shown to limit fibril diameter in vitro (Romanic et al., 1991), again probably due to steric effects. Similar reasoning applies to collagen I/V fibrils in cornea and collagen II/XI fibrils in cartilage. N-terminal processing of collagens V and XI is at best only partial, leaving large surface located N-terminal extensions which impede fibril growth, with the rest of the molecules being buried within the fibril interior (Fig. 2.6). The effects of

Fig. 2.6 The complex landscape of the corneal collagen fibril surface, showing N-terminal extensions of collagen V, surface bound FACITs and small leucine-rich proteoglycans (from (Birk and Bruckner, 2005), with permission)

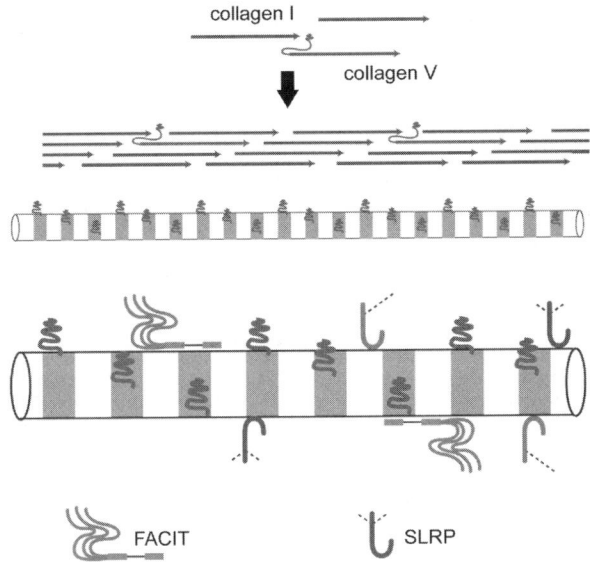

collagen I

collagen V

FACIT

SLRP

collagens V and XI have also been demonstrated in vitro (Birk et al., 1990; Blaschke et al., 2000). In vivo, targeted down-regulation of the expression of collagens III (Liu et al., 1997) or V (Wenstrup et al., 2006), or a natural mutation that prevents collagen XI expression (Li et al., 1995), also results in the formation of large diameter fibrils, consistent with the steric blocking mechanism of fibril growth.

Steric blocking by N-terminal extensions is not the only mechanism by which the minor fibrillar collagens might limit heterotypic fibril diameter. Another mechanism is through altering the rate of fibril nucleation. This has been shown most dramatically in the case of collagen V, where complete deficiency of the proα1(V) chain (which results in death at embryonic day 10) leads to the absence of procollagen V molecules and an almost total lack of collagen fibrils in the dermis, despite their being normal levels of collagen I (Wenstrup et al., 2004). The minor fibrillar collagens therefore seem to act as initiators of fibril formation, during the early nucleation stage. Large amounts of collagens V and XI would therefore lead to greater numbers of heterotypic fibrils (with collagens I and II, respectively). For a given amount of collagen, the presence of a greater number of fibrils would result in the average fibril diameter being smaller.

2.5.4 Interactions with Proteoglycans and Other Components of the Extracellular Matrix

A large body of evidence shows that interactions with the so-called small leucine-rich proteoglycans (SLRPs) have marked effects on collagen assembly (Iozzo, 1999;

McEwan et al., 2006). At present, 15 different types of SLRPs are known, each one consisting of a single polypeptide chain (or protein core) with an N-terminal cysteine-rich cap, followed by several leucine-rich repeat motifs and in most cases a C-terminal disulphide bonded cap. In proteoglycan form, a small number of gly-cosaminoglycan (GAG) chains, either chondroitin/dermatan sulphate or keratan sul-phate, are covalently attached to the protein core. The leucine-rich repeats (each one, 21–27 amino acid residues in length) give SLRPs on overall banana shape, as exemplified by the crystal structure of decorin (Scott et al., 2004). SLRPs known to affect collagen assembly include decorin, fibromodulin, lumican, biglycan, kerato-can and osteoglycin/mimecan. Early in vitro studies showed that decorin and fibro-modulin can markedly interfere with collagen fibril formation, resulting in delayed assembly and usually reduced fibril diameters. Removal of the GAG chains has little effect, showing that this is largely a property of the protein core. Lumican has similar effects in vitro, while biglycan appears to bind to collagen without affect-ing fibril assembly. In the case of osteoglycin/mimecan, it has recently been shown that processing by BMP1 potentiates the inhibiting effect on fibril assembly (Ge et al., 2004).

In vivo, further insights have been gained by targeted disruption (knockout) of the corresponding genes. The decorin knockout is characterized by abnormal skin fragility associated with collagen fibrils having a much broader range of diameters than in normal skin, with non-uniform axial mass distributions and irregular con-tours resulting from fibril fusion (Danielson et al., 1997). This is consistent with a role for decorin molecules in coating the fibril surface thereby keeping them apart. In the case of the lumican knockout (Chakravarti et al., 1998), fibrils in cornea which are normally thin (~30 nm) and uniform in diameter show a simi-lar loss of diameter control and organization resulting in opaque corneas (Fig. 2.7; see Chapter 13). The fibromodulin knockout results in tendon fibrils with reduced diameters, though interpretation of these data is complicated by a compensatory up-regulation of lumican expression (Svensson et al., 1999). Finally, the biglycan knockout leads to an osteoporosis-like defect in bone associated with large irregular fibrils (Xu et al., 1998). It should be noted that SLRPs have multiple biological effects independent of their effects on collagen assembly, so interpretation of the in vivo data is complex.

The interaction of decorin and collagen I has been studied in some detail (Graham et al., 2000). Unlike fibrils formed in vitro, fibrils isolated from tissues show diame-ter limitation for most of their length, as evidenced by a constant axial mass distribu-tion. This is thought to be due, at least in part, to interactions with decorin that coats the fibrils preventing further accretion. Interestingly, the amount of decorin appears to be less at the fibril tips. This is to be expected if the stoichiometry of binding of decorin to collagen is constant, since the number of molecules exposed on the surface, as a proportion of the total number of molecules in a fibril cross-section, increases as the tip diameter becomes smaller. Thus, tips are coated with a relatively small amount of decorin that will therefore no longer prevent fibril growth, until a limiting diameter is reached. This provides a mechanism for fibril to grow in length

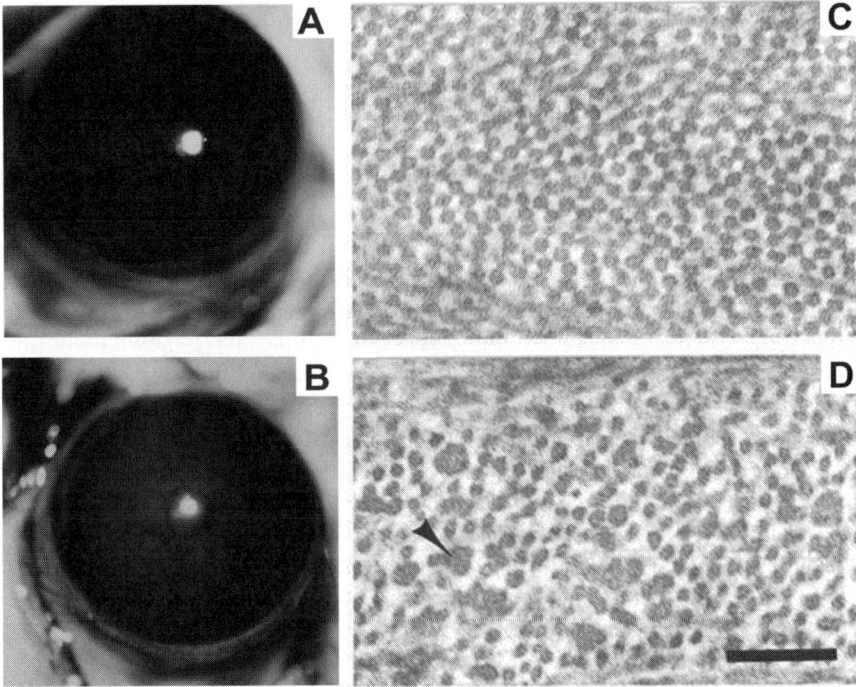

Fig. 2.7 Targeted disruption of lumican expression leads to corneal opacity as a result of loss of collagen fibril diameter control. **A** normal cornea and **C** corresponding electron micrograph of collagen fibrils (in cross-section) in the corneal stroma. **B** and **D** Lumican deficient cornea (from (Chakravarti et al., 1998), with permission)

whilst maintaining a constant diameter. It also permits early fibrils to fuse end to end, with the molecules pointing in the same direction, or in an antiparallel manner by C–C fusion, which creates bipolar fibrils with two N-terminal tips.

While SLRPs have usually been found to diminish the rate of fibril formation, this is not always the case. Recombinant lumican, for example, has been found to accelerate fibril formation in vitro (Neame et al., 2000), as have other extracellular matrix molecules, such as dermatopontin (MacBeath et al., 1993), tenascin-X (Minamitani et al., 2004), perlecan (Kvist et al., 2006), hevin (Sullivan et al., 2006) and cartilage oligomeric matrix protein (Halasz et al., 2007). It is likely that such interactions stabilize the initial nuclei that form during the lag phase prior to subsequent fibril growth. Fibronectin has also been found to accelerate collagen assembly in vitro (Speranza et al., 1987). Consistent with this observation, targeted disruption of fibronectin expression (Velling et al., 2002) or blocking fibronectin assembly with anti-α5β1 integrin antibodies (Li et al., 2003) prevents collagen assembly in cell culture.

2.5.5 Cell Interactions and Long-Range Order

It is now well established that in vivo, procollagen processing, secretion and fibril formation are intimately associated with membrane-enclosed compartments. Prior to secretion (see Section 2.4.4), there is evidence that both procollagen processing and early fibril formation can occur in Golgi to plasma membrane carriers. Subsequently, these carriers appear to fuse with the plasma membrane forming cell-surface invaginations (Canty et al., 2004). Such structures were first observed by Birk, Trelstad and colleagues (Birk et al., 1989; Birk and Trelstad, 1986)(Fig. 2.8). Originally considered to be recesses within the plasma membrane, more recent studies using three-dimensional reconstruction techniques have shown early fibrils within membranous extrusions projecting from the cell surface, called fibropositors (Canty et al., 2004) (Fig. 2.8). It is thought that such structures deliver newly formed fibrils to the extracellular matrix, where they join existing fibril bundles within channels formed between neighboring cells. These structures are readily observed in embryonic tendon when ECM synthesis is high, though not in adult tissues. Tendon is a tissue where fibril alignment is particularly important in order to withstand tensile stress. Recent observations have shown that fibroblasts within chick embryo fibroblasts are connected by cadherin-mediated cell–cell junctions, which appear to define extracellular spaces that facilitate fibril alignment (Richardson et al., 2007).

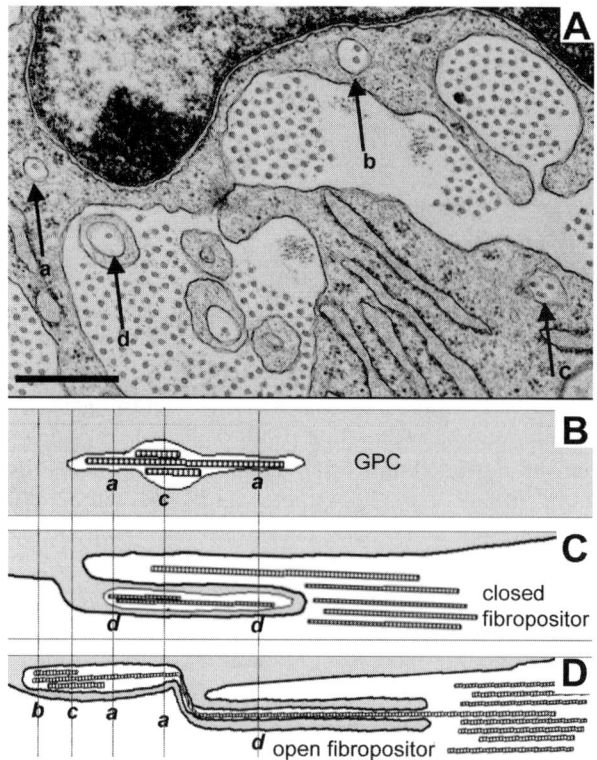

Fig. 2.8 Association of newly formed fibrils with cell membranes. A Electron micrograph of a transverse section through embryonic mouse tail showing bundles of extracellular collagen fibrils between adjacent cells, as well as membrane-bounded fibrils (or small groups of fibrils) within the cytoplasm (a, b, c) or in plasma membrane (PM) extensions called fibropositors (d). Scale bar 500 nm. B–D Schematic representations of collagen fibrils in a Golgi-to-PM transport compartment (B), as well as in closed (C) and open (D) fibropositors, with individual fibrils or small groups of fibrils indicated as in (A). From Canty and Kadler (2005), with permission

Early fibrils in 12–16-day chick embryo tendons have diameters of about 30 nm and lengths in the range 20–30 μm (Birk et al., 1995). Thereafter, there is a sharp increase in length and an increase in fibril diameter. There is evidence that subsequent fibril growth occurs by a process of fibril fusion involving tapered fibril tips. As discussed previously (Section 2.5.2), fibrils in vivo occur in two forms, either N–C unipolar or N–N bipolar, with a change in orientation somewhere in the middle. Since N-to-N fusion is never observed, subsequent fusion of bipolar fibrils cannot occur (Kadler et al., 1996). This limits possible modes of fibril fusion to unipolar–unipolar or unipolar–bipolar, thereby providing a mechanism for controlling fusion within fibril bundles. Another factor controlling fibril fusion is the presence of small leucine-rich proteoglycans, notably decorin and lumican. For example, the rapid rise in fibril length seen in chick embryo tendon at 17 days is correlated with a sharp drop in decorin content (Birk et al., 1995).

While fibril diameters in tendon increase with age and can reach up to 500 nm in rat tail tendon, fibril diameters in corneal stroma remain at about 30 nm throughout life. This is essential for the maintenance of optical transparency (see Chapter 13). Furthermore, unlike the aligned fibril bundles seen in tendon, fibrils in cornea are arranged in lamellae, within each of which fibrils are aligned but between which there is an abrupt change in orientation. In chick corneal, the angle is almost 90° giving rise to a twist in orientation between pairs of lamella throughout the depth of the stroma. Such changes in orientation between lamellae are also encountered in other connective tissues, notably compact bone and intervertebral disc, and are reminiscent of cholesteric liquid crystals (Giraud-Guille, 1996). This similarity raises the question of whether liquid crystalline organization might have a role to play in connective tissue morphogenesis. Clearly, once formed, collagen fibrils are too big to reorient to any major degree in tissues. Collagen molecules, on the other hand, can show liquid crystalline organization in acetic acid solution, resulting in cholesteric phases (Giraud-Guille, 1992). More recently, soluble procollagen molecules have been found to form pre-cholesteric phases in a physiological buffer, thus raising the possibility that long range order might arise prior to fibril formation (Martin et al., 2000). Whether such liquid crystalline ordering of procollagen occurs in vivo, or whether cells control long range order, is a matter of debate. The observation that orthogonal arrays of collagen fibrils form in the chick primary corneal stroma from procollagen molecules that must traverse a basement membrane after being secreted by an epithelial cell sheet (Linsenmayer et al., 1998), is an argument in favor of such a self-assembly process. Indeed, formation of fibropositors may be a consequence of collagen assembly, rather than vice versa (Birk and Bruckner, 2005).

2.6 Assembly of Collagen-Like Peptides

This chapter would not be complete without a discussion of a new and rapidly growing area of research, the use of self-assembling collagen-like peptides to make novel scaffolds for applications in tissue engineering and regenerative medicine. It is now well established that the extracellular matrix provides signals that control many cell activities such as differentiation, adhesion, migration, proliferation and apoptosis. These signals are orchestrated by the three-dimensional organization of the matrix, which provides both specific interactions and also structural clues that are only beginning to be understood. The possibility of synthesizing collagen-like peptides with the ability to self-assemble into nanofibrous structures permits the construction of well-characterized scaffolds for the study of cell–matrix interactions and for applications in tissue repair (Koide, 2005).

Some of the pioneering works in this area was carried out by Fields and his colleagues, who developed "peptide-amphiphiles" consisting of peptides with a collagen-like sequence attached to a long-chain dialkyl ester lipid tail (Fields et al., 1998). The collagen sequence conforms to the host–guest peptide model (Brodsky and Persikov, 2005), with two (Gly-Pro-Hyp)$_4$ sequences (for triple helix stability) straddling a real collagen sequence (in this case, from the collagen $\alpha 1$(IV) chain). These peptide-amphiphiles form stable triple helices and assemble into monolayers, due to the presence of the lipid tail. A related approach was that of Kaplan and colleagues (Martin et al., 2003), who synthesized "collagen-triblock" peptides consisting of the sequence (Gly-X-Hyp-Gly-Pro-Hyp)$_6$, where X = Pro, Ala, Val or Ser, straddled by a pentaglutamate sequence at either end. All four peptides formed stable triple helices. Furthermore, at high concentrations (> 20 mg/ml), all but the Ser-containing triple helices formed liquid crystals. The type of liquid crystal depended on the amino acid sequence: nematic when X = Val and cholesteric when X = Ala or Pro. In nematic phases, neighboring molecules are aligned in parallel, while in cholesteric phases, they are related by a twist. This shows that single amino acid substitutions can have important consequences for supramolecular assembly.

In order to form high molecular weight polymers of collagen-like peptides, two approaches have been used. In the first method, host–guest peptides containing an integrin-binding sequence were constructed with an N-terminal cysteine and a C-terminal thioester (Paramonov et al., 2005). When incubated in an aqueous solvent at neutral pH, these peptides undergo triple-helix formation and also spontaneous chemical ligation to form polymers 10–20 nm in diameter and several microns in length. The second approach is inspired by the notion of "sticky ends" in DNA ligation, and it involves the synthesis of a molecule with a triple-helical core, with two of the chains having N-terminal (Pro-Y-Gly)$_5$ extensions and the third a C-terminal (Pro-Y-Gly)$_5$ extension (Kotch and Raines, 2006) or a variant thereof (Koide et al., 2005) (Fig. 2.9). Molecules then associate by self-complementary interactions to form near-continuous triple helices that are hundreds of nanometer in length. Interestingly, polymers are longer when Y is Hyp rather than Pro, suggesting that hydroxyproline favors self-assembly. This has recently been verified using classical (Pro-Hyp-Gly)$_{10}$ peptides that assemble into branched filamentous triple-helical networks in a temperature and concentration-dependent manner, unlike (Pro-Pro-Gly)$_{10}$ triple helices that do not (Kar et al., 2006).

Fig. 2.9 Collagen-like peptides. **A** Amino acid sequences showing the central triple-helical zone and the overhanging "sticky ends". **B** Self-assembly of molecules shown in (**A**) each identified by a different color. From Kotch and Raines (2006), with permission

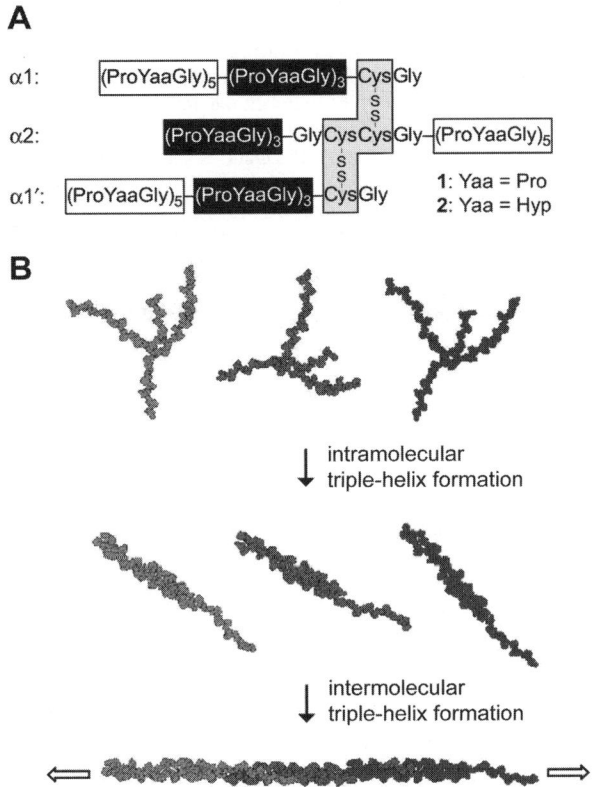

2.7 Conclusions

Less than 40 years ago, only one type of collagen was known. Since then, the number of different types has increased dramatically, in addition to the now large number of collagen-like proteins. The collagen triple-helical motif has turned out to remarkably be versatile in function, leading to the assembly of diverse supramolecular structures. The biosynthesis of collagens is a complex process, involving several post-translational modifications, the functions of which continue to be elucidated. While studies on collagen assembly have gradually switched from in vitro to in vivo, both approaches give complementary information on the molecular mechanisms involved. Armed with this vast amount of data, the time is ripe for developing new approaches to the use of collagens in tissue regeneration and repair.

References

Baldock C, Sherratt MJ, Shuttleworth CA, Kielty CM (2003) The supramolecular organization of collagen VI microfibrils. J Mol Biol 330: 297–307.
Barik S (2006) Immunophilins: for the love of proteins. Cell Mol Life Sci 63: 2889–2900.

Beck K, Brodsky B (1998) Supercoiled protein motifs: The collagen triple-helix and the alpha-helical coiled coil. J Struct Biol 122: 17–29.

Birk DE, Bruckner P (2005) Collagen superstuctures. Top Curr Chem 247: 185–205.

Birk DE, Trelstad RL (1986) Extracellular compartments in tendon morphogenesis: Collagen fibril, bundle, and macroaggregate formation. J Cell Biol 103: 231–240.

Birk DE, Fitch JM, Babiarz JP, Doane KJ, Linsenmayer TF (1990) Collagen fibrillogenesis in vitro: interaction of types I and V collagen regulates fibril diameter. J Cell Sci 95 (Pt 4): 649–657.

Birk DE, Nurminskaya MV, Zycband EI (1995) Collagen fibrillogenesis in situ: Fibril segments undergo post-depositional modifications resulting in linear and lateral growth during matrix development. Dev Dyn 202: 229–243.

Birk DE, Zycband EI, Winkelmann DA, Trelstad RL (1989) Collagen fibrillogenesis in situ: fibril segments are intermediates in matrix assembly. Proc Natl Acad Sci U S A 86: 4549–4553.

Blaschke UK, Eikenberry EF, Hulmes DJS, Galla HJ, Bruckner P (2000) Collagen XI nucleates self-assembly and limits lateral growth of cartilage fibrils. J Biol Chem 275: 10370–10378.

Bonfanti L, Mironov AA, Jr., MartÆnez-Mençrguez JA, Martella O, Fusella A, Baldassarre M, Buccione R, Geuze HJ, Mironov AA, Luini A (1998) Procollagen traverses the Golgi stack without leaving the lumen of Cisternae: Evidence for cisternal maturation. Cell 95: 993–1003.

Borel A, Eichenberger D, Farjanel J, Kessler E, Gleyzal C, Hulmes DJS, Sommer P, Font B (2001) Lysyl oxidase-like protein from bovine aorta. Isolation and maturation to an active form by bone morphogenetic protein-1. J Biol Chem 276: 48944–48949.

Bottomley MJ, Batten MR, Lumb RA, Bulleid NJ (2001) Quality control in the endoplasmic reticulum. PDI mediates the ER retention of unassembled procollagen C-propeptides. Curr Biol 11: 1114–1118.

Brittingham R, Uitto J, Fertala A (2006) High-affinity binding of the NC1 domain of collagen VII to laminin 5 and collagen IV. Biochem Biophys Res Commun 343: 692–699.

Brodsky B, Persikov AV (2005) Molecular structure of the collagen triple helix. Adv Protein Chem 70: 301–339.

Cabral WA, Chang W, Barnes AM, Weis M, Scott MA, Leikin S, Makareeva E, Kuznetsova NV, Rosenbaum KN, Tifft CJ, Bulas DI, Kozma C, Smith PA, Eyre DR, Marini JC (2007) Prolyl 3-hydroxylase 1 deficiency causes a recessive metabolic bone disorder resembling lethal/severe osteogenesis imperfecta. Nat Genet 39: 359–365.

Canty EG, Kadler KE (2005) Procollagen trafficking, processing and fibrillogenesis. J Cell Sci 118: 1341–1353.

Canty EG, Lu Y, Meadows RS, Shaw MK, Holmes DF, Kadler KE (2004) Coalignment of plasma membrane channels and protrusions (fibripositors) specifies the parallelism of tendon. J Cell Biol 165: 553–563.

Chakravarti S, Magnuson T, Lass JH, Jepsen KJ, LaMantia C, Carroll H (1998) Lumican regulates collagen fibril assembly: Skin fragility and corneal opacity in the absence of lumican. J Cell Biol 141: 1277–1286.

Chiquet M, Renedo AS, Huber F, Fluck M (2003) How do fibroblasts translate mechanical signals into changes in extracellular matrix production? Matrix Biol 22: 73–80.

Colige A, Beschin A, Samyn B, Goebels Y, van Beeumen J, Nusgens BV, Lapiere CM (1995) Characterization and partial amino acid sequencing of a 107-kDa procollagen I N-proteinase purified by affinity chromatography on immobilized type XIV collagen. J Biol Chem 270: 16724–16730.

Danielson KG, Baribault H, Holmes DF, Graham H, Kadler KE, Iozzo RV (1997) Targeted disruption of decorin leads to abnormal collagen fibril morphology and skin fragility. J Cell Biol 136: 729–743.

Ellgaard L, Ruddock LW (2005) The human protein disulphide isomerase family: Substrate interactions and functional properties. EMBO Rep 6: 28–32.

Eyre DR, Pietka T, Weis MA, Wu JJ (2004) Covalent cross-linking of the NC1 domain of collagen type IX to collagen type II in cartilage. J Biol Chem 279: 2568–2574.

Fields GB, Lauer JL, Dori Y, Forns P, Yu YC, Tirrell M (1998) Protein-like molecular architecture: Biomaterial applications for inducing cellular receptor binding and signal transduction. Biopolymers 47: 143–151.

Fleischmajer R, MacDonald ED, Perlish JS, Burgeson RE, Fisher LW (1990) Dermal collagen fibrils are hybrids of type I and type III collagen molecules. J Struct Biol 105: 162–169.

Franzke CW, Bruckner P, Bruckner-Tuderman L (2005) Collagenous transmembrane proteins: recent insights into biology and pathology. J Biol Chem 280: 4005–4008.

Fromme JC, Schekman R (2005) COPII-coated vesicles: flexible enough for large cargo? Curr Opin Cell Biol 17: 345–352.

Ge G, Seo NS, Liang X, Hopkins DR, Hook M, Greenspan DS (2004) Bone morphogenetic protein-1/tolloid-related metalloproteinases process osteoglycin and enhance its ability to regulate collagen fibrillogenesis. J Biol Chem 279: 41626–41633.

Ghai R, Waters P, Roumenina LT, Gadjeva M, Kojouharova MS, Reid KB, Sim RB, Kishore U (2007) C1q and its growing family. Immunobiology 212: 253–266.

Giraud-Guille MM (1992) Liquid crystallinity in condensed type I collagen solutions. A clue to the packing of collagen in extracellular matrices. J Mol Biol 224: 861–873.

Giraud-Guille MM (1996) Twisted liquid crystalline supramolecular arrangements in morphogenesis. Int Rev Cytol 166: 59–101.

Graham HK, Holmes DF, Watson RB, Kadler KE (2000) Identification of collagen fibril fusion during vertebrate tendon morphogenesis. The process relies on unipolar fibrils and is regulated by collagen–proteoglycan interaction. J Mol Biol 295: 891–902.

Grenard P, Bresson-Hadni S, El Alaoui S, Chevallier M, Vuitton DA, Ricard-Blum S (2001) Transglutaminase-mediated cross-linking is involved in the stabilization of extracellular matrix in human liver fibrosis. J Hepatol 35: 367–375.

Halasz K, Kassner A, Morgelin M, Heinegard D (2007) COMP acts as a catalyst in collagen fibrillogenesis. J Biol Chem, 282: 31166–31173.

Holmes DF, Capaldi MJ, Chapman JA (1986) Reconstitution of collagen fibrils in vitro; the assembly process depends on the initiating procedure. Int J Biol Macromol 8: 161–166.

Holmes DF, Chapman JA, Prockop DJ, Kadler KE (1992) Growing tips of type-I collagen fibrils formed in vitro are near-paraboloidal in shape, implying a reciprocal relationship between accretion and diameter. Proc Nat Acad Sci U S A 89: 9855–9859.

Holmes DF, Watson RB, Chapman JA, Kadler KE (1996) Enzymic control of collagen fibril shape. J Mol Biol 261: 93–97.

Holmskov U, Thiel S, Jensenius JC (2003) Collections and ficolins: humoral lectins of the innate immune defense. Annu Rev Immunol 21: 547–578.

Hopkins DR, Keles S, Greenspan DS (2007) The bone morphogenetic protein 1/Tolloid-like metalloproteinases. Matrix Biol, 26: 508–523.

Hudson BG, Tryggvason K, Sundaramoorthy M, Neilson EG (2003) Alport's syndrome, Goodpasture's syndrome, and type IV collagen. N Engl J Med 348: 2543–2556.

Hulmes DJS (2002) Building collagen molecules, fibrils, and suprafibrillar structures. J Struct Biol 137: 2–10.

Hulmes DJS, Bruns RR, Gross J (1983) On the state of aggregation of newly secreted procollagen. Proc Natl Acad Sci U S A 80: 388–392.

Hulmes DJS, Kadler KE, Mould AP, Hojima Y, Holmes DF, Cummings C, Chapman JA, Prockop DJ (1989) Pleomorphism in type I collagen fibrils produced by persistence of the procollagen N-propeptide. J Mol Biol 210: 337–345.

Iozzo RV (1999) The biology of the small leucine-rich proteoglycans – Functional network of interactive proteins. J Biol Chem 274: 18843–18846.

Ishida Y, Kubota H, Yamamoto A, Kitamura A, Bachinger HP, Nagata K (2006) Type I collagen in Hsp47-null cells is aggregated in endoplasmic reticulum and deficient in N-propeptide processing and fibrillogenesis. Mol Biol Cell 17: 2346–2355.

Kadler KE (1995) Extracellular matrix 1: fibril-forming collagens. Protein Profile 2: 491–619.

Kadler KE, Holmes DF, Trotter JA, Chapman JA (1996) Collagen fibril formation. Biochem J 316: 1–11.

Kadler KE, Baldock C, Bella J, Boot-Handford RP (2007) Collagens at a glance. J Cell Sci 120: 1955–1958.

Kar K, Amin P, Bryan MA, Persikov AV, Mohs A, Wang YH, Brodsky B (2006) Self-association of collagen triple helix peptides into higher order structures. J Biol Chem 281: 33283–33290.

Knupp C, Squire JM (2005) Molecular packing in network-forming collagens. Adv Protein Chem 70: 375–403.

Koch M, Schulze J, Hansen U, Ashwodt T, Keene DR, Brunken WJ, Burgeson RE, Bruckner P, Bruckner-Tuderman L (2004) A novel marker of tissue junctions: collagen XXII. J Biol Chem 279: 22514–22521.

Koide T (2005) Triple helical collagen-like peptides: engineering and applications in matrix biology. Connect Tissue Res 46: 131–141.

Koide T, Nagata K (2005) Collagen biosynthesis. Top Curr Chem 247: 85–114.

Koide T, Homma DL, Asada S, Kitagawa K (2005) Self-complementary peptides for the formation of collagen-like triple helical supramolecules. Bioorg Med Chem Lett 15: 5230–5233.

Kotch FW, Raines RT (2006) Self-assembly of synthetic collagen triple helices. Proc Natl Acad Sci U S A 103: 3028–3033.

Kuhn K (1987) The classical collagens: types I, II and III. In *Structure and Function of Collagen Types*, Mayne R, Burgeson RE (eds) pp. 1–42. Academic Press: Orlando.

Kvist AJ, Johnson AE, Morgelin M, Gustafsson E, Bengtsson E, Lindblom K, Aszodi A, Fassler R, Sasaki T, Timpl R, Aspberg A (2006) Chondroitin sulfate perlecan enhances collagen fibril formation. Implications for perlecan chondrodysplasias. J Biol Chem 281: 33127–33139.

Kwan APL, Cummings CE, Chapman JA, Grant ME (1991) Macromolecular organization of chicken type X collagen in vitro. J Cell Biol 114: 597–604.

Lees JF, Tasab M, Bulleid NJ (1997) Identification of the molecular recognition sequence which determines the type-specific assembly of procollagen. EMBO J 16: 908–916.

Le Goff C, Somerville RP, Kesteloot F, Powell K, Birk DE, Colige AC, Apte SS (2006) Regulation of procollagen amino-propeptide processing during mouse embryogenesis by specialization of homologous ADAMTS proteases: insights on collagen biosynthesis and dermatosparaxis. Development 133: 1587–1596.

Leighton M, Kadler KE (2003) Paired basic/Furin-like proprotein convertase cleavage of Pro-BMP-1 in the trans-Golgi network. J Biol Chem 278: 18478–18484.

Leikina E, Mertts MV, Kuznetsova N, Leikin S (2002) Type I collagen is thermally unstable at body temperature. Proc Natl Acad Sci U S A 99: 1314–1318.

Li Y, Lacerda DA, Warman ML, Beier DR, Yoshioka H, Ninomiya Y, Oxford JT, Morris NP, Andrikopoulos K, Ramirez F, Wardell BB, Lifferth GD, Teuscher C, Woodward SR, Taylor BA, Seegmiller RE, Olsen BR (1995) A fibrillar collagen gene, Col11a1, is essential for skeletal morphogenesis. Cell 80: 423–430.

Li S, Van Den DC, D'Souza SJ, Chan BM, Pickering JG (2003) Vascular smooth muscle cells orchestrate the assembly of type I collagen via alpha2beta1 integrin, RhoA, and fibronectin polymerization. Am J Pathol 163: 1045–1056.

Linsenmayer TF, Fitch JM, Gordon MK, Cai CX, Igoe F, Marchant JK, Birk DE (1998) Development and roles of collagenous matrices in the embryonic avian cornea. Prog Retin Eye Res 17: 231–265.

Liu X, Wu H, Byrne M, Krane S, Jaenisch R (1997) Type III collagen is crucial for collagen I fibrillogenesis and for normal cardiovascular development. Proc Natl Acad Sci U S A 94: 1852–1856.

Lucero HA, Kagan HM (2006) Lysyl oxidase: an oxidative enzyme and effector of cell function. Cell Mol Life Sci 63: 2304–2316.

MacBeath JR, Shackleton DR, Hulmes DJS (1993) Tyrosine-rich acidic matrix protein (TRAMP) accelerates collagen fibril formation in vitro. J Biol Chem 268: 19826–19832.

Martin R, Farjanel J, Eichenberger D, Colige A, Kessler E, Hulmes DJS, Giraud-Guille MM (2000) Liquid crystalline ordering of procollagen as a determinant of three-dimensional extracellular matrix architecture. J Mol Biol 301: 11–17.

Martin R, Waldmann L, Kaplan DL (2003) Supramolecular assembly of collagen triblock peptides. Biopolymers 70: 435–444.

McAlinden A, Smith TA, Sandell LJ, Ficheux D, Parry DAD, Hulmes DJS (2003) α-helical coiled-coil oligomerization domains are almost ubiquitous in the collagen superfamily. J Biol Chem 278: 42200–42207.

McEwan PA, Scott PG, Bishop PN, Bella J (2006) Structural correlations in the family of small leucine-rich repeat proteins and proteoglycans. J Struct Biol 155: 294–305.

Minamitani T, Ikuta T, Saito Y, Takebe G, Sato M, Sawa H, Nishimura T, Nakamura F, Takahashi K, Ariga H, Matsumoto K (2004) Modulation of collagen fibrillogenesis by tenascin-X and type VI collagen. Exp Cell Res 298: 305–315.

Mironov AA, Mironov AA, Jr., Beznoussenko GV, Trucco A, Lupetti P, Smith JD, Geerts WJ, Koster AJ, Burger KN, Martone ME, Deerinck TJ, Ellisman MH, Luini A (2003) ER-to-Golgi carriers arise through direct en bloc protrusion and multistage maturation of specialized ER exit domains. Dev Cell 5: 583–594.

Moali C, Font B, Ruggiero F, Eichenberger D, Rousselle P, Francois V, Oldberg A, Bruckner-Tuderman L, Hulmes DJS (2005) Substrate-specific modulation of a multisubstrate proteinase. C-terminal processing of fibrillar procollagens is the only BMP-1-dependent activity to be enhanced by PCPE-1. J Biol Chem 280: 24188–24194.

Molnar J, Fong KS, He QP, Hayashi K, Kim Y, Fong SF, Fogelgren B, Szauter KM, Mink M, Csiszar K (2003) Structural and functional diversity of lysyl oxidase and the LOX-like proteins. Biochim Biophys Acta 1647: 220–224.

Morello R, Bertin TK, Chen Y, Hicks J, Tonachini L, Monticone M, Castagnola P, Rauch F, Glorieux FH, Vranka J, Bachinger HP, Pace JM, Schwarze U, Byers PH, Weis M, Fernandes RJ, Eyre DR, Yao Z, Boyce BF, Lee B (2006) CRTAP is required for prolyl 3-hydroxylation and mutations cause recessive osteogenesis imperfecta. Cell 127: 291–304.

Mould AP, Hulmes DJS, Holmes DF, Cummings C, Sear CH, Chapman JA (1990) D-periodic assemblies of type I procollagen. J Mol Biol 211: 581–594.

Myllyharju J (2005) Intracellular post-translational modifications of collagens. Top Curr Chem 247: 115–247.

Myllyharju J, Kivirikko KI (2004) Collagens, modifying enzymes and their mutations in humans, flies and worms. Trends Genet 20: 33–43.

Myllyla R, Wang C, Heikkinen J, Juffer A, Lampela O, Risteli M, Ruotsalainen H, Salo A, Sipila L (2007) Expanding the lysyl hydroxylase toolbox: new insights into the localization and activities of lysyl hydroxylase 3 (LH3). J Cell Physiol 212: 323–329.

Neame PJ, Kay CJ, McQuillan DJ, Beales MP, Hassell JR (2000) Independent modulation of collagen fibrillogenesis by decorin and lumican. Cell Mol Life Sci 57: 859–863.

Paramonov SE, Gauba V, Hartgerink JD (2005) Synthesis of collagen-like peptide polymers by native chemical ligation. Macromolecules 38: 7555–7561.

Parkinson J, Kadler KE, Brass A (1994) Simple physical model of collagen fibrillogenesis based on diffusion limited aggregation. J Mol Biol 247: 823–831.

Parry DA (2005) Structural and functional implications of sequence repeats in fibrous proteins. Adv Protein Chem 70: 11–35.

Peinado H, Del Carmen Iglesias-de la Cruz M, Olmeda D, Csiszar K, Fong KS, Vega S, Nieto MA, Cano A, Portillo F (2005) A molecular role for lysyl oxidase-like 2 enzyme in snail regulation and tumor progression. EMBO J 24: 3446–3458.

Plumb DA, Dhir V, Mironov A, Ferrara L, Poulsom R, Kadler KE, Thornton DJ, Briggs MD, Boot-Handford RP (2007) Collagen XXVII is developmentally regulated and forms thin fibrillar structures distinct from those of classical vertebrate fibrillar collagens. J Biol Chem 282: 12791–12795.

Porter S, Clark IM, Kevorkian L, Edwards DR (2005) The ADAMTS metalloproteinases. Biochem J 386: 15–27.

Prockop DJ, Fertala A (1998) Inhibition of the self-assembly of collagen I into fibrils with synthetic peptides – Demonstration that assembly is driven by specific binding sites on the monomers. J Biol Chem 273: 15598–15604.

Rentz TJ, Poobalarahi F, Bornstein P, Sage EH, Bradshaw AD (2007) SPARC regulates processing of procollagen I and collagen fibrillogenesis in dermal fibroblasts. J Biol Chem 282: 22062–22071.

Ricard-Blum S, Dublet B, van der Rest M (2000) *Unconventional Collagens*. Oxford University Press: Oxford

Ricard-Blum S, Ruggiero F, van der Rest M (2005) The collagen superfamily. Top Curr Chem 247: 35–84.

Richardson SH, Starborg T, Lu Y, Humphries SM, Meadows RS, Kadler KE (2007) Tendon development requires regulation of cell condensation and cell shape via cadherin-11-mediated cell–cell junctions. Mol Cell Biol 27: 6218–6228.

Romanic AM, Adachi E, Kadler KE, Hojima Y, Prockop DJ (1991) Copolymerization of pNcollagen III and collagen I. J Biol Chem 266: 12703–12709.

Ruotsalainen H, Sipila L, Vapola M, Sormunen R, Salo AM, Uitto L, Mercer DK, Robins SP, Risteli M, Aszodi A, Fassler R, Myllyla R (2006) Glycosylation catalyzed by lysyl hydroxylase 3 is essential for basement membranes. J Cell Sci 119: 625–635.

Scott PG, McEwan PA, Dodd CM, Bergmann EM, Bishop PN, Bella J (2004) Crystal structure of the dimeric protein core of decorin, the archetypal small leucine-rich repeat proteoglycan. Proc Natl Acad Sci U S A 101: 15633–15638.

Seidah NG, Prat A (2005) Proprotein convertases in the secretory pathway, cytosol and extracellular milieu. Essays Biochem 38: 79–94.

Snellman A, Tuomisto A, Koski A, Latvanlehto A, Pihlajaniemi T (2007) The role of disulfide bonds and alpha-helical coiled-coils in the biosynthesis of type XIII collagen and other collagenous transmembrane proteins. J Biol Chem 282: 14898–14905.

Speranza ML, Valentini G, Calligaro A (1987) Influence of fibronectin on the fibrillogenesis of type I and type III collagen. Coll Relat Res 7: 115–123.

Stephan S, Sherratt MJ, Hodson N, Shuttleworth CA, Kielty CM (2004) Expression and supramolecular assembly of recombinant alpha1(viii) and alpha2(viii) collagen homotrimers. J Biol Chem 279: 21469–21477.

Sullivan MM, Barker TH, Funk SE, Karchin A, Seo NS, Hook M, Sanders J, Starcher B, Wight TN, Puolakkainen P, Sage EH (2006) Matricellular hevin regulates decorin production and collagen assembly. J Biol Chem 281: 27621–27632.

Svensson L, Aszùdi A, Reinholt FP, Fèssler R, Heinegard D, Oldberg A (1999) Fibromodulin-null mice have abnormal collagen fibrils, tissue organization, and altered lumican deposition in tendon. J Biol Chem 274: 9636–9647.

van der Slot ΛJ, Zuurmond ΛM, Bardoel ΛF, Wijmenga C, Pruijs HE, Sillence DO, Brinckmann J, Abraham DJ, Black CM, Verzijl N, DeGroot J, Hanemaaijer R, TeKoppele JM, Huizinga TW, Bank RA (2003) Identification of PLOD2 as telopeptide lysyl hydroxylase, an important enzyme in fibrosis. J Biol Chem 278: 40967–40972.

Velling T, Risteli J, Wennerberg K, Mosher DF, Johansson S (2002) Polymerization of type I and III collagens is dependent on fibronectin and enhanced by integrins alpha 11beta 1 and alpha 2beta 1. J Biol Chem 277: 37377–37381.

Verderio EA, Johnson TS, Griffin M (2005) Transglutaminases in wound healing and inflammation. Prog Exp Tumor Res 38: 89–114.

Walmsley AR, Batten MR, Lad U, Bulleid NJ (1999) Intracellular retention of procollagen within the endoplasmic reticulum is mediated by prolyl 4-hydroxylase. J Biol Chem 274: 14884–14892.

Wang WM, Lee S, Steiglitz BM, Scott IC, Lebares CC, Allen ML, Brenner MC, Takahara K, Greenspan DS (2003) Transforming growth factor-beta induces secretion of activated ADAMTS-2. A procollagen III N-proteinase. J Biol Chem 278: 19549–19557.

Ward NP, Hulmes DJS, Chapman JA (1986) Collagen self-assembly in vitro: electron microscopy of initial aggregates formed during the lag phase. J Mol Biol 190: 107–112.

Wenstrup RJ, Florer JB, Brunskill EW, Bell SM, Chervoneva I, Birk DE (2004) Type V collagen controls the initiation of collagen fibril assembly. J Biol Chem 279: 53331–53337.

Wenstrup RJ, Florer JB, Davidson JM, Phillips CL, Pfeiffer BJ, Menezes DW, Chervoneva I, Birk DE (2006) Murine model of the Ehlers–Danlos syndrome. col5a1 haploinsufficiency disrupts collagen fibril assembly at multiple stages. J Biol Chem 281: 12888–12895.

Williams BR, Gelman RA, Poppke DC, Piez KA (1978) Collagen fibril formation: optimal in vitro conditions and preliminary kinetic results. J Biol Chem 235: 6578–6585.

Xu T, Bianco P, Fisher LW, Longenecker G, Smith E, Goldstein S, Bonadio J, Boskey A, Heegaard AM, Sommer B, Satomura K, Dominguez P, Zhao C, Kulkarni AB, Robey PG, Young MF (1998) Targeted disruption of the biglycan gene leads to an osteoporosis-like phenotype in mice. Nat Genet 20: 78–82.

Yeowell HN, Walker LC (2000) Mutations in the lysyl hydroxylase 1 gene that result in enzyme deficiency and the clinical phenotype of Ehlers-Danlos syndrome type VI. Mol Genet Metab 71: 212–224.

Chapter 3
Collagen Fibrillar Structure and Hierarchies

T.J. Wess

Abstract Collagen is most commonly found in animals as long, slender generally cylindrical fibrillar structures with tapered ends that are most easily recognized by a 65–67 nm axial periodicity. Collagen fibrils are substantial constituents of skin, tendon, bone, ligament, cornea, and cartilage, where the fundamental tensile properties of the fibril are finely tuned to serve bespoke biomechanical, and less well understood structural signaling roles.

Many of these properties derive from the structural organization within a fibril, where the axial and lateral organization and topology of the collagen molecules ensure strong intermolecular interactions and cross-linkage. The presence of different collagen types within a single fibril is a structural prerequisite in many tissues. However, the necessity for heterotypic fibrillar structures may point to fine tuning of the structural properties in a composite, such as fibril size regulation, dispersion of crystallinity, and interfibrillar communication. Furthermore the specific properties of an individual tissue also rely on the suprafibrillar architecture at the mesoscopic level.

The chapter will discuss current information about fibrillar structure from both homo- and heterotypic fibrils from a structural and biochemical viewpoint. Fibrils rich in collagen I, II and III will be considered as will the contribution of minor fibrillar and FACIT collagens. The molecular organization in both axial and lateral senses will be reviewed for both helicoidal and quasi crystalline fibrillar structures. Current models that account for the dynamic behavior of collagen segments within the fibril will be reviewed and the basis for order and disorder within the fibril discussed. The discrete size and polydispersity of fibrils will be discussed in terms of tissue properties and characteristics. The surface features of fibrils will be considered which conveniently leads to the possible features of interfibrillar interactions.

The overall properties and morphology of a fibril are as important as its internal organization. For example, the fibril surface is a complex area that contains collagen molecules and a variety of proteoglycans. These dictate the interaction between fibrils and specify the environment of partner macromolecules. They are also important in restricting fibril growth and permitting fusion to occur. The defined diameter (or distribution of fibril diameters) and overall slender tapering of collagen fibrils has significance in determining the macroscopic mechanical properties of the tissues.

P. Fratzl (ed.), *Collagen: Structure and Mechanics*,
© Springer Science+Business Media, LLC 2008

The variety of local suprafibrillar and resultant architectures such as bundles, felt work, lamellae and fibers that are evidenced in tissues will be discussed.

3.1 Introduction and Background

Most of the collagen found in animals is in an insoluble fibrillar form where it is commonly found as long, slender, cylindrical, tapered structures. A feature that distinguishes them from other fibrillar forms of macromolecules is that they are most easily recognized by their axial 67 nm periodicity, which can be seen in atomic force microscopy, electron microscopy, and can also be inferred from X-ray diffraction data (Fig. 3.1).

Fig. 3.1 *Left* Electron micrograph of unstained tendon collagen fibrils showing the contrast within the 67 nm repeating unit. *Middle* Atomic force micrograph of collagen fibrils from tendon, where the ridges perpendicular to the fibril direction indicate the axial repeating structure and an aspect of depth to the image. Reprinted from Baselt et al. (1993), with permission from the Biophysical Society. *Right* X-ray diffraction image obtained from tendon. The tendon was held with its axis vertical in the X-ray beam. The diffraction peaks in the center of the image correspond to the lattice made by axial 67 nm periodicity

The collagen fibril therefore provides the key to scaffolding structures in the body from the nanoscopic to macroscopic length scales. Collagen fibrils along with water are substantial constituents of skin, tendon, bone, ligament, cornea, and cartilage. The interaction with other materials such as proteoglycan, minerals, and elastic fibers can lead to the modulation of the overall structural and, thence, the mechanical properties. Few tissues of the body do not require the tensile properties of collagen; the ubiquity of fibrillar collagen does however seem to require a number of different collagen gene products that combine to produce qualitatively different fibrillar forms that allow the modulation of the interactions within and between fibrils. However, the necessity for heterotypic fibrillar structures may point to fine-tuning of the structural properties, such as fibril size regulation, dispersion of crystallinity, and interfibrillar connectivity. These structure–function relationships are still being resolved.

The structural organization of collagen molecules within a fibril has long been the focus of attention, where the organization and topology of the collagen molecules are required to ensure strong intermolecular interactions. Although the axial intermolecular interactions seem to be well understood, the specificity of lateral interactions between collagen molecules appears to be far more variable. The presence of subfibrillar organizations in some fibril types may point to structural levels of organization that are required for the overall performance of a fibril within a composite. The required presence of different collagen types may be to give additional functionality or modulate local molecular packing as part of striking the balance between crystallinity and disorder within a biological polymer.

The overall properties and the morphology of a collagen fibril are as important as its internal organization. For example, the surface of a collagen fibril is a potentially complex area that contains collagen molecules and a variety of proteoglycans. The interfaces provided between the collagen on a fibril surface and the proteoglycan appear to have an importance in restricting fibril growth and permitting/restricting fibril fusion to occur. This is of importance since factors such as diameter, length, and tapering of collagen fibrils (probably governed by collagen–proteoglycan interactions) have significance in defining the macroscopic mechanical properties of the tissues. The basis of fibrils existing either as a fused mass in a tissue or as discontinuous discrete structures is still the subject of debate. The hierarchical organization of fibrils as bundles and macroaggregates (Birk and Trelstad, 1996) is also of paramount importance in providing bespoke mechanical properties. The interactions with other molecules in the composite are key to the transmission of tensile properties and especially viscoelastic properties from the nano- to mesoscopic levels. The molecular interactions that lead to these properties and the interfibrillar interactions that give bespoke tissue properties are the focus of this chapter.

3.2 The Fibril-Forming Collagens

Collagen fibrils contain a number of collagen gene products. This seems to be important to a number of factors in fibril development and especially in surface properties; therefore, minor contributions of collagen may contribute significantly to the interface behavior of fibrils. In addition, the specificity of links between collagen and other non-collagenous matrix components can be made by modulation of the composition of the fibril. Tissue-specific alterations to collagen chain structure as post-translational modifications also need to be considered in their own right.

There are over 30 specific collagen genes that produce over 25 collagen types (some collagen triplexes are made from two or more gene products); here we restrict our overview to the fibril-forming collagens. These can be conveniently divided into two classes: (a) fibril-forming collagens (Kadler, 1994) and (b) fibril-associated collagens with interrupted triple helices (FACIT collagens). In terms of incorporation into fibril structure, the fibril-forming collagens, types I, II, and III (van der Rest and Garrone, 1991), constitute the majority of the fibril, especially the internal structure. Types V and XI are minor components (Burgeson, 1988) but have similar structures. All fibril structure is dominated by the 300 nm triple-helical structure, though some differences in the size and complexity of the telopeptides (the ends of the molecule) occur with collagen type. It is also generally understood that the mechanical properties within a fibril will arise from the interactions between the fibril-forming collagens.

FACIT collagens (Shaw and Olsen, 1991) are a far more heterogeneous family of molecules (for example, types IX, XII, XIV, XVI, and XIX). They usually have additional structural complexity and frequently contain structural motifs that ensure control of interactions between each fibril (Eyre, 2002) and the cell/matrix environment. To that end, FACITs usually occupy the surface of fibrils, and some have a transient interaction with fibrils during development.

Types I and II collagens form two classes of fibrils, in which they are the major form of collagen present. Fibril-forming collagen types III and V are preferentially associated with type I-rich fibrils, whereas type II-rich fibrils frequently contain types IX and XI collagen. Fibrils comprising types II and III collagen have also been observed in cartilage (Young et al., 2000; Wu et al., 1996). Many fibrillar structures are therefore described as heterotypic.

3.3 Molecular Composition of Type I Collagen-Rich Fibrillar Structures

Fibrillar structures that consist primarily of type I collagen, such as those from tendon, appear to have distinct structural features, for example, a greater degree of internal crystallinity and a wide distribution of fibril diameters. Type I-rich fibrils in skin, however, are heterotypic and contain significant amounts of type III collagen, typically around 20% by mass. The length of the triple helix in type III collagen is slightly longer than in type I although, in contrast, the telopeptides are slightly shorter (Cameron et al., 2002). The presence of type III collagen is often associated with tissues, with regulated fibril diameters (Zhang et al., 2005). The absence of type II collagen is the basis of disease, such as Ehlers Danlos type IV syndrome (Pope et al., 1975). Although type III collagen can be buried within the fibril, slowly processed N-terminal propeptides may persist on its surface. Cross-links between type I and type III collagen have also been identified, indicating specific interactions within the fibril (Henkel and Glanville, 1982). These factors are of importance since fibril diameter distribution has been associated with overall mechanical properties (Parry, 1988).

Type V collagen is coassembled with type I collagen in the fibrils of many tissues but notably cornea, and it is thought to be one of the factors responsible for the small, uniform fibrillar diameter (25 nm) characteristic of this tissue (Parry et al., 1978). Suggested mechanisms of fibril diameter regulation are varied (Marchant et al., 1996); however, it may be related to the fact that the entire triple-helical domain of type V collagen molecules is believed to be buried within the fibril. The type I collagen molecules are thus present at the fibril surface (Chanut-Delalande et al., 2001), along with the retained N-terminal domains of the type V collagen. The latter are believed to extend outward through the gap zones.

A significant feature of the triple-helical domain of type V collagen is the high content of glycosylated hydroxylysine residues (10 times higher than for type I collagen); this must have a significant influence on both the intermolecular and interfibrillar interactions of the triple-helical domain of the type V collagen molecule (Mizuno et al., 2001). Further evidence of the type V collagen role in fibrillogenesis has been made (Wenstrup et al., 2004). A collection of other minor collagen components often join the family of proteins that associate with type I collagen in fibrillar forms, for example, collagen type XXIV is found in cornea and bone and is proposed to contribute to the regulation of fibril formation (Koch et al., 2003).

3.4 Molecular Composition of Type II Collagen-Rich Fibrillar Structures

Type II collagen and its associated minor collagens tend to be glycosylated to a greater extent than type I-rich fibrils. Type II collagen is found almost exclusively in cartilage, where the presence of additional minor collagens and non-collagenous glycoproteins is crucial for modulating fibril diameter, surface properties, and interfibrillar interactions (Eyre, 2002).

Collagen XI, a heterotrimeric molecule, is found predominantly in heterotypic cartilage fibrils (Mendler et al., 1989), where it is incorporated into type II-rich fibrils and is involved in the regulation of fibrillogenesis. The partly processed non-helical N-terminal regions are important in altering fibril surface properties (Blaschke et al., 2000). Coassembly of collagens I and XI has also been found in fibrils of several normal and pathologically altered tissues, including fibrous cartilage, bone (Yamazaki et al., 2005), and osteoarthritic joints (Hansen and Bruckner, 2003). These fibril structures appear to be significantly different from fibrils composed of collagen types II and XI.

Type IX collagen is one of the more widely studied FACIT collagens. Its molecular structure contains helical features that allow integration into the fibril, as well as globular and arm-like structures that allow interactions between fibrils (Eyre, 2002). Transgenic mice with mutations in a type IX gene develop normally, but show degenerative changes in articular cartilage after birth. In addition, evidence for the importance of collagen IX in human articular cartilage comes from the finding that a mutation in one of the collagen IX genes causes multiple epiphyseal dysplasia (Olsen, 1997). Other fibril-associated molecules such as collagen XVI represent minor FACIT component of fibrillar collagens (Kassner et al., 2004).

3.5 Collagen Molecular Packing in Fibrils

From the foregoing, it is clear that the collagen fibril is complex even in terms of the overall structure and composite nature. However, there are some central structural features of collagen packing that pertain more or less to all fibrillar structures.

The purpose of this section is to discuss the generic and specific modulation in molecular packing. The least contentious aspect of collagen molecular packing is the axial structure within a fibril. The frequently observed 67 nm density (D) step function repeat of fibrils is explained by the molecular stagger between molecules being 67 nm, or an integer multiple of this. Since the collagen molecular length of 300 nm corresponds to $4.4\,D$, the molecular stagger leads in projection to regions of high and low electron density, these being the overlap and gap regions (Hodge and Petruska, 1963). The association between collagen molecules is driven by electrostatic and hydrophobic interactions, where a 234-amino acid pseudoperiodicity observed within the collagen sequence of all fibril-forming types is the key to optimal electrostatic pairings between adjacent triple helices and maximizes the contact between hydrophobic regions (Hofmann et al., 1980; Hulmes et al., 1973; Itoh et al., 1998; Ortolani et al., 2000). Further small adjustments in the alignment of collagen molecules in the D stagger may be required for the optimization of interactions between triple helices (Cameron et al., 2007). The structural features of axial packing and the corresponding imaged data are shown in Fig. 3.2. These interac-

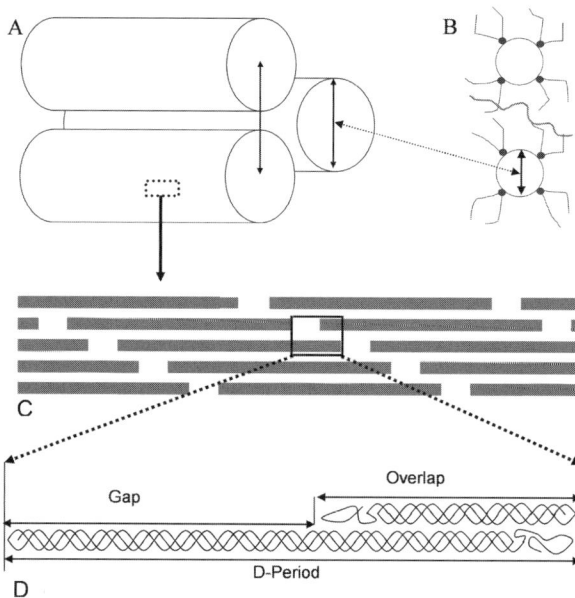

Fig. 3.2 Representation of collagen fibril packing and collagen molecule arrangement within a fibril. **A** Schematic representation of collagen fibril diameter and lateral packing. **B** An axial view of collagen fibrils (*large circles*), with associated proteoglycans and glycosaminoglycans (GAGs). The *small circle* is the proteoglycan protein core and the outward facing lines represent the GAG chains. **C** Each *filled block* represents a collagen molecule in the staggered array conformation found within a fibril. **D** The D-period of the collagen molecules represented by a gap and overlap area within the staggered array. This is representative of the electron density profile, which gives the characteristic step function of the collagen fibril and the strong meridional diffraction series in X-ray diffraction

tions are further stabilized by the development of molecular cross-links between the collagen molecules. These occur between sites in the short non-helical N- and C-terminal telopeptides of the collagen molecules and the main chain of the helix (for a review of collagen cross-linking see Bailey, 2001).

The axial association of heterotypic collagen molecules within a fibril is less well understood. Information pertaining to interactions has also been derived from determining the position of specific natural cross-links between heterotypic collagens in fibrils. This indicates a specificity of axial interaction between type I collagen and type III collagen. The interactions between type I and type V collagen have been determined in part by cross-link analysis (Niyibizi and Eyre, 1994). Electron microscope evidence for the D-periodic interaction of type IX collagen in a type II-rich fibril was established by Vaughan et al. (1988). The basis of the interaction has been elaborated by biochemical studies that show type II collagen forms extensive cross-links with type IX collagen (see, for example, Eyre et al., 2004). The heterotypic axial register has been studied, for example, in the electron microscopy study (Bos et al., 2001), where the simulation of the charge density of a type II fibril (with added contributions from glycosylation) was modulated with predicted contributions from type V/XI and IX collagen density profile. The study was partially successful; however other factors must be required to match the observed electron density. Further studies have also improved the fit for small fibrils from cartilage (Holmes and Kadler, 2006). Suggested interactions are shown schematically in Fig. 3.3.

Detailed analysis of X-ray diffraction data and electron microscope data corresponding to the axial structure of fibrillar collagens has been predominately carried out on type I collagen tissues where the presence of minor collagen components are minimal. These studies have indicated that the exact molecular periodicity contained 234.2 amino acids per D repeat (Meek et al., 1979). The non-helical ends of the collagen molecules are essential for fibril formation, however their molecular organization has been more difficult to ascertain. Attempts to understand telopeptide structure have relied on cross-linkage modeling in situ (Orgel et al., 2001, 2006) and structure determination of isolated telopeptides. The structural stability of the telopeptides is probably conferred in part by the triple-helical environment in which they are surrounded in a fibril. Modeling studies of telopeptides constrained (or docked) within a hexagonally packed triple-helical environment have been made by Jones and Miller (1991), in part by Vitagliano et al. (1995), and by Malone et al. (2004). These studies have indicated that the axial translation per residue within the telopeptides is less than that of the triple helix. This is in agreement with the interpretation of X-ray and neutron diffraction data made by Hulmes et al. (1977, 1980), where the N-terminal telopeptide sequence was shown to be contracted with a reduced axial rise per residue than seen in the triple helix. However, the C-terminal telopeptide contained a turn that causes the polypeptide chain to fold back on itself. A profile of the electron density at a resolution of approximately 0.5 nm was obtained by Orgel et al. (2000). This indicated that the folding point of the type I collagen $\alpha 1$ chain was between residues 13 and 14 of the C-terminal telopeptide. This feature also ensured that the telopeptide lysine residue involved in

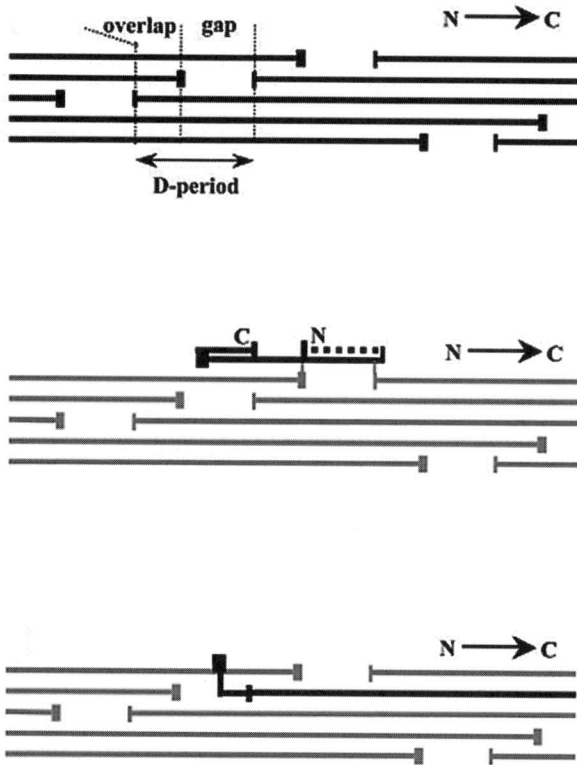

Fig. 3.3 *Top* Representation of the known axial arrangement of type II collagen molecules in a fibril. The triple helix is represented by the *horizontal lines*, and the telopeptides (extrahelical domains) are indicated by the *vertical lines*. *Middle* Representation of the axial location of type IX collagen (*black*) in the type II fibril. The IX–II intermolecular cross-links involve the type II telopeptides and are marked as *grey vertical lines*. The COL3 domain (indicated by *dotted line*) and amino-terminal NC4 domain of the type IX molecule have been represented in bent-back, axially aligned configuration in this preparation. *Bottom* Representation of the axial location of the type XI molecule (*black*) on the type II fibril. The main triple helix of the type XI is in an equivalent axial position to those of the type II molecules. The minor triple helix of the N-propeptides extends into the gap zone; the non-triple domains are shown extended perpendicular to the fibril axis. Reprinted from Holmes and Kadler (2006), Copyright (2006) National Academy of Sciences, USA

intermolecular cross-linking is in register with its acceptor sidechain on an adjacent helical segment. Schematic structures of telopeptides are shown in Fig. 3.4.

The heterotrimeric structure of types II and III collagen, and the lack of sequence similarity to type I collagen, ensures that the conformation of telopeptides in a type II or type III collagen chain must be different to that in type I collagen. In addition, the lack of stability in type I homotrimeric collagen was seen in the oim/oim mouse (McBride et al., 1997; Miles et al., 2002) A study of the type II collagen telopeptide structure utilized electron micrograph staining patterns that were

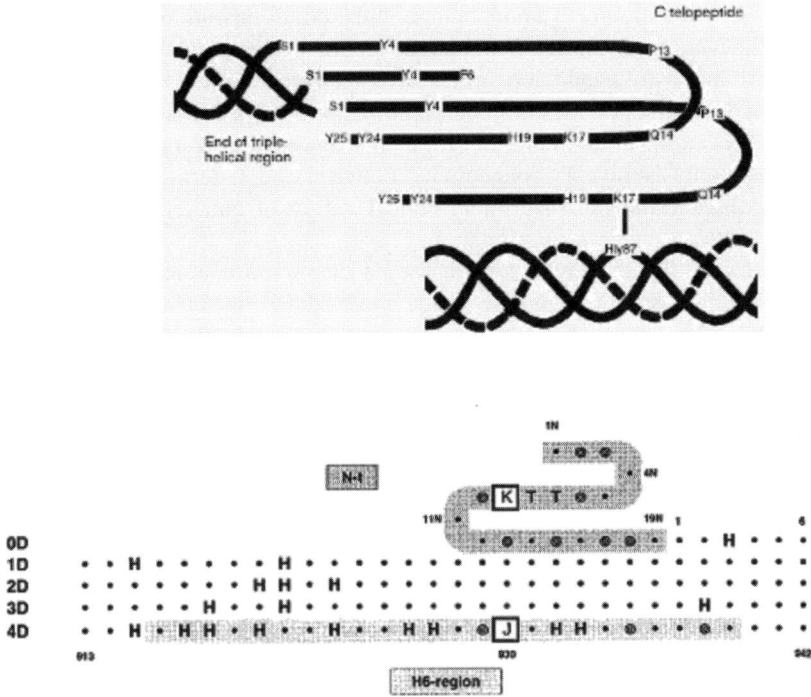

Fig. 3.4 *Top* C-telopeptide of type I collagen containing the proposed tight turn involving residues Pro13/Gln14. This would bring the Tyr24 residues into rough axial alignment with the Tyr4 residues, and bring Lys17 into a favorable position to form the lysine–hydroxylysine cross-link at Hly87 (shown) and help stabilize the microfibrillar structure. Reprinted from Structure, 8, Orgel et al. (2000), Copyright (2000), with permission from Elsevier. *Below* Schematic display of putative N-terminal telopeptide conformation of the type II collagen. Five 234 amino acid-staggered α(II)-chains spanning a 30-residue-long region are placed at a gap–overlap junction. Along the 4D-staggered chain, region H6 (darkened) contains cross-linking Hyl930 (framed). The 0D-staggered chain is preceded by the N-telopeptide (darkened), where Lys9 (framed) is placed. *H*: helix charged amino acids; *T*: telopeptide charged amino acids. *Circle-shaped symbols*: helix hydrophobic residues Leu929, Phe935, Leu938, and telopeptide hydrophobic residues Met2, Ala3, Tyr6, Ala10, Ala13, Met15N, Val17, Met18. Taken from Biopolymers, Vol. 54, 2000, 448–463. Copyright 2000 John Wiley & Sons, Inc. Reprinted with permission of John Wiley & Sons, Inc.

analyzed and interpreted in terms of amino acid sequence (Ortolani et al., 2000). A telopeptide model for both the N- and C-terminii was developed that contained molecular reversals at positions 10N–12N, 12C–14C, and 17C–19C for N- and C-telopeptides. This indicates that the telopeptides have an "S-fold" conformation, which can be interpreted as axial projections of a tridimensional conformation. In contrast, a study of the NMR solution structures of isolated telopeptides from type II collagen (Liu et al., 1993) indicated that the type II C-terminal telopeptide is extended. The recent findings that type II N-telopeptide interacts with Annexin V

(Lucic et al., 2003) points to a novel intercellular communication, which adds to the need for the local structure of this telopeptide region to be resolved. The structures of the type III and V collagen telopeptides have been less studied. However, the NMR study of type III telopeptides has been reported, and the 22-amino acid C-terminal telopeptide is extended with a tight turn involving residues 8–11 (Liu et al., 1993). Cross-link analysis reveals connectivity between the C-terminal telopeptide of type III collagen and the N-terminal helical region of another type III molecule (Henkel, 1996).

The telopeptides are characterized by a lack of the classical triplet Gly-Pro-Hyp. This characteristic signature is the most frequently occurring sequence in fibril-forming collagens, but it only accounts for about 10% of the amino acid triplets found. The significance of sequences with low Hyp/Pro occupancy has led to suggestions that real collagen sequences may contain local distortions from the accepted helical structure formed by "model" Gly-Pro-Pro peptides (Paterlini et al., 1995). Simple modeling of the electron density profile of collagen defined by the presence of amino acids at discrete locations produces an adequate fit with electron micrograph data and X-ray diffraction data. However, the expected position of amino acids requires fine-tuning, in order for an optimal fit with the data to be obtained (Brown et al., 1997; Orgel et al., 2000). This offers the possibility of regions of rarefaction and relative compression in the axial density profile. This may in part explain why crystallization of real collagen sequences has proven difficult. The presence of local distortions in the structure may be induced by the fibril formation; these areas could correspond to regions of differing molecular extensibility, thereby bestowing unique mechanical properties to the fibril (Silver et al., 2001). A detailed study of the local modulation of axial rise per residue has been made by Cameron et al. (2007); this indicates a correlation with sequence of the variability of axial rise per residue with the more stable Gly-Pro-Pro sequences exhibiting least variability.

3.6 Lateral Packing and Molecular Connectivities

The lateral association of collagen molecules within a fibril requires to be understood in terms of fibrillogenesis, fracture properties, cross-linking, and molecular connectivity. A great deal of emphasis has been put on understanding the internal structure of fibrils from tissues such as rat tail tendon. These are attractive to study as they exhibit long-range crystallinity and thus can provide diffraction data, which is rich in information (North et al., 1954; Wess et al., 1998). The long-range crystallinity can also be found in other tissues such as turkey tendon (Jesoir et al., 1981) and lamprey notochord, a type II fibrillar structure (Eikenberry et al., 1984). Determination of molecular packing from these tissues may indicate possible commonalities in molecular packing that may pervade all collagen fibrillar structures. Some dispute still remains in the acceptance of the models proposed from electron microscopic studies; these arguments are rehearsed by Bozec et al. (2007).

3.7 Evidence of Subfibrillar Structures

There is strong evidence that certain fibril types contain subdomains of structure that lie between molecular and fibrillar levels. Such nanoassemblies are of great interest, since they may reveal important information that relates to the overall mechanical properties of the fibrils by preventing crack propagation and also to the nucleation and growth processes of fibril formation.

The presence of pentameric microfibrillar structural units (Smith, 1968) is often viewed as a convenient link between the axial and lateral structures, direct evidence of microfibrils is scarce however many measurements have indicated the presence of a 4 nm lateral periodicity or growth phenomenon that points to evidence of a subfibrillar structure. Evidence for a microfibrillar structure has also been obtained using a number of other physical characterization techniques. For example, electron tomographic reconstructions in dry cornea fibrils revealed 4 nm microfibrillar type structures (Holmes et al., 2001). AFM was used to reveal microfibrils of type I collagen (Baselt et al., 1993).

X-ray diffraction data was used to modulate the rather simplistic pentameric model of collagen association into one arranged as compressed microfibrils on a triclinic unit cell lattice (Fraser et al., 1983; Wess et al., 1995), distortion of a regular pentagon to a compressed structure that allows quasi-hexagonal lateral packing (Trus and Piez, 1980). These units have been revealed in more detail recently by the determination of the structure of a single-collagen unit cell from the electron density map derived from X-ray diffraction studies (Orgel et al., 2006). A consensus of opinion from these studies points to a one-dimensional staggered microfibrillar structure

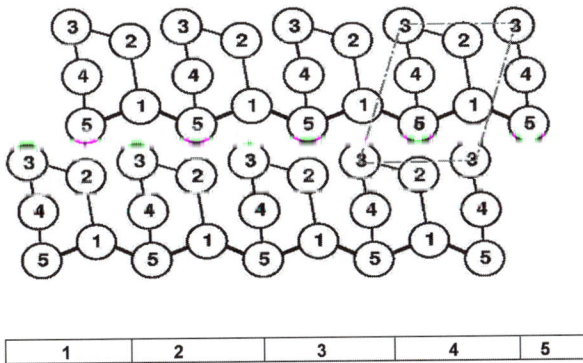

Fig. 3.5 *Top* Topology of interactions of collagen molecular segments within a crystalline region of a fibril. *Bottom* A collagen molecule can be represented by five pseudoperiodic "d" segments where the final C-terminal segment is approximately half the length of other segments. In cross-section through a fibril, the molecular packing allows the in-plane interactions of segments to be revealed. Here the segments of collagen molecules adopt a cyclic microfibril structure where the connectivity in the plane reflects the slow helical climb of contiguous segments within the fibril (see in addition Fig. 3.7). The proposed positioning of the cross-linkage (shown as a *thicker line*) between microfibril structures further stabilizes the collagen molecules. The compatibility between the base of a triclinic unit cell (axes shown in *grey*) and the microfibril topology is shown

with intermicrofibrillar cross-links. Such intermicrofibrillar links are important in the hierarchical connectivity at the supramicrofibrillar level and also provide the basis as to why individual microfibrils have proven so difficult to isolate. The molecular topology of compressed microfibrillar structure is shown in Fig. 3.5.

3.8 Order and Disorder in the Collagen Fibril

Much of the evidence for specificity of molecular packing of collagen in a fibril comes from crystallinity as judged by X-ray diffraction. However, the X-ray fiber diagram of fibrillar collagen also contains a significant amount of diffuse scatter, indicating that fibrils contain a large amount of static or even dynamic disorder. Fibrils that contain crystalline regions are also thought to contain significant levels of collagen molecules exhibiting liquid-like disorder. A large contribution to this could come from the lower density gap region of the fibril, where the structure is believed to be more disordered than in the overlap region (Fraser et al., 1987; Wess et al., 1998). Furthermore, the recent structure from Orgel et al. (2006) indicates that the electron density is clearest at the lateral level of the telopeptides. Here the packing density will be highest due to the accommodation of the folded telopeptides at the ends of the gap region. Thermal vibration and disorder may pervade in the overlap region.

The molecular packing of collagen has been likened to that of a liquid or liquid crystal, where only local molecular interactions are of significance (Fratzl et al., 1993; Cameron et al., 2003; Knight and Vollrath, 2002). In some fibril structures, this type of scatter is the only feature observed by X-ray diffraction, and the overall molecular packing can be described as liquid-like. These observations point to a variety of levels of lateral molecular organization in collagen fibrils, ranging from liquid-like to crystalline. The relationship between crystallinity, disorder, and supramolecular topology of crystalline packing within a fibril requires explanation and is related to fibril formation, overall morphology, and mechanical properties (Prockop and Fertala, 1998).

3.9 Partition of Structure in the Collagen Fibril

Hand in hand with the contrasting evidence of crystallinity and liquid-like order of collagen in a fibril is the evidence for partitioning of structures within the fibril. Beyond the level of the microfibril, AFM studies by Wen and Goh (2004) reveal fibrillar substructures, and work by Gutsmann et al. (2003) led to the concept that the collagen fibril is an inhomogeneous "tube-like" structure composed of a relatively hard shell and a softer, less dense core. In contrast, evidence from Franc (1993) showed central condensed material of 4 nm diameter in all collagen fibrils. The case for a generic fibrillar substructure therefore remains unresolved, and it is possible that bespoke accretion properties may be directed by fundamentally different subfibrillar architectures.

Long-range, organized lateral packing within a fibril also results in partition of substructures within a fibril; this can be understood simply enough in the observation that the fibril cross-section is a disc, which is incompatible with a crystalline lattice. However, the "squaring of the circle" can be made. Investigations by Hulmes et al. (1995) attempted to produce a model that unified (1) electron micrograph observations from lateral sections of fibrils (Hulmes et al., 1985), (2) X-ray diffraction fiber diagrams containing contributions from diffraction and scattering, (3) models of fibril growth that indicated quantal accretion at the fibril surface, and (4) evidence of substructures within a fibril.

The model developed within this particular study based on concentric rings of microfibrillar structures was found to have the best agreement with X-ray diffraction data. The structure contained sectors of crystalline order interfaced by disordered grain boundaries; this allowed crystallites to be accommodated systematically within a fibril with a circular cross-section. The model also provided a structural basis for growth with the incremental deposition of molecules or microfibrils at the fibrillar surface, as suggested by the 8 nm quantal variation in fibril diameter observed by Parry and Craig (1979). The cross-sectional structure of the proposed fibrillar structure is shown in Fig. 3.6. The reported topology of minor collagens and type II collagen in thin cartilage fibrils shows the possible diversity of lateral packing; however, the microfibril structure often remains a common theme (Holmes and Kadler, 2006).

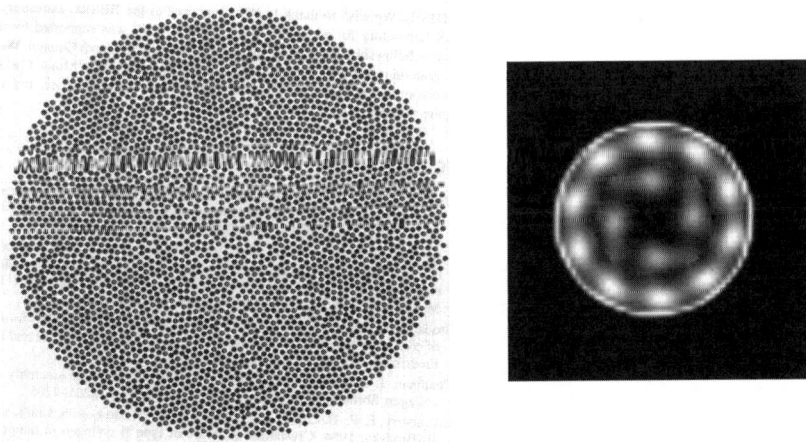

Fig. 3.6 *Left* The proposed structure of collagen molecules packing in a type I collagen fibril. Reprinted from Hulmes et al. (1995), with permission from the Biophysical Society. The structure demonstrates the location of crystalline regions of collagen lateral packing within a structure, which is circular in cross-section. The model was produced as an explanation of order and disorder in collagen fibrils. The center of the fibril (50 nm diameter) is also relatively disordered, which may explain some of the observations of "hollow fibrils" from AFM imaging

Right Experimental reconstructions from negatively stained thin cartilage fibrils exhibiting the 10+4 microfibril structure. This shows that fibril structure may be very tissue dependent. Reprinted from Holmes and Kadler (2006). The 10+4 microfibril structure of thin cartilage fibrils. PNAS, 103 (46): 17249–17254. Copyright (2006) National Academy of Sciences, USA

3.10 Molecular Kinking

Interpretation of X-ray diffraction data shows that in many fibrils, the collagen molecules are tilted with respect to the fibril axis (Fig. 3.7). Detailed interpretation from fibrils that give sufficient diffraction information showed that the sections of molecules within the "overlap" region have a common direction of tilt of magnitude,

Fig. 3.7 The proposed microfibrillar structure of collagen, showing the path of five collagen molecules as they pass around each other in successive D-periods. The common direction of molecular segments in the overlap regions are easily identified, and the molecular rearrangements that must occur to ensure consistency of molecular occupancy occur in the gap regions. This leads to an overall molecular crimped structure that can be compared to the macroscopic crimp structures seen in fibril architectures in Fig. 3.12. The structure was taken from Orgel et al. (2006). Microfibrillar structure of type I collagen in situ, PNAS, 103 (24): 9001–9005. Copyright (2006) National Academy of Sciences, USA. The axial structure has been compressed by a factor of 10 in order to visualize the crimp structure

approximately 5° (Miller and Tochetti, 1981). In order to ensure that the contents of all unit cells are identical in a crystallite, there must be a rearrangement of molecular connectivity in the "gap" region, and this implies that one collagen molecule contains several kinks, where the overall magnitude of its tilt is similar in both the gap and overlap regions. However, the azimuthal orientation of the tilt varies to ensure that each D segment passes through the correct topological position of the unit cell (Fraser et al., 1983; Orgel et al., 2006).

It has been proposed that such a tilt is typical of a structural class of fibril found in tissues that resist uniaxial stress, such as tendon, ligament, and bone (Ottani et al., 2001). A distinct class of fibril is one that exhibits a molecular tilt of approximately 18°. This can be seen in the fibrils from tissues such as skin (Brodsky et al., 1980), chordae tendinae (Folkhard et al., 1987a), cornea (Marchini et al., 1986; Yamamoto et al., 2000a), blood vessels, and nerve sheaths (Ottani et al., 2001). In these tissues, fibril diameters are relatively uniform, typically small (less than 100 nm), and the overall mechanical requisite of the tissue is to resist multidirectional stresses. This leads to a characteristic shortening of the axial repeat from 67 to 65 nm (67 cos 18°). Evidence from X ray diffraction and electron microscopy shows that the molecular tilt of the molecules is believed to be constant throughout the fibril (Holmes et al., 2001). This implies that such helicoidal arrangements must lead to the disruption of intermolecular or intermicrofibrillar interactions, and that the molecular strain at the center of a fibril is difficult to accommodate (Fig. 3.8). The basis for helicoid fibril formation is not clearly defined, although differences in the helical lengths of heterotypic fibril-forming collagens could play a part in defining the helical path of molecules within a fibril (Cameron et al., 2002). These tissues also typically demonstrate a lack of crystallinity, usually contain minor collagens, and demonstrate more liquid-like lateral packing. A detailed review of such fibrillar architectures is given in Ottani et al. (2001).

3.11 The Fibril Surface and Interface Properties

The surface of the fibril provides the "all important" interface with fibrils and other macromolecules. This means that the surface must be endowed appropriately with a complement of molecular species capable of mutual recognition to allow the specific properties of a tissue to be fulfilled. The surface, therefore, is the most complex area of the fibril in terms of molecular heterogeneity and structure.

The rope-like nature of the collagen molecule combined with its composition of relatively simple amino acids may restrict the potential for hetrologous interactions with other molecules. There are however important interactions that are based on the sequences held within the collagen helix. The sequence RGD (arginine, glycine, glutamic acid) is conserved in many collagen molecules and is recognized in D-periodic fibrils by the α2β1 integrin fibril receptor on cells (Yamamoto and Yamamoto, 1994), (Jokinen et al., 2004). Further interactions may then be facilitated by molecules that decorate the fibril surface. The telopeptides also act as a site for cell recognition for receptors such as Endo180 (Thomas et al., 2005). The minor collagens and FACIT collagens may have their most significant role at the fibril

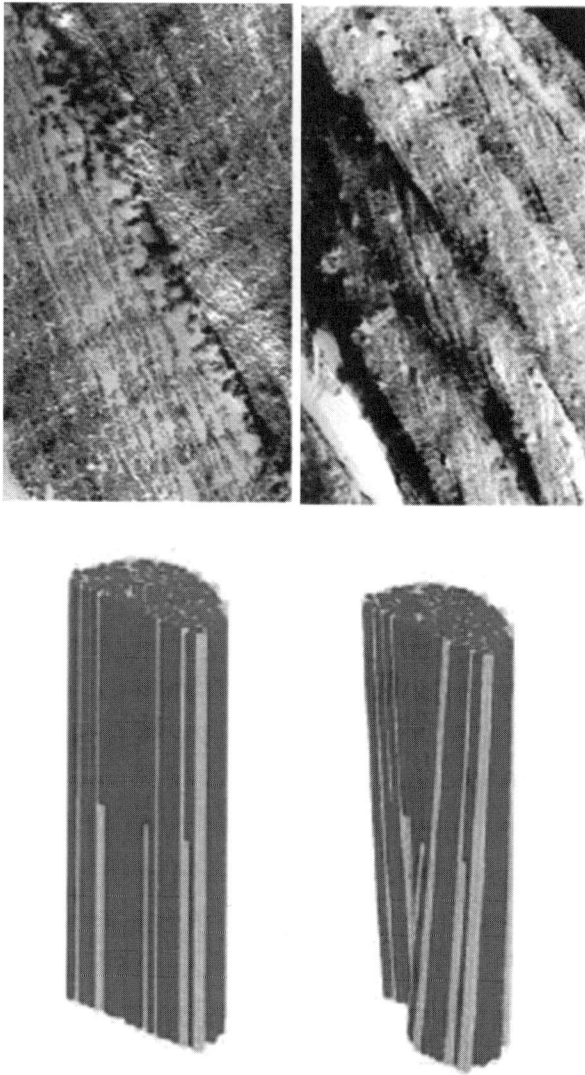

Fig. 3.8 *Top right* The almost parallel array of the collagen molecules in fibrils from rat tail tendon is clearly revealed by the freeze-etching process. *Top left* Thin, uniform fibrils that consistently exhibit the $\sim 16°$ helical winding of their molecules. *Bottom left* A cutaway of a simulated D-periodic structure within a 70-nm-diameter fibril construct, where type I collagen molecules are shown as dark grey rods and the type II molecules are in lights grey rods. The molecules are aligned in parallel to the fibril axis. *Bottom right* A simulated D-period of a fibril where all the molecules make a helicoid twist of 18° to the fibril axis. Reprinted from Journal of Structural Biology, 137, Cameron et al. (2002), Copyright (2002), with permission from Elsevier

surface; the specificity of interactions can be conferred by the presence of many non-helical sequences of FACIT collagens.

Type IX collagen is an exemplar of this class where its globular structural features that would clearly disrupt the internal structure of a collagen fibril are found on the fibril surface. Electron micrographs demonstrating type IX collagen protruding from the surface of fibrils establish this class of molecule as modulators of surface properties. Type IX collagen appears to shield type II collagen from exposure on the fibril surface. The N-terminal NC4 domain of type IX collagen is a globular structure projecting away from the surface of the cartilage collagen fibril. Several interactions have been suggested for this domain, reflecting its location on the fibril (Pihlajamaa et al., 2004). Binding assays showed that the NC4 domain of type IX collagen specifically binds heparin at a site located in the extreme N terminus containing a heparin-binding consensus sequence, whereas electron microscopy suggested the presence of at least three additional heparin-binding sites on full-length type IX collagen. The NC4 domain was also shown to bind cartilage oligomeric matrix protein. This is probably due to the presence of chondroitin sulphate sidechains, which will also play an important role in maintaining fibril spacing (Bishop, 2000). A model of type IX distribution on the fibril surface proposed by Eyre et al. (2004) can accommodate potential interfibrillar as well as intrafibrillar links between the type IX collagen molecules themselves, so providing a mechanism whereby type IX collagen can stabilize a collagen fibril network. Such links at the interfibrillar level are essential for the maintenance of structural integrity.

In addition to a number of collagen gene products crowding the fibril surface, contributions are made by several non-collagenous proteins. Principally, these are the proteoglycans such as decorin, lumican, and fibrimodulin. Their role is thought to involve intrafibrillar connectivity and matrix–cell communication. Most proteoglycans show a non-random axial distribution, with a clear preference for the gap region within the D-period (see, for example, Danielson et al., 1999; Hedlund et al., 1994; Scott and Thomlinson, 1998). Proteoglycans (PG) interact with collagen through globular protein cores that recognize specific sequences in the collagen composition. The extended and highly hydrated glycosaminoglycan components dictate the interfibrillar interactions either via PG to PG interactions or through interaction with other glycosaminoglycans, thus they are also essential modulators of the interfacial shear between fibrils.

Models of interfibrillar relationships usually assume non-specific interactions occurring through the proteoglycan-rich gel and hydrated milieu between fibrils. In the vitreous humor, thin heterotypic collagen fibrils have a coating of non-covalently bound macromolecules which, along with the surface features of the collagen fibrils themselves, probably play a fundamental role in maintaining gel stability. They are likely to both maintain the short-range spacing of vitreous collagen fibrils and link the fibrils together to form a network. A collagen fibril-associated macromolecule that may contribute to the maintenance of short-range spacing is opticin, a leucine-rich repeat protein. Collagen fibrils in extracellular matrices of connective tissues (tendon, cornea, etc.) are also bridged and linked by the anionic glycosaminogly-cans (AGAGs) of the small proteoglycans (decorin, etc.). Such bridges maintain the collagen fibril supramolecular architecture and were proposed as shape modules by Scott and Thomlinson (1998). Figure 3.9 shows an electron micrograph and schematic of the fibril surface. The close association of proteoglycans or glycosaminoglycans (PGs/GAGs) with maturing type VI collagens may provide a further indication to the link that associates D-periodic collagen fibrils via PGs/GAGs. Ruthenium red stainability on the surface of D-periodic collagen fibrils was also examined; the results show that these sites were also D-periodically associated (Watanabe et al., 1997).

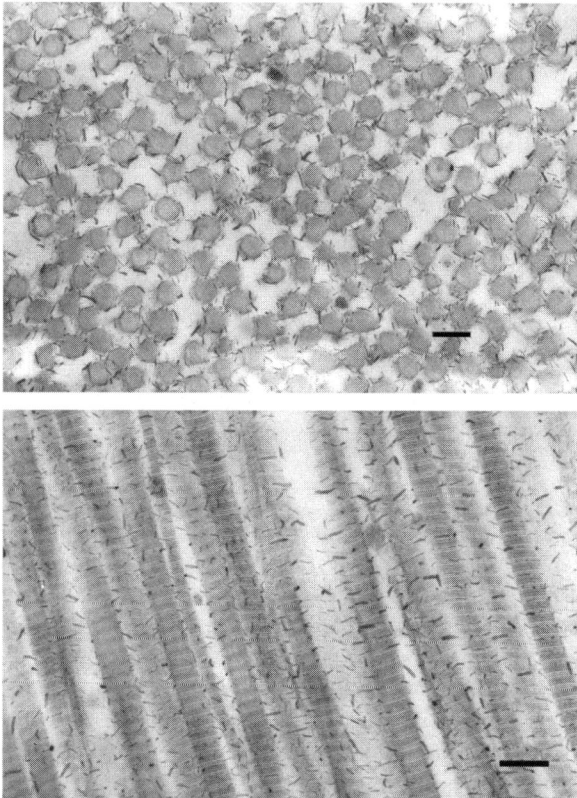

Fig. 3.9 Fibril surface properties: electron micrographs of transverse (*top*) and longitudinal (*bottom*) sections of collagen sclera. Bar = 50 nm. The proteoglycans that decorate the surface of the fibrils are shown as darker stained objects due to the use of a metal-based electron dense stain. The collapsed arms of the GAG chains can be seen to be connecting between fibrils. Images courtesy of Dr R Young, Cardiff University

Furthermore, specific interactions of collagen fibrils with the GAG-rich regions of several aggrecan monomers aligned within a proteoglycan aggregate have been identified. The fibril could therefore serve as a backbone in at least some of the aggrecan complexes (Hedlund et al., 1999). A highly specific pattern of cross-linking sites suggests that collagen type IX has evolved to function as an interfibrillar network-bonding agent. Interfibrillar coherence has been observed in some mineralized tissues, such as turkey leg tendon (Landis et al., 1996), where the axial banding pattern of several mineralized fibrils are in register. This may result from the growth of mineral between fibrils, but may be an association required prior to interfibrillar calcification. The significance of this and the possible specificity of interaction (or simple entanglement) of the glycosaminoglycan part of proteoglycans remains a rich source for future research.

3.12 Factors Involved in Fibril Growth and Size

Studies of the mechanical properties of tissues require knowledge of fibril size and interface properties; these factors vary widely among collagen-rich tissues with fibril diameters ranging from above a micron in tendon to below 40 nm in cornea, and ~10 nm fine filaments of cartilage. The distribution of fibril diameters is also tissue and age dependent (Fig. 3.10). Tissues such as tendon, which are predominantly type I collagen, contain a wide variety of fibril dimensions. This in part may relate to the tapered nature of fibrils and the in-filling that occurs between the close-packed cylindrical collagen fibrils. The interplay of heterotypic collagens, fibril-associated collagens, and co-fibrillar macromolecules that alter the accretion properties of available procollagen molecules may be required to form a typical collagen fibril with its characteristic cylindrical central shaft section and differential tip shapes. Although collagen fibril formation can be mimicked in vitro as an acellular system, the role of cellular processes cannot be underestimated. The challenge remains in research to understand the interplay that exists between extracellular macromolecules and cellular processes that produce the final fibrillar structure.

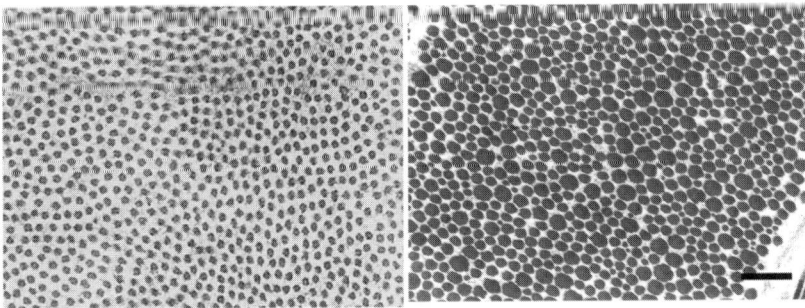

Fig. 3.10 A comparison of two tissues exhibiting differences in fibril diameter, diameter distribution, and spacing. *Left* Cornea fibrils are small with a distinct spacing between neighboring fibrils. The crystalline packing is however difficult to identify beyond nearest neighbors in even this tissue. *Right* Scleral tissue exhibits fibrils that are both larger and smaller in diameter than cornea. The fibrillar packing is more irregular and the spaces between larger fibrils is often filled with small fibrils. Images courtesy of Dr R. Young, Cardiff University. Bar = 50 nm

In the extracellular matrix, the procollagen C-propeptides ensure procollagen solubility prior to deposition, while it has been postulated that the persistence of the N-propeptides controls fibril shape (Hulmes, 2002). The N-terminal propeptides that persist to different extents in different collagen chain types play a role in determining fibril size distribution. Persistence of the N-propeptide from the slower processing of type III collagen or the partial removal of the bulky propeptide units of types V or XI collagen may result in a decoration of the fibril surface with globular protein domains. This may prevent further accretion, and various possibilities are discussed by Chapman (1989), Birk (2001), and Linsenmayer et al. (1993). Such surface interactions may be an important factor in why many fibrils are heterotypic in nature. The relative distribution of collagen types within fibril structures however remains poorly resolved.

Type V collagen integration in corneal tissue from both avian and mammalian sources points to heterotypic collagen interactions, providing some basis for fibril diameter regulation (White et al., 1997). The triple-helical structure of type V collagen alone has been shown to be an important factor in regulating the fibril diameter of type V/type I heterotypic fibrils in vitro (Adachi and Hayashi, 1986). It may therefore be possible that the disruption of molecular packing within a fibril caused by heterotypic interactions itself limits fibril growth.

The fibril surface proteoglycans also have an important role in the regulation of fibril structures. Molecules such as lumican (Chakravarti et al., 1998), decorin (Danielson et al., 1999), and fibromodulin (Svensson et al., 1999) coat the surface of fibrils in a specific manner and may restrict further growth. The competitive or sequential mechanisms that may be responsible for the coating of fibrils after nucleation or a certain amount of growth are still subject to speculation (MacBeath et al., 1993). Knockout gene experiments in mice have indicated that the loss of proteoglycans (such as decorin) causes alteration of fibrillar morphology, and fibril sizes are observed to be more polydisperse. However, upregulation and compensatory behavior of other associated proteoglycans make the results more complex to interpret than the removal of a specific gene product (Derwin et al., 2001).

3.13 Distribution of Fibril Diameter and Length

Fibril-forming collagens are typically 300 nm long and approximately 1 nm in diameter. These regular structures, however, form fibrillar structures of length and diameter that tend to vary depending on anatomical location. The basis for this wide distribution of fibrillar architectures is not well understood, but may lie in the macromolecular composite nature of many fibrils. The effective length of fibrils has been difficult to study, since in vitro fibril preparation is probably not reflective of the overall length of a fibril in vivo. Disruption and dispersion of fibril structures from tissues inevitably leads to breakage of fibrils, and resultant intact length may not be representative of the overall fibrillar length. Some attempts have been made to study fibril lengths in electron microscope transverse section. This is a painstaking and time-consuming pursuit; therefore, the number of fibrils that can be studied systematically in a tissue is low. The study by Craig et al. (1989) estimated the length scale of rat tail tendon fibrils to be from 0.3 mm to greater than 10 mm, depending heavily on developmental status. Studies of fibril length in development by Birk et al. (1997) failed to find both ends of any fibril. This indicated a possible fibril length longer than sequential microscope sections would allow, or that the fusion of fibrils produced a networked continuum at maturity. Furthermore, the lengths of the fibrils probably change with maturation. Fibril fusion and fibril-splitting events both seem to be possible descriptions of observed fibril bifurcation observed by electron microscopy (Christiansen et al., 2000; Graham et al., 2000; Hoshi et al., 1999), see Fig. 3.11. A useful resource on the comparison of fibril length and diameter measurements using different approaches is described in Redaelli et al. (2003).

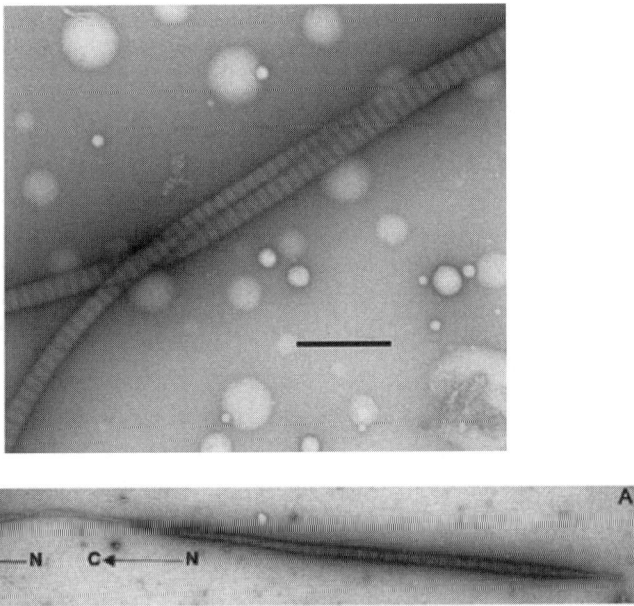

Fig. 3.11 *Top* A fibrillar structure that has either split or fused to produce a network of fibrils. The continuity between the fibrils can easily be evidenced by observation of the banded profile in both parent and progeny strands. Reprinted from Matrix Biology, 19, Kadler et al., Tip-mediated fusion involving unipolar collagen fibrils accounts for rapid fibril elongation, the occurrence of fibrillar branched networks in skin and the paucity of collagen fibril ends in vertebrates, 359–365, Copyright (2000), with permission from Elsevier. *Bottom* Collagen fibrils have tapered tips; these are also asymmetric. This is more easily seen in young growing fibrils. Reprinted from Journal of Molecular Biology, 295, Graham et al. (2000), Copyright (2000), with permission from Elsevier

Collagen fibril diameters vary widely, but the mechanical or developmental reasons for such a large variation among tissues remain poorly understood. It has been suggested that the mechanical properties of tendon are related to the fibril

diameter distribution; the large fibrils have a primary role in withstanding high tensile forces and the smaller fibrils have a special ability to resist creep (Parry and
Craig, 1984). Fibril diameters increase with maturation of tissue, but break down
at senescence (Parry et al., 1978). In terms of development, thinner fibrils may be
broken down more easily. However, experiments monitoring changes in fibril diameter with in vivo loading of tissue have resulted in little consensus (Michna, 1984;
Patterson-Kane et al., 1997). Tissues such as mature Achilles tendon contain a
bimodal distribution of diameters with thicker fibrils of diameters about 150–250 nm
and a population of thinner fibrils with diameters of about 50–80 nm (Svensson
et al., 1999).

In structures such as cartilage and vitreous humor, there is also a clear bimodal
distribution of fibril diameters; however, many tissues such as skin and submucosa
contain a more uniform distribution of diameters around 60 nm (Sanders and Goldstein, 2001). The rationale behind a discrete fibril diameter in load-bearing tissues
is not immediately apparent, since a priori a distribution of fibril sizes would be
thought to dissipate the load more evenly throughout the tissue. It has been argued
that these tissues are usually associated with tear resistance and higher compliance
than either tendon or ligament. The fibril diameter distribution in a tissue is possibly,
therefore, a balance between overall tensile strength and creep resistance, the former
being related to fibril size and the latter to the interfacial shear and thus the surface
area:volume ratio (for a review, see Ottani et al., 2001). A special case where the
optical properties of the tissue are paramount is that of cornea. Probably the most
highly regulated fibrillar diameter in animals is that found in adult human cornea
(33 nm; Meek and Fullwood, 2001). The relatively small and highly constrained
diameters, and the distribution on a semicrystalline lattice, are both essential features
for the transparency of the tissue (see Chapter 13).

Techniques such as small-angle X-ray scattering, which has been used to estimate fibril diameter without embedding or sectioning, tends to skew the distribution
to larger fibril sizes unless the scattering power from large fibrils up to 500 nm in
diameter is taken into account. An approach to determining fibril diameters with no
a priori knowledge of size is given by Goh et al. (2005).

3.14 Suprafibrillar Architectures

Although there is a strong molecular and nanoscopic driving force behind our understanding of the functionality of collagen fibrillar structures, the mesoscopic and
macroscopic properties of the tissue are where the structural integrity manifests
itself. Studies of collagen often focus on one specific level of structural organization
and more needs to be done to understand the linkages between different levels of
structural organization, for example see (Buehler, 2006). One of the differences
in mechanical resistive properties between tissues such as skin and tendon is the
feltwork nature of fibril distribution in skin. This allows resistance to strain to occur
within a two-dimensional plane. In contrast, tendon is required (normally) to resist
strain only along its axis; such properties of angular distribution and reinforcement
are discussed by Ottani et al. (2001).

Cornea is a tissue where collagen fibrils of uniform diameter are regularly arranged within a matrix of proteoglycans. The fiber direction in cornea, however, is not uniform and consists of a succession of superposed layers of fibrils with different orientations. The collagen fibrils in each layer are parallel to one another but are kept at a significant distance from one another. This is essential for the transparency of the tissue. Fibril distribution in the cellular cementum presents a lamellar packing pattern that conforms to the twisted plywood principle of bone lamellation, with a periodic rotation of matrix fibrils resulting in an alternating lamellar pattern (Yamamoto et al., 2000); further information on collagen interactions in bone at a variety of hierarchical levels can be found in Fratzl et al. (2004). In normal articular cartilage, the collagen fibrils in the superficial zone are compactly arranged into layers of decussating flat ribbons, mostly parallel to the artificial split lines used for specimen orientation (Hwang et al., 1992; Jeffery et al., 1991).

Collagen fibrils associate in many tissues to form discrete "macro-aggregate" structures (Birk and Trelstad, 1996). Within one collagen fiber, the fibrils are oriented not only longitudinally but also transversely and horizontally. The longitudinal fibers do not only run parallel but also cross each other, forming spirals, cross-plys, and some of the individual fibrils and fibril groups form spiral-type plaits (Kannus, 2000). Such local suprafibrillar architectures present organizational features of lamellae, twisting or sinusoidal "crimps" on a mesoscopic scale. Hierarchical structures of fibrils can be seen in Fig. 3.12.

It has been suggested that a possible commonality between many different collagen-based tissues is that the suprafibrillar organization resembles the cholesteric or precholesteric organization of liquid crystals (Giraud-Guille, 1996; Hulmes, 2002). Such liquid crystalline behavior was shown to occur in very high concentrations of collagen at low pH; this behavior has also been observed in procollagen molecules at high concentrations in a physiological buffer (Martin et al., 2000). Therefore, it may be possible that liquid crystalline ordering of collagen molecules may precede condensation into true fibrillar forms.

Relatively few observations have been made that relate to the specificity of intrafibrillar or larger scale interactions. In the main observation, fibrils are not required to have a regular axial spatial relationship that would lead to long-range coherence in the tissue. The distribution of fibrils as fibers or fibril bundles occurs as a common motif in many tissues (Silver et al., 1992), with the fiber direction being aligned to provide the bespoke mechanical characteristics of the tissue (Kastelic et al., 1978). Suprabundle structures of collagen often provide the interface of collagen with other biopolymers, such as elastin and fibrillin. These elastic components complement the stiff collagen fibers and produce the overall functional properties of the tissue. Periodontal ligament contains principal fibers developed from aggregates of fibrils that are allowed to form during root development in the space between ligamental cells (Yamamoto and Wakita, 1992). In the case of tendon, the suprafibrillar hierarchical structures are well characterized. Here a group of collagen fibrils forms a collagen fiber, the basic unit of a tendon. Fibers are bound together by a fine sheath of connective tissue called the endotenon. A bunch of collagen fibers forms a primary fiber bundle, and a grouping of primary fiber bundles results in a

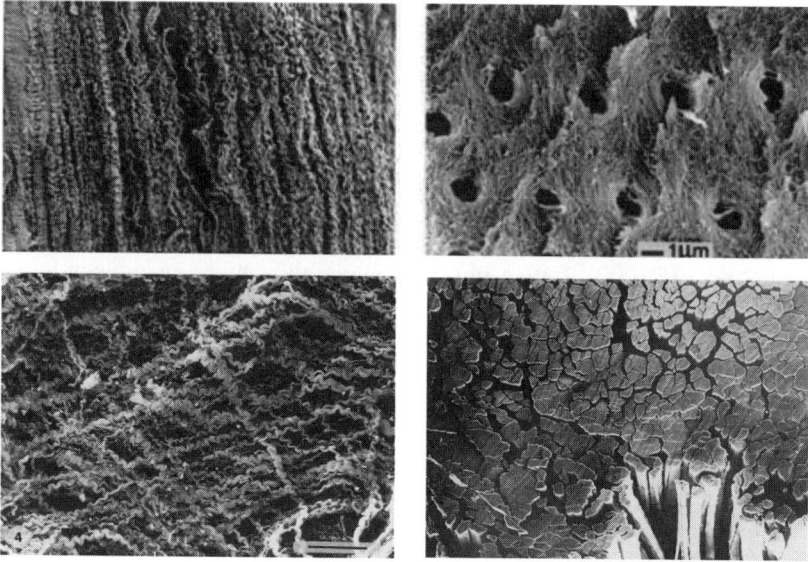

Fig. 3.12 Examples of suprafibrillar architectures. *Top left* SEM images of collagen fibril morphology during late development. The images show the formation of fibril bundles (fibers) that possess the characteristic crimp morphology seen in mature tendon and ligament. Reprinted from Matrix Biology, 25, Provenzano and Vanderby Jr. Collagen fibril morphology and organization: Implications for force transmission in ligament and tendon, 71–84, Copyright (2005), with permission from Elsevier. *Top right* Radial fibril arrays. SEM: human dentin fractured roughly parallel to the pulp cavity surface. The tubules (holes) are surrounded by collagen fibrils that are all more or less in one plane. Taken from Weiner et al. (1999). Reprinted, with permission, from the Annual Review of Materials Science, Volume 28 (c)1998 by Annual Reviews www.annualreviews.org. *Bottom left* The connectival stroma of the small intestine is made of slender collagen bundles running in all directions and forming a bidimensional isotropic web. SEM, scale bar indicates 50 μm. Reprinted from Micron, 32, Ottani et al. (2001), Copyright (2000), with permission from Elsevier. *Bottom right* Low magnification micrograph of a transected tendon; collagen bundles form thick, straight, parallel fascicles. SEM, scale bar indicates 50 μm. Reprinted from Micron, 32, Ottani et al. (2001), Copyright (2000), with permission from Elsevier

secondary fiber bundle. Secondary fiber bundles accumulate to form tertiary bundles that comprise the tendon (Kannus, 2000).

3.15 Relationships with Mechanical Properties of Collagen-Rich Tissues

The outstanding mechanical properties of tissues that rely on fibrillar collagen are due to the optimization of their structure on many hierarchical levels. The interplay between structures already described from the molecular to fibrillar, and thence to interfibrillar, is critical to the overall mechanical properties (Fratzl, 2003). Strain in the tissue results from two principal mechanisms – the molecular elongation and molecular shear within a fibril, and the shear deformation of the proteoglycan-rich matrix between fibrils. The correct supramolecular organization and cross-linking within each fibril is therefore essential. The response of the tissue to the stress imposed depends very much upon the strain rate, where high strain rates cause elongation of the collagen molecules and initiate shearing effects within a fibril. Slow shear rates result in shear of the matrix between the fibrils, leading to creep (Sasaki et al., 1999).

Synchrotron radiation studies of fibril behavior in tissues have been critical to investigations of structural alterations, since they allow transient structural features to be monitored in realistic time frames. Stress–strain experiments conducted synchronously with X-ray diffraction reveal changes in the fibril axial D-period from 67 nm to just under 69 nm before failure, however only 40% of this lengthening is a contribution from the extension of the collagen helix. The rest of the fibril extension must result from some form of molecular rearrangement. This appears with the change in the relative intensity of the second- and third-order Bragg peaks of the meridional series, indicating a severe distortion in the usual gap/overlap step function that is characteristic of fibrillar collagens (Sasaki and Odajima, 1996). In situ synchrotron X-ray scattering experiments suggest that several different processes could dominate, depending on the amount of strain.

While at small strains there is a straightening of kinks in the collagen structure (first at the fibrillar and then at the molecular level), higher strains are believed to lead to molecular gliding within the fibrils and ultimately to a disruption of the fibril structure. Preliminary attempts to quantify these distortions in terms of alteration of the collagen axial structure met with limited success (Folkhard et al., 1987b). Here, in similar studies, three elementary models for molecular elongation and rearrangement of collagen within a fibril were proposed. In the first model, molecular elongation arises through an alteration of the helical pitch. In the second model, there is an increase in the length of the gap region. The third model shows that a relative slippage of laterally adjoining molecules occurs

The shortcomings of such investigations can be interpreted as (a) the starting fibrillar model is of insufficient accuracy to mimic the changes imposed by the mechanisms of distortion or (b) the emphasis, or even the entire basis, for the mechanisms of distortion is incorrect. The relationship of collagen molecular mechanical properties with fibrillar properties is described in models by Buehler (2006), where there are two length scales of interactions characterized by first the deformation changes from homogeneous intermolecular shear and secondly by covalent bond fracture. The latter factor is not as yet incorporated into published structural models.

It has also been noted that the strain within collagen fibrils is always considerably smaller than in the whole tendon. This phenomenon is still very poorly understood, but points toward the existence of additional gliding processes occurring at the interfibrillar level (Fratzl et al., 1998). Once again, this leads to the fibril surface and interfibrillar interactions having a more prominent role than generally appreciated. In turn, this deserves more attention than it currently receives. Mechanisms for interfibrillar stress transfer almost certainly involve the surface and interfibrillar proteoglycans, where molecular entanglement and electrostatic interactions may provide the basis for shear resistivity. The glycosaminoglycans bound to decorin act like bridges between contiguous fibrils connecting adjacent fibrils every 64–68 nm. This architecture would suggest their possible role in providing the mechanical integrity of the tendon structure. Such interactions have been investigated (among others) by Redaelli et al. (2003).

In conclusion, the collagen fibril is a complex structure with a fundamental D-repeat of approximately 65–67 nm. This provides a framework on how different molecular species interact with one another, to provide fibrillar structures with features suited to both their roles as biomechanical tensile materials and as a source of intermolecular connectivity. The ubiquity of collagen in most animal tissues shows that a high degree of functionality can be married by modulations on a relatively stable framework. The hierarchical organization of collagen-based materials from the molecular to the functional tissue allows for the intervention and interplay of a series of design features that ensure that the "triple-helical rope" is put to best use in each case.

References

Adachi, E. and Hayashi, T. (1986). In vitro formation of hybrid fibrils of type-V collagen and type-I collagen – Limited growth of type-I collagen into thick fibrils by type-V collagen. Connect. Tissue Res. 14, 257–266.

Bailey, A. J. (2001). Molecular mechanisms of ageing in connective tissues. Mech. Ageing Dev. 122, 735–755.

Baselt, D. R., Revel, J. P., and Baldeschwieler, J. D. (1993). Subfibrillar structure of type-I collagen observed by atomic-force microscopy. Biophys. J. 65, 2644–2655.

Birk, D. E. (2001). Type V collagen: Heterotypic type I/V collagen interactions in the regulation of fibril assembly. Micron 32, 223–237.

Birk, D. E. and Trelstad, R. L. (1996). Extracellular compartments in tendon morphogenesis: collagen fibril, bundle, and macroaggregate formation. J. Cell Biol. 103, 231–240.

Birk, D. E., Zychband, E. I., Woodruff, S., Winkelmann, D. A., and Trelstad, R. L. (1997). Collagen fibrillogenesis in situ: Fibril segments become long fibrils as the developing tendon matures. Dev. Dyn. 208, 291–298.

Bishop, P. N. (2000). Structural macromolecules and supramolecular organisation of the vitreous gel. Prog. Retin. Eye Res. 19, 323–344.

Blaschke, U. K., Eikenberry, E. F., Hulmes, D. J. S., Galla, H. J., and Bruckner, P. (2000). Collagen XI nucleates self-assembly and limits lateral growth of cartilage fibrils. J. Biol. Chem. 275, 10370–10378.

Bos, K. J., Holmes, D. F. Meadows, R. S., Kadler, K. E., McLeod, D., and Bishop, P. N. (2001). Collagen fibril organisation in mammalian vitreous by freeze etch/rotary shadowing electron microscopy. Micron 32, 301–306.

Bozec L, van der Heijden G, and Horton, M. (2007). Collagen fibrils: nanoscale ropes. Biophys J. 92, 70–75.

Brodsky, B., Eikenberry, E. F., and Cassidy, K. (1980). An unusual collagen periodicity in skins. Biochim. Biophys. Acta. 621, 162–166.

Brown, E. M., King, G., and Chen, J. M. (1997). Model of the helical portion of a type I collagen microfibril. J. Am. Leather Chem. As. 92, 1–7.

Buehler, M. K. (2006). Nature designs tough collagen: Explaining the nanostructure of collagen fibrils. PNAS 103, 12285–12290.

Burgeson, R. E. (1988). New collagens, new concepts. Ann. Rev. Cell Biol. 4, 551–577.

Cameron, G. J., Alberts, I. L., Laing, J. H., and Wess, T. J. (2002). Structure of type I and type III heterotypic collagen fibrils: An X-ray diffraction study. J. Struct. Biol. 137, 15–22.

Cameron, G. J, Cairns, D. E, and Wess, T. J. (2007). The variability in type I collagen helical pitch is reflected in the D periodic fibrillar structure. J Mol Biol. 28;372(4),1097–1107.

Chakravarti, S., Magnuson, T., Lass, J. H., Jepsen, K. J., LaMantia, C., and Carroll, H. (1998). Lumican regulates collagen fibril assembly: Skin fragility and corneal opacity in the absence of lumican. J. Cell Biol. 141, 1277–1286.

Chanut-Delalande, H., Fichard, A., Bernocco, S., Garrone, R., Hulmes, D. J. S., and Ruggiero, F. (2001). Control of heterotypic fibril formation by collagen V is determined by chain stoichiometry. J. Biol. Chem. 276, 24352–24359.

Chapman, J. A. (1989). The regulation of size and form in the assembly of collagen fibrils in vivo. Biopolymers 28, 1367–1382.

Christiansen, D. L., Huang, E. K., Silver, F. H., and Christiansen, U. (2000). Assembly of type I collagen: fusion of fibril subunits and the influence of fibril diameter on mechanical properties. Matrix Biol. 19, 409–420.

Craig, A. S., Birtles, M. J., Conway, J. F., and Parry, D. A. D. (1989). An estimate of the mean length of collagen fibrils in rat tail tendon as a function of age. Connect. Tissue Res. 19, 51–62.

Danielson, K. G., Siracusa, L. D., Donovan, P. J., and Iozzo, R. V. (1999). Decorin, epiphycan, and lumican genes are closely linked on murine chromosome 10 and are deleted in lethal steel mutants. Mamm. Genome 10, 201–203.

Derwin, K. A., Soslowsky, L. J., Kimura, J. H., and Plaas, A. H. (2001). Proteoglycans and glycosaminoglycan fine structure in the mouse tail tendon fascicle. J. Orthop. Res. 19, 269–277.

Eikenberry, E. F., Childs, B., Sheren, S. B., Parry, D. A. D., Craig, A. S., and Brodsky, B. (1984). Crystalline fibril structure of type-II collagen in lamprey notochord sheath. J. Mol. Biol. 176, 261–277.

Eyre, D. R. (2002). Collagen of articular cartilage. Arthritis Res. 4, 30–35.

Eyre, D. R., Pietka, T., Weis, M. A., and Wu, J. J. (2004). Covalent cross-linking of the NC1 domain of collagen type IX to collagen type II in cartilage. J. Biol. Chem. 279, 2568–2574.

Folkhard, W., Christmann, D., Geercken, W., Knorzer, E., Koch, M. H., Mosler, E., Nemetschek-Gansler, H., and Nemetschek, T. (1987a). Twisted fibrils are a structural principle in the assembly of interstitial collagens, chordae tendinae included. Z. Naturforsch. [C] 42, 1303–1306.

Folkhard, W., Mosler, E., Geercken, E., Knorzer, H., Nemetschek-Gansler, H., and Nemetschek, T. (1987b). Quantitative analysis of the molecular sliding mechanism in native tendon collagen – Time resolved dynamic studies using synchrotron radiation. Int. J. Biol. Macromol. 9, 169–175.

Franc, S. (1993). Ultrastructural evidences of a distinct axial domain within native rat tail tendon collagen fibrils. J. Submicrosc. Cytol. Pathol. 25, 85–91.

Fraser, R. D. B., MacRae, T. P., Miller, A., and Suzuki, E. (1983). Molecular conformation and packing in collagen fibrils. J. Mol. Biol. 167, 497–521.

Fraser, R. D. B., MacRae, T. P., and Miller, A. (1987). Molecular packing in type I collagen fibrils. J. Mol. Biol. 193, 115–125.

Fratzl, P. (2003). Cellulose and collagen: from fibres to tissues. Curr. Opin. Colloid Interface Sci. 8, 32–39.

Fratzl, P., Fratzl-Zelman, N., and Klaushofer, K. (1993). Collagen packing and mineralization. An X-ray scattering investigation of turkey leg tendon. Biophys. J. 64, 260–266.

Fratzl, P., Gupta, H. S., Paschalisb, E. P., and Roschgerb, P. (2004). Structure and mechanical quality of the collagen–mineral nano-composite in bone. J. Mater. Chem. 14, 2115–2123.

Fratzl, P., Misof, K., Zizak, I., Rapp, G., Amenitsch, H., and Bernstorff, S. (1998). Fibrillar structure and mechanical properties of collagen. J. Struct. Biol. 122, 119–122.

Giraud-Guille, M.-M. (1996). Twisted liquid crystalline supramolecular arrangements in morpho-genesis. Int. Rev. Cytol. 166, 59–101.

Goh, K. L., Hiller, J., Haston, J. L., Holmes, D. F., Kadler, K. E., Murdoch, A., Meakin, J. R., and Wess, T. J. (2005). Analysis of collagen fibril diameter distribution in connective tissues using small-angle X-ray scattering. Biochim. Biophys. Acta. 1722, 183–188.

Graham, H. K., Holmes, D. F., Watson, R. B., and Kadler, K. E. (2000), Identification of collagen fibril fusion during vertebrate tendon morphogenesis. The process relies on unipolar fibrils and is regulated by collagen-proteoglycan interaction. J. Mol. Biol. 295, 891–902.

Gutsmann, T., Fantner, G. E., Venturoni, M., Ekani-Nkodo, A., Thompson, J. B., Kindt, J. H., Morse, D. E., Fygenson, D. K., and Hansma, P. K. (2003). Evidence that collagen fibrils in tendons are inhomogeneously structured in a tubelike manner. Biophys. J. 84, 2593–2598.

Hansen, U., and Bruckner, P. (2003). Macromolecular specificity of collagen fibrillogenesis. Fibrils of collagens I and XI contain a heterotypic alloyed core and a collagen I sheath. J. Biol. Chem. 278, 37352–37359.

Hedlund, H., Hedbom, E., Heinegard, D., Mengarelli-Widholm, S., Reinholt, F. P., and Svensson, O. (1999). Association of the aggrecan keratan sulfate-rich region with collagen in bovine articular cartilage. J. Biol. Chem. 274, 5777–5781.

Hedlund, H., Mengarelli-Widholm, S., Heinegard, D., Reinholt, F. P., and Svensson, O. (1994). Fibromodulin distribution and association with collagen. Matrix Biol. 14, 227–232.

Henkel, W. (1996). Cross-link analysis of the C-telopeptide domain from type III collagen. Biochem. J. 318, 497–503.

Henkel, W., and Glanville, R. W. (1982). Covalent crosslinking between molecules of type I and type III collagen. The involvement of the N-terminal, nonhelical regions of the alpha 1 (I) and alpha 1 (III) chains in the formation of intermolecular crosslinks. Eur. J. Biochem. 122, 205–213.

Hodge, A. J., and Petruska, J. A. (1963). Recent studies with the electron microscope on ordered aggregates of the tropocollagen molecule. In "Aspects of Protein Chemistry" (G. N. Ramachan-dran, Ed.), pp. 289–300. Academic Press, London.

Hofmann, H., Fietzek, P. P., and Kuhn, K. (1980). Comparative analysis of the sequences of the three collagen chains a1(I), a2 and a1(III); Function and genetic aspects. J. Mol. Biol. 141, 293–314.

Holmes, D. F., Gilpin, C. J., Baldock, C., Ziese, U., Koster, A. J., and Kadler, K. E. (2001). Corneal collagen fibril structure in three dimensions: Structural insights into fibril assembly, mechanical properties, and tissue organization. PNAS 98, 7307–7312.

Holmes, D. F. and Kadler, K. E. (2006). The 10+4 microfibril structure of thin cartilage fibrils. PNAS 103, 17249–17254.

Hoshi, K., Kemmotsu, S., Takeuchi, Y., Amizuka, N., and Ozawa, H. (1999). The primary calcifi-cation in bones follows removal of decorin and fusion of collagen fibrils. J. Bone Miner. Res. 14, 273–280.

Hulmes, D. J. S. (2002). Building collagen molecules, fibrils, and suprafibrillar structures. J. Struct. Biol. 137, 2–10.

Hulmes, D. J. S., Holmes, D. F., and Cummings, C. (1985). Crystalline regions in collagen fibrils. J. Mol. Biol. 184, 473–477.

Hulmes, D. J. S., Miller, A., Parry, D. A. D., Piez, K. A., and Woodhead-Galloway, J. (1973). Analysis of the primary structure of collagen for the origins of molecular packing. J. Mol. Biol. 79, 137–148.

Hulmes, D. J. S., Miller, A., White, S. W., and Brodsky-Doyle, B. (1977). Interpretation of the meridional diffraction pattern from collagen fibres in terms of the known amino acid sequence. J. Mol. Biol. 110, 643–666.

Hulmes, D. J. S., Miller, A., White, S. W., Timmins, P. A., and Berthet-Colominas, C. (1980). Interpretation of the low angle meridional neutron diffraction patterns from collagen fibres in terms of the amino acid sequence. Int. J. Biol. Macromol. 2, 338–345.

Hulmes, D. J. S., Wess, T. J., Prockop, D. J., and Fratzl, P. (1995). Radial packing, order and disorder in collagen fibrils. Biophys. J. 68, 1661–1670.

Hwang, W. S., Li, B., Jin, L. H., Ngo, K., Schachar, N. S., and Hughes, G. N. F. (1992). Collagen fibril structure of normal, aging, and osteoarthritic cartilage. J. Pathol. 167, 425–433.

Itoh, T., Kobayashi, M., and Hashimoto, M. (1998). The role of intermolecular electrostatic interaction on appearance of the periodic band structure in type I collagen fibril. Jpn. J. Appl. Phys. 37, L190–L192.

Jeffery, A. K., Blunn, G. W., Archer, C. W. and Bentley, G. (1991). Three-dimensional collagen architecture in bovine articular cartilage. J. Bone Joint Surg. Br. 73-B, 795–801.

Jesoir, J. C., Miller, A., and Berthet-Colominas, C. (1981). Crystalline three dimensional packing is a general feature of type I collagen fibrils. FEBS Lett. 13, 238–240.

Jokinen, J., Dadu, E., Nykvist, P., Käpylä, J., White, D. J., Ivaska, J., Vehviläinen, P., Reunanen, H., Larjava, H., Häkkinen, L., and Heino, J. (2004). Integrin-mediated cell adhesion to type I collagen fibrils. J. Biol. Chem. 279, 31956-31963.

Jones, E. Y., and Miller, A. (1991). Analysis of structural design features in collagen. J. Mol. Biol. 218, 209–219.

Kadler, K. (1994). Extracellular matrix. 1: fibril-forming collagens. Protein Profile 1, 519–638.

Kannus, P. (2000). Structure of the tendon connective tissue. Scand. J. Med. Sci. Spor. 10, 312–320.

Kassner, A., Tiedemann, K., Notbohm, M., Ludwig, T., Morgelin, M., Reinhardt, D. P., Chu, M. L., Bruckner, P., and Grassel, S. (2004). Molecular structure and interaction of recombinant human type XVI collagen. J. Mol. Biol. 339, 835–853.

Kastelic, J., Galeski, A., and Baer, E. (1978). Multicomposite structure of tendon. Connect. Tissue Res. 6, 11–23.

Knight, D. P., and Vollrath, F. (2002). Biological liquid crystal elastomers. Philos. Trans. R. Soc. Lond., B, Biol. Sci. 357, 155–163.

Koch, M., Laub, F., Zhou, P., Hahn, R. A., Tanaka, S., Burgeson, R. E., Gerecke, D. R., Ramirez, F., and Gordon, M. K. (2003). Collagen XXIV, a vertebrate fibrillar collagen with structural features of invertebrate collagens selective expression in developing cornea and bone. J. Biol. Chem. 278(44), 43236–43244.

Landis, W. J., Hodgens, K. J., Song, M. J., Arena, J., Kiyonaga, S., Marko, M., Owen, C., and McEwen, B. F. (1996). Mineralization of collagen may occur on fibril surfaces: Evidence from conventional and high-voltage electron microscopy and three dimensional imaging. J. Struct. Biol. 117, 24–35.

Linsenmayer, T. F., Gibney, E., Igoe, F., Gordon, M. K., Fitch, J. M., Fessler, L. I., and Birk, D. E. (1993). Type-V collagen: Molecular-structure and fibrillar organization of the chicken alpha-1(V) NH2-terminal domain, a putative regulator of corneal fibrillogenesis. J. Cell Biol. 121, 1181–1189.

Liu, X, H., Otter, A., Scott, P. G., Cann, J. R. and Kotovych, G. (1993). Conformational analysis of the type II and type-III collagen alpha-1 chain C-telopeptides by H- 1-nmr and circular-dichroism spectroscopy. J. Biomol. Struct. Dyn. 11, 541–555.

Lucic, D., Mollenhauer, J., Kilpatrick, K. E., and Cole, A. A. (2003). N-telopeptide of type II collagen interacts with annexin V on human chondrocytes. Connect. Tissue Res. 44, 225–239.

MacBeath, J. R., Shackleton, D. R., and Hulmes, D. J. S. (1993). Tyrosine-rich acidic matrix protein (TRAMP) accelerates collagen fibril formation in vitro. J. Biol. Chem. 268, 19826–19832.

McBride, D. J., Choe, V., Shapiro, J. R., and Brodsky, B. (1997). Altered collagen structure in mouse tail tendon lacking the alpha 2(I) chain. J. Mol. Biol. 270, 275–284.

Malone, J. P., George, A., and Veis, A. (2004). Type I collagen N-telopeptides adopt an ordered structure when docked to their helix receptor during fibrillogenesis. Proteins 54, 206–215.

Marchant, J. K., Hahn, R. A., Linsenmayer, T. F., and Birk, D. E. (1996). Reduction of type V collagen using a dominant-negative strategy alters the regulation of fibrillogenesis and results in the loss of corneal-specific fibril morphology. J. Cell Biol. 135, 1415–1426.

Marchini, M., Morocutti, M., Ruggeri, A., Koch, M. H. J., Bigi, A., and Roveri, N. (1986). Differences in the fibril structure of corneal and tendon collagen. An electron microscopy and X-ray diffraction investigation. Connect. Tissue Res. 15, 269–281.

Martin, R., Farjanel, J., Eichenberger, D., Colige, A., Kessler, E., Hulmes, D. J. S., and Giraud-Guille, M. M. (2000). Liquid crystalline ordering of procollagen as a determinant of three-dimensional extracellular matrix architecture. J. Mol. Biol. 301, 11–17.

Meek, K. M., Chapman, J. A., and Hardcastle, R. A. (1979). The staining pattern of collagen fibrils. Improved correlation with sequence data. J. Biol. Chem. 254, 10710–10714.

Meek, K. M., and Fullwood, N. J. (2001). Corneal and scleral collagens – A microscopist's perspective. Micron 32, 261–272.

Mendler, M., Eich-Bender, S. G., Vaughan, L., Winterhalter, K. H. and Bruckner, P. (1989). Cartilage contains mixed fibrils of collagen types II, IX, and XI. J. Cell Biol., 108, 191–197.

Michna, H. (1984). Morphometric analysis of loading-induced changes in collagen fibril populations in young tendons. Cell Tissue Res. 236, 465–470.

Miles, C. A, Sims, T. J, Camacho, N. P, and Bailey, A. J. (2002). The role of the alpha2 chain in the stabilization of the collagen type I heterotrimer: a study of the type I homotrimer in oim mouse tissues. J. Mol. Biol. 321, 797–805.

Miller, A., and Tochetti, D. (1981). Calculated X-ray diffraction pattern from a quasihexagonal model for the molecular arrangement in collagen. Int. J. Macromol. 3, 9–18.

Mizuno, K., Adachi, E., Imamura, Y., Katsumata, O., and Hayashi, T. (2001). The fibril structure of type V collagen triple-helical domain. Micron 32, 317–323.

Niyibizi, C. and Eyre, D. R. (1994). Structural characteristics of cross-linking sites in type V collagen of bone. Chain specificities and heterotypic links to type I collagen. Eur. J. Biochem. 224, 943–950.

North, A. C. T., Cowan, P. M., and Randall, J. T. (1954). Structural units in collagen fibrils. Nature 174, 1142–1143.

Olsen, B. R. (1997). Collagen IX. Int. J. Biochem. Cell Biol. 29, 555–558.

Orgel, J. P. R. O., Irving, T. C., Miller, A., and Wess, T. J. (2006). Microfibrillar structure of type I collagen in situ. PNAS 103, 9001–9005.

Orgel, J. P. R. O., Miller, A., Irving, T. C., Fischetti, R. F., Hammersley, A. P., and Wess, T. J. (2001). The in situ supermolecular structure of type I collagen. Structure 9, 1061–1069.

Orgel, J. P., Wess, T. J., and Miller, A. (2000). The in situ conformation and axial location of the intermolecular cross-linked non-helical telopeptides of type I collagen. Struct. Fold. Des. 8, 137–142.

Ortolani, F., Giordano, M., and Marchini, M. (2000). A model for type II collagen fibrils: Distinctive D-band patterns in native and reconstituted fibrils compared with sequence data for helix and telopeptide domains. Biopolymers 54, 448–463.

Ottani, V., Raspanti, M., and Ruggeri, A. (2001). Collagen structure and functional implications. Micron 32, 251–260.

Parry, D. A. (1988). The molecular and fibrillar structure of collagen and its relationship to the mechanical properties of connective tissue. Biophys. Chem. 29, 195–209.

Parry, D. A. D., Barnes, G. R. G., and Craig, A. S. (1978). A comparison of the size distribution of collagen fibrils in connective tissues as a function of age and a possible relationship between fibril size distribution and mechanical properties. Proc. R. Soc. Lond. B. 203, 305–321.

Parry, D. A. D., and Craig, A. S. (1979). Electron microscope evidence for an 80 Angstrom unit in collagen fibrils. Nature 282, 213–225.

Parry, D. A. D., and Craig, A. S. (1984). Growth and Development of Collagen Fibrils in Connective Tissues. In "Ultrastructure of the Connective Tissue Matrix" (A. Ruggeri and P. M. Motta, Eds.), pp. 34–64. Martinus Nijhoff, Netherlands.

Paterlini, M. G., Nemethy, G., and Scheraga, H. A. (1995). The energy of formation of internal loops in triple-helical collagen polypeptides. Biopolymers 35, 607–619.

Patterson-Kane, J. C., Wilson, A. M., Firth, E. C., Parry, D. A. D., and Goodship, A. E. (1997). Comparison of collagen fibril populations in the superficial digital flexor tendons of exercised and non-exercised thoroughbreds. Equine Vet. J. 29, 121–125.

Pihlajamaa, T., Lankinen, H., Ylostalo, J., Valmu, L., Jaalinoja, J., Zaucke, F., Spitznagel, L., Gosling, S., Puustinen, A., Morgelin, M., Peranen, J., Maurer, P., Ala-Kokko, L., and Kilpelainen, I. (2004). Characterization of recombinant amino-terminal NC4 domain of human collagen IX: Interaction with glycosaminoglycans and cartilage oligomeric matrix protein. J. Biol. Chem. 279, 24265–24273.

Pope, F. M., Martin, G. R., Lichtenstein, J. R., Penttinen, R., Gerson, B., Rowe, D. W., and McKusick, V. A.. (1975). Patients with Ehlers-Danlos syndrome type IV lack type III collagen. PNAS 72, 1314–1316.

Prockop, D. J., and Fertala, A. (1998). The collagen fibril: the almost crystalline structure. J. Struct. Biol.122, 111–118.

Redaelli, A., Vesentini, S., Soncini, M., Vena, P., Mantero, S., and Montevecchi, F. M. (2003). Possible role of decorin glycosaminoglycans in fibril to fibril force transfer in relative mature tendons: A computational study from molecular to microstructural level. J. Biomech. 36, 1555–1569.

Sanders, J. E., and Goldstein, B. S. (2001). Collagen fibril diameters increase and fibril densities decrease in skin subjected to repetitive compressive and shear stresses. J. Biomech. 34, 1581–1587.

Sasaki, N., and Odajima, S. (1996). Elongation mechanism of collagen fibrils and forcestrain relations of tendon at each level of structural hierarchy. J. Biomech. 29, 1131–1136.

Sasaki, N., Shukunami, N., Matsushima, N., and Izumi, Y. (1999). Time resolved X-ray diffraction from tendon collagen during creep using synchrotron radiation. J. Biomech. 32, 285–292.

Scott, J. E., and Thomlinson, A. M. (1998). The structure of interfibrillar proteoglycan bridges ('shape modules') in extracellular matrix of fibrous connective tissues and their stability in various chemical environments. J. Anat. 192, 391–405.

Shaw, L. M., Olsen, B. R. (1991). FACIT collagens: diverse molecular bridges in extracellular matrices. Trends Biochem. Sci. 1, 191–194.

Silver, F. H., Christiansen, D. L., Snowhill, P. B., and Chen, Y. (2001). Transition from viscous to elastic-based dependency of mechanical properties of self-assembled type I collagen fibers. J. Appl. Polymer Sci. 79, 134–142.

Silver, F. H., Kato, Y. P., Ohno, M., Wasserman, A. J. (1992). Analysis of mammalian connective tissue: relationship between hierarchical structures and mechanical properties. J. Long Term Eff. Med. Implants. 2(2–3),165–98.

Smith, J. W. (1968). Molecular pattern in native collagen. Nature 219, 157–158.

Svensson, L., Aszodi, A., Reinholt, F. P., Fassler, R., Heinegard, D., and Oldberg, A. (1999). Fibromodulin-null mice have abnormal collagen fibrils, tissue organization, and altered Lumican deposition in tendon. J. Biol. Chem. 274, 9636–9647.

Thomas, E. K., Nakamura, M., Wienke, D., Isacke, C. M., Pozzi, A., and Liang, P.. (2005). Endo180 binds to the C-terminal region of type I collagen. J. Biol. Chem. 280, 22596–22605.

Trus, B. L., and Piez, K. A. (1980). Compressed microfibril models of the nativecollagen fibril. Nature 286, 300–301.

van der Rest, M., and Garrone, R. (1991). Collagen family of proteins. FASEB J 51, 2814–2823

Vaughan, L., Mendler, M., Huber, S., Bruckner, P., Winterhalter, K. H., Irwin, M. I., and Mayne, R. (1988). D-periodic Distribution of collagen type IX along cartilage fibrils. J. Cell Biol. 106, 991–997.

Vitagliano, L., Nemethy, G., Zagari, A., and Scheraga, H. A. (1995). Structure of the type-I collagen molecule based on conformational energy computations: The triple-stranded helix and the n-terminal telopeptide. J. Mol. Biol. 247, 69–80.

Watanabe, M., Kobayashi, M., Fujita, Y., Senga, K., Mizutani, H., Ueda, M., and Hoshino, T. (1997). Association of type VI collagen with D-periodic collagen fibrils in developing tail tendons of mice. Arch. Histol. Cytol. 60, 427–434.

Weiner, S., Traub, W., and Wagner, H. D. (1999). Lamellar bone: structure-function relations. J. Struct. Biol. 126, 241–255.

Wen, C. K., and Goh, M. C. (2004). AFM nanodissection reveals internal structural details of single collagen fibrils. Nano. Lett. 4, 129–132.

Wenstrup, R. J., Florer, J. B., Brunskill, E. W., Bell, S. B., Chervoneva, I., and Birk, D. E. (2004). Type V collagen controls the initiation of collagen fibril assembly. J. Biol. Chem. 279, 53331–53337.

Wess, T. J., Hammersley, A., Wess, L., and Miller, A. (1995). Type I collagen packing conformation of the triclinic unit cell. J. Mol. Biol. 248, 487–493.

Wess, T. J., Hammersley, A. P., Wess, L., and Miller, A. (1998). A consensus model for molecular packing of type I collagen. J. Struct. Biol. 122, 92–100.

White, J., Werkmeister, J. A., Ramshaw, J. A. M., and Birk, D. E. (1997). Organization of fibrillar collagen in the human and bovine cornea: Collagen types V and III. Connect. Tissue Res. 36, 165–174.

Wu JJ, Murray J, Eyre DR (1996). Evidence for copolymeric crosslinking between types II and III collagens in human articular cartilage. Trans. Orthop. Res. Soc. 21, 42.

Yamamoto, T., Domon, T., Takahashi, S., Islam, N., and Suzuki, R. (2000b). Twisted plywood structure of an alternating lamellar pattern in cellular cementum of human teeth. Anat. Embryol. 202, 25–30.

Yamamoto, S., Hashizume, H., Hitomi, J., Shigeno, M., Sawaguchi, S., Abe, H., and Ushiki, T. (2000a). The subfibrillar arrangement of corneal and scleral collagen fibrils as revealed by scanning electron and atomic force microscopy. Arch. Histol. Cytol. 63, 127–135.

Yamamoto, T. and Wakita, M. (1992). Bundle formation of principal fibers in rat molars. J. Periodont. Res. 27, 20–27.

Yamamoto K, and Yamamoto, M. (1994). Cell adhesion receptors for native and denatured type I collagens and fibronectin in rabbit arterial smooth muscle cells in culture. Exp. Cell Res. 214, 258–263.

Yamazaki, M., Majeska, R. J., Yoshioka, T. H., Moriya, H., Thomas, A., and Einhorn, T. A. (2005). Spatial and temporal expression of fibril-forming minor collagen genes (types V and XI) during fracture healing. J. Orthop. Res. 15, 757–764.

Young, R. D., Lawrence, P. A., Duance, V. C., Aigner, T., and Monaghan, P. (2000a). Immunolocalization of collagen types II and III in single fibrils of human articular cartilage. J. Histochem. Cytochem. 48, 423–432.

Zhang, G., Young, B. B., Ezura, Y., Favata, M., Soslowsky, L. J., Chakravarti, S., and Birk, D. E. (2005). Development of tendon structure and function: Regulation of collagen fibrillogenesis. J. Musculoskelet. Neuronal. Interact. 5, 5–21.

Chapter 4
Restraining Cross-Links Responsible for the Mechanical Properties of Collagen Fibers: Natural and Artificial

N.C. Avery and A.J. Bailey

Abstract The mechanical properties of collagen fibers primarily depend on the formation of head to tail Schiff base cross-links between end-overlapped collagen molecules within the fiber induced by the enzyme lysyl oxidase. Inhibition of these cross-links results in the complete loss of mechanical strength of the fiber. During maturation these initial divalent cross-links react further with molecules in register from an adjacent fiber forming stable trivalent cross-links and further increasing its mechanical strength. This system of cross-linking is well established and exists throughout the animal kingdom from sponges to man, but there remain a number of unidentified cross-links known to be present in some tissues. In addition, there are some unusual cross-links in certain invertebrates. The nature of the collagen cross-linking is tissue specific rather than species specific and depends on the extent of hydroxylation of both the telopeptide and triple helical lysines involved in the cross-link and on the rate of collagen metabolism. The cross-link profile of collagen fibers therefore varies considerably within and between tissues, for example in different bones. Recent studies indicate that pyrrole cross-links rather than pyridinoline cross-links correlate with mechanical strength of avian bones. The profile can change between normal loading and extreme exercise and these differences appear to relate to their particular function, but further studies to identify whether a particular cross-link is responsible remain to be carried out.

Following maturation the low turnover of collagen allows the non-enzymic random accumulation of glucose oxidation products, some of which form intermolecular cross-links, ultimately rendering the fiber too stiff for normal function. The significance of the major glycation cross-link, believed to be glucosepane, remains to be confirmed. The successful use of inhibitors of this glycation reaction and of specific glycation cross-link breakers should lead to a reduction in this deleterious effect in both aging and diabetes mellitus.

The high mechanical strength and resistance to heat and bacterial degradation of collagen fibers has been utilized industrially. Additional chemical cross-linking in vitro has been employed historically to attain specific mechanical and thermal properties, for example tanning skin to leather, and more recently in medical and cosmetic products with low cytotoxic effects. The resultant increases in denaturation temperature have recently been correlated with reduced water content of the fiber.

P. Fratzl (ed.), *Collagen: Structure and Mechanics*,
© Springer Science+Business Media, LLC 2008

A wide range of cross-linking agents are available for modification of collagen to provide a product with specific properties.

4.1 Introduction

The mechanical properties of collagenous tissues predominantly depend on the formation of intermolecular cross-links between the collagen molecules within the fibers to prevent slippage under load. The mechanical properties of tissues are further refined by the varied alignment of these cross-linked fibers within a tissue. For example, the parallel alignment of fibers in tendons enhances longitudinal strength, the random layered organization in the skin maximizes compliance, the laminated layers in intervertebral disc provide flexibility, the concentric layers in bone lend tensile strength, while the precise layered organization in the cornea facilitates both strength and transparency. Heterotypic fibers are normal in most tissues (Hulmes Chap. 3) but the different collagen types are cross-linked by the same mechanism.

The cross-link profile varies with the tissue, for example skin, bone and tendon, all have different profiles, but it also depends on the relative stress these tissues experience, for example compressor versus extensor tendons, and also on the rate of turnover and the age of the tissue. In general the mechanical strength of collagenous tissues increases with age and this can now be accounted for by changes in cross-link profile. Similarly, functionally different bones vary in their cross-link profile, but we need to know which cross-link(s) is functionally the most important and whether the cross-linking can be modified. An understanding of the role of individual cross-links related to the particular physical properties of a tissue under sedentary and extreme physical exercise is now being achieved. For example mechanically strong bone correlates with strong connecting tendons, (Goodship Chap. 10) but which are the significant cross-links in each of these tissues remains to be elucidated.

The cross-linking process occurs in two stages, the initial stage to give an optimum functioning tissue involves the enzymic formation of divalent and then trivalent intermolecular cross-links at precisely defined sites linking the molecules head to tail and ultimately forming a network within the fiber. The second stage involves the non-enzymic adventitious reaction with glucose and its products to form additional intermolecular cross-links through the linking of lysine and arginine residues in the triple helical region of the molecule. These cross-links render the fibers stiffer and ultimately brittle, thus decreasing the optimal efficiency of the fiber. The chemistry of these glycation cross-links is slowly being unraveled.

The close packing and cross-linking of the collagen molecules in fibers ensure that the fibers are virtually inextensible. However, the fibers do have some elasticity (like steel wire) due to about 5% extension and this is important in the locomotion of animals. The stiff, cross-linked fibers have sufficient elasticity to store energy, which on release ensures efficiency of movement in animals as diverse as humans and kangaroos by allowing them to reuse the stored energy on rebound (Alexander 1988). Can this be property be related to the type of cross-link?

The high mechanical strength of native collagen fibers provides a good basis for its use in industry and through its low antigenicity such products have found widespread use in medicine. A common feature of these products is the requirement to control and reproduce particular mechanical properties and this is generally achieved by additional specific chemical cross-linking.

4.2 Enzyme Cross-Linking (Lysyl Oxidase)

The major cross-links of collagen are based on the oxidative deamination of the ε-amino groups of specific lysine residues by the enzyme lysyl oxidase to form lysine-aldehydes. The lysyl oxidases act on the molecular aggregates in the fibril, not single molecules, oxidatively deaminating only specific, non-triple helical telopeptide lysines and hydroxylysines. Surprisingly five isoforms of lysyl oxidase have been identified to date but the specific function of each has not yet been elucidated (Lucero and Kagan 2006). However, it is likely that the different amino acid sequences at the amino and carboxy terminals of collagen, the different collagen types and even different tissues may require a number of specific lysyl oxidases.

The aldehydes so formed then react with an opposing ε-amino group of hydroxylysine in the highly conserved lysyl oxidase binding site within the triple helical region (Hyl-Gly-His-Arg) of an adjacent molecule quarter-staggered in relation to the first, thereby allowing the formation of a Schiff base (aldimine) intermolecular cross-link. Lysyl oxidase is inhibited by copper deficiency or by β-aminopropionitrile, a lathyritic agent. In the absence of these cross-links collagen fibers have little or no strength, and even bone is extremely fragile.

4.2.1 Immature Tissues

Aldimines. In tissues such as skin and rat tail tendon where the telopeptide lysines are barely hydroxylated the predominant cross-link is the aldimine formed between lysine-aldehyde and hydroxylysine, dehydro-hydroxylysinonorleucine (deH-HLNL) (Bailey and Peach 1968) (Fig. 4.1). The cross-link is stable under physiological conditions but can be readily cleaved in vitro by dilute acetic acid or by mild heat, thus accounting for the high solubility of immature skin collagen. The presence of lysine instead of hydroxylysine in the sequence Hyl-Gly-His-Arg would produce dehydro-lysinonorleucine (deH-LNL), but this lysine is generally hydroxylated in skin, consequently the lysine cross-link is barely detectable.

A second type of aldimine reported to be present in skin and tendon is the tetravalent dehydro-histidino-hydroxymerodesmosine (deH-HHMD) formed by the reaction of histidine across the carbon–carbon double bond of the dehydrated aldol condensation product and reaction of the free aldehyde of the aldol with ε-amino group of hydroxylysine in the triple helix. Borohydride reduction of the Schiff base yields histidino-hydroxymerodesmosine (HHMD). However, the existence of this Schiff base in vivo is controversial. It has been suggested that the reaction of the histidine residue across the double bond of the aldol is driven by the conditions of the borohydride reduction (Robins and Bailey 1973). This proposal has been refuted

Immature cross-link   Mature cross-link

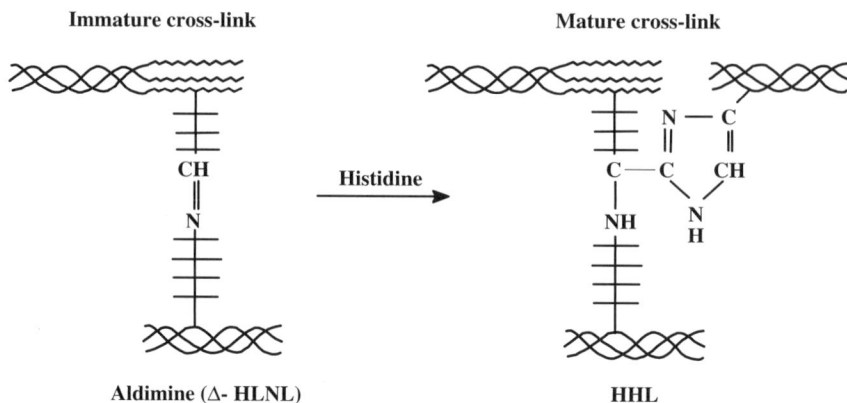

Aldimine (Δ- HLNL)   HHL

Fig. 4.1 Reaction of telopeptide lysine-aldehyde with triple helical hydroxylysine to form the divalent aldimine cross-link dehydro-hydroxylysinonorleucine (deH-HLNL). This cross-link subsequently reacts with triple helical histidine to form the stable mature cross-link histidino-hydroxylysinonorleucine (HHL), further increasing the mechanical strength of the fiber

by Bernstein and Mechanic (1980) although they have only been able to demonstrate the presence of HHMD following borohydride reduction. Further alternative techniques are required to clarify the in vivo existence of this potential tetravalent cross-link.

Keto-amines. If the telopeptide lysine is hydroxylated, as in bone collagen, the hydroxylysine-aldehyde formed reacts with the ε-amino group of hydroxylysine in the Hyl-Gly-His-Arg sequence of the triple helix to form a Schiff base, which spontaneously undergoes an Amadori rearrangement to form hydroxylysino-keto-norleucine (HLKNL) (Fig. 4.2). The keto-amine is stable to acid and heat thus accounting for the insolubility of bone and cartilage collagen even at the immature fetal stage.

In bone collagen a cross-link may form between a hydroxylysine-aldehyde and a lysine in the triple helix (Robins and Bailey 1975), the Schiff base formed undergoing the Amadori rearrangement to form lysino-keto-norleucine (LKNL). Unlike skin collagen this cross-link is present in significant quantities because of the lower overall hydroxylation of lysines in the triple helix of bone collagen. However, because the reduced form is a structural isomer of HLNL and therefore co-elutes with the reduced deH-HLNL, it is rarely considered in the literature.

4.2.2 Mature Tissues

The presence of stable non-reducible cross-links derived from the immature cross-links was suspected from the decline in the latter with increasing maturity despite an increase in tensile strength of the fibers (Bailey and Shimokomaki 1971) (Fig. 4.3a).

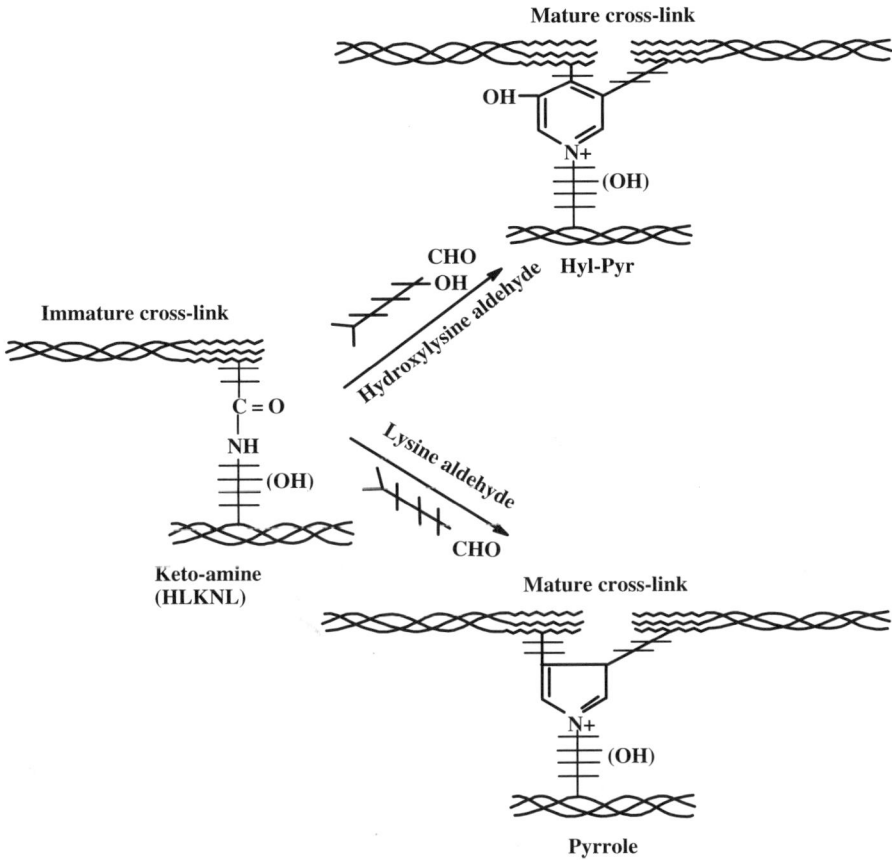

Fig. 4.2 Reaction of telopeptide hydroxylysine-aldehyde with triple helical hydroxylysine to form an aldimine which spontaneously rearranges to form the divalent keto-amine cross-link, hydroxylysino-keto-norleucine (HLKNL). Further reaction can occur with either a telopeptide hydroxylysine-aldehyde to form the trivalent hydroxylysino-pyridinoline or with telopeptide lysine-aldehyde to form the trivalent hydroxylysino-pyrrole. Alternatively the divalent cross-link lysino-keto-norleucine (LKNL) is the precursor of the lysyl derivatives of pyridinoline and pyrrole trivalent cross-links

4.2.2.1 Histidino-hydroxylysinonorleucine (HHL)

DeH-HLNL reacts spontaneously with histidine to form HHL, a trivalent cross-link (Yamauchi et al. 1987) and the major mature cross-link in adult skin and some tendons (Fig. 4.1).

Surprisingly HHL is not present in mature mouse or rat skin, although the aldimine decreases with maturation and the fiber increases in mechanical stiffness, indicating the presence of an unknown cross-link.

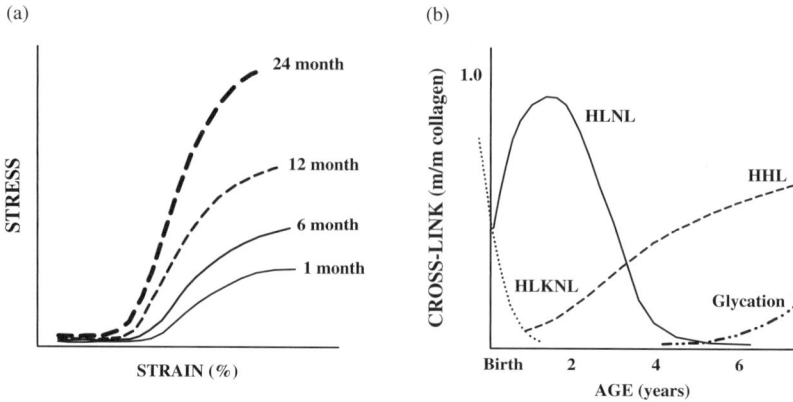

Fig. 4.3 (a) Increasing mechanical strength of rat tail tendon fibers with age, even after maturation at 6 months (—) due to glycation cross-links (- - - -). (b) Changing cross-link profile of bovine skin collagen with age showing the rapid disappearance of HLKNL from fetal skin, the increase in HLNL with growth and conversion of the latter to HHL during maturation, followed by glycation cross-linking (From Jackson, Avery, Tarlton, Eckford, Abrams & Bailey "Changes in metabolism of collagen in genitourinary prolapse" in The Lancet published by Elsevier 1996.)

4.2.2.2 Hydroxylysyl-pyridinoline (Hyl-Pyr)

It is a common cross-link present in bone, cartilage and tendon. It is predominant in highly hydroxylated collagens such as cartilage and at the same time in generally poorly hydroxylated collagens such as bone collagen, but which has highly hydroxylated lysine in the telopeptide regions. Hydroxylysyl-pyridinoline was first identified by Fujimoto et al. (1978) and lysyl–pyridinoline was later identified by Ogawa et al. (1982) (Fig. 4.2). The nomenclature deoxy-pyridinoline is incorrect since it suggests a condensation product by loss of water, whereas in fact it is simply a lysine rather than a hydroxylysine side-chain.

The precise mechanism of formation of the pyridinolines is debatable. Eyre and Oguchi (1980) proposed that they are formed from two keto-amines with the release of hydroxylysine thereby linking three collagen molecules, while Robins and Duncan (1983) proposed a condensation reaction between the keto-amine and an additional hydroxylysine-aldehyde within the same molecules thereby only linking two type II collagen molecules. Similarly Light and Bailey (1985) reported that pyridinoline only cross-linked two type I molecules and that it was not the major mature cross-link in bone and tendon collagen.

4.2.2.3 Pyrroles

The presence of pyrroles has been convincingly demonstrated in mature tendons by Kupyers et al. (1992) who proposed they were derived by a similar mechanism to the pyridinolines, from the keto-amine (HLKNL) with a lysine-aldehyde. Alternatively by the reaction of the keto-amine and the aldimine (deH-HLNL) (Hanson and Eyre 1996) (Fig. 4.2). The structure of the pyrroles have been confirmed as the biotinylated derivative employing mass spectrometry and, as in the case of the pyridinolines, there are both hydroxylysyl-pyrrole and lysyl-pyrrole present in some tissues, but they have not yet been isolated (Brady and Robins 2001).

4.2.3 Changing Cross-Link Profiles of Different Tissues

4.2.3.1 Skin

Fetal skin possesses the keto-amine due to the high turnover and high lysyl hydroxylase activity, which then disappears rapidly as infant growth occurs and the tissue is stabilized by the aldimine deH-HLNL (Bailey and Robins 1972) and HHL in the mature tissue (Yamauchi et al. 1987) (Fig. 4.3b). These changes are consistent with the alterations in the mechanical properties of the fiber (Fig. 4.3a). High keto-amine levels are again present in the initial stages of dermal wound healing and the collagen goes through the same immature to mature change in cross-link profile during healing. Other soft tissues such as cornea and sclera also contain the divalent aldimine as the predominant cross-link.

4.2.3.2 Bone

The keto-amine is the major cross-link in young bone collagen and is present at about 1 cross-link per collagen molecule. In mature bone the level of the pyridinolines is only approximately 0.2 pyridinolines per collagen molecule, while the pyrroles are about 0.4 per collagen molecule. Clearly there appears to be a deficiency in bone cross-linking. Following isolation and characterization of the C-terminal telopeptide (ICTP) from type I bone Eriksen et al. (2004) reported that more than half of the trivalent cross-links are neither pyridinoline nor pyrrole indicating the presence of unknown cross-link structures.

The pyridinolines are spread equally between the N and C termini while the pyrroles appear to be concentrated at the N-terminal of the molecule (Knott et al. 1995; Hanson and Eyre 1996; Brady and Robins 2001). This suggests a different function for the pyrrole and the pyridinoline cross-links. We have correlated the pyrrole with tensile strength of both avian and human bone which is in contrast to the absence of a strength correlation with the pyridinolines (Knott and Bailey 1998). We therefore proposed that the pyrroles may act as interfibrillar cross-links involving three different molecules and thereby account for the mechanical properties of compact bone, while the pyridinolines stabilize the fibers by only cross-linking two collagen molecules within a single fibril and therefore do not stabilize the fiber over and above that of the divalent keto-amine. The latter proposal is consistent with the findings of Robins and Duncan (1983) described above. However, there appear to be variations in the predominant cross-link in different types of avian bone (Knott and Bailey 1999) and further studies are required. Similarly there does not appear to be a pyrrole/compressive strength correlation with cancellous bone. This may be due to the higher turnover rate and increased lysyl hydroxylase activity of cancellous bone so that compressive strength (not tensile strength) is consequently related to the pyridinolines (Banse et al. 2002a). On the other hand, the size and shape of the trabeculae in cancellous bone, determined histomorphometrically, correlates with the pyrrole cross-link (Banse et al. 2002b). A significant feature of this study was that the cross-link profile was highly variable between the bones of individual human subjects and these differences influenced the mechanical competence of the bones.

4.2.3.3 Tendons and Ligaments

These tissues possess a cross-link profile intermediate between skin and bone possessing significant amounts of both the aldimine and the keto-amine immature cross-links and subsequently the mature cross-links HHL, Hyl-Pyr and Pyrrole, respectively. The proportion of each depends on the particular mechanical function of the tissue, for example, the compressor tendon in the rabbit foot possesses the keto-amine while the aldimine is the predominant cross-link in the extensor tendon (Kent et al. 1985). Ligaments tend to have more keto-amine while the tendons have higher aldimine content (Amiel et al. 1984). This difference may be due to variation in the turnover rate and lysyl hydroxylase activity.

Mineralization of tendons in some animals, e.g., chickens and turkeys, does not involve mineralization of the existing type I collagen, the latter is replaced by a less hydroxylated type I bone collagen before mineralization (Knott et al. 1997).

4.2.3.4 Cartilage

The compressive strength of cartilage is due to the network of type II collagen fibers in a highly hydrated proteoglycan matrix. The type II fibers appear to be cross-linked through the type IX collagen present on the surface of these fibers, thereby conferring mechanical strength across the cartilage (Miles et al. 1998). Type II collagen is highly hydroxylated hence the cross-links are predominantly keto-amine (HLKNL) and its maturation product Hyl-Pyr. The turnover of human cartilage is very slow, approximately 100 years (cf. skin is about 20 years) and therefore high levels of glycation cross-links accumulate (see below) which leads to deleterious stiffening and fragility of the cartilage.

4.2.3.5 Basement Membrane

The non-fibrous type IV of basement membranes is highly hydroxylated and the major cross-link is the keto-amine but pyridinoline is absent in the mature tissue. These cross-links only bind the anti-parallel molecule in the N-terminus, while the large non-triple helical domains at the C-terminal are cross-linked by di-sulphide cross-links and a newly identified cross-link S-hydroxylysyl-methionine (Vanacore et al. 2005).

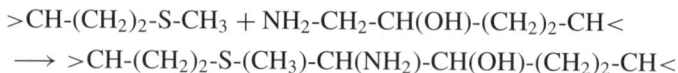

$$>\text{CH-(CH}_2)_2\text{-S-CH}_3 + \text{NH}_2\text{-CH}_2\text{-CH(OH)-(CH}_2)_2\text{-CH}<$$
$$\longrightarrow \ >\text{CH-(CH}_2)_2\text{-S-(CH}_3)\text{-CH(NH}_2)\text{-CH(OH)-(CH}_2)_2\text{-CH}<$$

The mechanical properties of basement membrane are important, for example, in the lens capsule of the eye, the capillaries and the kidney glomeruli. These membranes become more permeable, rigid and brittle with age, which is believed to be primarily due to non-enzymic glycation cross-linking (see below).

It is interesting to note that the demonstration of high levels of cross-links in fibers does not always lead to high tensile strength tissues. The cuverian tubules of the sea cucumber can be extensively stretched due to the presence of a proteoglycan sheath surrounding the fine, highly cross-linked fibrils (Bailey et al. 1982).

4.2.4 Importance of Lysine Hydroxylation

It is clear from the above that the nature of the cross-links depends on the post-translational modification of the specific lysines to hydroxylysine. The extent of hydroxylation varies considerably between collagen types and tissues, and the variation occurs both in the triple helical lysines and in the amino and carboxy telopeptides. Low levels of total hydroxylysine can lead to thicker fibers (Notbohm et al. 1999) than found with high levels of hydroxylysine (Torre-Blanco et al. 1992) indicating hydroxylysine may play a role in fibril formation.

Cartilage and basement membrane are highly hydroxylated in both the helix and the telopeptide, skin collagen less so in the helix and virtually absent from the telopeptide, while in contrast bone collagen is poorly hydroxylated in the triple helix but possesses highly hydroxylated telopeptide lysines.

The differences in lysine hydroxylation of the triple helical lysines and the N and C telopeptide lysines is probably accounted for by the activity of the isomeric lysyl hydroxylases. Three lysyl hydroxylases (LH 1–3) have been identified to date (Passoja et al. 1998), and there is some evidence that one is specific for the C-terminal telopeptide lysine and one for the N-terminal lysine, and the third for the triple helical lysine prior to folding of the triple helix. More recent studies indicate that the isoenzymes are not quite so specific as all three can hydroxylate the helical sequences, but does confirm that LH-2 hydroxylates the N-telopeptide (Takaluoma et al. 2007). Although this conveniently accounts for the observed variation in the hydroxylation of these domains it remains to be confirmed and assumes no more lysyl hydroxylases are subsequently identified.

Some genetic and acquired disorders reveal modifications of the mechanical properties of the collagenous tissues due to changes in the lysyl hydroxylase activity and consequently can provide important new data on the role of hydroxylation. Importantly, lysine hydroxylation deficiency can lead to reduced mechanical strength. For example, *Ehlers–Danlos syndrome (EDS) type VIA* is distinguished by marked tissue weakness resulting from under-hydroxylation of lysine residues due to mutations in the lysyl 1 hydroxylase gene (PLOD1). Biochemically these changes in the cross-link profile of bone reveal a virtual absence of Hyl-Pyr, due to non-hydroxylation of the specific receptor lysines in the triple helix (Eyre et al. 2002). Conversely over-hydroxylation is seen in several forms of *osteogenesis imperfecta* (OI) and can lead to retarded triple helix formation and compromised bone strength (Kirsch et al. 1981) (see Comacho, Chapter 16). Increases in lysyl hydroxylation occur during rapid growth in fetal bone, in wound healing and bone fracture repair (Glimcher et al. 1980) and in bone disorders such as osteoarthritis and osteoporosis. Biochemically *osteoarthritis* reveals thickening of the subcondral bone of joints, very high collagen turnover and lysyl hydroxylase activity (Mansell and Bailey 1998) again leading to mechanically weaker bone (Li and Aspden 1997). *Osteoporosis* involves overall loss of bone, but again the collagen turnover is increased leading to weaker bone (Bailey et al. 1992; Batge et al. 1992). TRAP knock-out mice display mild *osteopetrosis* in the limbs, and biochemically have increased collagen turnover and enzymic cross-links resulting in increased mechanical strength (Roberts et al. 2007).

4.2.5 Cross-Linking and Tissue Adaptation to Mechanical Force

An understanding of the fundamental mechanism of "mechanoregulation", that is, the manner in which tissue adapts to acute or long-term mechanical loading, could lead to strategies to enhance normal tissue function. The process by which mechanical forces are converted into changes in cellular physiology or "mechanotransduction" may result in a change in the cross-link profile and achieve enhanced mechanical properties. Studies on the changes arising from extreme exercise demonstrate an increase in tensile strength of bone and tendon (Viidik 1986; Woo et al. 1982; Birch et al. 1999). Curwin et al. (1988) reported a change in the rate of collagen turnover in exercised chickens with a consequent reduction in the pyridinoline cross-links, presumably due to an increase in the immature cross-links. However, a detailed study correlating the change in strength and the change in cross-link profile in both young and mature tissues has not yet been carried out. An understanding of these processes could lead to better control of exercise regimes and improvements in recovery from injury.

4.2.6 Determination of the Cross-Links

The immature cross-links are determined by initially stabilizing the aldimine and keto-amine bonds by reduction with a mild reducing agent such as sodium borohydride followed by acid hydrolysis and subsequent separation from other amino acids by ion-exchange chromatography. The mature cross-links can be determined similarly, although prior reduction is not necessary. Pyridinoline can also be determined by fluorescence following separation using an HPLC column. The pyrrole is determined colorimetrically by reaction with dimethylaminobenzaldehyde as the Ehrlich chromogen following partial enzyme digestion of the collagen (Sims et al. 2000).

More recently attempts are being made to determine the cross-links spectroscopically employing FT-infrared (Paschalis et al. 2001) but, as yet, the resolution of the amide 1 peaks of the spectra is not sufficiently accurate. In vivo analysis of bone separating mineral and collagen through the skin has been achieved by Raman emission excited by pulsed laser (Draper et al. 2005). An extension of this technique to determine the collagen cross-links would be a very exciting prospect.

To summarize, the mechanism of stabilization of the collagen fiber to provide optimal functionality occurs through the enzyme lysyl oxidase and the spontaneous conversion of the initial divalent aldimine and keto-amine cross-links to the mature trivalent cross-links is now fairly well established. The chemical isolation of the cross-link peptides confirmed the quarter-stagger end-overlap alignment deduced from electron microscopy, and their head to tail polymerization of the molecules by cross-linking accounts for the high tensile strength of the fiber. To account for the increased fiber strength on maturation we have proposed that the trivalent cross-links form interfibril cross-links.

Some cross-link questions remain to be solved: confirmation of the presence and identification of the unknown cross-links in mature bone collagen, the location of the known trivalent cross-links, the nature of the mature cross-links in rat and mouse skin and in basement membranes. Our inability to determine the mature cross-links in mouse tissues is currently a problem when following the metabolic changes in knock-out mice. The different roles of the pyridinoline and pyrrole cross-links in determining the material strength of various bones, and the effect of exercise on the cross-link profile, are also current questions that need to be answered. The high proportion of apparently uncharacterized cross-links in mature bone is certainly a very important question to be investigated.

4.2.7 Non-enzymic Cross-Linking (Glycation)

The concentration of mature cross-links tends to plateau following maturation but the tensile properties of the collagen fibers continue to increase indicating a second process of cross linking (Fig. 4.3a). These cross-links have now been shown to be the non-enzymic, adventitious reactions of collagen with glucose and other aldehydes from glyoxal to malondialdehyde. The initial process is the Maillard reaction but further oxidative reactions occur and the process is now referred to as glycation, ultimately leading to advanced glycation end-products (AGEs) (Fig. 4.4). The reaction with collagen has been extensively investigated and is now well established primarily due to the interest in diabetes mellitus, but also in normal age changes, where the functional properties of vulnerable tissues such as renal basement membranes, the cardiovascular system and the retinal capillaries are readily altered. Glycation is therefore a major contributor to the change in mechanical properties with age.

Direct evidence for the role of AGEs in aging was achieved by the administration of AGEs (a mixture of AGEs formed by glycation of BSA) to normal rats. The rats subsequently revealed typical changes found in diabetes and aging, i.e., basement membrane thickening, glomerula hypertrophy and an increase in mesangial volume in the absence of hyperglycemia (Vlassara et al. 1994). The AGEs employed in these

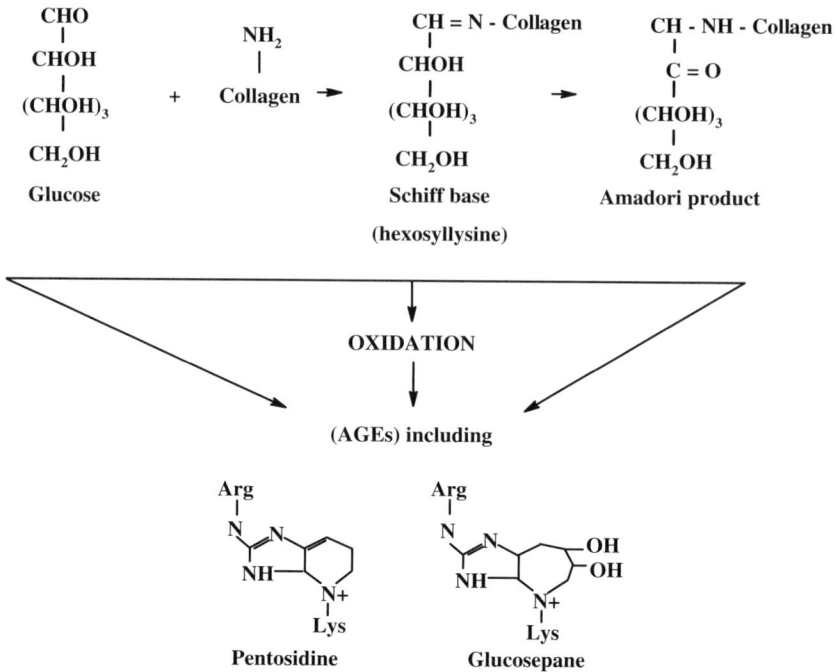

Fig. 4.4 Initial reaction of glucose with collagen, and subsequent oxidation of all three products to reactive fragments, results in advance glycation end-products (AGEs) some of which are glycation cross-links such as pentosidine and glucosepane involving arginine–lysine reactions. Other cross-links derived from lysine–lysine reactions have been identified, e.g., GOLD, MOLD, vesperlysine and crosslines, but they are only present in minute quantities

studies are clearly not stable, inert, end-products but some must contain complex compounds possessing groups that react directly with tissues. It would be interesting to investigate the relative importance of some of these potential cross-linking compounds.

The subsequent oxidation processes lead to many products, including intermolecular cross-links, and these AGE cross-links lead to increases in breaking load and stiffness, denaturation temperature, and a decrease in solubility and susceptibility to degradative enzymes during aging (Andreassen et al. 1981; Monnier and Cerami 1981; Kent et al. 1985) (Fig. 4.5). Glycation cross-links when added to the enzymic cross-links, already producing a functional fiber, clearly have a deleterious effect on the properties of the fiber. The increase in failure stress and the failure to absorb elastic strain energy when stressed results in a loss of the compliant nature of the fiber, which is crucial to the normal functioning of all collagenous tissues.

The nature of these cross-links is complex but is slowly being unraveled (for reviews see Avery and Bailey 2005; Monnier et al. 2005; Baynes 2003). Similar changes occur in diabetes mellitus but at a much faster rate than normal aging.

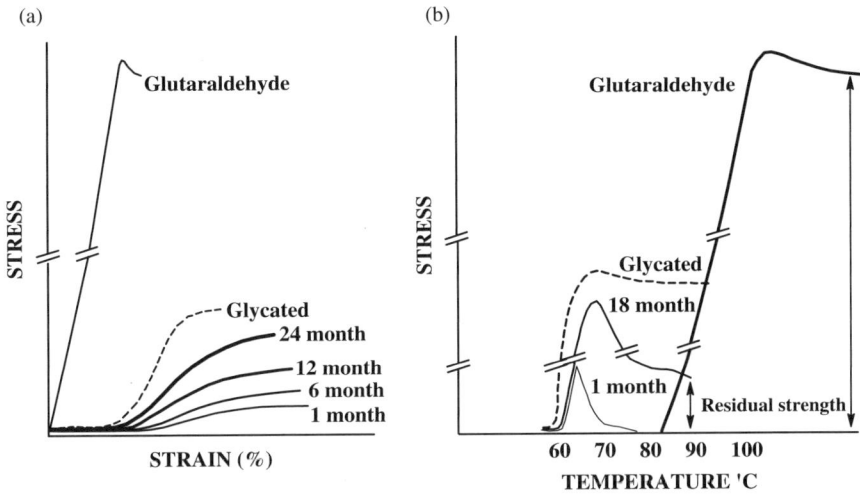

Fig. 4.5 Effect of glycation cross-linking and artificial chemical cross-linking with glutaraldehyde on the physical properties of collagen fibers. (a) Typical stress/strain curves for immature, mature, glycated and glutaraldehyde-treated fibers. (b) Typical hydrothermal isometric tensile tension curves for immature, mature, glycated and glutaraldehyde-treated collagen fibers

The initial route to glycation cross-links involves the reaction of glucose with the ε-amino group of lysine to form a Schiff base, glucosyl-lysine (Robins and Bailey 1972), which undergoes an Amadori rearrangement to fructose-lysine. Both these adducts undergo oxidation to deoxyglucose and fragmentation to smaller sugar aldehydes, such as methylglyoxal, which is several thousand times more reactive than glucose but present in minute quantities (Brinkmann et al. 1998). The inter-molecular cross-links are formed by further reaction between the modified lysines and arginines of adjacent collagen molecules.

Most reported cross-links have been obtained by model reactions in vitro, only a few have been subsequently isolated from tissue. The best characterized cross-link is pentosidine (Sell and Monnier 1989) formed between fructose-lysine and arginine (Fig. 4.4) but is present in very small quantities in most tissues and, at about 1 cross-link per 200 collagen molecules, is therefore unlikely to have a significant effect on the physical properties of the fiber. MOLD and GOLD (Odani et al. 1998) are derived from the initial reaction with methylglyoxal and glyoxal, respectively, with lysine but are again only present in minute quantities. On the other hand glucosepane, a lysine–arginine cross-link (Biemel et al. 2001) formed under non-oxidative conditions is reported to be present at 20 times the concentration of pentosidine, increasing with age to 1 cross-link per 5 collagen molecules and up to 1 per 2 molecules in diabetic tissues (Sell et al. 2005) (Fig. 4.4). Glucosepane is therefore the only current potential glycated cross-link likely to have a significant effect on the mechanical properties of the collagen fiber.

Methylglyoxal, is also formed from triosephosphate, acetone from ketone bodies, degradation of unsaturated lipids in addition to the oxidative fragmentation of fructose-lysine in collagen. The relative importance of the triosephosphate pathway appears to be significantly greater than the minor amount from fructose-lysine (Thornalley 1996). Another product of the breakdown of lipids is malondialdehyde which could account for the stiffening of large arteries (Slatter et al. 2000) where collagen and lipid are in close contact. Previously believed to act through the two aldehydes to form a double Schiff base we have shown that the reaction results in a dihydropyridine cross-link (Slatter et al. 1999). We have also shown that glucose reaction with soluble BSA or insoluble fibrous collagen, under near physiological conditions, produces a much faster reaction with BSA and some differences in the nature of the glycation products formed (Slatter et al. 2008). We attribute these differences to the inaccessibility of side-chain residues within the tightly packed collagen fiber to glucose, and therefore the reactions with BSA cannot be used as a model for fibrous proteins.

The location of glycation cross-links has not been determined but clearly must cross-link adjacent molecules via the triple helical region (Fig. 4.8). Whether there are preferential lysine and arginine residues as cross-link sites is unknown, but it is possible in view of the preferential sites for the glucose modification of BSA (Hinton and Ames 2006) and the predominant methylglyoxal modification of arginine. Attempts have been made to predict glycation sites from the amino acid sequence (Johansen et al. 2006). More recently Doblar (2008) employing pKa values and bio-informatics predicted that the arginines of the cell receptor sites, RGD and GFOGER would be preferentially glycated by methylglyoxal. This was confirmed using LC-MS/MS. It will be difficult to assign a specific cross-link to a change in mechanical properties since the conditions for the formation of these cross-links has been shown to be dependent on environmental factors that can vary considerably between body tissues.

The contribution of glycation products to pathological changes in the physical properties during aging is often challenged on the basis that the compounds formed are only detectable in trace quantities. Certainly the non-enzymic and therefore random nature of the reaction results in several different pathways leading to a diversity of products at low concentration and therefore difficult to detect. However, evidence of the modifications in physical properties and the ability to specifically inhibit these changes provide incontrovertible evidence for the role of glycation in aging.

4.2.7.1 Reported Effect of AGEs on Lysyl Oxidase Cross-Links

The formation of additional lysine-aldehydes by AGE-induced Strecker degradation of the ε-amino group of lysine has been reported with levels of lysyl-aldehyde in diabetic rat serum 4/5 times higher than controls (Akagawa et al. 2002). These lysine-aldehydes and their subsequent cross-links would have to be formed between two regions of the triple helix of adjacent collagen molecules (instead of the telopeptides involved in normal enzyme cross-links) and would make a difference to the physical properties of the fiber. Although an interesting proposal, the reaction only appears to be effective in the presence of copper and oxygen which suggests superoxide and

H_2O_2 are involved. The mechanism may therefore be a simple Fenton-type reaction rather than the Strecker reaction proposed as operating in vivo. Unfortunately no lysine-aldehyde-derived cross-links have been detected to support this proposal.

Although generally agreed that the concentration of mature, enzyme-induced cross-links (lysyl oxidase) plateaus following maturity, there is a report of an increase of lysyl oxidase-derived cross-links in the dermis of subjects with diabetes mellitus and that the level of HLKNL (Buckingham and Reiser 1990) correlated with the tightening and thickness of the skin. This cross-link is not normally present in collagen fibers from skin, the major cross-link being deH-HLNL and its maturation product HHL, although it could be present in high turnover tissues but in that case there would be little accumulation of the glycation cross-links normally observed in diabetic skin.

4.2.7.2 AGE Cross-Linking Induced by Diet

As discussed above, injection of AGEs into rats produced typical age-related changes and diabetic manifestations in collagenous tissues. In an extension of these studies several recent reports have dealt with the ingestion of AGEs formed during the heat processing of foods (Henle 2005) and from tobacco smoke (Cerami et al. 1997). The dietary AGEs appear to contribute significantly to the body's AGE pool and at these levels may overload the natural defense systems against AGE accumulation, such as the AGE receptor system on cell surfaces (Vlassara et al. 1985). As expected processed foods high in lipid and protein, including collagen, exhibit the highest AGE levels, e.g., fat and meat which contain 10/20 fold that of carbohydrate food such as bread and fruit (Goldberg et al. 2004).

Some of these food-derived AGEs could clearly display collagen cross-linking activity and consequently affect the mechanical properties of sensitive collagenous tissues.

4.2.7.3 Inhibition of Glycation Cross-Links

Lysyl oxidase produces native fibers with optimal mechanical properties, whereas the subsequent glycation cross-links are deleterious to normal function. Attempts have therefore been made to inhibit the formation of AGEs. Various types of inhibitors have been studied, metal chelators (Price et al. 2001), chemical and natural antioxidants such as green tea, garlic and carnosine (Hipkiss 2005). Aldehyde competitive reactors have been reasonably successful but aminoguanidine (Brownlee et al. 1986) was finally withdrawn from clinical trials due to adverse side-reactions. Pyridoxamine (Voziyan and Hudson 2005) is now being trialed. The mechanism of action of the latter is as yet unknown but may involve scavenging reactive carbonyls thus inhibiting post-Amadori stages, or possibly by inhibiting the effects of ROS (Price et al. 2001). Inhibition may also take place in the animal body itself, the AGEs being targeted by the specific receptors on macrophages (Vlassara et al. 1985) known as RAGEs, although they have now been identified on a number of cell types. RAGE is a multi-ligand receptor and a member of the immunoglobulin super-family of cell surface molecules. Although RAGE is the best characterized AGE receptor for CML and hydroimidazolones it also binds non-specifically to many other diverse proteins (Goldin et al. 2006). Other recent studies have shown the multiplicity of AGEs and cellular receptors is very complex and they elicit multiple effects (Bierhaus et al. 2005; Hudson and Schmidt 2004). Hence the role of RAGE as a key defense against accumulation of AGEs is now debatable.

Since glycation is already an on-going reaction during aging and diabetes before patient presentation at clinics an important alternative approach is the cleavage of existing cross-links. Phenyl-thiazolium and related compounds have been reported to cleave these cross-links through the dicarbonyl group (Vasan et al. 2003). However, the mechanism is controversial, there is no direct evidence for dicarbonyl AGEs, which actually would be highly reactive and form more complex compounds. It has also been reported that the hydrolysis products of these cross-link breakers are potent metal chelators (Price et al. 2001). On the other hand these compounds have been shown to be very effective in vivo in restoring the flexibility of large arteries in aged animals and animals with diabetes (Wolfenbuttle et al. 1998), but their mode of action is unknown. Further compounds are likely to be developed and their mechanisms elucidated in this important approach.

To summarize, the slow adventitious accretion of AGE cross-links alters the mechanical properties of fibers and amorphous basement membranes ultimately resulting in a rigid and brittle tissue that cannot fulfill its specific functional role. Glycation cross-linking is therefore a true aging effect, rather than the maturation of the enzymic cross-links as often reported. To date only one potential cross-link, glucosepane, is present in sufficient quantity to account for the observed increase in mechanical properties.

Biochemical markers of the extent of glycation are important in diabetes. Glycated hemoglobin has been used as a marker for the management of diabetes, but other markers of the changes in collagenous tissues are clearly required. Fluorescent changes in the skin have been correlated with retinopathy and arterial stiffness, but specific tissue cross-links would be more valuable to ascertain risk.

The study of AGEs is proving to be more diverse and complex than previously thought. Further research is important and should concentrate on the molecular effects, rather than histological changes, and elucidation of the cross-linking mechanisms involved in order to identify the products responsible and then specifically inhibit these deleterious changes.

4.2.8 Unusual Cross-Linking Mechanisms in Native Collagen

Animal collagens from sponge to human are predominantly cross-linked by lysyl oxidase-derived cross-links described above but Nature has evolved alternative mechanisms for specialized environmental conditions.

4.2.8.1 Nε($γ$-Glutamyl)-Lysine Cross-Links

These cross-links have been long established in keratin and fibrin, and more recently in fibrillin (Kielty et al. 2003), but they do occur in some collagenous structures. The cross-links are generated by transglutaminase (TGase) for which seven different genes have been characterized to date and transglutaminase is now known to play a diverse role in biology (Aeschlimann and Thomazy 2000). TGase activity is Ca^{2+} dependent and catalyses the amine $γ$-glutamyl transferase reaction, which leads to the formation of Nε($γ$-glutamyl)-lysine isopeptide cross-link.

Transglutaminase

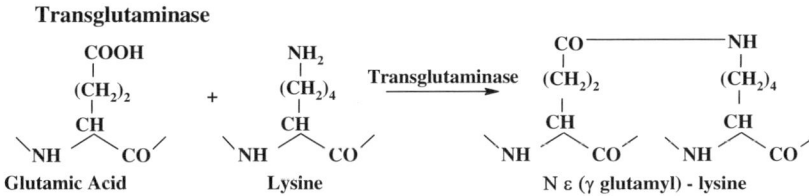

Glutamic Acid   +   Lysine $\xrightarrow{\text{Transglutaminase}}$ N ε ($γ$ glutamyl) - lysine

The isopeptide is hydrolyzed by both acid and alkali hence isolation and identification involves extensive enzyme digestion. Its presence can be detected by antibodies to the cross-link and indirectly by amino acid analyses before and after fluorodinitrobenzaldehyde (FDNB) derivativization to distinguish free and bound ε-amino lysine residues. This cross-link has been reported (Aeschlimann and Paulsson 1991) to be present in type IV basement membrane at the dermal–epidermal junction. There is reasonable evidence that types V and XI are assembled as a core for collagens I and II and therefore play a role in the fibrillogenesis of these major collagens. The type V/XI collagen fibrils have been shown to be insoluble even following treatment with lathyritic agents. Synthesis of these collagens from A204 rhabdomyosarcoma cells revealed that their insolubility was due to the formation of Nε($γ$-glutamyl)-lysine by transglutaminase (Kleman et al. 1995). Various collagens have been reported to be stabilised by TGase in human liver fibrosis (Grenard et al. 2001).

4.2.8.2 Catechol Oxidase: Quinones

Byssus threads, which attach mussels to rocks, possess a gradient of mechanical properties based on a chimeric collagen that encompasses, within the primary structure domains corresponding to collagen, polyhistidines and either elastin or dragline spider silk (Waite et al. 2003). The threads act as shock absorbers and are attached to rock by a water-proof polypeptide glue secreted from the foot (Waite et al. 1985). The byssus threads contain peptide-bound 3,4-dihydroxyphenylalanine (DOPA) and these are converted by catechol oxidase to quinones which can then react with lysine to form a cross-link (Fig. 4.6a). McDowell et al. (1999) on the other hand reported that these DOPA residues formed a 5,5 diDOPA cross-link (Fig. 4.6b). The extent of diDOPA cross-linking correlates with the stress on the thread by the flow of seawater

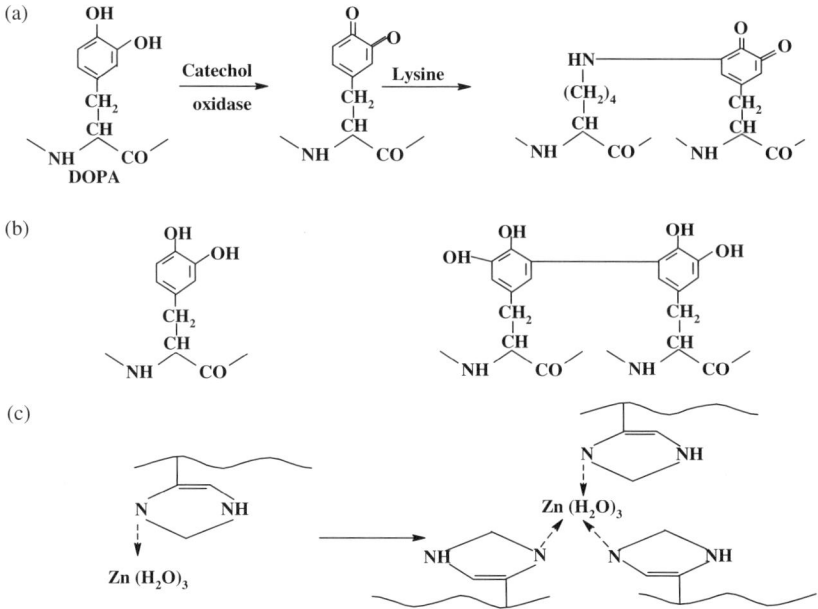

Fig. 4.6 Unusual cross-links identified in collagen fibers from invertebrates. (**a**) Oxidation of DOPA to quinones by catechol oxidase and subsequent reaction with lysine side-chains. (**b**) Reaction of DOPA to form a diDOPA cross-link. (**c**) Formation of a co-ordinate complex by interaction of zinc and histidine

and these threads have been shown to undergo an increase in tensile strength of 4–5 times due to aeration by seawater.

4.2.8.3 Co-ordinate Metal–Ion Complex

The byssus threads may initially be cross-linked by metal–ion bonds and later reinforced by oxidation products involving the catechol oxidase. Histidine-rich regions occur at both termini of the chimeric collagen and can form di- and tri-histidyl zinc complexes which act as non-covalent stabilizing cross-links (Waite et al. 1998) (Fig. 4.6c).

Sea urchin egg shell membranes must be refractory to chemical, enzymic and mechanical disruption to allow the embryo to develop in a protected environment. The stabilization of the membrane is an excellent example of the sequential formation of different cross-links for particular conditions. Initially the covalent cross-link Nε(γ-glutamyl)-lysine is formed. Protoliaisin then attaches ovoperoxidase to the membrane in turn producing hydrogen peroxide free radicals resulting in the formation of di-tyrosine between the various proteins thereby hardening the egg shell membrane within 10 min of fertilization. The process is finely regulated, the ovoperoxidase only becoming active after fertilization and subsequent exposure to seawater (Foerder and Shapiro 1977).

4.2.9 Stabilization by Chemical Cross-Linking for Bioengineering Tissues

Collagen is abundant in the animal body and has a high mechanical strength plus resistance to enzymes and these properties are utilized in a wide range of industrial and medical products. However, there is generally a need to further stabilize the fibers by chemical cross-linking to modify and control their mechanical properties, their residence time in the body and to decrease their immunogenicity.

Chemical cross-linking has a long history, for example, to convert skin to leather with specialized properties (Covington 1997) and more recently for medical and cosmetic applications.

The reactive groups, lysine, glutamic acid and hydroxyl groups project radially from the rod-like collagen molecules, which are in parallel alignment in the fiber thus ensuring intermolecular or interfibrillar cross-linking can occur. Cross-links involving these groups further prevent the molecules/fibers sliding past each other under load and thereby increase the mechanical strength of the fibers and also increase the denaturation temperature. Such artificial cross-linking also renders the collagen fibers less susceptible to enzyme degradation, due to decreased accessibility for the enzyme and reduced hydration of the fibers.

4.2.9.1 Mechanical Properties

In a standard stress–strain curve of a collagenous tissue cross-linking results in an increase in breaking strength and stiffness with increasing age, but this is dramatically increased by chemical cross-linking, the toe region shortens, the linear region becomes steeper indicating a stiffer fiber and the breaking strength increases (Fig 4.5). The increase is variable dependent on the chemical nature of the cross-link (Covington 1997; Paul and Bailey 2003).

It is interesting to note that during air drying of collagen fibers, in addition to increasing the denaturation temperature, the fiber changes from a wet flexible fiber to a dry and very stiff fiber. This poses the question of how far the mechanical properties are modified by the resistance to slippage due to intermolecular and interfibrillar cross-linking versus that induced by the dehydration of the fibers.

4.2.9.2 Denaturation Temperature

The almost crystalline structure of the collagen molecules collapses on heating, at a sharply pronounced denaturation temperature, to random chains of gelatin. The temperature can range from $15°C$ for cod skin to $50°C$ for *Ascaris* cuticle, while mammalian collagens denature at about $40°C$ and the differences have been correlated with the proline and hydroxyproline content of the collagen (Burjanadze and Kisirya 1982). The stability of the triple helix is believed to be due to water hydrogen-bonded to the hydroxyproline residue (Ramachandran et al. 1973; Bella et al. 1995). Alternatively, Holmgren et al. (1998) have proposed that stabilization occurs through an inductive effect of the hydroxyl group of hydroxyproline on the peptide backbone thus stabilizing the *trans*-configuration of the prolyl peptide bond of the triple helix. The relative importance of these two effects is currently unknown.

The aggregated fibers of mammals possess a denaturation temperature of about 65°C, but cross-linking with glutaraldehyde can increase the denaturation temperature to almost 100°C.

4.2.9.3 Hydrothermal Isometric Tension

During denaturation at 65°C the collagen fiber shrinks to about one quarter of its original length. Maintaining the fiber at constant length generates a tension which can be measured using a strain gauge (Fig 4.5b). The tension profile of this process alters with age as the immature cross-links mature to thermally stable bonds. Additional changes occur following chemical cross-linking which are valuable in assessing the properties of the modified tissue. The method also provides an estimate of the proportion of the total and thermally stable cross-links since both the ultimate and residual strength of the denatured fiber can be determined (Kopp et al. 1977).

4.2.10 Mechanisms of Some Common Chemical Cross-Link Reactions

4.2.10.1 Reaction with ε-Amino Groups of Lysine

Glutaraldehyde is the commonest aldehyde cross-linking agent since it forms chemically stable bonds and significantly increases the physical properties of the fiber (Covington 1997). The chemical nature of the bonds is complex since glutaraldehyde contains aldols and forms polymeric complexes during storage. These complexes are believed to react with lysine to form heterocyclic compounds which subsequently undergo oxidation to pyridine rings (Fig. 4.7a). Although extensively used to stabilize prostheses for medical application in vivo its use has been questioned due to its propensity to induce cytotoxic effects, and a variety of alternative cross-linking agents are currently being investigated as a potential replacement.

Isocyanates are also commonly used for example, hexamethylene di-isocyanate (HMDIC) (Fig 4.7b) has been successfully used (Damink et al. 1995) and raises the denaturation temperature of sheep skin from 54 to 74°C.

4.2.10.2 Reaction with Carboxyl Groups

Carbodiimides ($NH=C=NH$) and acyl azides have been used to react with carboxyl groups without themselves being incorporated into final linkages:

$2 \times$ Collagen–COOH + R-NC=N-R $\longrightarrow$ Collagen-CO –NH-CO-Collagen.

Ethyl-3(3-dimethylamino)propyl carbodiimide (EDC) does not induce the cytotoxic side-effects associated with glutaraldehyde, and the extent and rate of

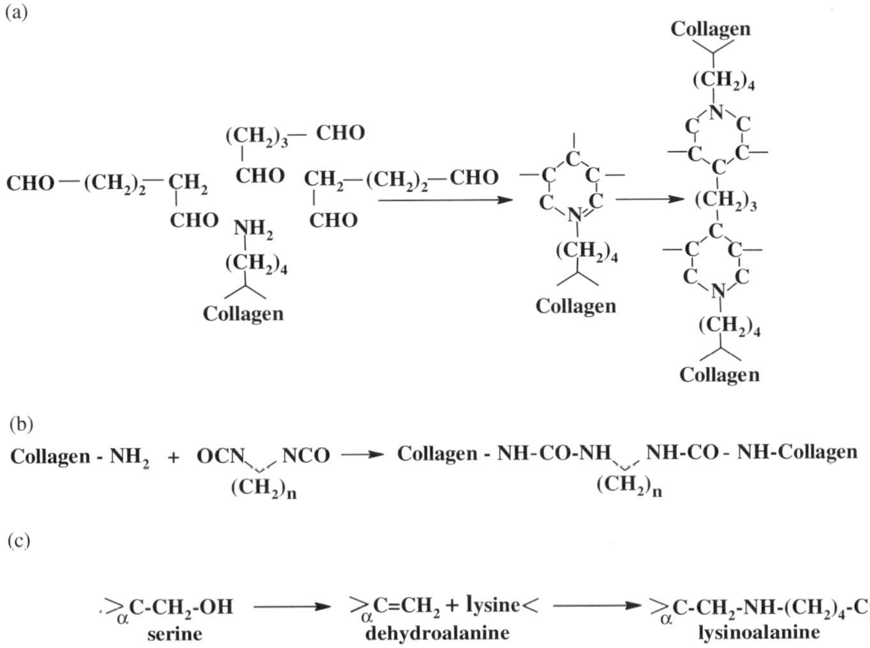

(a)

Collagen

$(CH_2)_4$

$(CH_2)_3$— CHO

CHO—$(CH_2)_2$—CH_2 CHO CH_2—$(CH_2)_2$—CHO

CHO NH_2 CHO

$(CH_2)_4$

Collagen

$(CH_2)_4$

Collagen

$(CH_2)_3$

$(CH_2)_4$

Collagen

$(CH_2)_4$

Collagen

(b)

Collagen - NH_2 + OCN NCO ⟶ Collagen - NH-CO-NH NH-CO- NH-Collagen

$(CH_2)_n$ $(CH_2)_n$

(c)

$\gtrdot_{\alpha}$C-CH_2-OH ⟶ $\gtrdot_{\alpha}$C=CH_2 + lysine$\lessdot$ ⟶ $\gtrdot_{\alpha}$C-CH_2-NH-$(CH_2)_4$-C$\lessdot_{\alpha}$

serine   dehydroalanine   lysinoalanine

Fig. 4.7 Mechanisms of the different reactions utilized in the modification of collagen fibers to provide specific properties: (**a**) proposed reaction with glutaraldehyde polymers; (**b**) reaction with isocyanates; (**c**) β-elimination, a non-chemical dehydrothermal treatment

cross-linking can be enhanced by the addition of *N*-hydroxysuccinimide (NHS), the combination results in a significant increase in tensile properties (Damink et al. 1996).

Similarly, amide bond formation can be achieved by acyl azides, the initial reaction being with carboxy groups and further reaction with ε-amino groups has been shown to give a level of cross-linking comparable with glutaraldehyde but with a much reduced cytotoxic effect (Petite et al. 1995).

4.2.10.3 Enzyme-Generated Cross-Links

Transglutaminase can be used to generate isopeptide cross-links between glutamic acid and lysine to form stable γ-glutamyl-lysine cross-links (Chen et al. 2005) in native collagen as described above. The cross-links will be located between residues in the helices of adjacent molecules/fibers and thereby increase the mechanical properties of the fiber.

4.2.10.4 Radiation

X-ray and γ-ray irradiation of collagen results in two competing reactions: degradation through peptide bond cleavage is predominant in the dry state, while intermolecular cross-link formation is predominant in the wet state. Degradation increases the solubility and reduces tensile strength, while irradiation in the wet state decreases solubility, increases the denaturation temperature, decreases enzyme susceptibility and produces a significant increase in hydrothermal isometric tension due to the formation of thermally stable cross-links (Bailey 1968). The nature of the cross-links has not been elucidated but presumably is initiated by free radicals formed in the water resulting in random cross-links.

UV Radiation. It has been suggested that aging in terms of the stiffening of skin is accelerated by UV radiation. However, both cross-linking and degradation occur, the relative proportion depending on the presence of oxygen, pH, type of collagen and wavelength of the UV light. The change in mechanical properties is therefore variable depending on the conditions. UV light produces similar effects to X-rays and γ-rays since free radicals are produced in the water by all these types of radiation. Rapid degradation of phenylalanine and tyrosine occurs and it has been proposed that cross-linking occurs through energy absorption by these aromatic groups, resulting in the formation of a di-tyrosine cross-link and a DOPA cross-link (Kato et al. 1995), but the presence of these cross-links have not been confirmed by other workers. Evidence for cross-links has included increased stiffness, enzyme resistance and denaturation temperature in skin (Yamauchi et al. 1991; Sionkowska and Kaminska 1999). UV irradiation has been employed in conjunction with riboflavin to stiffen the weakened cornea in keratoconus (Spoerl et al. 2004).

UV irradiation of collagen in solution initially produces a stiffened gel, but continued irradiation results in a loss of viscosity. Clearly the effect of cross-linking is predominant in the early stages but as the increase in peptide cleavage of the triple helix increased degradation predominated with the loss of viscosity. Both cross-linking and degradation are occurring simultaneously. The early degradation produced a damaged intermediate state of the triple helix with lower denaturation temperature due to the cleavage of peptide bonds, but as this reaction increased on further irradiation there was a rapid loss of the triple helix (Miles et al. 2000). In these studies on tropocollagen, skin and cornea no evidence for the formation of di-tyr and DOPA cross-links was obtained.

4.2.10.5 Borohydride Reduction

The aldimine cross-links present in immature skin are acid and heat labile, but can be stabilized by reduction with sodium borohydride, rendering the collagen insoluble with increased mechanical strength.

However, the reduction of the existing labile cross-links does not increase the denaturation temperature since there is no change in the number of cross-links or in the hydration of the fiber (see below).

4.2.10.6 Dehydrothermal Cross-Links

Extensive dehydration, particularly under high vacuum is used industrially to produce an insoluble cross-linked collagen with a high denaturation temperature. The cross-link is believed to be between lysine and alanine formed by β-elimination of serine to dehydro-alanine which then reacts with lysine to form lysinoalanine (Bohak 1968) (Fig. 4.7c). This non-chemical treatment can be combined with UV irradiation to increase mechanical strength and avoid associated cytotoxicity.

4.2.11 Location of Enzymic, Glycation and Chemical Cross-Links

The location of divalent enzymic cross-links is well established by analysis of cyanogen bromide-digested cross-linked peptides and occurs between the telopeptide lysines and hydroxylysines in the end-overlap domain of adjacent molecules (Fig. 4.8a), thus confirming the molecular alignment previously identified by electron-micrographs. These cross-links place restrictions on the models for the organization of the molecules in the fiber (Bailey et al. 1980). Based on the increase in physical properties it has been proposed that the mature enzymic cross-links form interfibrillar cross-links (Fig. 4.8b). Glycation cross-linking by glucose and its oxidation products involves predominantly lysine and arginine residues, while cross-linking with reactive chemical reagents involves lysine and glutamic acid. All these groups are radially exposed along the length of the rod-like collagen molecules and

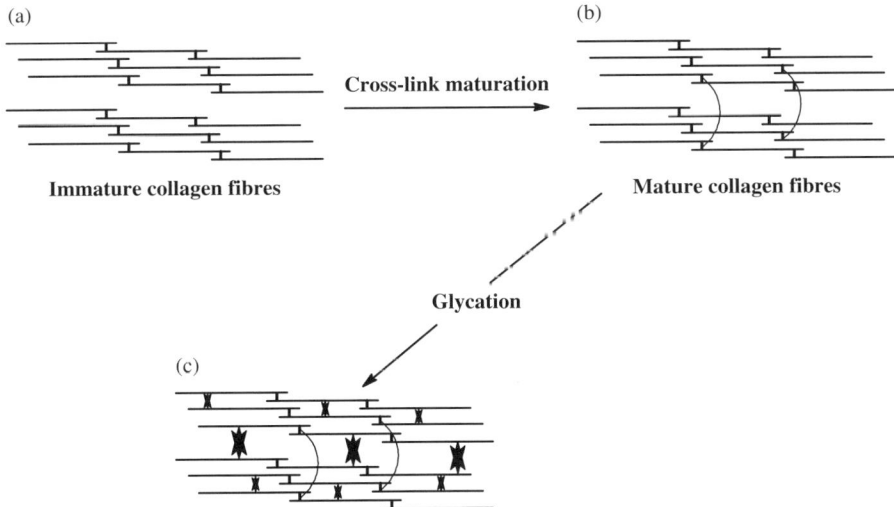

Fig. 4.8 Proposed locations of cross-links in collagen fibers; (**a**) divalent immature cross-linking within the fiber [I], (**b**) trivalent mature cross-links linking adjacent fibers [)] and (**c**) glycation cross-linking within and between fibers [✗]. Chemical cross-linking would also be located within and between collagen fibers as indicated for the glycation cross-linking (From Avery & Bailey "Enzymic and non-enzymic cross-linking mechanisms in relation to turnover of collagen: relevance to aging and exercise" in Scandinavian Journal of Medicine & Science in Sports published by Blackwell 2005.)

are therefore conveniently positioned to form interhelical cross-links, either within or between fibrils (Fig. 4.8c), resulting in altered physical properties of the tissue.

4.2.12 Mechanism of Increased Denaturation Temperature by Cross-Linking

The accumulation of mature intermolecular cross-links is associated with an increase in thermal stability and hydrothermal isometric tension and these properties are increased further by glycation reactions during aging. Dramatic increases in the denaturation temperature occur following artificial cross-linking with aldehydic reagents such as glutaraldehyde (from 65°C to about 100°C). The mechanism involved has been variously assigned to the stability of the particular cross-links or a change in the co-operative regions of the collagen molecule. However, we recently reported that the increased thermal stability is determined by the water content of the fiber (Miles et al. 2005), the cross-link reducing the axial separation of the molecules leading to a loss of water between the molecules. Employing cross-linking agents of different lengths between the functional molecules we found that all the samples yielded the same denaturation temperature at the same hydration demonstrating that the denaturation temperature was not affected by the nature of the cross-link other than through its effect on hydration of the fiber. The dehydration does not affect the bound water of the collagen molecule hence the enthalpy of denaturation is therefore unaffected. The denaturation temperature of collagen in bone is very high (about 155°C) due to the reduced water content induced by mineralization, but on demineralization the collagen denatures at normal temperature (65°C) for type I fibrous collagen.

The "polymer in a box" mechanism of stabilization of polymers in which the unfolded molecule is confined to a reduced fiber lattice explains this data (Miles and Ghelashvili 1999). The rate of unfolding is depressed by the surrounding molecules which act like the walls of a box to reduce the number of molecular configurations and hence the entropy. We have previously proposed the presence of a thermally labile region in the molecule which has to be denatured before the whole molecule can unzip. The entropy of this activated state is reduced by the confinement of the molecules in a smaller box due to the dehydration caused by the cross-links drawing the collagen molecules closer together.

The denaturation temperature decreases with increased hydration and ionic composition, but increases with mechanical load (Chen and Humphrey 1998). The importance of mechanical loading, and hence the mechanism by which tension delays shrinkage, has not received the attention it deserves.

To summarize, collagen biomaterials have a wide variety of applications in industry and medicine. A specific increase in mechanical strength, denaturation temperature, enzyme resistance and biocompatibility can be achieved by a correct choice from a wide range of available cross-linking agents. The increases are achieved by cross-linking between the fibers resulting in tighter binding of the fibers and loss of water.

4.3 Future Prospects

The mechanism of formation of the enzymic cross-links is now well established, but despite the extensive knowledge gained over the past few decades there are still several significant problems to solve. The nature of the mature cross-link in mouse skin is unknown and would be valuable in following the maturation of the cross-links in studies of knock-out mice. The recent report that bone collagen contains unknown cross-links, not involving lysyl oxidase, at a level equivalent to the combined pyridinoline and pyrrole cross-links is clearly an important question. We are beginning to correlate particular cross-links with the mechanical properties of bone and tendon collagen, but we need to know how this varies with particular bones and tendons, and the effect of the load exerted on them, both under normal circumstances and under strenuous exercise.

It is generally accepted that the changes in mechanical properties with age following maturation are the result of glycation cross-linking, but the nature of these cross-links is only just beginning to be unraveled. Several structures have been proposed but have either not been confirmed in vivo or shown to be present in insufficient quantities to affect the mechanical properties. Only glucosepane appears to satisfy these criteria but its presence and location in collagen needs to be confirmed.

Chemical and physical methods of cross-linking native and synthetic collagens have been utilized to increase the mechanical strength, denaturation temperature and resistance to degradative enzymes for industrial, medical and cosmetic applications. Further new sophisticated cross-linking agents that are both as efficient as glutaraldehyde but do not generate cytotoxic effects are now being developed.

The increases in denaturation temperature have been correlated with the reduced water content of the fiber, resulting from the tighter binding of the molecules in the fiber by the cross-linking agent. Although the denaturation temperature is increased by mechanical load on the fiber the mechanism remains to be elucidated, particularly so since most tissues operate under tension both in vivo and in vitro. The loss of water also stiffens the fiber, but the nature of the correlation between mechanical strength and hydration of the fiber has not yet been investigated in detail.

References

Aeschlimann D and Paulsson M (1991) Cross-linking of laminin-nidogen complexes by tissue transglutaminase – a novel mechanism for basement membrane stabilization. J Biol. Chem. 266 15308–15317.

Aeschlimann D and Thomazy V (2000) Protein cross-links in assembly and remodelling of extracellular matrices. The role of transglutaminase. Connect. Res. 41 1–27.

Akagawa M, Sasaki T and Suyama K (2002) Oxidative deamination of lysine residues in plasma proteins of diabetic rats. Novel mechanism via the Maillard reaction. Eur. J. Biochem. 269 5451–5420.

Alexander R McN (1988) Elastic Mechanisms in Animal Movement. Cambridge University Press, Cambridge UK.

Amiel D, Frank C, Harwood F, Fronek J and Akeson W (1984) Tendons and ligaments; a morphological and biochemical comparison. J. Ortho. Res. 1 257–265.

Andreassen T T, Seyer-Hansen K and Bailey A J (1981) Thermal stability, mechanical properties and reducible cross-links of rat tail tendon in experimental diabetes. Biochim. Biophys. Acta. 677 313–317.

Avery N C and Bailey A J (2005) Enzymic and non-enzymic cross-linking mechanisms in relation to turnover of collagen: relevance to ageing and exercise. Scand. J. Med. Sci. Sports. 15 231–240.

Bailey A J (1968) Effects of ionizing radiation on connective tissue components. Int. Rev. Connect. Tissue Res. 4 233–281.

Bailey A J and Peach C M (1968) Isolation and structural identification of a labile intermolecular cross-link in collagen. Biochem. Biophys. Res. Commun. 33 812–819.

Bailey A J and Shimokomaki M S (1971) Age related changes in the reducible cross-links of collagen. FEBS Lett. 16 86–88.

Bailey A J and Robins S P (1972) Embryonic skin collagen replacement of the type of aldimine cross-link during the early growth period. FEBS Lett. 21 330–334.

Bailey A J, Gathercole L J, Dlugosz J, Keller A and Voyle C A (1982) Proposed resolution of the paradox of extensive cross-linking and low tensile strength of the cuvierian tubules collagen from the sea cucumber *Holothuria forskali*. Int. J. Biol. Macromol. 4 329–334.

Bailey A J, Light N D and Atkins E D T (1980) Chemical cross-linking restrictions on models for the molecular organization of the collagen fibre. Nature (Lond) 288 408–410.

Bailey A J, Wotton S F, Sims T J and Thompson P W (1992) Post-translational modification in the collagen of human osteoporotic femoral head. Biochem. Biophys. Res. Commun. 185 801–805.

Banse X, Sims T J and Bailey A J (2002a) Mechanical properties of vertebral cancellous bone: Correlation with intermolecular cross-links. J. Bone Miner. Res. 17 1621–1628.

Banse X, Devogelaer J P, Lafasse A. Grynpas M, Sims T J and Bailey A J (2002b) The cross-link profile of bone collagen correlates with the structural organization of the trabeculae. Bone 31 70–76.

Batge B, Diebold J, Stein H, Bodo M and Muller P K (1992) Compositional analysis of the collagenous bone matrix: a study on adult normal and osteopenic bone tissue. Eur. J. Clin. Invest. 22 805–812.

Baynes J W (2003) Chemical modification of proteins by lipids in diabetes. Clin. Chem. Lab. Med. 41 1159–1165.

Bella J, Brodsky B and Berman H M (1995) Hydration structure of a collagen peptide. Structure 3 893–966.

Bernstein P H and Mechanic G L (1980) A natural histidine-based imminium cross-link in collagen and its location. J. Biol. Chem. 255 10414–10422.

Biemel K M, Reihl O, Conrad J and Lederer M O (2001) Formation pathways for lysine-arginine cross-links derived from hexoses and pentoses by the Maillard process. J. Biol. Chem. 276 23405–23412.

Bierhaus A, Humpert P M, Morcos M, Wenet T, Chavakis T, Arnold B, Stern D M and Nawroth P P (2005) Understanding RAGE, the receptor for advanced glycation end-products. J. Mol. Med. 83 876–886.

Birch H L, McLaughlin L, Smith R K and Goodship A E (1999) Treadmill exercise-induced tendon hypertrophy: assessment of tendons with different mechanical function. Equine Vet. J. 30 222–236.

Bohak Z (1968) N (dl-2-amino-2-carboxyethyl)-1-lysine: a new amino acid formed on alkaline treatment of proteins. J. Biol. Chem. 239 2878.

Brady J D and Robins S P (2001) Structural characterization of pyrrole cross-links in type I collagen of human bone. J. Biol. Chem. 276 18812–18818.

Brinkmann E, Degenhardt T P, Thorpe S R and Baynes J W (1998) Role of the Maillard reaction in aging of tissue proteins. Advanced glycation end-product-dependent increase in imidazolium cross-links in human lens capsules. J. Biol. Chem. 273 18714–18719.

Brownlee M, Vlassara H, Kooney A, Ulrich P and Cerami A (1986) Aminoguanidine prevents diabetes-induced arterial wall protein cross-linking. Science. 232 1629–1632.

Buckingham B and Reiser K M (1990) Relationship between the extent of lysyl oxidase-dependent cross-links in skin collagen, non-enzymatic glycosylation and long-term complications of type I diabetes mellitus. J. Clin. Invest. 86 1946–1054.

Burjanadze T V and Kisirya E L (1982) Dependence of thermal stability on the number of hydrogen-bonds in water-bridged collagen structure. Biopolymers 21 1695–1701.

Cerami C, Founds H, Nicholl I D, Mitsuhashi T, Giordano D, Vanpatten S, Lee A Al-Abed Y, Vlassara H, Bucala R and Cerami A (1997) Tobacco smoke is a source of toxic reactive glycation products. Proc. Natl. Acad. Sci. USA 94 13915–13920.

Chen S S and Humphrey J D (1998) Heat-induced change in the mechanics of a collagenous tissue: pseudo elastic behaviour at 37°C. J. Biomech. 31 211–216.

Chen R N, Ho H O and Sheu M T (2005) Characterization of collagen matrices cross-linked using microbial transglutaminase. Biomat. 26 4229–4235.

Covington A D 1997 Modern tanning chemistry. Chem. Soc. Rev. 26 111–126.

Curwin S L, Vailas A C and Wood J (1988) Immature tendon adaptation to strenuous exercise. J. Appl. Physiol. 65 2297–2301.

Damink L H H O, Dijkstra P J, vanLuyn M J A, Vanwachem P B, Nieuwenhuis P and Feijen J (1995) Cross-linking of dermal sheep collagen using hexamethylene diisocyanate. J. Mater. Sci. Mater. MED. 6 429–434.

Damink L H H O, Dijkstra P J, vanLuyn M J A, Vanwachem P B, Nieuwenhuis D and Feijen J (1996) Cross-linking of dermal sheep collagen using a water soluble carbodiimide. Biomaterials 16 1003–1008.

Doblar D (2008) PhD Thesis. University of Essex, UK.

Draper E R C, Morris M D, Camacho N P, Matousek P, Towrie M, Parker A W and Goodship A E (2005) Novel assessment of bone using time-resolved transcutaneous Raman Spectroscopy. J. Bone Miner. Res. 20 1968–1972.

Eriksen H A, Sharp C A, Robins S P, Sassi M-L, Risteli L and Risteli J (2004) Differently cross-linked and uncross-linked carboxy-terminal telopeptides of type I collagen in human mineralised bone. Bone 34 720–727.

Eyre D R and Oguchi H (1980) The hydroxypyridinolinium cross-links of skeletal collagen: their measurement, properties and proposed pathway of formation. Biochem. Biophys. Res. Commun. 92 403–410.

Eyre D R, Shao P, Weis M A and Steinmann B (2002) The kyphoscoliotic type of Ehlers-Danlos syndrome (type VI); differential effects on hydroxylation of lysines in collagens I and II revealed by analysis of cross-linked telopeptides from urine. Mol. Genet. and Metab. 76 211–216.

Foerder C A and Shapiro B M (1977) Release of ovoperoxidase from sea-urchin eggs hardens the fertilization membrane with di-tyrosine cross-links Proc. Natl. Acad. Sci. USA 74 4214–4218.

Fujimoto D, Ishida T and Hayashi H (1978) The structure of pyridinoline, a collagen cross-link. Biochem. Biophys. Res. Commun. 84 52 57.

Glimcher M J, Shapiro F, Ellis R D and Eyre D R (1980) Changes in tissue morphology and collagen composition during the repair of cortical bone in the adult chicken. J. Bone Joint Surg. 62A 964–973.

Goldberg T, Cai W, Peppa M, Dardaine V, Baliga B S, Uribarri J and Vlassara H (2004) Advanced glycoxidation end-products in commonly consumed foods. J. Am. Diet Assoc. 104 1287–1291.

Goldin A, Beckman J A, Schmidt A M and Creager M A (2006) Advanced glycation end-products. sparking the development of diabetic vascular injury. Circulation 114 597–605.

Grenard P, Bresson-Hadni S, El Alaoui S, Chevallier M, Vuitton DA and Ricard-Blum S (2001) Transglutaminase-mediated cross-linking is involved in the stabilization of the extra-cellular matrix in human liver fibrosis. J. Hepatol. 35 367–375.

Hanson D A and Eyre D R (1996) Molecular specificity of pyridinoline and pyrrole cross-links in type I collagen of human bone. J. Biol. Chem. 271 26508–26516.

Henle T (2005) Protein-bound advanced glycation end-products (AGEs) as bioactive amino acid derivatives in foods. Amino Acids 29 313–322.

Holmgren S K, Taylor K M, Bretscher L E and Raines R T (1998) Code for collagen's stability deciphered. Nature 392 666–667.

Hinton D J S and Ames J M (2006) Site specificity of glycation and carboxymethylation of BSA by fructose Amino Acids. 30 425–433.

Hipkiss A R (2005) Glycation ageing and carnosine: are carnivorous diets beneficial? Mech. Ageing Dev. 126(10) 1034–1039

Hudson B G and Schmidt A M (2004) RAGE: a novel target for drug intervention in diabetic vascular disease. Pharm. Res. 21 1079–1086.

Johansen M B, Kiemer L and Brunak S (2006) Analysis and prediction of mammalian protein glycation. Glycobiology 16 844–853.

Kato Y, Nishikawa T and Kawakishi S (1995) Formation of protein-bound 3,4 dihydroxyphenylalanine in collagen types I and IV exposed to ultraviolet light. Photochem. Photobiol. 61 367–372.

Kent M J C, Light N D and Bailey A J (1985) Evidence for glucose-mediated covalent cross-linking of collagen after glycosylation in vitro. Biochem. J. 225 745–752.

Kielty C M, Baldock, C, Sherratt M J, Rock M J, Lee D and Shuttleworth C A (2003) Fibrillin; from microfibril assembly to biomechanical function. In Elastomeric Proteins (Eds. P R Shewry, A S Tatham A S and A J Bailey) Cambridge University Press, Cambridge UK. pp 94–114.

Kleman J-P, Aeschlimann D, Paulson M and van der Rest M (1995) Transglutaminase-catalysed cross-linking of fibrils of collagen V/XI in A204 Rhabdomyosarcoma cells. Biochemistry 34 13768–13775.

Kirsch E, Kreig T, Remberger K, Fendel H, Bruckner P and Muller P K (1981) Disorder of collagen-metabolism in a patient with osteogenesis imperfecta (lethal type)- increased degree of hydroxylation of lysine in collagen type I and type III. Eur. J. Clin. Invest. 11 38–47.

Knott L, Whitehead C C, Fleming R H and Bailey A J (1995) The biochemistry of the collagenous matrix of osteoporotic avian bone. Biochem. J. 310 1045–1051.

Knott L, Tarlton J F and Bailey A J (1997) The chemistry of collagen cross-links. Biochemical changes in collagen during partial mineralization of turkey leg tendon. Biochem. J. 322 535–542.

Knott L and Bailey A J (1998) Collagen cross-links in mineralising tissues: a review of their biochemistry, function and clinical relevance. Bone 22 181–187.

Knott L and Bailey A J (1999) The collagen biochemistry of avian bone: a comparison of bone type and skeletal site. Brit. J. Poultry Sci. 40 371–379.

Kopp J, Sale P and Bonnet Y (1977) Apparatus for measuring contraction devised for studying physical properties of collagen fibres- isometric tension, extent of cross-linking, relaxation. Can. I. Food Sc. Tech. J. 10 69–72.

Kupyers R, Tyler M, Kurth L B, Jenkins I D, and Horgan D J (1992) Identification of the loci of the collagen associated Ehrlich chromogen in type I collagen confirms its role as a trivalent cross-link. Biochem. J. 283 129–136.

Li B H and Aspden R M (1997) Composition and mechanical properties of cancellous bone from the femoral head of patients with osteoporosis and osteoarthritis. J. Bone Mineral Res. 12 541–651.

Light N D and Bailey A J (1985) Collagen Cross-links: location of pyridinoline in type I collagen. FEBS Lett. 182 503–508.

Lucero HA and Kagan HM (2006) Lysyl oxidase; an oxidative enzyme and effector of cell function. Cell. Mol. Life Sci. 63 2604–2316.

Mansell J P and Bailey A J (1998) Abnormal cancellous bone collagen metabolism in osteoarthritis. J Clin. Invest. 101 1596–1603.

McDowell L M, Burzio L A, Waite J H and Schaefer J (1999) Rotational echo double resonance detection of cross-links formed in mussel byssus under high flow stress. J. Biol. Chem. 274 20293–20295.

Miles C A, Knott L, Sumner I G and Bailey A J (1998) Differences between the thermal stabilities of the three triple helical domains of type IX collagen. J. Mol. Biol. 27 135–144.

Miles C A, Sionkowska A, Hulin S, Sims T J, Avery N C and Bailey A J (2000) Identification of an intermediate state in the helix coil degradation of collagen by UV light. J. Biol. Chem. 275 33014–33020.

Miles C A and Ghelashvili M (1999) Polymer-in-a-box mechanism for the thermal stabilization of collagen molecules in fibres. Biophys. J. 76 3243–3252.

Miles C A, Avery N C, Rodin V and Bailey A J (2005) The increase in denaturation temperature following cross-linking is caused by dehydration of the fibres. J. Mol. Biol. 346 551–556.

Monnier V M and Cerami A (1981) Non-enzymatic browning in vivo, possible process for aging of long-lived proteins. Science 211 491–493.

Monnier V M, Mustata T G, Biemel K L, Reihl O, Lederer M O, Zhenyo D and Sell D R (2005) Cross-linking of the extracellular matrix by Maillard reaction in aging and diabetes: an update on "a puzzle nearing resolution". Ann. NY Acad. Sci. 1043 533–544.

Notbohm H, Nokelainen M, Myllyharju J, Fietzek P P, Muller P K and Kivirikko K I (1999) Recombinant human type II collagens with low and high levels of hydroxylysine and its glycosylated forms show marked differences in fibrillogenesis in vitro. J. Biol. Chem. 274 8988–8993.

Odani H, Shinzato T, Usami J, Matsumoto Y, Brinkmann Frye E, Baynes JW and Maeda K (1998) Imidazolium crosslinks derived from reaction of lysine with glyoxal and methylglyoxal are increased in serum proteins of uremic patients; evidence of increased oxidative stress in uremia. FEBS Lett. 427 (3) 381–385

Ogawa T, Ono T, Tsuda M, Kawanishi Y (1982) A novel fluorphore in insoluble collagen: a cross-linking moiety in collagen molecule. Biochem. Biophys. Res. Commun. 107 1252–1257.

Paul R G and Bailey A J (2003) Chemical stabilization of collagen as a biomimetric. Sci. World J. 3 138–155.

Paschalis E P, Verdelis K, Doty S B, Boskey A L, Mendelsohn R and Yamauchi M (2001) Spectoscopic characterisation of collagen cross-links in bone. J. Bone Mineral. Res. 16 1821–1828.

Passoja K, Rautavuoma K, Ala-Kokko L, Kosonen T and Kivirikko K I (1998) Cloning and characterization of a third human lysyl hydroxylase isoform. Proc. Natl. Acad. Sci. USA 95 10482–10486.

Petite H, Duval L L, Frei V, Abdulmalak N, Sigotluizard M F and Herbage D (1995) Cytocompatability of calf pericardium treated by glutaraldehyde and by acyl azide methods in an organotypic culture model. Biomaterials 16 1003–1008.

Price DL. Rhett P M, Thorpe S R and Baynes J W (2001) Chelating activity of advanced Glycation End-product inhibitors J. Biol. Chem. 276 48967–48972.

Ramachandran G N, Bansal M, and Bhatnaga R S (1973) A hypothesis on the role of hydroxyproline in stabilizing collagen structure. Biochem. Biophys. Acta 322 166–171.

Roberts H C, Knott L, Avery N C, Cox T M, Evans M J and Hayman A R (2007) Altered collagen in tartrate-resistant acid phosphatase deficient mice: a role for TRACP in bone collagen metabolism. Calcif. Tissue Intl. 80 400–410.

Robins S P and Bailey A J (1972) Age-related changes in collagen: the identification of reducible lysine-carbohydrate condensation products. Biochem. Biophys. Res. Communs. 48 76 84.

Robins S P and Bailey A J (1973) The chemistry of the collagen cross-links. The characterization of Fraction C, a possible artifact produced during the reduction of collagen fibres with borohydride. Biochem J. 135 657–665.

Robins S P and Bailey A J (1975) The chemistry of collagen cross-links. The mechanism of stabilization of the reducible intermediate cross-links. Biochem J. 149 381–385.

Robins S P and Duncan A (1983) Location of pyridinoline in bovine articular cartilage at two sites of the molecules Biochem. J. 215 175–182.

Sell D R and Monnier V M (1989) Structure elucidation of a senescent cross-link from human extracellular matrix. Implications of pentoses in the aging process. J. Biol. Chem. 264 21597–21602.

Sell D R, Biemel K M, Reihl O, Lederer M O, Strauch C M and Monnier V M (2005) Glucosepane is a major protein cross-link of the senescent human extracellular matrix. J. Biol. Chem. 280 12310–12315.

Sims T J, Avery N C and Bailey A J (2000) Quantitative determination of collagen cross-links. In Methods in Molecular Biology (Eds. C. Strueli and M E Grant) Vol 139 Extracellular Matrix Protocols. Humana Press, Totowa, NJ.

Sionkowska A and Kaminska A (1999) Thermal helix-coil transition in UV irradiated collagen from rat tail tendon. Int. J. Biol. Macromol. 24 337–340.

Slatter D A, Paul R G, Murray M and Bailey A J (1999) Reaction of lipid-derived malondialdehyde with collagen. J. Biol. Chem. 274 19661–19669.

Slatter D A, Bolton C H and Bailey A J (2000) The role of lipid-derived malondialdehyde in diabetes mellitus. Diabetologia 43 550–557.

Slatter D A, Avery N C and Bailey A J (2008) Collagen in the fibrillar state is protected from glycation. Int. J. Biochem. Cell Biol. In press.

Spoerl E. Wollensak G and Seiler T (2004) Increased resistance of cross-linked cornea against enzymatic digestion. Current Eye Res. 29 35–40.

Takaluoma K, Lantto J, and Myllyharju J (2007) Lysyl hydroxylase-2 is a specific telopeptide hydroxylase, while all three isoenzymes hydroxylate collagenous sequences. Matrix Biol. 26 396–403.

Thornalley P J (1996) Advance glycation and development of diabetic complications: unifying the involvement of glucose, methyl glyoxal and oxidative stress. Endocrin. Metabol. 3 149–166.

Torre-Blanco A, Adachi E, Hojima Y, Wotton J A M, Minor R R and Prockop D J (1992) Temperature induced post-translational over-modification of type I collagen. Effects of over modification of the protein on the rate of cleavage by procollagen N-proteinase and on self-assembly of collagen into fibrils. J. Biol. Chem. 267 2650–2655.

Vanacore R M, Friedman D B, Ham A J L, Sundaramoorthy M and Hudson B G (2005) Identification of S-hydroxylysyl-methionine as the covalent crosslink of the non-collagenous (NC1) hexamer of the alpha 1 alpha 1alpha 2 collagen IV network – A role for the post-translational modification of lysine 211 to hydroxylysine 211 in hexamer assembly. J.Biol.Chem. 280 (32) 29300–29310.

Vasan S, Foiles P and Founds H (2003) Therapeutic potential of breakers of advanced glycation end-products – protein cross-links. Arch. Biochem. Biophys. 419 89–96.

Viidik A (1986) Adaptability of connective tissue. in Biochemistry of Exercise: Human Kinetics (Ed. A Salton) Champaign Ill. USA.

Vlassara H, Brownlee M and Cerami A (1985) High-affinity-receptor-mediated uptake and degradation of glucose modified proteins: a potential mechanism for the removal of senescent macromolecules. Proc. Natl. Acad. Sci. USA 82 5588–5592.

Vlassara H, Striker L J, Teichberg S, Fuh H, Li Y M and Steffes M (1994) Advanced glycation end-products induce glomerular sclerosis and albuminurea in normal rats. Proc. Natl. Acad. Sci. USA. 91 11704–11708.

Voziyan P A and Hudson B G (2005) Pyridoxamine. The many virtues of a Maillard Reaction inhibitor. Ann. NY Acad. Sci. USA 1043 807–816.

Waite J H, Housley T J and Tanzer M L (1985) Peptide repeats in mussel glue protein. Theme and Variations Biochem 24 5010–5014.

Waite J H, Qin X X and Coyne K J (1998) The peculiar collagens of mussel byssus. Matrix Biol 17 93–106.

Waite J H, Vaccaro E , Sun C and Lucas J (2003) Collagens with elastin and silk-like domains. In Elastomeric Proteins (Eds. P R Shewry, A S Tatham and A J Bailey) Cambridge University Press, Cambridge, UK, pp 189–212.

Wolfenbuttle B H R, Boulanger C M, Crjins F R L, Huijberts M S P, Poitevin P, Swennen G N M, Vasan S, Egan J J, Ulrich P, Cerami A and Levy B I (1998) Breakers of advanced glycation end-products restore large artery properties in experimental diabetes. Proc. Natl. Acad. Sci. USA 95 4630–4640.

Woo S L Y, Gomez M A, Woo Y K and Akeson W K (1982) Mechanical properties of tendons and ligaments II. The relationships of immobilization and exercise on tissue remodelling. Biorheology 19 397–408.

Yamauchi M. London R E, Guenat C, Hashimoto F and Mechanic G L (1987) Structure and formation of a stable histidine-based trifunctional cross-link in skin. J Biol. Chem. 262 11428–11434.

Yamauchi M, Prisayanh P Haque Z and Woodley D T (1991) Collagen cross-linking in sun exposed and unexposed sites of aged human skin. J. Invest. Dermatol. 97 938–941.

Chapter 5
Damage and Fatigue

R.F. Ker

Abstract Collagenous tissues can, inevitably, be damaged and broken by over-load, but they are also susceptible to the prolonged application of a lesser load. The majority of studies are for bone with tendon in the second place. This chapter therefore concentrates on these tissues. Test variables which can effect the time-to-rupture include the applied stress, the frequency and the temperature. A knowl-edge of the stresses which arise in life helps to put the results of in vitro tests into perspective.

The concepts and quantities used to characterize the fracture and fatigue behavior of materials are based on an understanding of the initiation and propagation of cracks. These concepts and quantities were first applied to metals, but their use has been extended to other materials, including, among collagenous tissues, bone and dentin. However, the anisotropy and inhomogeneity of these tissue mean that caution must be used in assessing the results. Tendons are far more strikingly anisotropic and the standard techniques of Fracture Mechanics for studying crack propagation are not appropriate.

The fatigue behavior of tendon and bone illustrates the idea that load-bearing biological structures are built to be only just adequate for their function. "Just adequate" includes allowance for routine repair of non-symptomatic damage. This balance of damage and repair seems to be part of the control mechanism by which biological tissues are maintained by their cells in a viable state throughout life.

5.1 Introduction

Dissect a tendon from a vertebrate. Hang a weight from it, while keeping it moist, and wait. In due time, the tendon will rupture. That is more-or-less what Wang Xiao Tong said to me and Neill Alexander some 20 years ago. I was skeptical. What had he done wrong? It just did not seem a "sensible" way of making a structure whose purpose in life is to transmit tensile forces. But he was right (Wang and Ker 1995). Bone behaves in a rather similar way (Carter and Caler 1983, Caler and Carter 1989). Given sufficient time, the specimen ruptures, even when the stress is

P. Fratzl (ed.), *Collagen: Structure and Mechanics*,
© Springer Science+Business Media, LLC 2008

well below the ultimate tensile stress (UTS). Details of times and stresses will be given in Section 5.4.2. Since time-dependent damage and, ultimately, rupture occurs with a constant load, it seems inevitable it will also occur with a cyclic load. This has been demonstrated for tendon (Wang et al. 1995; Schechtman and Bader 1997; Pike et al. 2000; Wren et al. 2003) and bone (Evans and Lebow 1957, which I think is the earliest reference; Carter and Caler 1983; Caler and Carter 1989; Zioupos et al. 1996 and many others). Dentin behaves much like bone (Nalla et al. 2003). Among the other collagen-containing tissues, rather few papers explicitly demonstrate susceptibility to fatigue rupture. Exceptions are rabbit ligaments in tension (Thorton et al. 2007; Schwab et al. 2007) and human articular cartilage in tension (Bellucci and Seedhom 2001). However, it is widely assumed that collagenous tissues, in general, are susceptible to fatigue. Indeed, this seems to be a property of the collagen fibril itself (Eppell et al. 2006).

Bone and tendon are not unusual. Susceptibility to fatigue rupture is widespread among structural materials. Metal fatigue has been of particular concern since the nineteenth century. With structural metals, at everyday temperatures, fatigue only occurs when the loading is oscillatory. Therefore, the science of fatigue and its nomenclature were developed in relation to cyclic testing. The word "fatigue" conventionally implies time-dependent damage through cyclic loading. Similarly, the word "creep", which is used to describe time-dependent change in dimensions of a specimen under load, conventionally implies a constant load. I prefer to use the words more generally, regardless of the pattern of loading. There should be no confusion, for I aim to specify in each case, where it is relevant, whether the loads are "constant" or "cyclic" or whatever. Metals and ceramics only creep at high temperatures. Polymers creep and ultimately rupture under constant load at normal temperatures. Research on fatigue in metals has, of course, used all possible modes of loading, but tension predominates. Metals are more susceptible to fatigue in tension and failure often occurs through tensile loading. Note that failures in bending are usually failures in tension and initiate where the tensile stress is greatest. Bone is stronger in compression than in tension. On the other hand, some bones are subject to substantial compressive stresses. When a bone fails under compressive loading, shear is likely to be involved (Taylor et al. 2007). In an isotropic and homogeneous material, with a linear relation between stress and strain, the maximum shear stress is at $45°$ to the principal stresses. Though bone is none of these things, shear stresses are developed within the material. If a line of weakness in shear is found, sliding will occur within the material. Taylor et al. (2003b) and Winwood et al. (2006) have investigated fatigue of cortical bone in torsion. Tendons are subject exclusively to longitudinal tension (except in the fibrocartilagenous regions found where a tendon passes round a bony prominence). All fatigue tests on tendon have been in longitudinal tension. A longitudinal compressive load cannot be applied to a tendon: it just buckles. Cartilage is loaded in compression, but tensile stresses arise within the tissue, especially near the free surface. Failure, if it comes, is in tension, which is why Bellucci and Seedhom (2001) studied the fatigue of cartilage in tension. For these reasons, this chapter has a disproportionate emphasis on loading in tension.

In general, when a structure breaks in two, it is because a crack has spread across it. Fracture Mechanics is therefore concerned with the mechanics of cracks. The starting point came in 1921 when Griffith propounded his theory of crack growth. I will not deal with the full mathematics of the Griffith crack, but I aim to explain the concepts involved fairly fully. The key idea of balancing the energy available for crack growth against the energy required remains central to Fracture Mechanics. I will introduce some of the material properties, such as critical crack length, critical stress intensity and the R-curve, used in Fracture Mechanics. However, I will not attempt to fully explain the relevance, implications and inter-relationships of these quantities. To fill in the details see one of the numerous books on Fracture Mechanics and Fatigue. Notable are Anderson (2005) and Suresh (1998).

In the last 40 years, Fracture Mechanics has grown into a major scientific discipline, which can deal successfully with real structural materials of use in both the man-made and natural worlds. The next section is an outline of Fracture Mechanics. Subsequent sections deal with experimental observations on the creep and fatigue of bone and tendon and with some of the biological implications.

5.2 Cracks

The starting point assumes an isotropic, homogeneous and elastic material with a linear stress–strain plot (Hooke's law). The resulting body of theory is called linear elastic fracture mechanics (LEFM). LEFM applies directly only to brittle materials, but the concepts introduced under LEFM find much wider application.

5.2.1 The Griffith Crack: Material Resistance, Energy Release Rate and Stress Intensity Factor

A crack causes a stress concentration. Inglis, in 1913, considered an elliptical hole in an infinite thin sheet. An external load is applied to the sheet, so that, far from the hole, the tensile stress is uniform in a direction parallel to the shorter axis of the ellipse (Fig. 5.1). The stress concentration can be illustrated by drawing stress trajectories, which go round the ellipse. Inglis obtained expressions for the stresses throughout the sheet. The maximal stress is at the edge of the hole. If the ellipse is made very narrow it becomes a crack and the stress concentration becomes large. Note that, because of the symmetry about the center line in Fig. 5.1, Inglis' expressions also apply to a semi-infinite sheet with a half-ellipse cut into its edge. Inglis' result carries the important message that sharp corners are to be avoided, both in engineering design and in living organisms.

The stress concentration depends only on the ratio of the length of the ellipse to the minimum radius of curvature. Thus geometrically similar ellipses all cause

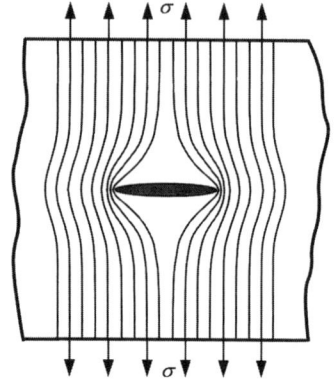

the same stress concentration, however small the ellipse. Does this mean that tiny
scratches are as serious as bigger cracks? Everyday experience shows that the
answer is "No". The key to understanding why small cracks do not propagate was
provided by Griffith (1921).[1] Griffith pointed out that rupture requires energy. At the
ultimate level, molecules (and, indeed, atoms) have to be moved apart: displacement
is involved as well as force. The new surfaces have surface energy. This is a famil-
iar concept with liquids, which applies equally to solids, though it is less obvious
because the solid cannot deform to minimize surface area. For a crack of length $2a$
in a plate of thickness B (Fig. 5.2), the surface energy, W_S, is

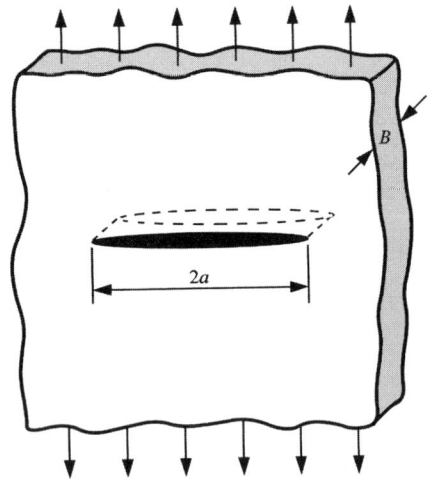

[1] The date is often stated as 1920. This is because Griffith read his paper to The Royal Society of
London in 1920. The printed version appeared in 1921.

$$W_S = 4aB\gamma_s \tag{5.1}$$

where γ_s is the surface energy per unit area (equivalent to surface tension).

Griffith calculated the energy available for the relatively simple case of an elliptical crack in a very large thin sheet. By making the plate "thin", a two-dimensional calculation is sufficient. Griffith imagined the hole appearing, while the remote stress is held at σ. The plate with the hole stretches more easily than the intact plate, so when the hole arrives the external clamps, which apply forces to the plate, move apart. Using Inglis' stress distribution, Griffith showed the energy available for crack building, W_P, to be

$$W_P = -\frac{\pi a^2 B \sigma^2}{E} \tag{5.2}$$

where E is the elastic modulus.[2]

The question of interest is whether an existing crack will grow, which depends on the changes in energy accompanying changes in area. Differentiating equations (5.1) and (5.2) with respect to the area $A\ (=2aB)$ gives:

$$\mathcal{G} = \frac{dW_P}{dA} = -\frac{\pi a \sigma^2}{E} \tag{5.3}$$

and

$$R = \frac{dW_S}{dA} = 2\gamma_s \tag{5.4}$$

$\mathcal{G}$, introduced by Irwin (1956), is the energy release rate. (Note that "rate" here refers to changes with area, not time.) R, the energy absorption rate, describes the resistance of the material to crack extension. $\mathcal{G}$ is referred to as the "driving force", though it is not strictly a force.

At the critical crack length, a_c, $\mathcal{G} + R = 0$ and:

$$a_c = \frac{2E\gamma_s}{\pi\sigma^2} \tag{5.5}$$

A crack longer than a_c can grow because by doing so the total energy is reduced. A shorter crack cannot grow by Griffith's mechanism alone. The critical stress, σ_c, for a crack of length a is given by:

---

[2] Because the clamps move apart, while the load is constant, external energy is fed into the plate. Half this energy goes into increasing the elastic energy of the plate. The other half is W_P. The division into two halves arises because, with linear elasticity, the elastic strain energy stored in a structure depends on (final force)/2.

$$\sigma_c = \sqrt{\frac{2E\gamma_S}{\pi a}} \qquad (5.6)$$

If an attempt is made to increase the stress above σ_c, the crack will grow and the specimen breaks.

The geometry of the Griffith/Inglis crack, an ellipse in an infinite thin sheet, does not lend itself to experimental study of the behavior of cracks in a material. Fracture Mechanics took a step forward with the derivation of expressions for the stress field in the neighborhood of a sharp crack tip in a specimen of a shape, which can be realized in practice (Irwin 1957). The expressions are written in polar coordinates (r, θ), with the origin at the tip of the crack. A point is specified by its distance r from the origin and the angle θ between the line joining the origin to the point and the direction of the length of the crack. Any component of stress at the point (r, θ) can be written in the form:

$$\sigma(r, \theta) = \frac{K}{\sqrt{2\pi r}} f(\theta) \qquad (5.7)$$

The form of the function $f(\theta)$ depends on which component of stress is being considered. The parameter K is called the stress intensity factor. It has units of stress multiplied by $\sqrt{\text{length}}$. The stress at any selected point, and therefore what might happen there, is determined by K. In particular, the crack will grow when K reaches a critical value, K_c.

K is proportional to the force applied to the specimen and otherwise depends solely on the geometry of the specimen, including the length of the crack. For a Griffith crack, which, in this case, is a super-thin ellipse,

$$K = \sigma\sqrt{\pi a} \qquad (5.8)$$

Combining equations (5.3) and (5.8) gives $K = \sqrt{\mathcal{G}E}$. This equation applies not only for the Griffith crack, but under LEFM in general.

Expressions for K with selected geometries are listed in the Handbooks of Fracture Mechanics. Figure 5.3 shows a compact tension specimen, which is among the selected geometries.

In "classical" mechanics, rupture occurs when the stress reaches a critical value, which, in tension, is called the ultimate tensile stress (UTS). In fracture mechanics, with a crack deliberately made in the test specimen, crack propagation occurs when stress intensity reaches a critical value K_c. For the Griffith crack, $K_c = \sqrt{2E\gamma_S}$. K_c, like UTS, is a property of the material. K_c is measured by preparing a specimen to one of the standard geometries listed in the handbooks. The specimen includes a deliberately manufactured crack of known length. The load is increased until the crack grows. The critical value of K thus reached is calculated, in essence from

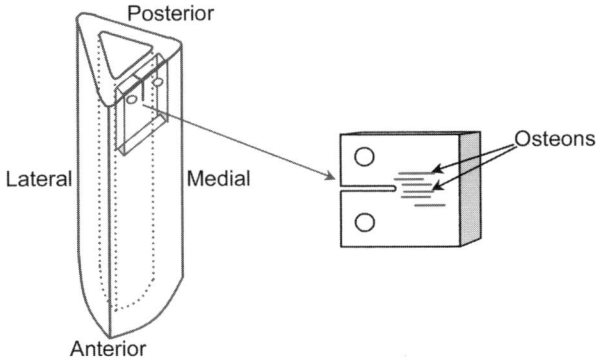

Fig. 5.3 A compact tension specimen machined from a human humerus. From Nalla et al. (2005), reprinted from Biomaterials with permission from Elsevier

equation (5.8), but with adjustment for the actual geometry of the test specimen using the appropriate expression.[3]

5.2.2 Tough Materials

Griffith confirmed his theory by experiments on glass, which is the archetypal brittle material. Tough materials have additional mechanisms to sink energy during crack growth. This can be allowed for by modifying the Griffith expression (equation (5.6)) to read:

$$\sigma_c - \sqrt{\frac{2E(\gamma_s + \gamma_p)}{\pi a}} \tag{5.9}$$

where γ_p covers the energy per unit surface area for all other energy sinks. The subscript "p" is used because, for some materials, notably metals, the most important energy sink is plastic yielding. A stress–strain curve for a bone is shown in Fig. 5.4. Energy per unit volume is represented by the area under the curve, which increases disproportionately beyond the yield region. Thus the energy balance is shifted to discourage crack growth. Tough materials resist impact fracture, in which the energy supplied by the external source is limited.

I chose a bone plot for Fig. 5.4, rather than metal, to fit in with the present context of collagenous materials. The analogy between bone and metal must not be pushed too far. Although energy absorption applies to both, the mechanisms are utterly different. For a metal, the region of the stress–strain plot after yielding is

---

[3] Stress intensity is often denoted by K_I. The "I" refers to Mode I loading in which the load is applied normal to the crack plane. Modes II and III are orthogonal to Mode I and to each other. By combining all three, any mode of loading can be covered. However, Mode I is usually the most relevant in fracture, because it tends to open the crack.

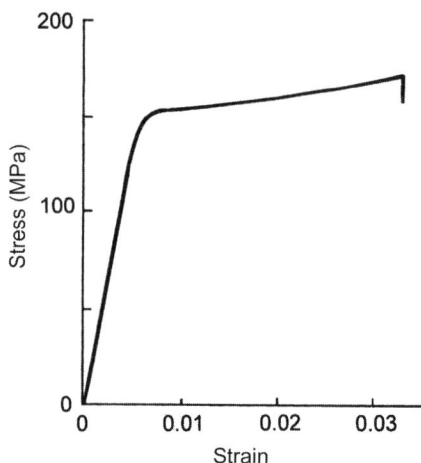

Fig. 5.4 A stress–strain plot for longitudinal tension for bovine cortical bone. From Currey (2002), who derived the figure from data in Reilly and Burstein (1975). With permission from Princeton University Press (Currey 2002) and from Elsevier (Reilly and Burnstein 1975, J Biomech)

described as plastic. When a metal is stretched into the plastic region, dislocations move and distortion is permanent, but the material is not thereby damaged. With bone, the permanent distortion is less, but the material is damaged (Currey 2002). This damage can only be repaired by cellular action. The extent of the post-yield region in bone varies greatly and correlates (negatively) with the calcium content (Currey 2004). Tendon shows no comparable yield. In this respect, and this respect only, tendon behaves as a brittle material.

K_c, obtained as described in Section 4.2.2, is widely used as a measure of toughness even though tough materials are not linearly elastic. Irwin (1956) proved that LEFM expressions listed in handbooks adequately allow for the shape of the specimen, so long as the region of plastic deformation is small, with the vast majority of the material still linearly elastic. A bigger snag with bone is that it is not isotropic or homogeneous. Nonetheless, critical stress intensities for bone have often been reported: the earliest reference is Bonfield and Datta (1976) and others include Norman et al. (1995) and Taylor et al. (2003a), who use the word "threshold" instead of "critical". Obviously, with a complex material, the meaning of critical (or threshold) stress intensity may not be fully defined. Nonetheless, for comparisons between fairly similar systems, it is hopefully meaningful. For example, Zioupos and Currey (1998) noted a (small) reduction in K_c with age in human bones.

5.2.3 The R-Curve

With a brittle material, crack growth, once started, is unstable, because the driving force, G, increases with crack length but the resistance, R, does not. With a tough material, other mechanisms for absorbing energy may cause the resistance to increase with crack length. If the driving force is increased by increasing the applied stress, the crack will grow until again $G = R$. Thus a range of crack lengths are stable and the crack grows with increasing stress. The process continues until the rate of increase of R with a becomes too small to match the increase in $\mathcal{G}$. The requirement for stability is:

$$\frac{d\mathcal{G}}{da} \leq \frac{dR}{da} \qquad (5.10)$$

These are second derivatives of energy, as is usual in questions of stability.

Since $\mathcal{G} = R$ whenever the crack is stable, the R-curve can be plotted as G (for a stable crack) against a. Often stress intensity (for a stable crack) is plotted against Δa (or $\sqrt{\Delta a}$). The slope of the line is a measure of toughness. R-curves with bone are, for example, given by Nalla et al. (2006) and Vashishth (2004).

5.2.4 The J-Integral

The J-integral was introduced by Rice (1968). Like $\mathcal{G}$, J is an energy release rate but applicable more broadly to include non-linear materials and large strains. For more details, see, for example, Anderson (2005). I know of only two papers which use the J-integral with a collagenous tissue: Zioupos and Currey (1998) and Yan et al. (2007), both of which are concerned with the toughness of cortical bone. In the wider biological world, the J-integral has been used by Bertram and Gosline (1986) to assess the toughness of horse hoof keratin and by Mach et al. (2007a, 2007b: Section 5.3 below) in relation to the propagation of fatigue cracks, with special reference to seaweeds.

5.3 Fatigue Cracks

Time-dependent damage does not arise within Fracture Mechanics as I have described it so far. The progress of a crack involves adjustments to stresses and strains throughout a structure. This happens at about the speed of sound and, for anything but the vastest of structures, is effectively instantaneous. This explains the very rapid spread of the final fatal crack. But fatigue rupture takes a long time. It may be minutes or hours, even years. What is happening during all this time? Why does it take so long? In some sense, viscosity may be involved. Solid ice flows in a glacier – very slowly. However, I am cautious about involving viscosity directly. The material of a liquid is not damaged when it flows, nor is ice. However, damage is central to fatigue. Ultimately, the randomness of molecular motions must be involved. Very occasionally a molecule will receive far, far more than the average thermal energy and something will give. This implies a strong temperature dependence for fatigue. Carter and Hayes (1976) and Rimnac et al. (1993) for cortical bone and Wang and Ker (1995) for tendon found that time-to-rupture is indeed strongly influenced by temperature. Rimnac et al. (1993) showed that the strain rates under constant load for cortical bone fits with the Arrhenius equation for thermally activated phenomena, such as chemical reactions, which is of the general form:

$$\text{Rate} = Ae^{-Q/RT} \tag{5.11}$$

where T is the absolute temperature, A is a constant, R is the gas constant and Q is a constant, termed the activation energy. Analysis of data in Wang and Ker (1995) shows that, for tendon, the fit to this equation is good for rates such as the reciprocal of time-to-rupture, initial strain rate and initial damage rate. Davison (1989) links tendon rupture and a chemical reaction by suggesting strain hydrolysis as a damage mechanism. Within collagenous structures and, indeed within collagen itself, there are many sites of chemical equilibria between bound states and dissociated (hydrolyzed) states. Under load, the equilibrium may be shifted toward dissociation, because the two sides, when dissociated, tend to spring apart and are therefore less likely to bind together again.

During fatigue, the time-to-rupture is the sum of the time for crack initiation and for crack growth. Crack growth is accommodated by the empirical power law due to Paris et al. (1961). In the secondary creep region (or a tertiary region of very low acceleration, as above), the rate of crack growth is given by

$$\mathrm{d}a\,/\mathrm{d}t = AK^n \tag{5.12}$$

or, for oscillatory loading,

$$\mathrm{d}a\,/\mathrm{d}t = C(\Delta K)^m \tag{5.13}$$

where K is the stress intensity and ΔK is the stress intensity range and A, n and C, m are constants. The use of ΔK, with no mention of the absolute value of stress intensity is appropriate for metals, but less so for collagenous materials. Wright and Hayes (1976) were the first to use the Paris law with bone. Numerous reports have appeared since about 1998: for example, Nalla et al. (2005) and Shelton et al. (2003); reviewed by Kruzic and Ritchie (2008) and Ritchie et al. (2005). Below a threshold of ΔK no crack growth occurs. This threshold is between about $0.5\,\text{MPa mm}^{1/2}$ to about $5\,\text{MPa mm}^{1/2}$. Just above the threshold growth rates as low as $10^{-9}\,\text{m s}^{-1}$ have been measured. Growth rate increases very sharply with ΔK, with about $10^{-4}\,\text{m s}^{-1}$ being reached at, perhaps, 10 times the threshold value of ΔK. Wright and Hayes (1976) and Nalla et al. (2005) are among those who prepared their specimen from long bones with the crack propagation direction parallel to the length of the original bone. The anisotropy of bone makes this the easiest way to propagate a crack. Shelton et al. (2003), working with horse metacarpal bones, oriented their compact tension specimen so that the initial crack was transverse to the bone. However, they were unable to persuade the specimens taken from the lateral side of the tibia to propagate in the direction of the initial crack. Instead the crack turned through $90°$ and ran in the longitudinal direction of the original bone. Given the long initial crack (about 10 mm), this is a powerful testimony to the ability of bone to deflect a transverse crack. This is a major factor in bones' toughness and fatigue resistance.

Another example, from biology, of slow macrocrack growth is provided by sea-weed and is the subject of a pair of papers by Mach et al. (2007a, 2007b). The first is a primer of fatigue and fracture mechanics, leading up to a description of the way in which the J-integral can be used when investigating slow crack growth in materials, such as elastomers, which are non-linear and undergo large strains. These are not the primary ways in which collagenous materials depart from the ideals of LEFM, which instead have the problems of being anisotropic and inhomogeneous. The second paper gives experimental data for the growth of macrocracks, analyzes this by the methods of the primer and establishes the consequences of the forces which arise in life. We will return to macroalgae in Section 5.6.3.2, because, although the materials are utterly different, the ultimate biology has similarities.

5.4 Creep and Fatigue in Tendon and Bone

In this section, we temporarily leave cracks and consider some observations which have been made regarding creep and fatigue in tendon and bone.

5.4.1 Creep

Figure 5.5 shows three creep curves – plots of strain against time for longitudinal tension. In Fig. 5.5a (tendon), the line is spread into a band because the load was oscillating. For Fig. 5.5b (tendon) and c (bone) the load was constant. The strain (or mean strain, for Fig. 5.5a) shows a similar pattern of progressive increase in all three plots. Immediately after the initial elastic extension, the creep is rapid but decreases to a point of inflexion (minimum velocity of creep). Thereafter, the creep

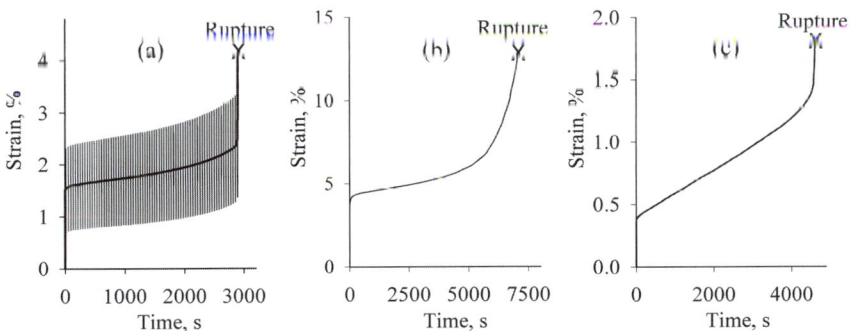

Fig. 5.5 Creep curves (strain v. time). (**a**) Wallaby tail tendon: cyclic loading, stress range 2–30 MPa; frequency 2 Hz; temperature 37°C. One cycle out of every 40 was recorded (more near the beginning and end) and each of these contributes to the mean strain, thick line. Only one in every 80 cycles is individually plotted to avoid the lines merging. New data. (**b**) Wallaby tail tendon: constant load, stress 30 MPa; temperature 37°C. From Wang and Ker (1995). (**c**) Bovine femoral bone: constant load, stress 100 MPa. From Sedman (1993) with permission

Fig. 5.6 Slope of the mean strain line from Fig. 5.5a. The *line* indicates the trend of the points. It is the best-fit line (*F*–ranked) according to Tablecurve2D from Systat Software Inc, San Jose, California, USA: $r^2 = 0.993$

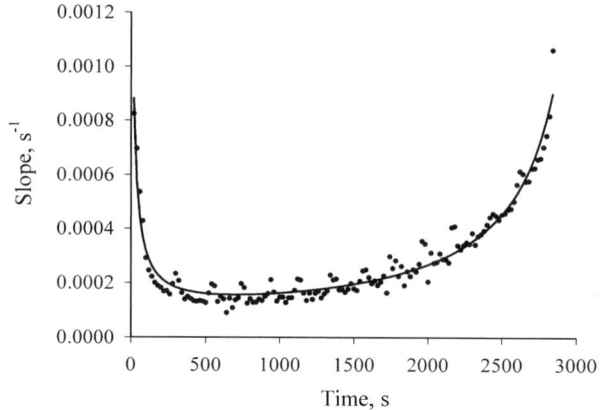

accelerates, slowly to start with and then more rapidly. This is illustrated by Fig. 5.6, in which slope of the mean line in Fig. 5.5a (i.e., the velocity of creep, expressed as change in % strain per unit time) is plotted against time. It is usual, in Materials Science, to recognize three stages of creep: "primary" with decelerating strain, "secondary" with a constant strain rate for the majority of the time and "tertiary" as the strain accelerates to rupture (Teoh and Cherry 1984; Klompen et al. 2005). However, my description of the plots of Fig. 5.5 has primary and secondary creeps only. The distinction between a broad "point" of inflexion and a region of constant strain rate is obviously somewhat arbitrary, but Fig. 5.6 shows change occurring throughout and no sustained region of constant velocity. Figure 5.5c indicates that this viewpoint is also tenable for bone, though the "point" of inflexion is even shallower.

For bone and tendon, I consider primary and tertiary creeps to be superimposed, rather than successive. Both continue throughout, with primary creep tending to an asymptote. Tertiary creep is associated with a reduction in stiffness due to damage by cracking. This reduction starts at the beginning of the test (Wang et al. 1995, for tendon: Zioupos et al. 1996, for bone). With damage occurring throughout, tertiary creep is to be expected at all times because damage is progressive: each addition to damage makes the material less able to withstand the applied load (Ker and Zioupos 1997).

It is reasonable to consider primary creep as viscoelastic and occurring without accompanying damage. Wang and Ker (1995; Fig. 12) plot stiffness ratio against creep strain. This plot has a positive intercept on the strain axis, showing that some strain (primary strain) occurs without reduction in stiffness.

5.4.2 Time-to-Rupture as a Function of Stress

Materials normally have a stress threshold below which fatigue does not occur, which is sometimes called the "endurance limit". The position with bone and tendon is unclear. Zioupos et al. (1996) consider that a threshold is more likely with laminar bone than with osteonal bone. Whether or not a threshold strictly exists, times-to-rupture are very long at stresses below about 50 MPa (Carter and Caler 1983: human) or 65 MPa (Zioupos et al. 1996: bovine femur; antler), which is to be compared with a strength achieved in a single pull of about 160 MPa (Currey 2002). The corresponding figures for tendon are $\leq$10 MPa and about 140 MPa (Wang and Ker 1995: wallaby tail tendons). (Again, it is not known whether 10 MPa is really a threshold. In a single test, Wang and Ker (1995) found no sign of rupture after 15 days at 10 MPa, but this has not been confirmed.) Beyond these stresses, the time-to-rupture decreases with stress extremely rapidly: with bone, to the power of -15 or even more dramatically (Carter and Caler 1983, Caler and Carter 1989, Zioupos et al. 1996). With wallaby tail tendons at constant load, the power is about -3.2 (Wang and Ker 1995). With cyclic loading at 5.3 Hz this becomes -5.6 (data of Wang et al. 1995, re-analyzed in terms of a power law). Schechtman and Bader (1997) obtained similar results with the human extensor digitorum longus. However, Wren et al. (2003), working with the human Achilles tendon, observed no clear trend of time-to-rupture with stress. I find this puzzling. Even more puzzling is Deakin (2006). She endeavored to investigate the fatigue rupture of human hand tendons, using stresses, which should, according to Ker et al. (2000), have been sufficient to cause rupture in a fairly short time. But rupture did not occur.

5.4.3 Cyclic Loads Compared to a Constant Load: Times-to-Rupture

The mechanisms of fatigue damage in metals, which are fairly well understood (see, for example, Anderson 2005), depend on a changing load. A constant load causes no damage. Collagenous materials are different. Constant loads cause damage. But, the question still arises as to whether cycling the load increases the damage rate. This question has been tackled, for bone, by Carter and Caler (1983, 1985), Caler and Carter (1989), Zioupos et al. (2001) and Nalla et al. (2005) and, for tendon, by Wang et al. (1995).

Two related methods are used. (i) Compare times-to-failure in constant load tests with the time-to-failure in cyclic tests. (ii) Compare times-to-failure at very different frequencies. The rationale behind (ii) is that if, at one extreme, cycles were all that mattered, the cycles-to-rupture would be independent of frequency, whereas, at the other extreme, if cycling per se was irrelevant, the time-to-rupture would be independent of frequency. With very different test frequencies, these are very different outcomes. In general, something in between is to be expected.

Method (i) is not entirely straightforward. Behavior under constant stress sets a maximum value for the time-to-rupture under changing stresses. This maximum can be assessed by integration, using data from constant stress tests and taking account of the times for which each stress is applied. Carter and Caler describe this maximum time (i.e., minimum damage rate) as being due to "time-dependent damage". Any extra damage rate, which may be observed in a cyclic test, they describe as being due to "cycle-dependent damage". This is a particular and technical use of the terms "time-dependent" and "cycle-dependent". In this section, I follow the Carter and Caler usage. Elsewhere, if the technical meaning is intended I state that explicitly.

The next two sub-sections will review the response of first tendon and then bone to these methods.

5.4.3.1 Tendon

Time-dependent damage is qualitatively similar for both modes of loading. Wang and Ker (1995) and Ker et al. (2000) used constant loads; Wang et al. (1995), Pike et al. (2000), Schechtman and Bader (1997) and Wren et al. (2003) used cyclic loading. Quantitatively, cycling reduces the time-to-rupture (Fig. 5.7).

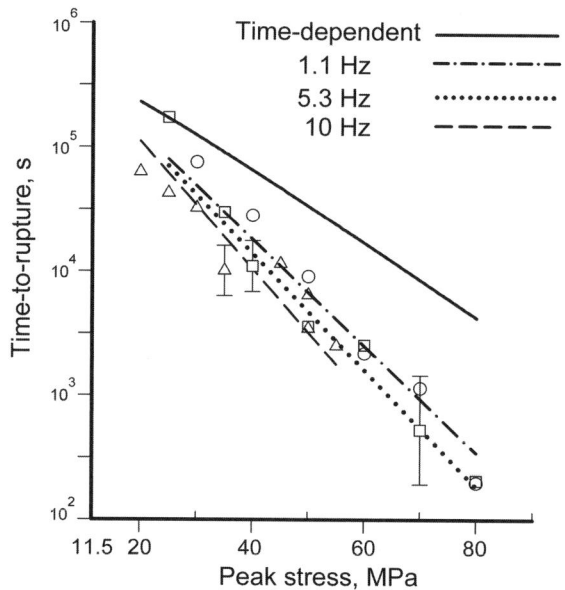

Fig. 5.7 The effect of frequency on time-to-rupture for wallaby tail tendons. The time-dependent line was obtained by integration using the results of constant load tests. From Wang et al. (1995)

This is an awkward section for me to write, because of the change in my preferred nomenclature. The line labeled "Time-dependent" in Fig. 5.7, is labeled "Creep" in the same figure in Wang et al. (1995).

5.4.3.2 Bone

Results, obtained with human bone, by Caler and Carter (1989) are shown in Fig. 5.8, which may be compared with Fig. 5.7 for tendon since both are in tension. Figure 5.8 is strikingly different. The time-dependent plot and both cyclic plots coincide. Cycling the load makes no difference. All that matters, for time-to-failure, are the magnitudes of the loads and the times of application. Tests involving compression, for which there is no tendon analog, give more complicated results. For details see Caler and Carter (1989).

Fig. 5.8 The effect of
frequency on time-to-rupture
for human cortical bone
under tension. The minimum
stress was 0. The abscissa is
stress range ($\Delta\sigma$), normalized
by the elastic modulus (E^*).
The asterisk indicates that the
modulus was measured at the
same stress rate (28 MPa s^{-1})
in each case. Reprinted, after
rotation of the figure through
90°, from Caler and
Carter (1989), J Biomech
with permission from
Elsevier

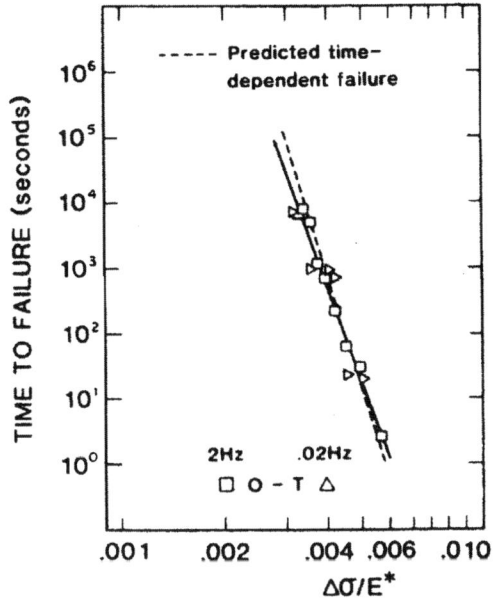

Zioupos et al. (2001), noting that the 1989 conclusion had not penetrated far into the literature, repeated part of Caler and Carter (1989) with human bone and also with bovine bone of somewhat different structure. Their result was the same. Cycles are irrelevant for the fatigue of bone in tension.

In 1983, Carter and Caler had used tests with a mean load of zero, on the assumption that no time-dependent damage then arises. By 1989, they had realized that this assumption is dubious. They therefore separated tension (0-T) and compression (0-C) tests, which enabled them to obtain reliable results. The 1989 results caused Caler and Carter to discard their earlier suggestion (Carter and Caler 1985) that there is a crossover, as to which category of damage is most important according to the number of cycles to failure. As the authors have dropped the idea, I will not explain further. The interested reader should compare the 1985 and 1989 papers.

5.5 Crack Stopping in Bone and Tendon

5.5.1 Bone

Cracks can be restricted either by a high resistance, R, achieved through energy sinks (Section 5.2.2) or by reducing the driving force, $\mathcal{G}$, acting at the crack-tip. Bone uses both approaches in a multitude of ways.

Vashishth (2007) reviews the formation of cracks at all scales throughout the hierarchical structure of bone. An important energy sink, with tensile loading, is the formation of diffuse cracking: regions with innumerable, very small cracks, which do not coalesce into more dangerous cracks. Vashishth considers diffuse damage to be time dependent in the technical sense of Carter and Caler.

One of the mechanisms for restricting the stress intensity at the crack tip is crack-bridging. This is particularly likely to occur when propagating a transverse crack in an anistropic material, which has strong and tough elements in the longitudinal direction. The occurrence of uncracked "ligaments" in bone in the wake of the crack tip is reviewed by Ritchie et al. (2005). These ligaments carry part of the stress and, thus, reduce the stress intensity at the crack tip. Crack-deflection is another mechanism used by bone. In an inhomogeneous material, when a growing crack encounters a particularly tough inclusion it may be deflected along the boundary. Once deflected, stress is carried less directly to the crack tip and growth may cease (Taylor et al. 2007).

5.5.2 Tendon

Tendon is a fiber-reinforced composite, in which more-or-less parallel fibers lie within a less stiff matrix. In such a material, longitudinal cracks can be formed easily and crack-deflection becomes the rule. Tendon is an extreme example of a fiber-reinforced composite, in that the ratio of moduli (matrix/fiber) is extreme. It is therefore exceptionally non-homogeneous and anisotropic. The theories of Fracture Mechanics, as outlined in Section 5.2 above, were intended for homogeneous materials (Anderson 2005, p. 271). Caution should therefore be exercised in applying them to bone, but with tendon, they are hardly relevant. However, crack formation and propagation remain crucial. For rupture to occur, a crack, however convoluted, must traverse the structure. Just as the theories are different, experiments on crack propagation also have very different outcomes.

Experiments in fracture mechanics usually involve pulling on a specimen containing an artificial crack, such as that illustrated in Fig. 5.3. Figure 5.9, from Ker (2007), shows what happens when a tendon, with an artificially inserted trans-

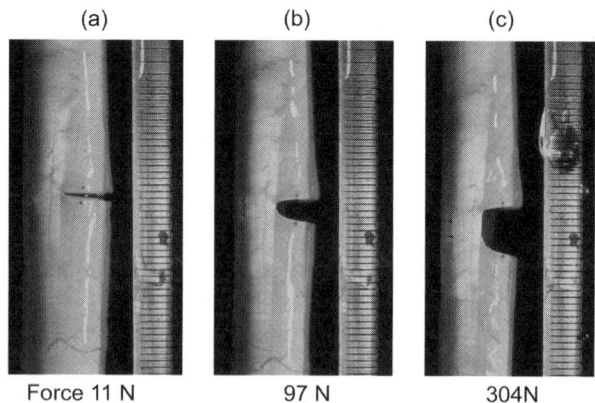

(a)   (b)   (c)

Force 11 N   97 N   304N

Fig. 5.9 The effect of tensile force on a slit in a sheep plantaris tendon. (**a**) Force 11 N. (**b**) 97 N. (**c**) 304 N. From Ker (2007)

verse crack, is pulled longitudinally. Figure 5.9a shows the starting position. Figure 5.9b shows the position with a modest load. The crack has opened and the tip is curved. The slope of the curve is equal to the longitudinal shear strain at that point. With a continuum, the shear strain would be infinite at the crack tip. With structural units of a finite size, infinity is avoided, even in the absence of any other stress-diffusing mechanism. However, it is clear from Fig. 5.5b that the tendon can maintain a very high longitudinal shear strain. But there is a limit and by Fig. 5.5c failure in shear has led to a longitudinal crack. Rupture starts at the crack tip and may run along the whole length of the specimen. With a 50% initial slot, half the tendon remains intact, while the other half is unloaded.

Although the shear strain at rupture is high, the ultimate shear stress may be quite low, because the longitudinal shear modulus is orders of magnitude less than the tensile modulus. For the tendon of Fig. 5.9, I estimate the ratio of moduli (longitudinal shear/tensile) to be about 10^{-3} (Ker 2007). Easy propagation of longitudinal cracks leads to a mode of rupture which might be called "interdigitation". On rupture, numerous fine fingers, of a wide range of lengths, extend from each end. They still interdigitate, so that the tendon appears to be intact. Further movement causes the fingers to slide out past each other, with negligible force. With a suitably long specimen, the overlap can be long, as is illustrated in Ker (2007). Even with the longest specimens and a break centered on the middle, it is hard to be certain that stress concentrations in the region of the clamps did not contribute to the rupture.

5.6 Biological Aspects: Evolution, Growth and Adaptation

5.6.1 Tendons

All biological tissues are, necessarily, adequate for their function. But, what is "adequate" depends on the circumstances. Those tendons which act as energy-saving springs in locomotion are subject to much higher stresses in life than the majority of tendons (Ker et al. 1988). A resistance to fatigue damage that was just adequate for one of the majority of tendons would be far from adequate for a spring-tendon. Wang et al. (1995) calculated that an Achilles tendon made of a material with the same properties as a toe flexor tendon would break after running for about 1 hour. Ker et al. (2000) noted a correlation between stress-in-life and resistance to fatigue damage, such that all tendons are about equally liable to damage in a living animal. This liability is high, so that damage seems likely to be part of daily life. It will be non-symptomatic, so long as it can be routinely repaired by the cells. Ker et al. (2000) hypothesized that damage acts as a trigger for repair. On this hypothesis, damage is part of the control mechanism of tendon maintenance. The mechanism of routine repair is unknown, in contrast to the situation with bone: Section 5.6.2.

Pike et al. (2000) studied limb tendons from lambs over a range of ages from birth to nearly full-grown. They compared a toe flexor tendon with a toe extensor tendon. At birth the two tendons were fairly similar in their resistance to fatigue damage and were subjected to similar stresses by their respective muscles. During growth, the stress on the flexor tendon increased more than the stress on the extensor tendon. Improvements in resistance to fatigue damage matched the increasing stresses, again with the flexor outpacing the extensor tendon. This observation is consistent with adaptation to different stresses. But, of course, it may be that the tendons are genetically programmed to change differently during growth. These "nurture v. nature" questions are never easy to resolve! The chemical and structural differences underlying differences in fatigue quality are not known.

5.6.2 Bones

Remodeling is a major part of bone biology. Bone is absorbed in localized regions and entirely new bone is laid down in concentric cylinders around central blood vessels. This is the Haversian system, which despite much research still generates many unanswered questions – see Currey (2002; Chapter 11).

Martin (1995) and others have proposed that remodeling serves to repair fatigue microcracks. This seems a drastic way of tackling small cracks, but, because of calcification, it may be impossible to carry out repair, except following a complete clearance. Prendergast and Taylor (1994) and Martin (1995) have suggested that damage is a trigger to the bone cells both for remodeling and for the deposition of additional bone, so that the strains and stresses experienced in the structure are maintained at a desirable level. The control of cell activity is a complex matter involving the interaction of chemical as well as mechanical influences, but it would certainly make sense if damage is part of the mix.

Taylor et al. (2007) take the view that bone always has a tendency toward fatigue damage in life. This damage needs to be routinely repaired to keep the right balance of damage and fatigue at the right level for tissue maintenance. The difference, from the corresponding situation with tendon, is that all bones seem to "target" about the same normal peak stresses. With tendons, as noted above, peak stresses vary widely from tendon to tendon. Fatigue quality appears to be adjustable and is adjusted in a way that does not seem to happen with bone. Instead, the stress level is adjusted by altering the thickness of the bone until the proper rate of fatigue damage is occurring. Tendons can also adjust their thickness, but have the additional option of adjusting fatigue quality.

5.6.3 Other Materials

5.6.3.1 Heart Valves (Collagen with Elastin)

Prosthetic heart valves are prone to serious fatigue problems. Wells et al. (2005) suggest that the reason for the better performance of the native heart valve is that it is subject to routine repair in response to fatigue damage. However, some doubt is introduced by two subsequent papers by the same group. Grashow et al. (2006) and

Stella et al. (2007) studied the creep of native human heart valves using, respectively, mitral and aortic valve leaflets. They conclude that in both cases creep after 3 h is negligible. In these papers, negligible means "at most 1%". The reason for interpreting "at most 1%" as negligible is that the comparison the authors are making is with the observed rate of stress relaxation. In the context of this chapter, the comparison required is between the time for significant tertiary creep and the lifetime of the heart valve, say 70 years. This is much more demanding. I do not consider that the suggestion of Wells et al. (2005) is seriously undermined by the later papers.

5.6.3.2 Seaweed (No Collagen)

As mentioned in Section 5.3, Mach et al. (2007b) studied the fatigue of blades of macroalgae and assessed the lifetime to be expected, under the action of wave forces. This proved to be less than the observed lifetime of living macroalgae. They therefore conclude that routine repair takes place. This mirrors the situation found with collagenous materials.

5.6.4 Generalizations

All the tissues, which have been mentioned in this section, appear to live on the edge. Their fatigue quality is only just sufficient for the loads they encounter. Sufficiency includes allowing for ongoing repair. The situation is clearest with tendon because it is loaded in a simple and predictable way: longitudinal tension only and at a stress, which is limited by the weakness of the attached muscle.

If damage is balanced by repair, the appropriate fatigue quality will be determined by the availability of repair and this will vary with the tissue. But, even when ongoing repair is not possible, some damage can still be tolerated, so long as parts can be periodically replaced. This is the case for tissues which are beyond the reach of living cells, notably the keratinous tissues (hair, feathers, hooves) and insect cuticle. Their maintenance is closer to that of engineering structures. Collagenous tissues do it more subtly.

References

Anderson TL (2005) Fracture mechanics: fundamentals and applications. 3rd ed. Taylor & Francis, Boca Raton.

Bellucci G, Seedhom BB (2001) Mechanical behaviour of articular cartilage under tensile cyclic load. Rheumatology 40:1337–1345

Bertram JEA, Gosline JM (1986) Fracture toughness design in horse hoof keratin. J Exp Biol 125:29–47

Bonfield W, Datta PK (1976) Fracture toughness of compact bone. J Biomech 9:131–134

Caler WE, Carter DR (1989) Bone creep-fatigue damage accumulation. J Biomech 22:625–635

Carter DR, Caler WE (1983) Cycle-dependent and time-dependent bone fracture with repeated loading. J Biomech Eng 105:166–170

Carter DR, Caler WE (1985) A cumulative damage model for bone fracture. J Orthop Res 3:84–90

Carter DR, Hayes WC (1976) Fatigue life of compact bone. I. Effects of stress amplitude, temperature and density. J Biomech 9:27–34

Currey JD (2002) Bones: structure and mechanics. 2nd ed. Princeton University Press, Princeton.

Currey JD (2004) Tensile yield in compact bone is determined by strain, post-yield behaviour by mineral content. J Biomech 37:549–556

Davison PF (1989) The contribution of labile crosslinks to the tensile behaviour of tendons. Conn Tissue Res 18:293–305

Deakin AH (2006) The mechanical and morphological properties of normal and rheumatoid human forearm tendons. PhD thesis, The University of Strathclyde

Eppell SJ, Smith BN, Kahn H, Ballarini R (2006) Nano measurement with micro-devices: mechanical properties of hydrated collagen fibrils. J R Soc Interface 3:117–121

Evans FG, Lebow M (1957) Strength of human compact bone under repetitive loading. J App Physiol 10:127–130

Grashow JS, Sacks MS, Liao J, Yoganathan AP (2006) Planar biaxial creep and stress relaxation of the mitral valve anterior leaflet. Annals Biomed Eng 34:1509–1518

Griffith AA (1921) The phenomena of rupture and flow in solids. Philos Trans R Soc A 221: 163–191

Irwin GR (1956) Onset of fast crack propagation in high strength steel and aluminium alloys. In Proceedings of the second Sagamore Conference, Syracuse University, New York, 2:289–305

Irwin GR (1957) Analysis of stresses and strains near the end of a crack traversing a plane. J App Mech 24:361–364

Ker RF (2007) Mechanics of tendon, from an engineering perspective. Int J Fatigue 29:1001–1009

Ker RF, Zioupos P (1997) Creep and fatigue damage of mammalian tendon and bone. Comments Theor Biol 4:151–181

Ker RF, Alexander RMcN, Bennett MB (1988) Why are tendons so thick? J Zool Lond. 216: 309–324

Ker RF, Wang XT, Pike AVL (2000) Fatigue quality of mammalian tendons. J Exp Biol 203: 1317–1327

Klompen ETJ, Engels TAP, Van Breeman LCA, Schreurs PJG, Govaert LE, Meijer HEH (2005) Quantitative prediction of long-term failure of polycarbonate. Macromolecules 38:7009–7017

Kruzic JJ, Ritchie RO (2008) Fatigue of mineralized tissues: cortical bone and dentin. J Mech Biomed Mat 1:3–17

Mach KJ, Nelson DV, Denny MW (2007a) Techniques for predicting the lifetimes of wave-swept macroalgae: a primer on fracture mechanics and crack growth. J Exp Biol 210:2213–2230

Mach KJ, Hale BB, Denny MW, Nelson DV (2007b) Death by small forces: a fracture and fatigue analysis of wave-swept macroalgae. J Exp Biol 210:2231–2243

Martin B (1995) Mathematical model for repair of fatigue damage and stress fracture in osteonal bone. J Orthop Res 13:209–316.

Nalla RK, Imbeni V, Kinney JH, Staninec M, Marshall SJ, Ritchie RO (2003) In vitro fatigue behaviour of human dentin with implications for life prediction. J Biomed Mat Res Part A 66A:10–20

Nalla RK, Kruzic JJ, Kinney JH, Ritchie RO (2005) Aspects of in vitro fatigue in cortical bone: time and cycle dependent crack growth. Biomaterials 26:2183–2195

Nalla RK, Kruzic JJ, Kinney JH, Balooch M, Ager JW, Ritchie RO (2006) Role of microstructure in the aging-related deterioration of the toughness of human cortical bone. Mat Sci Eng C 26:1251–1260

Norman TL, Vashishth D, Burr DB (1995) Fracture toughness of human bone under tension. J Biomech 28: 309–329

Paris PC, Gomez, MP, Anderson WP (1961) A rational analytic theory of fatigue. Trends Eng 13: 9–14

Pike AVL, Ker RF, Alexander RMcN (2000) The development of fatigue quality in high- and low-stressed tendons of sheep (Ovis Aries). J Exp Biol 203:2187–2193

Prendergast PJ, Taylor D (1994) Prediction of bone adaptation using damage accumulation. J Biomech 27:1067–1076

Rice JR (1968) A path independent integral and the approximate analysis of strain concentrations by notches and cracks. J App Mech 35: 379–386

Reilly DT, Burstein AH (1975) The elastic and ultimate properties of compact bone tissue. J Biomech 8:393–405.

Rimnac CM, Petko AA, Santner TJ, Wright TM (1993) The effect of temperature, stress and microstructure on the creep of compact bone. J Biomech 26:219–228

Ritchie RO, Kinney JH, Kruzic JJ, Nalla RK (2005) A fracture mechanics and mechanistic approach to the failure of cortical bone. Fatigue Fract Engng Mater Struct 28: 345–371

Schechtman H, Bader DL (1997) In vitro fatigue of human tendons. J Biomech 30:829–835

Schwab TD, Johnston CR, Oxland TR, Thornton GM (2007) Continuum damage mechanics (CDM) modelling demonstrates that ligament fatigue damage accumulates by different mechanisms than creep damage. J Biomech 40:3279–3284

Sedman AJ (1993) Mechanical failure of bone and antler: the accumulation of damage. D.Phil thesis, University of York.

Shelton DR, Martin RB, Stover SM, Gibeling JC (2003) Transverse crack propagation behaviour in equine cortical bone. J Mat Sci 38:3501–3508

Stella JA, Liao J, Sacks MS (2007) Time-dependent biaxial mechanical behaviour of the aortic valve leaflet. J Biomech 40:3169–3177

Suresh (1998) Fatigue of Materials. 2nd edit. Cambridge University Press, Cambridge.

Taylor D, Hazenberg JG, Lee TC (2003a) A cellular transducer in damage-staimulated bone remodelling: a theoretical investigation. J Theor Biol 225:65–75

Taylor D, O'Reilly P, Valet L, Lee TC (2003b) Fatigue strength of compact bone in torsion. J Biomech 36:1103–1109

Taylor D, Hazenberg JG, Lee TC (2007) Living with cacks: damage and repair in human bone. Nat Mater 6:263–268

Teoh SH, Cherry BW (1984) Creep rupture of a linear polythene. I. Rupture and pre-rupture phenomena. Polymer 25:727–734

Thorton GM, Schwab TD, Oxland TR (2007) Fatigue is more damaging than creep revealed by modulus reduction and residual strength. Ann Biomed Eng 35:1713–1721

Vashishth D (2004) Rising crack-growth-resistance behaviour in cortical bone: implications for toughness measurements. J Biomech 37:943–946

Vashishth D (2007) Hierarchy of bone microdamage at multiple length scales. Int J Fatigue 29:1024–1033

Wang XT, Ker RF (1995) The creep rupture of wallaby tail tendons. J Exp Biol 198:831–845

Wang XT, Ker RF, Alexander, RMcN (1995) Fatigue rupture of wallaby tail tendons. J Exp Biol 198:847–852

Wells SM, Sellaro T, Sacks MS (2005) Cyclic loading response of bioprosthetic heart valves: effects of fixation state on the collagen fiber architecture. Biomaterials 26:2611–2619

Winwood, K, Zioupos P, Currey JD, Cotton JR, Taylor M (2006) Strain patterns during tensile, compressive and shear fatigue of human cortical bone and implications for bone biomechanics. J Biomed Mat 79A:289–297

Wren TAL, Lindsey DP, Beaupré GS (2003) Effects of creep and cyclic loading on the mechanical properties and failure of human Achilles tendon. Ann Biomed Eng 31:710–717

Wright TM, Hayes WC (1976) The fracture mechanics of fatigue crack propagation in compact cortical bone. J Biomed Mat Res 10:637–648

Yan J, Mecholsky JJ, Clifton KB (2007) How tough is bone? Applciation of elastic-plastic fracture mechanics to bone. Bone 40:279–484

Zioupos P, Wang XT, Currey, JD (1996) Experimental and theoretical quantification of the development of damage in fatigue tests of bone and antler. J Biomech 29:989–1002

Zioupos P, Currey, JD (1998) Changes in the stiffness, strength and toughness of human cortical bone with age. Bone 33:57–66.

Zioupos P, Currey JD, Casinos A (2001) Tensile fatigue in bone: are cycles-, or time to failure, or both, important? J Theor Biol 210: 389–399

Chapter 6
Viscoelasticity, Energy Storage and Transmission and Dissipation by Extracellular Matrices in Vertebrates

F.H. Silver and W.J. Landis

Abstract The extracellular matrix (ECM) of vertebrates is an important biological mechanotransducer that prevents premature mechanical failure of tissues and stores and transmits energy created by muscular deformation. It also transfers large amounts of excess energy to muscles for dissipation as heat, and in some cases, the ECM itself dissipates energy locally. Beyond these functions, ECMs regulate their size and shape as a result of the changing external loads. Changes in tissue metabolism are transduced into increases or decreases in synthesis and catabolism of the components of ECMs. Viscoelasticity is an important feature of the mechanical behavior of ECMs. This parameter, however, complicates the understanding of ECM behavior since it contains both viscous and elastic contributions in most real-time measurements made on vertebrate tissues.

The purpose of this chapter is to examine how time-dependent (viscous) and time-independent (elastic) mechanical behaviors of an ECM are related to the hierarchical structure of vertebrate tissues and the macromolecular components found in specific tissues. In most ECMs, energy storage is believed to involve elastic stretching of collagen triple helices found in the cross-linked collagen fibrils comprising vertebrate connective tissues, and energy dissipation is believed to involve sliding of such collagen fibrils by each other during tissue deformation. It may be concluded that viscoelasticity differs markedly among different ECMs and is related to ECM hierarchical structure at the molecular and supramolecular levels of any particular vertebrate tissue.

6.1 Introduction

Energy storage, transmission and dissipation are some of the many key mechanical functions provided by ECMs in vertebrate tissues and are required for efficient movement, effective locomotion, and tissue regeneration and repair through mechanochemical transduction events (Silver et al., 2001b–e, 2002a–b, 2002d–e, 2003a, 2006). To achieve and maintain the processes involved in movement and locomotion, vertebrates must be able to develop muscular forces, store

P. Fratzl (ed.), *Collagen: Structure and Mechanics*,
© Springer Science+Business Media, LLC 2008

elastic energy and transfer this energy to their joints attached to muscles. In addition, the energy remaining after movement must be transferred from the joints to the muscles where it can be dissipated as heat (Silver, 2006).

During the normal gait cycle in vertebrates, potential energy is stored as strain energy in tendons that are stretched during the impact with the ground. The same potential–strain energy relation in tendons can be said in general of movement by vertebrates through different environmental media, including trees, air and water. Elastic recoil primarily by tendons converts most of the stored energy to kinetic energy (Alexander, 1983, 1984). The manner by which strain energy is stored, transmitted and dissipated is important to the understanding of how ECMs function to allow efficient locomotion without premature failure of their components. Decreased energy storage and dissipation by ECMs are associated with debilitation, disease and normal aging in vertebrates (Silver et al., 2001f, 2002e, 2003a), and these parameters are intimately associated with molecular and higher order changes in ECM structure, including its constituent collagen molecules, fibrils, and fibers; elastic and smooth muscle fibers; proteoglycans and water (Silver, 2006).

In this chapter, the mechanisms of energy storage, transmission and dissipation in vertebrates are examined at the molecular and other levels of structural hierarchy and are related to the viscoelastic behaviors of a variety of vertebrate tissues. In this context, insight is provided for mineralized and unmineralized tendon, self-assembled collagen fibers, skin, cartilage and the vessel wall. Beyond these aspects relative to tissues, mechanochemical transduction and the relationship between viscoelastic behavior and hierarchical structure are investigated and used to suggest the modeling techniques for oriented, orientable and composite ECMs.

6.1.1 Concept of Energy Storage, Transmission and Dissipation

Energy is generated in the vertebrate musculoskeletal system by the contraction of muscle through the conversion of ATP to ADP. This energy is used to stretch tendons connected to skeletal muscles, and this results in energy storage (Alexander, 1983, 1984; Silver, 2006). Tendon stretching leads to joint movement and locomotion. Not all the energy generated by muscular contraction leads to movement. Approximately 10–25% is dissipated within the tendon because of the viscous sliding of collagen fibrils (Dunn and Silver, 1983), the rearrangement of water molecules, and other processes, and energy is transmitted back to the muscles at the end of each gait cycle to be dissipated as heat (Wilson et al., 2001). Results of studies with race horses indicate that the muscle–tendon unit not only stores energy during its extension but also dissipates the remaining energy after extension is complete (Wilson et al., 2001). These observations suggest that energy dissipation by the tendon–muscle unit may help prevent tendon tearing and stress fractures in bone. Since tendon and bone are ECMs composed of dense fibrous connective tissue, it is important to examine at the molecular and tissue levels how energy storage, transmission and dissipation occur.

The ability to store and transmit energy in ECM in musculoskeletal tissues is related to the rod-like structure of the fibrous collagens (types I, II and III) that form cross-linked networks in these tissues (Silver et al., 2002d). The predominant components of these ECMs are type I and II collagens, while types III, V and others are found in smaller amounts (Silver et al., 2006). While the type I collagen molecule is rod-like, it contains numerous bends and points of flexibility (Silver and Birk, 1984), and modeling studies based on the sequences of the types I, II and III collagen molecules suggest that these structural elements are composed of alternating sequences of rigid and flexible domains (Silver et al., 2001b–d, 2001f, 2001d; Landis et al., 2006) (Fig. 6.1). The flexible domains approximately coincide with positively stained bands found in the collagen D period (Silver et al., 2001b–d, 2002d; Landis et al., 2006). These domains are poor in the imino acids, proline and hydroxyproline, and they appear to be deformed when an external mechanical force is applied to the collagen fibrils (Freeman and Silver, 2004a, b).

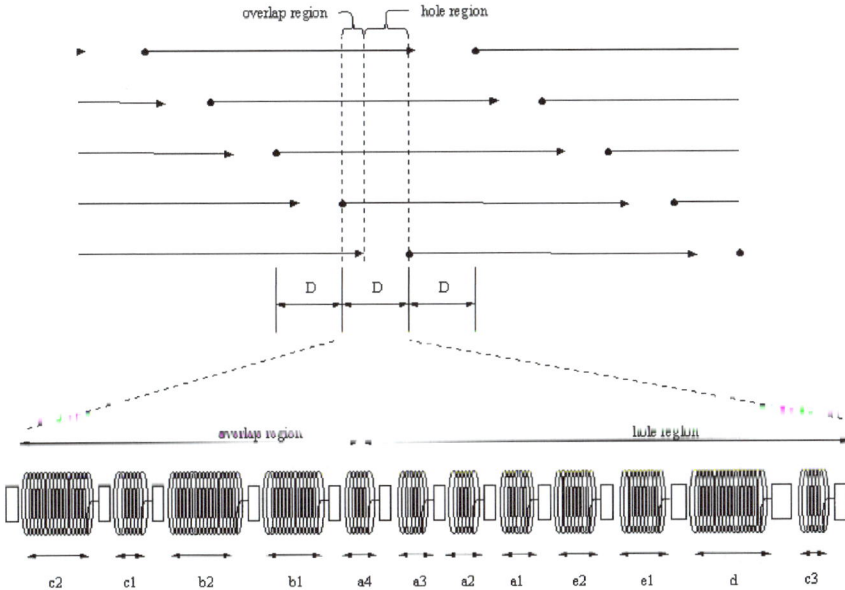

Fig. 6.1 Total (top line), elastic (middle line), and viscous (bottom line) stress-strain curves for rat tail tendon fibers tested in uniaxial tension. In the quarter-stagger packing pattern, collagen molecules, 4.4D long, are staggered with respect to their neighbors. In tendon, the collagen molecules are shifted with respect to each other by a distance D equal to 67 nm. When collagen molecules are stained with metal ions and then viewed in the electron microscope, a series of light and dark bands are observed across the axis of the fibril and are designated b2, b1, a4, a3, a2, a1, e2, e1, d, c2 and c1. The distance D is made up of a hole region of about 0.6 D and an overlap region of about 0.4 D. The D period is the characteristic fingerprint of fibrous collagen. The bands in the D period are enlarged at the bottom of the figure, and the flexible charged regions are denoted by springs, while the rigid regions are depicted by small rectangles

As discussed further below, energy storage during stretching is thought to be associated with conformational changes in the collagen triple helix caused by repulsion between sets of like charges found in the collagen α-chain sequences (Freeman and Silver, 2004b). In contrast, energy dissipation appears to be associated with the viscous sliding of collagen fibrils by each other during tensile deformation (Silver et al., 2001a–d, 2002b). Thus, the long collagen fibrils and fibers in tendon maximize energy storage and transmission, while minimizing energy dissipation during locomotion (Silver et al., 2001a–d). In the same manner, energy storage and transmission are putatively maximized in bone. However, this is not the case for all ECMs. In skin, for example, the fraction of energy dissipated during tensile deformation is higher than that in tendon and bone (Dunn and Silver, 1983). The skin protects the inner organs of vertebrates from mechanical injury by energy dissipation during stretching (Dunn and Silver, 1983).

6.1.2 Molecular Basis of Energy Storage and Dissipation

When ECMs are subjected to tension, stress is developed by the stretching of collagen fibril bundles (fibers), which in turn causes stretching of individual collagen fibrils and molecules (Silver et al., 2001a–d, 2001f). Increases in the collagen D period are reported to exceed increases in the h spacing (axial displacement per amino acid residue of collagen) for stretched rat tail tendon (Mosler et al., 1985), and the h spacing in tendon is found to increase only about 1% for every 10% increase in the macroscopic strain of this tissue (Mosler et al., 1985; Silver et al., 2001b–c). Therefore, 90% of the macroscopic strain applied to the tendon causes fibrillar and molecular slippage, while only 10% of the macroscopic strain causes direct stretching of the collagen triple helix. In this instance, if the slope of a stress–strain curve of tendon is corrected to reflect that (1) only 10% of the strain occurs at the molecular level of the collagen triple helix and (2) tendons are composed of only about 50% by weight of collagen, then the corrected slope of such a stress versus strain curve would serve as an estimate of the elastic modulus of the collagen molecule (Silver et al., 2001b–c) (Fig. 6.2). An estimate of the elastic modulus for tendon and type I collagen of between 7 and 8 GPa may be calculated as a result (Fig. 6.2). In this calculation, it is assumed that the stress is stored elastically, that is, no loss in energy occurs by viscous processes (Silver et al., 2001b–c). The assumption is true if the stress–strain curve is constructed from measurements made when the stress has decayed to an equilibrium value.

A molecular modeling program has been used to calculate the change in free energy if there were a 1–3% increase in the h spacing of a type I collagen triple helix packed into a five-membered unit. The resulting free energy calculation was compared to the change in energy under the molecular stress versus strain curve (Freeman and Silver, 2004b). From these considerations, it was determined that changes in free energy during stretching of a collagen microfibril were proportional to the energy changes under the experimental elastic stress–strain curve determined for tendon (Freeman and Silver, 2004b). Furthermore, other results indicated that

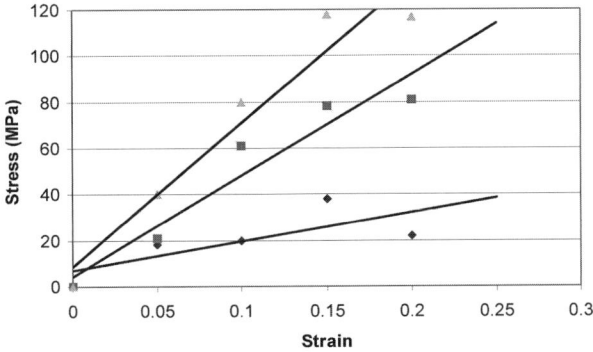

Fig. 6.2 Stress–strain curves for tendon. Total (*triangles*), elastic (*squares*) and viscous (*diamonds*) stress–strain curves for rat tail tendon fibers tested in uniaxial tension. The elastic modulus of collagen is estimated to be between 7 and 8 GPa from the slope of the elastic stress–strain curve after correction for the ratio of macroscopic to molecular strain and correction for the collagen content. The viscous stress is related to the shape factor and the energy dissipated as heat from Silver et al., 2001b-c

some regions of the molecule devoid of imino acids (proline and hydroxyproline) required lower energies to stretch than did regions rich in imino acids (Freeman and Silver, 2004b). The regions without imino acids have been predicted to form folds in the collagen triple helix (Paterlini et al., 1995), appear to open or uncoil when stress is applied to an ECM, and under these circumstances serve as sites that can store energy (Silver et al., 2001c). Molecular modeling results suggest that stretching increases steric energy of the triple helix that is attributable to van der Waals and electrostatic interactions between amino acids that are charged (Freeman and Silver, 2004b).

As described above, an estimate of the elastic modulus of the collagen triple helix may be obtained from the slope of an elastic stress–strain curve corrected for the actual molecular strain and collagen content. The approach assumes that measurements on tendon tissue in tension are made after all viscous relaxation has occurred. The actual measurement is made by allowing the stretched tissue to relax at constant length until the stress remains constant (Silver et al., 2001c–d, 2001f) (Fig. 6.3). The relaxation process may take a matter of several hours (Silver et al., 2001b–d, 2002a–b, 2002e; Snowhill and Silver, 2005). The stress at a particular strain at equilibrium, that is, when the stress no longer decreases with time, is the elastic or time-independent stress (Silver et al., 2001c), providing a modulus estimate of the collagen molecule on relevant stress–strain curve correction (Silver et al., 2001c). The difference between initial stress at a fixed strain and equilibrium stress is the stress dissipated as heat (Silver et al., 2001c).

Figure 6.2 shows the initial (total), equilibrium (elastic) and dissipated (viscous) stress–strain curves for tendon. As mentioned previously, much of the energy unutilized for locomotion is dissipated in muscles attached to tendons and causes the muscles to increase in temperature (Wilson et al., 2001).

When two rod-like elements slide past each other in a solution containing small solvent molecules, one can calculate the element length, in this case the collagen

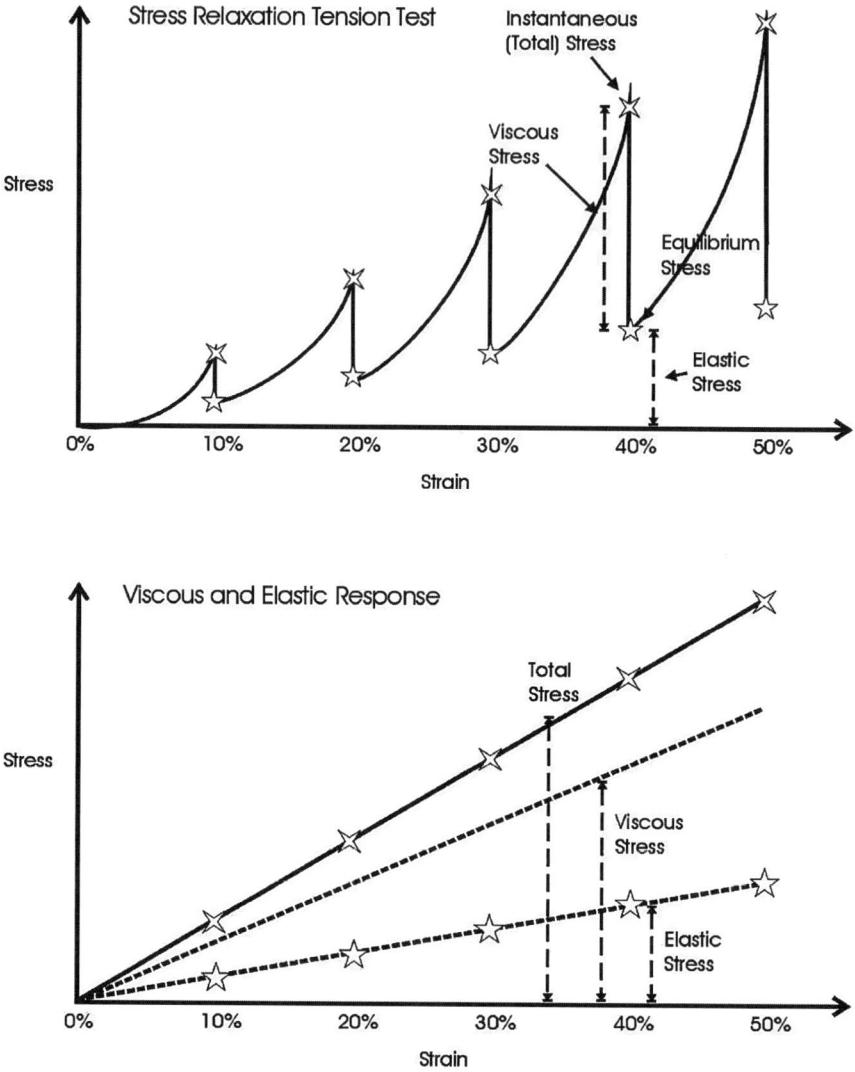

Fig. 6.3 Diagram of incremental stress–strain testing. (*Top*) Illustration of the tensile loading program used to generate elastic and viscous stress–strain curves. A sample is elongated to a fixed length in tension, and the stress is allowed to relax until it does not decrease with increasing time. The equilibrium or lowest stress is proportional to the energy stored, and the difference between the initial and equilibrium stresses is proportional to the energy dissipated (viscous stress) as heat. After the stress reaches an equilibrium value, another strain increment is applied, and the process is repeated until the sample fails in tension. (*Bottom*) The total, viscous and elastic stresses are plotted versus strain to construct the incremental stress–strain curves from Seehra and Silver, 2006

fibril length, knowing viscous stress and fibril diameter (Silver et al., 2001c; Silver, 2006). Thus, fibril lengths calculated from viscous stress for developing tendons in chicks (Silver et al., 2003c) coincide well with fibril lengths measured for the same tendons with electron microscopy (Birk et al., 1989). These observations suggest that energy dissipation in tendon and other ECMs occurs through the viscous sliding of fibrils and bundles of fibrils by each other during tensile deformation (Silver et al., 2003c). Energy dissipation in tendon is minimized by maximizing collagen fibril diameter since the frictional factor of rods decreases as the diameter increases (Silver, 2006). This result would explain in part the fact that collagen fibrils in tendon are much larger in diameter than those collagen fibrils comprising ECMs of other tissues (Silver et al., 2001c, 2002a). A smaller average collagen fibril diameter is associated with increased energy dissipation in ECMs, such as skin and vessel wall (Silver and Christiansen, 1999).

6.1.3 Viscoelastic Behavior of Tendon

Much of the current understanding of the relationship between hierarchical structure and viscoelastic behavior of ECMs is based on the studies of the mechanical properties of developing tendon (McBride et al., 1985, 1988; McBride, 1984). The properties of developing tendon rapidly change just prior to the onset of locomotion. For example, the maximum total stress of about 2 MPa that can be borne by a 14-day-old embryonic chick leg extensor tendon increases 2 days after birth to 60 MPa (McBride et al., 1985, 1988; McBride, 1984). This rapid increase in tensile stress by tendon occurs without changes in the tissue hierarchical structure (McBride et al., 1985, 1988; McBride, 1984). In this case, the collagen fibril length appears to be more important for energy storage than fibril diameter, but the two parameters are linked together since fibrils have been shown to grow in length by lateral fusion of fibril bundles (McBride et al., 1988; Birk et al., 1989, Silver et al., 2003c).

Viscoelastic properties of ECMs have been obtained by constructing incremental stress–strain curves for a variety of tissues including tendon (Silver, 2006). Such incremental stress–strain curves are derived for tendon and other ECMs by stretching the tissues in a series of strain increments and then allowing the stress to relax to an equilibrium value at each strain increment before another strain increment is added (Silver et al., 2001c) (Fig. 6.3). By subtracting the elastic stress (equilibrium stress value) from the initial or total stress value, the viscous stress (stress that is converted into heat) is determined. By plotting equilibrium stress versus strain and total stress minus equilibrium stress versus strain (Fig. 6.2), elastic and viscous stress–strain curves are derived (Silver et al., 2001c).

As discussed above, the slope of the elastic stress–strain curve is proportional to the elastic modulus of the collagen molecule (Silver et al., 2001c), while the viscous stress at a particular strain is a measure of collagen fibril length (Silver et al., 2001c). An estimate of the elastic modulus of the collagen molecule is obtained by dividing the slope of the elastic stress–strain curve by the collagen content and by the ratio of the molecular strain (change in h spacing) divided by the macroscopic strain (0.1). A value of between 7 and 8 GPa for the elastic modulus of the collagen molecule (Silver et al., 2001c) has been found for rat tail tendon. Collagen fibril lengths calculated from the viscous stress and hydrodynamic theory (Silver, 2006) range from about 20 μm for developing tendon to in excess of 1 mm for adult tendons (Silver et al., 2003c, 2006).

6.1.4 Viscoelasticity of Self-Assembled Type I Collagen Fibers

A great deal of information concerning the viscoelasticity of ECMs can be derived from understanding the behavior of model systems, such as self-assembled type I collagen fibers developed from solubilized rat tail tendon collagen (Christiansen et al., 2000; Silver et al., 2000, 2001a, 2002c). The fibers are self-assembled under conditions that produce D-banded collagen fibrils similar to those seen in rat tail tendons (Christiansen et al., 2000; Silver et al., 2000). The purified type I collagen fibrils produced by self-assembly are much narrower, for example between about 20 and 40 nm in diameter, as compared to those as large as 1 μm in diameter in normal rat tail tendon (Christiansen et al., 2000; Silver et al., 2000). Incremental stress–strain curves for self-assembled purified type I collagen are linear for uncross-linked collagen fibers (Silver et al., 2001a). However, unlike the incremental stress–strain curves for rat tail tendon, the viscous stress–strain curve for tendon lies above the elastic stress–strain curve (Silver et al., 2001a). This result suggests that, in the absence of cross-links, the ability of collagen fibers to store elastic energy is impaired, and elastic energy storage appears increased by the formation of cross-links (Silver et al., 2001a). When self-assembled collagen fibers are cross-linked by aging at room temperature, the elastic stress–strain curve lies above the viscous curve (Silver et al., 2000). On comparison of elastic stress–strain curves for tendon and self-assembled collagen fibrils, the slope of the elastic stress–strain curve for cross-linked self-assembled collagen fibrils is much closer to that of tendon than is the slope for uncross-linked collagen fibers (Silver et al., 2001a). This result underscores the need for end-to-end cross-links between collagen molecules, in order to promote energy storage during stretching (Silver et al., 2001a). The viscoelastic behavior of tendon and self-assembled collagen fibers is very similar. The energy storage capability of tendon is attributable to direct stretching of the triple helix, and energy dissipation occurs through the sliding of neighboring fibrils and bundles of fibrils during tensile deformation. However, the behavior of other ECMs is a bit more complicated as discussed below for skin.

6.1.5 Viscoelasticity of Skin

As just noted, when tendon is stretched and the collagen fibrils are aligned with the tendon axis, the elastic and viscous behaviors reflect the stretching and sliding

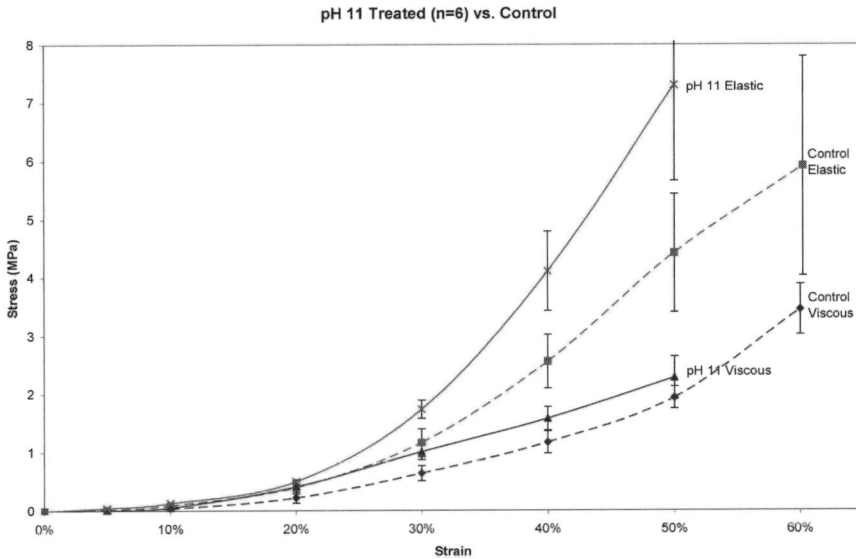

Fig. 6.4 Incremental stress–strain curve for skin. Elastic and viscous stress–strain curves for decellularized human dermis at pH 7.0 (control) and pH 11.0 stretched in tension. At pH 7.0 and 11.0, the slope of the elastic and viscous stress–strain curves increases with increasing strain. Note the elastic stress is increased at pH 11.0 compared to pH 7.0, suggesting that the elastic modulus is proportional to the net charge on the surface of the collagen triple helix from Seehra and Silver, 2006

of collagen fibrils (Silver et al., 2000). Incremental stress–strain behavior of skin is different from a tendon model as shown in Fig. 6.4 (Silver et al., 2001b). This typical curve shows that stress is almost immeasurable until a strain of about 20% is reached. Beyond this value, slopes of the elastic and viscous curves increase considerably (Dunn and Silver, 1983; Silver et al., 2001c; Seehra and Silver, 2006). In this instance, the slope of the initial part of the elastic stress–strain curve for skin, when corrected for its elastic tissue content, is similar to the elastic modulus reported for ligamentum nuchae (0.4 MPa), itself rich in elastic tissue (Dunn and Silver, 1983; Silver et al., 2001c). The slope of the final part of the elastic stress–strain curve for skin, corrected for its collagen content, is about 4 GPa, a value closer to that for the elastic modulus of normal type I collagen (Silver et al., 2001c, 2002a). The value of 4 GPa for the elastic modulus at high strains of skin versus 7–8 GPa for the elastic modulus of tendon may be attributable to the differences between the type of cross-links found in skin versus tendon or to the fact that the collagen fibers in skin are biaxially oriented and not oriented along a single axis (Silver et al., 2002a). At low strains (below 20%), the elastic and viscous stresses are about the same when one expands the low axis of the stress–strain curve (Silver et al., 2001c, 2002a), while at high strains, the elastic stress is much greater than the viscous stress (Silver et al., 2002a). This result may suggest that two different mechanisms underlie the viscoelastic character and behavior of skin. It has been suggested that, at low strains, the viscoelasticity of skin is caused by stretching of the elastic fibers that normally surround its composite collagen fibers. Elastic fibers are easily stretched but offer little energy storage or dissipation (Dunn and Silver, 1983; Silver et al., 2001c, 2002a). At high strains, the collagen fibers, which form a network independent of elastic fibers, bear loads and store energy that prevent premature mechanical failure of the skin (Dunn and Silver, 1983; Silver et al., 2001c, 2002a). The mechanical behavior of skin is less complex compared to the behavior of other ECMs such as articular cartilage described below, in which collagen fiber orientation changes in successive hierarchical tissue layers.

6.1.6 Viscoelastic Behavior of Cartilage

Cartilage from a variety of sources in the human body is characterized in part by having its component collagen fibrils and fibers oriented in different directions depending on the hierarchical structure of the particular tissue. Thus, normal human meniscal tissue is found to consist of distinct layers or zones of cartilage, each with collagen fibrils of specific, but different, orientation with respect to its adjacent layers (Arnoczky, 1992; Schoenfeld et al., 2007). Likewise, normal human articular cartilage consists of type II collagen fibrils secreted in different orientations in the successively deeper zones of this tissue (Silver et al., 2001f). The stress–strain behavior of the cartilage of such tissues reflects their respective structures. As an example, incremental stress–strain curves have been evaluated for articular cartilage from humans (Fig. 6.5), in order to assess the energy storage, transmission and dissipation ability of elastic cartilage (Silver et al., 2001f, 2002b, 2004; Silver and Bradica, 2002). After an initial non-linear region, the elastic and viscous stress–strain curves for normal adult cartilage are almost linear (Silver et al., 2001f). Elastic moduli for collagen molecules in normal adult cartilage approach the value of 7 GPa for the superficial zone, where the constituent collagen fibrils and fibers are aligned with the cartilage surface (Silver et al., 2001f, 2002b, 2004). The value falls to about 2 GPa for collagen fibrils and fibers that are found in deeper layers and oriented at angles to the cartilage tensile axis (Silver et al., 2002f). The viscous

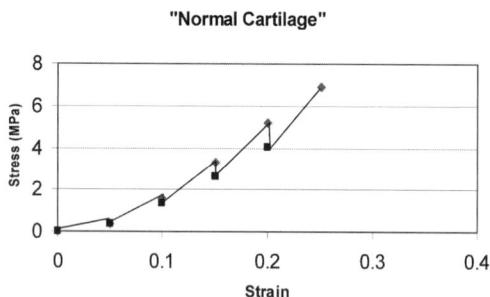

Fig. 6.5 Incremental stress–strain curve for articular cartilage. Total (*diamonds*) and elastic (*squares*) stress–strain curves for human articular cartilage taken from the femoral head. Note the linearity between stress and strain as the strain is increased. The incremental stress–strain curve of articular cartilage in tension is similar to that of tendon Silver and Bradica, 2002

loss by adult cartilage stretched in tension is very small, similar to tendon, and calculated collagen fibril length approaches 1 mm (Silver et al., 2002f). Analysis of incremental stress–strain curves determined for normal adult articular cartilage suggests that cartilage stores energy during deformation (Silver et al., 2004). Most of the energy is transmitted back to the muscle after joint deformation because of the low energy loss by normal tissue (Silver et al., 2004). This result indicates that a normal function of articular cartilage is to help transmit energy from tendon to bone, where it is transferred into changes in potential and kinetic energy. After joint movement occurs, the remaining energy is dissipated in muscle as heat.

Analysis of incremental stress–strain curves for osteoarthritic cartilage suggests that both elastic and viscous stress–strain curves for this tissue are decreased in magnitude with respect to normal cartilage (Silver et al., 2004). Such a conclusion leads to the inference that energy storage and dissipation mechanisms are impaired in this pathology, a situation that induces tears in the cartilage and subsequent wear and loss of its articular surfaces (Silver et al., 2004). While the mechanical behavior of articular cartilage is dominated by the character of collagen fibers as related to their structural hierarchy, other tissues such as vessel wall have more complicated behaviors.

6.1.7 Viscoelastic Behavior of Vessel Wall

The viscoelasticity of vessel wall is more difficult to understand than that of other ECMs (Snowhill et al., 2004). Incremental stress–strain curves for human aorta, for instance, are similar in shape to the curves for skin, but the magnitude of the maximum stress for aorta is only about 2% of that of skin and about 0.2% that of tendon. The differences may be explained by examining the curves in more detail. If one corrects the slope of the elastic stress–strain curve for aorta at low strains by the elastic tissue content of vessel wall, the elastic modulus is about that found for skin at low strains (Silver and Siperko, 2003; Silver et al., 2003b; Snowhill et al., 2004). This observation implies that elastic tissue may bear the loads at low strains in aorta similar to that occurring in skin (Snowhill et al., 2004, Snowhill and Silver, 2005). At high strains, the elastic modulus is much lower than that of the skin, and this result indicates that collagen alone in aorta does not bear the loads at high strains as in skin (Snowhill and Silver, 2005). Underlying this behavior, collagen is supported by other structures or by changes in its hierarchical assembly in the vessel wall. Such additional components, including smooth muscle, are necessary at high strains since the vessel wall is dynamically involved in mechanochemical transduction induced by variations in blood pressure in the human body (Silver and Siperko, 2003).

6.1.8 Determination of Elastic and Viscous Properties of Mineralized Tendon and Type I Collagen

Mineralized avian tendon and its constituent type I collagen have also been studied to determine the particular elastic and viscous properties of this tissue (Silver et al., 2001c; Landis and Silver, 2002; Landis et al., 2006). Normally mineralized avian tendon can bear higher stresses compared to unmineralized avian tendon (Silver et al., 2001c). There is an increase in the slope of elastic stress–strain curves for this tissue, and the elastic modulus of collagen increases as well (Silver et al., 2001c; Landis and Silver, 2002; Landis et al., 2006). From typical viscous stress–strain curves derived from tendons having low mineral content, the stress lost through viscous processes is small but increases upon mineralization. Mineralized tendons bear much higher loads but are less efficient at storing energy (Freeman and Silver, 2004a, 2005). Thus in avian tendon, mineralization may be a process by which tendon failure is prevented but at the expense of lowering the efficiency of locomotion of the animal.

6.1.9 Effects of Strain Rate and Cyclic Loading

As noted above, interpretation of viscoelastic behavior of an ECM is complicated by several factors, including its normal or pathological condition, reflected in its structural integrity, and the possible presence of mineral deposition that in some tissues serves a critical role in load bearing and prevention of mechanical failure. In addition, viscoelastic behavior is affected by the sampling rate at which measurements are made. An important consideration in this circumstance is the effect of strain rate on viscoelastic properties of ECMs. While the shape of the stress–strain curve for various ECMs may appear to be similar for several different tissues, the elastic and viscous stress–strain behaviors may be different for each ECM and, as discussed previously in the example of cartilage, are dependent on the hierarchical structure of each tissue (Silver, 2006). Whereas the elastic behavior of tendon is approximately linear because of the direct stretching of its constituent collagen fibrils and fibers that are oriented almost in parallel with the loading direction (Silver et al., 2003c), the behavior of cartilage, skin and vessel wall are more complex (Silver, 2006). Furthermore, the elastic behavior and modulus of tendon and skin are approximately strain-rate independent at least at high strains (Silver, 2006), while those of cartilage and vessel wall are strain-rate dependent (Silver et al., 2004; Snowhill and Silver, 2005). At high strain rates, the presence of proteoglycans (PGs) appears to increase the elastic modulus of cartilage collagen, distinct from the behavior of tendon collagen (Silver et al., 2004). PGs may modify cartilage collagen in this manner by binding to the flexible regions of the molecule and sterically hindering their deformation (Silver et al., 2004).

The cyclic dependency of the stress–strain curve for ECMs has been well documented in the literature (Silver, 2006). If a rat tail tendon is cycled more than ten times through tensile loading and unloading cycles, its continuously loaded stress–strain curve continues to change after each cycle (Rigby, 1964). However, if the elastic stress–strain curve is measured at low strain rates, the elastic modulus does not change for tendon and also for skin (Silver, 2006). Variations in the viscous stress–strain curve probably occur because of water loss and irreversible slippage of collagen fibrils (Silver, 2006). This result suggests that preconditioning a sample before cyclic testing may only affect the collagen fibril orientation and the energy lost attributable to viscous processes. It may not affect the elastic behavior of the tissue if collagen fibrils are the only load-bearing elements in a tissue (Silver, 2006).

6.2 Concept of Mechanochemical Transduction and Changes in Tissue Metabolism and Aging

Energy storage and dissipation attributable to internal and external mechanical loading of tissues induce changes in gene expression and protein synthesis in ECMs. Thus, unlike earlier conceptions, ECMs are not passive biomechanical elements (Silver et al., 2003a–b, 2004), and they modulate the mechanical forces imparted to them into changes in biochemical signals in a series of phosphorylation events in a process termed mechanochemical transduction (Silver et al., 2003a; Silver, 2006). An example of such a process may be described by lifting weights repeatedly for several months. In this situation, both the size of body muscles and the amount of skin increase (Silver et al., 2003a; Silver 2006). These changes occur because most of the ECMs in the human body are under internal tension so that the balance between external forces resulting from lifting weights against the force of gravity and internal forces present in ECMs is altered during musculoskeletal movement (Silver and Siperko, 2003; Silver et al., 2003a). Alteration of the balance between external and internal mechanical forces acting on resident cells found in ECMs leads to changes in gene expression and subsequent production of ECM proteins (Silver and Siperko, 2003). In concert with larger muscle and skin content, the weight lifting also develops tendons and bones that increase in size and weight (Silver et al., 2003a). In contrast to such tissue stimulation by weight lifting, prolonged bed rest and a reduction in external mechanical loading result in tissue catabolism. This leads to loss of muscle and atrophy of the musculoskeletal system (Silver, 2006). Therefore, any changes in the equilibrium between external and internal mechanical forces acting on ECMs cause variations in mechanochemical transduction at the cellular level. These changes appear to be important mechanisms by which mammals and vertebrates in general adjust their needs to store, transmit and dissipate energy that is required for bodily movements (Silver, 2006).

The ability of cells in ECMs to facilitate locomotion in a gravitational field dictates that cells must be able to adapt their responses to changing energy requirements (Silver, 2006). This means that information in the form of external mechanical forces must somehow set in motion a series of biological steps that produce an increase in muscle mass and associated muscle tissues when external loads are increased (Silver, 2006). Increased muscle mass is required to amplify the amount of work that can be done by an organism in response to the need for growth or as a result of environmental demand. This change is equivalent to conversion of mechanical energy to chemical energy in the form of high molecular weight components found inside and outside ECM cells (Silver, 2006).

Mechanochemical transduction is thought to involve several different macromolecular components and biochemical processes. Among them, constituents and events may include direct stretching of protein–cell surface integrin binding sites that are present on all eukaryotic cells (integrin-dependent mechanisms). Such stretching occurs by means of direct collagen fibril–integrin interactions (Silver et al., 2003a). Stress-induced conformational changes in ECMs may alter integrin structure and lead to activation of several secondary messenger pathways within the cell (Fig. 6.6). These pathways in turn may effect altered regulation of genes that are critical to the ultimate secretion of ECM proteins, as well as to altered regulation of cell division (Fig. 6.7). Mechanochemical transduction may also be initiated through deformation of gap junctions containing calcium-sensitive stretch receptors (Silver and Siperko, 2003). Once activated, these channels may trigger secondary messenger signals through pathways similar to those involved in integrin-dependent activation and allow communication between cells with similar or different phenotypes. Mechanochemical transduction may additionally occur through activation of ion channels in cellular membranes. Mechanical forces have been shown to alter cell membrane ion channel permeability associated with Ca^{+2} and other ion fluxes (Silver and Siperko, 2003). The application of mechanical forces to cells promotes

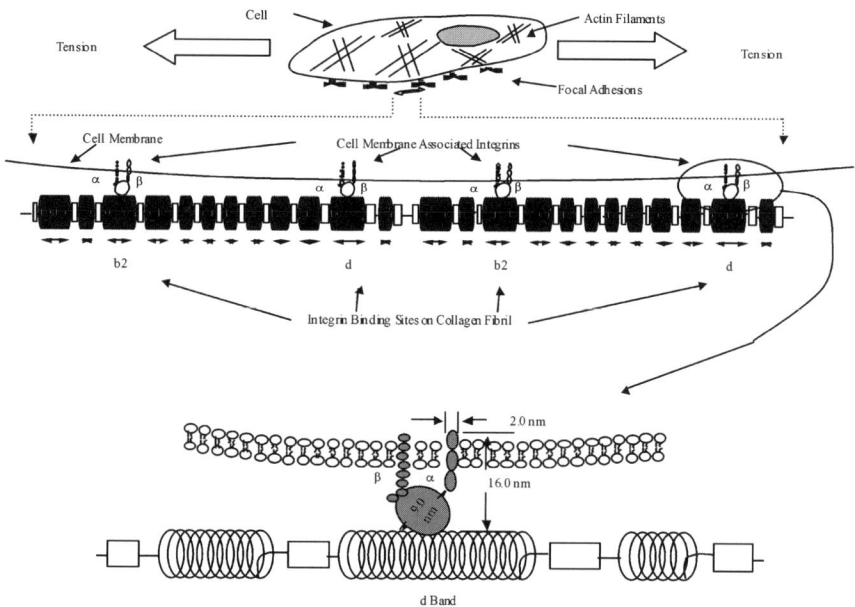

Fig. 6.6 Diagram illustrating how mechanical stresses are transduced into changes in mechanochemical transduction in ECM. Tensile forces applied to collagen fibrils are putatively transduced into activation of phosphorylation systems through integrin-dependent and integrin-independent events. (*Top*) Illustration demonstrating how deformation of collagen fibrils leads to stretching of attached cell membranes and integrin subunits α and β. (*Bottom*) Enlarged view of contact between α and β integrin subunits and the flexible regions on the collagen fibril that are denoted by springs. When collagen fibrils are deformed, the flexible regions become extended, an action that putatively causes a conformational change in attached integrin subunits

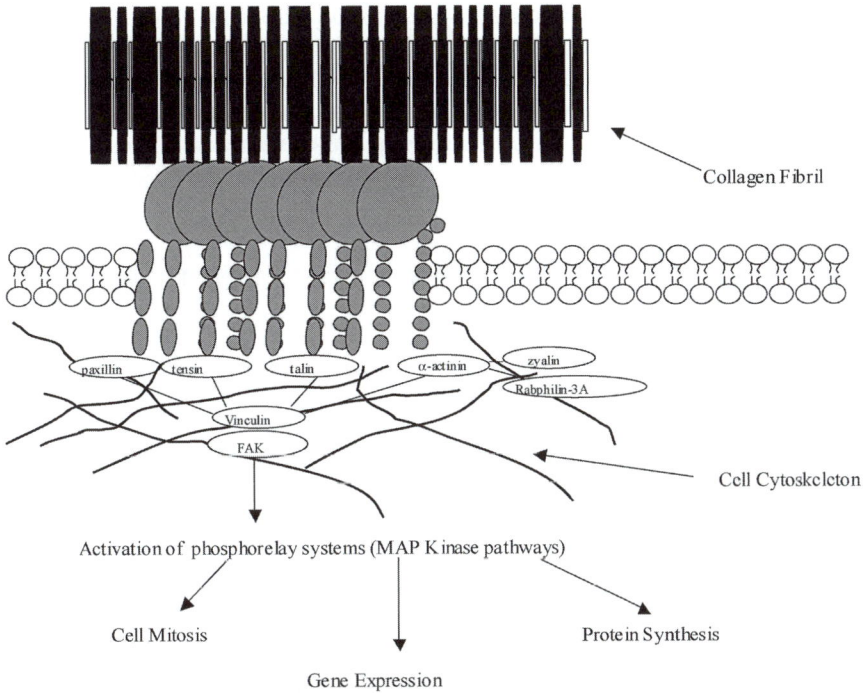

Fig. 6.7 Illustration showing the possible relationship between cellular and ECM components involved in mechanochemical transduction and regulation of mitosis, gene expression and protein synthesis. This diagram shows the proximity of collagen fibrils (*top*), integrin subunits α and β and the neighboring cytoskeletal components. Tension applied to collagen fibrils causes deformation of integrin subunits and cytoskeletal components putatively leading to activation of FAK and upregulation of the phosphorylation pathways that regulate cell division, gene expression and protein synthesis

as well as upregulation of growth factor and hormone receptors on cell surfaces, even in the absence of ligand binding, as recently reviewed (Silver and Siperko, 2003).

6.3 Relationship Between Viscoelasticity and Hierarchical Structure

Mechanochemical transduction is an important process that results in changes in ECM hierarchical structure. In this context, relationships between viscoelastic mechanical behavior and vertebrate tissue structure should be considered in more detail. The mechanical behavior of ECMs may be subdivided on the basis of the directional alignment of their component collagen fibers with respect to the loading axis in tension, as well as to the content and arrangement of ECM constituents other than collagen (Silver, 2006). To a first approximation, the behavior of a simple macromolecular network found in an ECM depends principally on collagen fiber orientation. While this result is an oversimplification, it can be used to model the behavior of aligned collagen networks such as those found in tendon (Silver et al., 2003c). In general, the physical structure and its relation to the mechanical properties of an ECM are important to understand since otherwise an analysis of the viscoelastic behavior of the ECM becomes a curve-fitting process without a biochemical, biomechanical or biological frame of reference.

6.3.1 Aligned Collagen Networks and Mechanical Models of Tendon

To expand upon the considerations of tendon noted above, a first approximation of the behavior of this tissue reflects stretching of its constituent collagen fibers. The behavior can be modeled as an elastic spring in parallel with a viscous element (Silver, 2006). The elastic spring represents the elastic stretching of collagen triple helices, while the viscous element represents sliding of collagen fibrils past each other in the network (Silver et al., 2001a–b). Tendon behavior is best modeled as a collection of elastic springs, one for each collagen fibril, aligned in parallel with a collection of viscous elements that denote slippage of each collagen fibril. The elastic modulus for collagen molecules and fibrils in tendon has been estimated from the corrected slope of its elastic stress–strain curve (Silver et al., 2001 a–b). The values of the elastic modulus of a collagen molecule have been determined for rat tail tendon and mineralized turkey leg tendon and are listed in Table 6.1 (Silver et al., 2001a; Landis and Silver, 2002). These values fall between 4.2 and 7.69 GPa, depending on the tissue investigated. Values for the elastic modulus of self-assembled collagen fibers are also given in Table 6.1. The values of the elastic modulus for cross-linked self-assembled type I collagen approach those reported for normal tendon, a result suggesting, as noted earlier, that the elastic behavior of tendon to a first approximation reflects that of its collagen fibers. Since the diameters

Table 6.1 Estimated elastic moduli for ECM components based on elastic stress–strain measurements (Silver, 2006)

Molecule	Tissue	Elastic modulus (GPa)
Type I collagen	Self-assembled	6.51
Type I collagen	Rat tail tendon	7.69
Type I collagen	Turkey tendon (no mineral)	4.20
Type I collagen	Turkey tendon (0.245 mineral content)	7.22
Types I and III collagen	Skin	4.4
Type II collagen	Articular cartilage (surface parallel)	7.0
Type II collagen	Articular cartilage (surface perpendicular)	2.21
Type II collagen	Articular cartilage (whole parallel)	4.91
Type II collagen	Articular cartilage (whole perpendicular)	1.52
Type II collagen	Osteoarthritic cartilage (whole perpendicular)	0.092
Elastin	Skin	0.040
Elastin	Vessel wall	0.01

Table 6.2 Estimated collagen fibril lengths based on mechanical measurements of viscous loss in different ECMs (Silver, 2006)

Tissue	Fibril length (mm)
Rat tail tendon	0.860
Self-assembled collagen fibers	0.0373
Turkey tendon (no mineral)	0.108
Turkey tendon (0.245 mineral content)	0.575
Human skin	0.0548
Articular cartilage (surface parallel)	1.265
Articular cartilage (surface perpendicular)	0.688
Articular cartilage (whole parallel)	0.932
Articular cartilage (whole perpendicular)	0.696
Osteoarthritic cartilage (whole perpendicular)	0.164

of self-assembled collagen fibers are much smaller than those of rat tail tendon fibrils, fibril diameter is not critical in determination of elastic modulus. However, the presence of cross-links between molecules is important (Silver et al., 2000) and greatly influences the value of the viscous loss (related to the fibril length) compared to elastic modulus (Silver, 2006). Uncross-linked collagen fibrils are shorter and dissipate more energy than cross-linked fibrils as shown in Table 6.2. Diseases of the ECM that reduce cross-link density or modify cross-link characteristics dramatically reduce energy storage and dissipation in ECMs and lead to premature mechanical failure of the tissue involved.

6.3.1.1 Mechanical Models of Mineralized Tendon

Mechanical models of mineralized tendon parallel the analysis presented above for unmineralized tendon. Elastic moduli of mineralized turkey tendons have been calculated from experimental elastic incremental stress–strain curves. For mineralized tendons, the stress–strain curves are linear at a mineral content of about 0.3, and the elastic modulus is about 7 GPa and increases with mineral content (Landis and Silver, 2002). Fibril lengths calculated for mineralized turkey tendon increase with mineral content from 0.108 to 0.575 mm (Silver, 2006). As fibril lengths increase, so does the elastic modulus, but energy storage may decrease as a result of loss of extensibility.

Molecular modeling results suggest that the increased elastic modulus of mineralized turkey tendon compared to unmineralized tendon is a result of the effect of calcium and phosphate ions increasing the resistance of the flexible regions in collagen to axial deformation (Silver et al., 2001b; Freeman and Silver, 2004a–b, 2005). These ions appear to act as bridges or cross-links between collagen α-chains and collagen molecules, and they thereby make the flexible charged regions on collagen more rigid. It is also possible that the onset and progression of mineralization stiffens collagen by preventing deformation of its regions comprised of charged amino acid residues.

6.3.2 Mechanical Models of Orientable ECMs

The analysis of behavior of tendon as presented above is rather straightforward, but it is more complicated for other ECMs. When one considers skin, for example, a tissue containing a collagen network that is closely oriented biaxially, it is clear that its elastic and viscous stress–strain curves are highly non-linear, as noted previously. However, results of experimental stress–strain studies suggest that at high strains under uniaxial tension, the stress–strain behavior of skin is almost linear (Dunn and Silver, 1983). In this case, skin would appear to contain two independent networks, elastic tissue and collagen fibers, that behave independently. If elastic behavior is examined separately from viscous behavior, two straight lines approximate the elastic stress–strain curve, one representing the low strain elastic region of the tissue and the other representing its high strain elastic region (Silver et al., 2001c, 2002a). This approximation is consistent with earlier literature suggesting that the low strain region reflects behavior of the elastic fiber network of skin, while the high strain region reflects behavior of the collagen fibers (Dunn and Silver, 1983). The result makes modeling skin more difficult since an elastic spring and viscous element (Voigt element) occur respectively for the elastic tissue and collagen network comprising this tissue (Silver, 2006). Further, the two Voigt elements representing elastic tissue and collagen fibers must be connected in parallel with each other since loss of elastic fiber content associated with aging of skin shifts its stress–strain behavior in such a manner as to indicate the absence of elastic fibers (Silver, 2006). The elastic modulus of skin increases significantly with strain, and for this reason its collagen fibers are modeled in a wave form until its elastic fibers are fully stretched (Silver et al., 2001c, 2002a).

The elastic fiber contribution of skin may be analyzed by examining the low strain behavior of its elastic stress–strain curve. The collagen fiber contribution of the tissue may be obtained by assessing its high strain elastic behavior. From incremental stress–strain curves for skin utilizing this approach (Silver et al., 2001b, 2002a), the elastic modulus for elastic fibers after correction for the elastin content of the tissue was found to be between 0.01 and 0.04 GPa, while the elastic modulus for its collagen fibers was measured to be 4.4 GPa. Values are listed in Table 6.1. The low strain behavior of skin was reported to be strain-rate dependent, while its high strain behavior was not (Seehra and Silver, 2006). This result suggests that elastic fiber stretching in skin may involve a change in the structure of elastin or fibrillin, the microfibrillar component of elastic fibers, while that of collagen remains unaltered (Silver, 2006).

6.3.3 Mechanical Models of ECMs Comprised Only of Collagen Fibers

Analysis of ECM behavior increases in complexity as tissue hierarchical structure becomes more complicated. From a mechanical perspective, tissues such as articular cartilage are composites of layers of regular (aligned) and irregular (unaligned) collagen networks (Silver and Christiansen, 1999). Although there are PGs and other components in cartilage besides collagen fibers, these constituents appear to reinforce the fibers and do not drastically alter their viscoelastic behavior. This result permits cartilage modeling based on known collagen fiber orientation in each tissue layer. Thus, the surface zone of articular cartilage contains aligned collagen fibers and can be modeled in a manner similar to tendon, using a spring in series with a viscous element. Deeper articular cartilage zones containing collagen fibers that run at angles to the fibers found in the surface can be modeled as a spring in series with a viscous element, but the stress generated in these regions must be corrected for the collagen fiber orientation. This factor may be accounted for by the spatial (angular) relation between composite collagen fibers and the tensile axis of the tissue.

Incremental elastic and viscous stress–strain curves of articular cartilage have been reported in the literature (Silver et al., 2004). The elastic modulus of its superficial zone measured in tension parallel to its collagen fibrils is estimated to be 7.0 GPa while that perpendicular to the collagen fibers of the superficial zone is determined as 2.21 GPa (Table 6.1). Values of the elastic moduli for collagen fibers in normal intact articular cartilage range between 2 and 7 GPa, depending on fiber orientation in a specific zone. These measurements suggest the behavior of the tissue can be modeled as a composite of layers of collagen fibers with different collagen fibril orientations (Silver et al., 2004).

Collagen fibril lengths within articular cartilage superficial, intermediate and deep zones have been determined from the stress values of incremental viscous stress–strain curves. Lengths of up to 1.265 mm are reported (Table 6.2) (Silver, 2006). Fibril lengths for mature articular cartilage are similar to those found for tendon.

6.3.4 Mechanical Models of Composite ECMs Containing More Than Collagen Fibers

Composite ECMs, such as vessel walls, do not behave mechanically as a series of layers of simple collagen fibers (Snowhill et al., 2004; Snowhill and Silver, 2005). Analysis of incremental stress–strain curves of vessel walls suggests that their low and high strain behaviors cannot simply be analyzed like skin (Snowhill and Silver, 2005). Their composition consists of elastic fibers, collagen and smooth muscle, and each component appears to contribute to aspects of vessel wall strain behavior. While the low strain behavior of vessel wall is to a first approximation a reflection of the behavior of elastic tissue, the high strain behavior of this tissue reflects the behavior of both collagen and smooth muscle fibers (Snowhill and Silver, 2005). Maximum elastic stress measured in vessel wall is much lower than that observed for skin, and this observation has been explained by postulating that the collagen fibers and smooth muscle comprising the wall structure interact in series with respect to viscoelasticity (Snowhill and Silver, 2005). Based on this hypothesis, it was proposed that the behavior of vessel wall may be modeled as a spring in series with a dashpot, a Voigt element, representing the behavior of the elastic fibers, connected in parallel to two other Voigt elements in series, representing the behavior of collagen and smooth muscle fibers (Snowhill and Silver, 2005).

The values for the elastic modulus of collagen in different vessel walls are significantly lower than 1 GPa and vary depending on the collagen and smooth muscle content for each particular wall (Snowhill et al., 2004, Snowhill and Silver, 2005). The elastic modulus for elastic tissue has been determined to be about 0.013 GPa as listed in Table 6.1 and appears to vary markedly depending on the elastin content for each vessel wall (Snowhill and Silver, 2005).

6.4 Conclusions

ECMs act as biological mechanotransducers that prevent premature mechanical failure of tissues. ECMs store and transmit energy created by muscular deformation and amplify cell division and subsequent protein synthesis as applied stress and loads are increased to tissues internally or externally. Changes in tissue stress, load and metabolism are transduced into increases or decreases in cellular synthesis and catabolism of the biochemical and structural components of ECMs. Thus, ECMs regulate their size and shape as a result of the constantly modulating local mechanical demands on the tissues they comprise. In the cellular synthesis of high molecular weight macromolecules, energy may be stored by ECMs in the form of covalent bonds and later released through catabolic reactions as may be required for continued tissue growth and development.

While viscoelasticity is an important feature of ECM behavior, it complicates the understanding of mechanical behavior of an ECM since most of its real-time measurements are characterized by both elastic and viscous contributions. Energy storage and elastic behavior primarily involve the stretching of collagen triple helices found in cross-linked collagen fibrils that constitute many ECMs. Energy dissipation or viscous loss is believed to involve sliding of collagen fibrils by each other during tissue deformation. It has been concluded that viscoelasticity of collagenous tissues differs markedly among different ECMs and is related to the hierarchical structure of each particular tissue. Since tissue architecture varies among different ECMs, so too does the amount of energy that each may store and dissipate. Attempts to model the viscoelastic behavior of different ECMs suggest that use of springs to represent the stretching of the collagen triple helix, elastic fibers and smooth muscle placed in series with viscous elements that represent the sliding of collagen and other fibrous ECM components may offer a molecular explanation for the complex viscoelastic properties of ECMs.

Acknowledgments This work was supported in part by grant AR 41452 from the National Institutes of Health (to WJL).

References

Alexander R M (1983) Animal Mechanics (2nd ed.). Blackwell Scientific, Oxford, UK.

Alexander R M (1984) Elastic energy stores in running vertebrates. A. Zool. 24, 85–94.

Arnoczky S P (1992) Gross and vascular anatomy of the meniscus and its role in meniscal healing, regeneration, and remodeling. In: Mow V C, Arnoczky S P, and Jackson D W (Eds.), Knee Meniscus: Basic and Clinical Foundations, Raven Press, New York, pp. 1–14.

Birk D E, Zycband E I, Winkelmann D A, Trelstad R L (1989) Collagen fibrillogenesis in situ: Fibril segments are intermediates in matrix assembly. Proc. Natl. Acad. Sci. U S A 86: 4549–4553.

Christiansen D L, Huang E K, Silver F H (2000) Assembly of type I collagen: Fusions of fibril subunits and the influence of fibril diameter on mechanical properties. Matrix Biol. 19: 409–420.

Dunn M G, Silver F H (1983) Viscoelastic behavior of human connective tissue: Relative contribution of viscous and elastic components. Connect. Tissue Res. 12: 59–70.

Freeman J W, Silver, F H (2004a) Analysis of mineral deposition in turkey tendons and self-assembled collagen fibers using mechanical techniques. Connect. Tissue Res. 45: 131–141.

Freeman J W, Silver F H (2004b) Elastic energy storage in unimineralized and mineralized extracellular matrices (ECMs): A comparison between molecular modeling and experimental measurements. J. Theor. Biol. 229: 371–381.

Freeman J W, Silver F H (2005) The effects of prestrain on in vitro mineralization of self-assembled collagen fibers. Connect. Tissue Res. 46: 107–155.

Landis W J, Silver, F H (2002) The structure and function of normally mineralizing tendons. Comp. Biochem. Physiol. A, 133: 1135–1157.

Landis W J, Silver F H, Freeman J (2006) Collagen as a scaffold for biomimetic mineralization of vertebrate tissues. J. Mater. Chem. 16: 1495–1503.

McBride D J (1984) Hind Limb Extensor Tendon Development in the Chick: A Light and Transmission Electron Microscopic Study. M.S. Thesis in Physiology, Rutgers University, Piscataway, NJ.

McBride D J, Hahn R, Silver F H (1985) Morphological characterization of tendon development during chick embryogenesis: Measurement of birefringence retardation. Int. J. Biol. Macromol. 7: 71–76.

McBride D J, Trelstad R L, Silver, F H (1988) Structural and mechanical assessment of developing chick tendon. Int. J. Biol. Macromol. 10: 194–200.

Mosler E, Folkhard W, Knorzer E., Nemetschek-Gansler H, Nemetschek T H, Koch M H (1985) Stress-induced molecular arrangement in tendon collagen. J. Mol. Biol. 182: 589 596.

Paterlini M G, Nemethy G, Scheraga H A (1995) The energy of formation of internal loops in triple-helical collagen polypeptides. Biopolymers 35: 607–619.

Rigby B J (1964) Effect of cyclic extension on the physical properties of tendon collagen and its possible relation to biological aging of collagen. Nature 202: 1072–1074.

Schoenfeld A, Landis W J, Kay D. (2007) Meniscal tissue engineering. Am. J. Orthop. 36: 614–620.

Seehra G P, Silver F H (2006) Viscoelastic properties of acid- and alkaline-treated human dermis: A correlation between total surface charge and elastic modulus. Skin Res. Technol. 12: 190–198.

Silver, F H (2006) Mechanosensing and Mechanochemical Transduction in Extracellular Matrix: Biological, Chemical, Engineering and Physiological Aspects, Springer, New York.

Silver F H, Birk D E (1984) Molecular structure of collagen in solution: Comparison of types I, II, III, and V. Int. J. Biol. Macromol. 6: 125–132.

Silver F H, Bradica G (2002) Mechanobiology of cartilage: how do internal and external stresses affect mechanochemical transduction and elastic energy storage? Biomechanics & Modeling in Mechanobiology 1: 1–19.

Silver F H, Christiansen D L (1999) Biomaterials Science and Biocompatibility, Chapter 6, Springer, New York.

Silver F H, Siperko L M (2003) Mechanosensing and mechanochemical transduction. Crit. Rev. Biomed. Eng. 31: 255–331.

Silver F H, Christiansen D L, Snowhill P, Chen Y (2000) Role of storage on changes in the mechanical properties of tendon and self-assembled collagen fibers. Connect. Tissue Res. 41: 155–164.

Silver F H, Christiansen D L, Snowhill P B, Chen Y (2001a) Transition from viscous to elastic-based dependency of mechanical properties of self-assembled type I collagen fibers. J. Appl. Polym. Sci. 79: 134–142.

Silver F H, Freeman J W, Horvath I, Landis W J (2001b) Molecular basis for elastic energy storage in mineralized tendon. Biomacromolecules 2: 750–756.

Silver F H, Freeman J, DeVore D (2001c) Viscoelastic properties of human skin and processed dermis. Skin Res. Technol. 7: 18–25.

Silver F H, Horvath I, Foran D (2001d) Viscoelasticity of the vessel wall: Role of collagen and elastic fibers. Crit. Rev. Biomed. Eng. 29: 279–301.

Silver F H, Bradica G, Tria A (2001e) Relationship among biomechanical, biochemical and cellular changes associated with osteoarthritis. Crit. Rev. Biomed. Eng. 29: 373–391.

Silver F H, Bradica G, Tria A (2001f) Viscoelastic behavior of osteoarthritic cartilage. Connect. Tissue Res. 42: 223–233.

Silver F H, Seehra P, Freeman J W, DeVore D (2002a) Viscoelastic properties of young and old human dermis: Evidence that elastic energy storage occurs in the flexible regions of collagen and elastin. J. Appl. Polym. Sci. 86: 1978–1985.

Silver F H, Bradica G, Tria A (2002b) Elastic energy storage in human articular cartilage: Estimation of the elastic spring constant for type II collagen and changes associated with osteoarthritis. Matrix Biol. 21: 129–137.

Silver F H, Ebrahimi A, Snowhill P B (2002c) Viscoelastic properties of self-assembled type I collagen fibers: Molecular basis of elastic and viscous behaviors. Connect. Tissue Res. 43: 1–12.

Silver F H, Horvath I, Foran D J (2002d) Mechanical implications of the domain structure of fibril forming collagens: Comparison of the molecular and fibrillar flexibility of α-chains found in types I, II and III collagens. J. Theor. Biol. 216: 243–254.

Silver F H, Siperko L M, Seehra G P (2002e) Mechanobiology of force transduction in dermis. Skin Res. Technol. 8: 1–21.

Silver F H, DeVore D, Siperko L M (2003a) Invited Review: Role of mechanophysiology in aging of ECM: Effects of changes in mechanochemical transduction. J. Appl. Physiol. 95: 2134–2141.

Silver F H, Snowhill P B, Foran D (2003b) Mechanical behavior of vessel wall: A comparative study of aorta, vena cava, and carotid artery. Ann. Biomed. Eng. 31: 793–803.

Silver F H, Freeman J, Seehra G P (2003c) Collagen self-assembly and development of matrix mechanical properties. J. Biomech. 36: 1529–1553.

Silver F H, Bradica G., Tria A (2004) Do changes in mechanical properties of articular cartilage alter mechanochemical transduction and promote osteoarthritis? Matrix Biol. 23: 467–476.

Snowhill P B, Silver F H (2005) A mechanical model of porcine vascular tissues-Part II: Stress–strain and mechanical properties of juvenile porcine blood vessels. Cardiovasc. Eng. 5: 157–169.

Snowhill P B, Foran D J, Silver F H (2004) A mechanical model of porcine vascular tissues-Part I. Determination of macromolecular component arrangement and volume fractions. Cardiovasc. Eng. 4: 281–294.

Wilson A M, McGuigan M P, Su A, van den Bogert A J (2001) Horses damp the spring in their step. Nature 414: 895–899.

Chapter 7
Nanoscale Deformation Mechanisms in Collagen

H.S. Gupta

Abstract The mechanical properties of collagen Type I tissues are, like many biological connective tissues, crucially dependent on the hierarchical architecture at the nanometer and micron length scale. Triple helical collagen molecules aggregate into ordered fibrils of $\sim$ 100–200 nm diameter, which in turn can form parallel-fibered fiber bundles or lamellae at the micron level in tendon, bone and other tissues. To determine quantitatively the structural response of elements at each level in the structural hierarchy to applied mechanical stresses, in situ methods, which combine a high-resolution structural determination tool like x-ray scattering along with micromechanical testing, are a unique tool. These methods have recently provided a range of information and insights into the actual deformation processes occurring at the molecular, fibrillar and fiber bundle level in both unmineralized and mineralized collagen tissue types. In this chapter, we provide an overview of our current understanding of the nanoscale deformation processes in tendon, bone and related tissues.

7.1 Introduction

Collagen (principally type I and II) plays a major structural role in a range of vertebrate tissues from hard and stiff tissues like bone, tendon, ligament and cartilage to more compliant materials such as cornea. The unique hierarchical structure (Chapter 1) where molecules are self-assembled into microfibrils and fibrils, which aggregate into fiber bundles and fascicles (in tendon) (Kastelic et al. 1978) or into lamellae and compact or spongy bone tissue types (in vertebrate bone) (Weiner and Wagner 1998) has long been considered to be crucial in determining the mechanical properties. The mechanical properties of the different types of collagen-containing tissues (bone, tendon, etc.) have been extensively investigated and are reported in Chapters 2, 3 and 4 of this volume. However, relatively little quantitative understanding exists of the deformation mechanisms at the fibrillar and fiber bundle level, since such measurements require special techniques to probe the deformation exclusively at a specific length scale. For example, to measure the change in pitch of the collagen molecules, when the tissue in which it is present is subjected to an external load, requires an x-ray diffraction image of the helical pitch before and

P. Fratzl (ed.), *Collagen: Structure and Mechanics*,
© Springer Science+Business Media, LLC 2008

after the load. Since following the structural changes has to occur in real time, nanoscale and microscale deformation mechanisms in collagen require the use of special in situ setups, which are essentially micromechanical testing devices coupled to a microscopic, structural or spectroscopic characterization tool. For investigating structural changes at the level of 100 nm and smaller, x-ray scattering and diffraction are the preferred methods, while changes at the level of 1 μm and above have been investigated with light and confocal microscopy. This chapter summarizes the work done by numerous groups over the past years on understanding quantitatively the nanoscale and microscale deformation of collagen in collagen-containing tissues. For reasons of experimental choice, most of the work on uncalcified collagen has been carried out on the parallel-fibered rat tail tendon collagen model system.

7.2 Deformation at the Fiber Bundle Level

Some of the earlier works on the fiber–matrix interactions in tendon focused on the macroscopic crimp observed at the scale of 10–100 μm, as a possible example of fiber buckling inside a hydrated, polysaccharide-rich matrix. Using synthetic composite model systems (nylon/poly(ethyl) acrylate) with different shrinkage properties of the two components, buckled and wavy fiber forms were observed, qualitatively similar to the crimp pattern seen in polarized light microscopy of tendons (Dale and Baer 1974). Using native and reconstituted collagen fibrils mounted on substrates deformed to different strain levels, electron microscopy of these fixed fibrils showed that deformation at the fibril level was homogeneous across the entire D-period of the fibril. However, the effect of staining on the fibril mechanics was pronounced, with unstained bands appearing more deformable (Barenberg et al. 1978). An interesting suggestion by these authors was that the degree of cross-linking within the fibril was much more limited than previously supposed, and not present in the triple helical regions, due to the observed large deformability of the fibrils. The role of noncovalent (ionic or hydrogen) bonds was suggested to be more relevant in maintaining fibrillar integrity (Barenberg et al. 1978).

Structurally, tendon fascicles are composed of 50–100 μm diameter fibers, which are in turn composed of 100–200 nm diameter fibrils, held together by inter- and intra-fibrillar cross-links and a hydrated proteoglycan-rich matrix between fibrils and also between fibers (Fig. 7.1). In situ measurement of deformation in tendon under load observed at the microscopic length scale (10–100 μm) would therefore provide quantitative information on the deformation mechanisms between fibers. Since the surface contrast of tendon tissue is low, techniques such as laser confocal microscopy are needed to visualize the fiber structure and associated tenocytes (tendon cells). Screen and coworkers have in recent years pioneered this technique, using the shift of the tendon cells within the fibers as markers of intra-fiber strain and studying the effect of incubation and partial removal of the matrix phase on the fiber level mechanics (Screen et al. 2003, 2004a,b, 2005a,b,c, 2006).

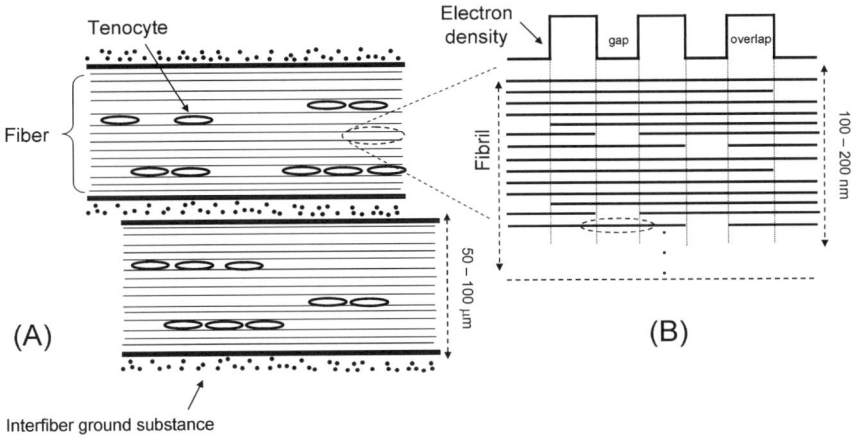

Fig. 7.1 Schematic of the tendon structure at the (**a**) microscale and (**b**) nanoscale. In (**a**), fiber bundles of typically 50 μm lie parallel to each other, with tendon cells (tenocytes) in parallel rows within the fiber bundle. Thin lines denote the fibrils (50–200 nm in diameter). In (**b**) a single fibril is schematically indicated as a staggered array of Type I collagen molecules (shown as rods). The step profile above the schematic indicates the periodic 1D electron density profile arising from this staggered arrangement

Screen et al. (2003) used confocal microscopy of tendons, whose cells (tenocytes) were fluorescently labeled, to show that the fiber strain was significantly less than the macroscopic strain. The variation of fiber strain was not linear with applied tissue strain, but showed a sharp increase between 2 and 3% tissue strain. However, strain *between* fibers was also significant. Comparison of the inter- and intra-fiber strains showed that the intra-fiber strains were less (<1–1.5%) than inter-fiber strains (1.5–5.0%), implying that sliding between fibers (not just between fibrils) also played a significant role in tendon deformation (Screen et al. 2004a,b). Possibly, proteoglycans can coat the fiber surfaces as well as the internal fibril surfaces. It will be critical for future studies to clearly differentiate between the noncollagenous organic molecules that are found *within* fibers and between fibrils versus the smaller fraction found at the larger length scale *between* fibers, if the question of intra- versus inter-fiber sliding is to be answered. When the proteoglycan chondroitin sulfate (CS) is removed enzymatically from the tendons, it was observed that the *intra*-fiber strain was reduced relative to the control case and the *inter*-fiber strain was increased, suggesting that mechanically connective proteoglycan molecules were removed preferentially between the fibers and not within them (Screen et al. 2006). However, the effect of the incubating medium was also found to be significant – when tendon fascicles were stored for a long time in PBS (phosphate-buffered saline, the standard medium for testing samples mechanically), there was significant swelling of the fibrils (Screen et al. 2006). Control of the storage media and careful attention to its ionic concentration are no doubt crucial to maintaining tendons in a state as close to "native" as possible, since the anionic glycosaminoglycan component can exhibit significant swelling and electrostatic repulsion, which

will coincidentally enable the diffusion of smaller molecules in the created spaces
as well (Scott 2003). The physiological relevance toward understanding collagen
micromechanics is brought out by the result that cyclic tensile strain upregulates the
synthesis of collagen (although the glycosaminoglycan synthesis and cell proliferation remains unaltered) (Screen et al. 2005a,b,c).

Structurally, the location and nature of the proteoglycan binding, both to the collagen as well as to each other, is crucial in determining the mechanical properties of
the material as a whole (Scott 1991, 2003, Fratzl et al. 2004a, b). The initial concept
of a randomly entangled meshwork of proteoglycans has been refined to an "ordered
aggregation model" (in the words of J. E. Scott (1992)), which can especially
explain the cohesiveness of very dilute meshworks of collagen found in umbilical
cord, vitreous humor (Scott 1992). Small proteoglycans like decoran bind nonco-
valently but specifically to the collagen fibril (linking identical d/e bands across
fibrils) (Scott 2003). These "shape modules" (Scott 2003) can deform elastically
(i.e., no rigid connections between fibrils) and are found localized along the collagen fibril, linking adjacent fibrils together. Molecular mechanics approaches have
been used to model the transfer of force between collagen fibrils and decorin-bound
glycosaminoglycans (Redaelli et al. 2003). Aggrecan and biglycan may also exist
and bind to decoran and weakly to collagen in the tendon (Screen et al. 2005a,b,c,
2006, Scott 1991, Raspanti et al. 1997, Cribb and Scott 1995, Yanagishita 1993).

The conclusion from these numerous studies on deformation at the fiber bun-
dle/fiber level is that it is still unclear what the hierarchical partition of strain from
the macroscopic to the molecular level is. Unlike in bone, where recent studies can
at least separate the hierarchical deformation at the fibril and mineral particle level
from all length scales above (Gupta et al. 2006a, b), the apparently simpler case
of tendon collagen still does not deliver an unequivocal answer as to how much
of the reduction from tissue to fibril strain comes about at the intermediate fiber
bundle/fiber level. Techniques to look at the two levels simultaneously are expected
to be difficult, because of the technical challenge of combining confocal microscopy
with synchrotron x-ray diffraction, but comparative studies between similar sets of
samples using the two methods would no doubt deliver very interesting results.

7.3 Fibrillar and Molecular Deformation Mechanisms

Much of our current understanding of the mechanics of collagenous tissues at the
ultrastructural (sub-micron) level arises from the use of time-resolved x-ray diffraction and scattering methods. At the molecular and supramolecular levels, the hierarchical structure of collagen gives rise to a rich structure in reciprocal space, where
wide-angle x-ray diffraction (WAXD) and small-angle x-ray scattering (SAXS) can
be used to infer changes in intermolecular spacing (Fratzl et al. 1993), axial fibril
stagger and on the existence of microfibrillar units inside fibrils (Orgel et al. 2006)
and fibril self-assembly (Wess et al. 1998), as described in Chapter 3 in this volume. A corollary is that the ultrastructural changes associated with mechanical

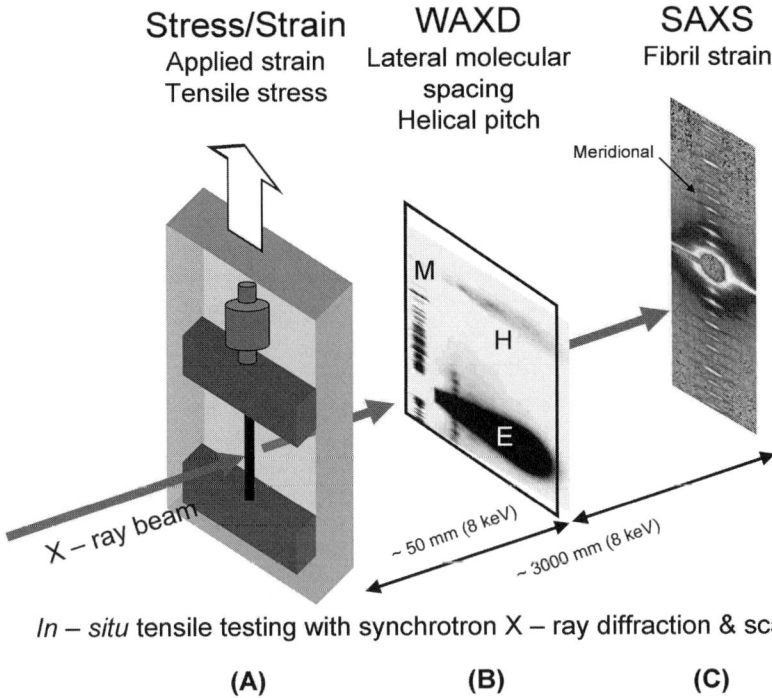

Stress/Strain
Applied strain
Tensile stress

WAXD
Lateral molecular
spacing
Helical pitch

SAXS
Fibril strain

In – situ tensile testing with synchrotron X – ray diffraction & scattering

(A)   **(B)**   **(C)**

Fig. 7.2 Schematic of the in situ tensile testing with synchrotron x-ray diffraction and scattering. The tissue is mounted in a tensile tester enclosed in a fluid chamber filled with physiological saline, and the incident x-ray beam (transmission geometry) enters through x-ray transparent windows. Depending on the distance of the x-ray detector to the sample, the (**B**) small length scale structures like helical collagen pitch and lateral molecular packing or (**C**) large-scale ordering such as the fibril D stagger ($D = 65$–67 nm) can be visualized, as seen in the representative spectra

deformation can be measured quantitatively from the changes in the WAXD and SAXS spectra. A typical spectrum of collagen is shown in Fig. 7.2, where the structural parameters associated with each range of reciprocal space (helical pitch, axial fibril stagger and lateral molecular spacing) are also shown. Most, but not all, of the work described here has been carried out on the model system of rat tail tendon, due to its simple, one-dimensional fibrillar arrangement, ease of extraction and mechanical testing.

The application of x-ray diffraction and scattering techniques to mechanically deformed collagenous tissues was pioneered in the 1970s by Th. Nemetschek and coworkers. They observed that under mechanical stress the meridional spectrum split into two pairs of Bragg reflections, indicating either fibrillar splitting into sub-fibrils or kinking of the fibrils along the stress direction (Hosemann et al. 1974). A model of alternately soft and hard regions along the fibril was proposed, with the amorphous, nonhelical regions undergoing the maximum deformation, and a kink structure, where the helical versus nonhelical regions are tilted by small but opposite degrees to the strain axis. Such a model bears similarities to the latest proposed microfibril structure proposed by Orgel et al. (2006).

Nemetschek and coworkers also pioneered the application of high-brilliance synchrotron x-ray radiation to the in situ investigation of changes in collagen ultrastructure (Folkhard et al. 1987a, b, Mosler et al. 1985a, b, c, Knorzer et al. 1986). Synchrotron sources, which enable the generation of very intense beams of light through the relativistic acceleration of electrons, can be used to produce x-ray spectra of biological connective tissues at time-scales orders of magnitude shorter (10–100 s) than laboratory sources (1–10 h per spectra). Such high-brilliance beams make it possible to follow the real-time ultrastructural changes occurring during collagen deformation, as shown in Fig. 7.2. Usually, the sample is maintained in a physiological saline solution (like phosphate-buffered saline (PBS)) in an x-ray sample chamber with x-ray transparent windows. An x-ray beam is used to take spectra in transmission geometry. By varying the sample to detector distance, the diffracted x-ray signal can provide information about the molecular (short sample to detector distance or WAXD) or fibrillar structure (long sample to detector distance or SAXS).

Under application of tension, these workers were the first to show that the fibrillar D-period increases when macroscopic load is applied, implying a stretching of the fibrils (Mosler et al. 1985a, b, c). However, because the relative intensities of the different orders in the meridional SAXS spectra changed as well, they also inferred that there must be a change in the relative length of the "gap" versus the "overlap" regions. Hence, under mechanical loading, the fibril strain does not arise solely due to homogeneous molecular elongation (which would keep the gap/overlap ratio constant), but is also comprised of a component of sliding between the triple helical molecules (Mosler et al. 1985a, b, c). Model calculations showed that the initial deformation (from 67 to 67.6 nm) corresponded to a stretching of the triple helices, while for larger strains sliding of triple helices relative to each other is relevant (Folkhard et al. 1987a, b). Because the macroscopic tissue strain was always larger than the fibril strain, strain in the interfibrillar space (which takes up 10–20% of the area fraction in tendon) must take up the remaining deformation, and the fibrils shift axially relative to each other (Folkhard et al. 1987a, b).

By carrying out very fast loading and unloading experiments to simulate the dynamic or in vivo loads experienced during sports injuries or accidents, Knorzer et al. (1986) showed that macroscopic rupture was preceded, on the timescale of a few seconds, by *intra*-fibrillar sliding and damage, as evidenced from changes in the SAXS spectrum just before failure. It was also observed that the change induced in D-periodicity by application of stress was reversible, as long as the final D-period did not exceed 68.4 nm (from an initial value of 67 nm) (Folkhard et al. 1987a, b).

Sasaki and coworkers extended the above picture by considering the deformation of the triple helical collagen molecule as well as that of the fibril (Sasaki and Odajima 1996a, b), thereby considering three different levels of the structural hierarchy. The change in the helical pitch of the collagen molecule was taken as a measure of the molecular strain ε_M, compared to the fibrillar strain ε_F. Using a different system, bovine Achilles tendon, from the rat tail tendon of Nemetschek and coworkers, and a laboratory x-ray diffraction system, they obtained a value of 2.9 ± 0.1 GPa. This "static" modulus was compared to that obtained by Brillouin light scattering and dilute solution measurements by other workers, and found to be lower. The authors proposed the (somewhat implausible) suggestion that the lower static modulus of the collagen molecule meant that viscoelastic processes occur within the triple helix itself. It is more likely that due to the longer time for measurement of x-ray spectra in a lab source, there was considerable fibrillar level relaxation at the level studied by Nemetschek and coworkers above (Folkhard et al. 1987a, b, Mosler et al. 1985a, b, c, Knorzer et al. 1986), leading to a lower fibril strain and molecular stretch. Using these results combined with measures of the fibrillar and tissue strain and a quantitative (two dimensional) model of molecular packing in a single fibril, Sasaki and Odajima (1996a, b) attempted to separate the effect of molecular elongation, gap elongation and molecular slippage. Their conclusion was that molecular elongation was the dominant mode, but that each mode had to operate concurrently, since a single mode without the others would lead to a tilt in the meridional reflections, which is not observed experimentally. However, it should be noted that due to transverse isotropy in the distribution of fibrils around the tensile axis, a tilt could indeed occur in the fibrils, which would manifest itself as a projected increase in meridional width of the SAXS reflections. The width of the peaks was not considered by Sasaki and coworkers.

Fratzl and colleagues investigated the deformation at the fibrillar level in tendon collagen by separating the stress–strain curve into four separate regions and considering the changes in fibrillar structure (as inferred from the SAXS pattern) at each stage (Fratzl et al. 1998, Misof et al. 1997a, b, Puxkandl et al. 2002). As described in the introductory chapter by Fratzl, the tensile stress–strain curve for tendon collagen has four regions: macroscopic crimp (toe region), the heel region with concave upward curvature, elastic (linear) regime and late linear regime just prior to fiber failure. In situ mechanical testing with SAXS and WAXD can be used to show that the predominant mechanisms involved at each stage involve different scales in the structural hierarchy, as described in the following.

In the first stage, the removal of macroscopic crimp corresponds to the straightening out of fibril bundles at the level of 10 μm and above. This straightening results in the very low effective modulus region known as the "heel" region, at the beginning of the stress–strain curve. Due to different degrees of pre-straightening or tautening while clamping the tendon, the length of this region is variable, and depends to a large degree on the specific experimental protocol used to clamp the samples.

In the second "toe" region, the effective elastic modulus (as defined by the tangent to the stress–strain curve) slowly increases, and results in a concave upward stress/strain curve. In this region, Misof et al. (1997a, b) looked at the changes of the intermolecular spacing and degree of order in the gap zone as the tissue was elongated. They found that, while there was no net reduction in intermolecular spacing, the degree of entropic disorder dramatically reduced. This is observed by considering the intensity of the equatorial Bragg peak arising from the lateral spacing of molecules within the fibril. On application of tensile strain, the equatorial intensity was found to be proportional to the strain applied.

Assuming a model where the molecules vibrate laterally due to thermal activation, the mean energy of vibration $<E>$ is proportional to the mean-squared displacement $u^2(<E> = kT = (2f/l)<u^2>$, where f is the force along the molecule). Relating the force f and the lateral displacement u^2 to the macroscopic stress σ and strain ε, respectively, a simple model calculation reproduces the upward curvature of the stress/strain curve in the heel region. The mechanism is essentially a form of rubber elasticity, due to loss of freedom of motion of the gap zone segments of the collagen molecules (Fig. 7.3). It is noteworthy that an explicit temperature dependence was introduced into the constitutive equation between stress and strain (Eq. (5) in (Misof et al. 1997a, b)), enabling a test of this model at different temperatures, although this has not been done.

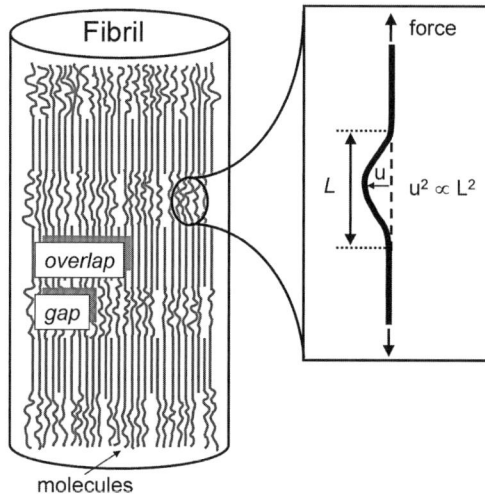

Fig. 7.3 Schematic of the lateral thermally wiggling of collagen molecules in the "gap" zones, which can be modeled as described in Misof et al. (1997a, b). Sketch on the right follows Misof et al. (1997a, b)

When the kinks have been removed and the fibrillar structure straightened out in the linear elastic regime, the strain in the fibrils was observed to be consistently less than the tissue strain (Fratzl et al. 1998), consistent with the work of Folkhard et al. (1987a, b) and implying interfibrillar sliding and shear in the proteoglycans-rich matrix. As the strain rate increased, the fraction of the strain taken up in the fibrils increased, consistent with the viscoelastic behavior of a 2-phase system, consisting of the (mostly elastic) fibrils and (mostly viscous) proteoglycan-rich matrix (Puxkandl et al. 2002). For large strains of greater than 5–7%, the intensity of the meridional reflections decreased, implying that the degree of axial ordering was reduced, perhaps due to a progressively smoothing out of the sharp gap/overlap interface due to axial disorder in the collagen packing (Fratzl et al. 1998) (Fig. 7.4).

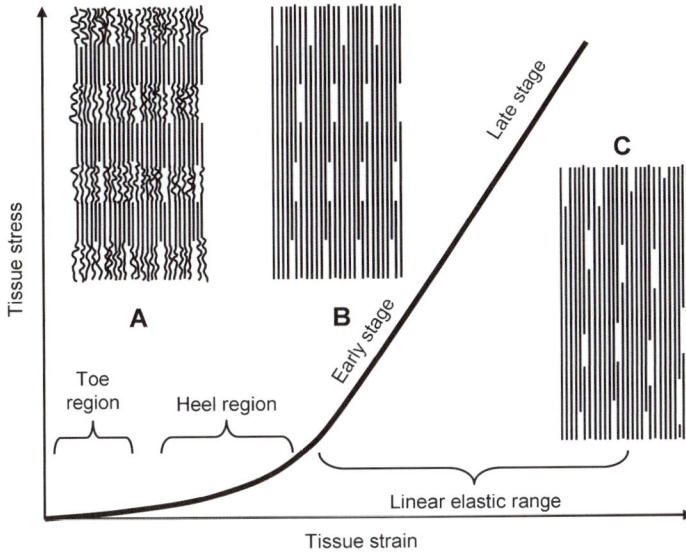

Fig. 7.4 Following Fratzl et al. (1998), the changes at the fibrillar level with applied tensile stress in tendon collagen. Stress/strain curve shows an upward curvature in the "heel" region, where molecular kinks in the gap region are progressively straightened out. For larger strains, the stress–strain relationship is linear, with stretching of the triple helices and gliding of the molecules. The interface between the gap and overlap zones becomes progressively less well-defined with increasing strain

By comparing the fibrillar deformation in tendons with normal cross-linking and in tendons where cross-linking was inhibited due to a diet of β-aminoproprionitrile fed to the test animals, Puxkandl et al. (2002) showed that the intermolecular cross-links are vital for the normal mechanical deformation of tendon collagen. Their experiments showed that cross-link-deficient collagen has a much reduced maximum tensile strength (10–20% that of normal collagen) as well as, in many cases, a long plateau region where the stress remains constant and the strain increases continuously. Such behavior is typical of very young or embryonal collagen from rats, where cross-links do not have the time to form, as shown in a comparison curve of mechanical properties by Bailey (2001).

The lack of intermolecular cross-links manifests itself in the mechanical response of the fibrils as well, as can be directly seen from in situ tensile experiments. In contrast to normal collagen, the strain ratio $\varepsilon_F/\varepsilon_T$ decreases with increasing strain rate. This reversal of the normal behavior can be easily explained when the fibrils and proteoglycan-rich matrix are modeled as two viscoelastic systems in series. While in normal collagen, the fibril component has a high elasticity (storage modulus) and low viscosity (loss modulus), inhibition of cross-linking between molecules in the fibril reverses the situation – in the cross-link-deficient case, the fibril is more viscous than the matrix (Fig. 7.5). A simple calculation shows that the effect of increasing strain rate in this case is to reduce the effective strain fraction carried by the fibrils, which is precisely what is observed in experiments (Puxkandl et al. 2002).

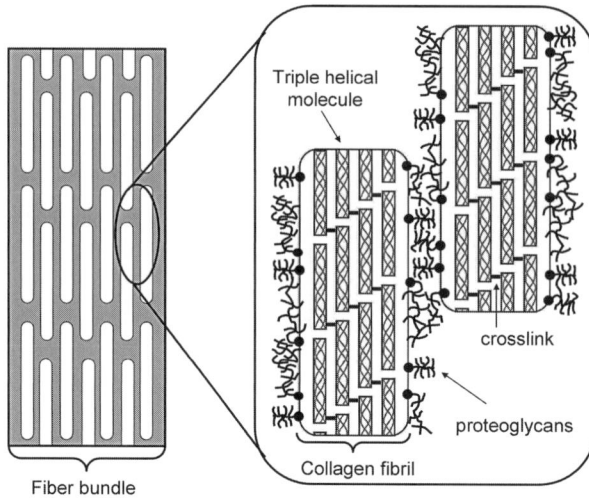

Fig. 7.5 Collagen fibrils and proteoglycan-rich matrix, showing the thin fibrils separated by viscous matrix. Tensile strains in the collagen fibrils are transmitted via viscous shearing in the hydrated matrix (which is negatively charged). The right hand inset shows the structure in more detail, with covalent cross-links (horizontal lines) between collagen molecules. Proteoglycan molecules bind specifically to the collagen fibrils and are entangled with each other in the matrix (Figure after (Puxkandl et al. 2002))

Relatively little is known about the processes involved in fibril and tendon failure, apart from the work by Knorzer referenced above (Knorzer et al. 1986), and some investigations by P. Purslow and coworkers. While unexplored experimentally, several suggestions can be put forward based on our knowledge of fibrillar and microfibrillar structures. Close to fracture, the fibrils could break up into microfibrils, splitting laterally into the basic units. There could be disruption within the microfibrils themselves, resulting in bond disruption in the collagen network.

The majority of work on fibrillar and molecular deformation has been carried out on rat tail tendon, due to its ease of use, ready availability and simple geometry lending itself to ready interpretation in x-ray diffraction and scattering experiments. The mechanical integrity and functioning of numerous other uncalcified connective tissues like skin, cartilage and ligament, which contain collagen as a principal component, are expected to depend on the fibrillar microstructure and deformation. Relatively little is known about the fibrillar deformation mechanisms in such tissues. For articular cartilage, numerous groups working in the field of biomechanical engineering have attempted to correlate macroscopic compression properties with deformation at the fibril–proteoglycan gel level (for typical examples, see (Li et al. 1999, 2000)), but relatively little is known quantitatively about the actual fibrillar deformation, as measured, for example, with in situ x-ray techniques. The relative mechanical importance of the negatively charged proteoglycan gel is larger in cartilage than in tendon, and poroelastic models which consider fluid flow in the open fibril–gel network have been found to give good agreement with experiment

(Li et al. 2000). Purslow and coworkers investigated the stress relaxation in rat skin collagen and bovine intramuscular tissue at the fibrillar level, using in situ x-ray diffraction with synchrotron radiation (Purslow et al. 1998). They find that, in contrast to tendon, the viscoelastic processes associated with creep cannot be related to changes in fibrillar ultrastructure. Neither fibril D-period nor the degree of angular orientation of fibrils was found to change significantly over the time course of stress relaxation or creep. However, the authors note that their applied stresses of 0.01–0.05 MPa are orders of magnitude less than those applied by Mosler et al. (1985a, b, c), who observed fibril D-period reduction during stress relaxation, which could perhaps explain the observed constant fibrillar ultrastructure.

7.4 Mineralized Collagen Deformation

In contrast to noncalcified tissues other than tendon, the fibrillar deformation and mechanics of the collagen in bone and related mineralized tissues has been the focus of extensive study, partially due to the growing socioeconomic importance of the treatment of bone diseases like osteoporosis. In bone, the Type I collagen fibrils are laid down by bone-forming cells (osteoblasts), in initially unmineralized osteoid packets. During an initial fast (primary) and later slow (secondary) mineralization process, poorly crystalline calcium hydroxyapatite particles deposit within (Landis et al. 1993, 1996a, b) and on the surface of (Landis et al. 1996a, b) the collagen fibrils, leading eventually to a fully mineralized organic/inorganic composite which nonetheless shows residual traces of the underlying D-periodicity when viewed at the ultrastructural level (Hassenkam et al. 2004). As expected from the dramatically different moduli of collagen (2–4 GPa) versus apatite ($\sim$ 100 GPa), this process of biomineralization is accompanied by an order of magnitude increase in the stiffness of the collagen/mineral composite, but without a significant reduction in toughness (Wainwright et al. 1982) (Fig. 7.6).

The deformation of the collagen/mineral composite (mineralized collagen fibril) is expected to play a crucial role in determining the material properties of the bone organ as a whole (Fratzl et al. 2004a, b). Progressive decalcification of bone was found to change the effective modulus and yield strength, but leave the slope of the post-yield curve unaffected (Reilly and Burstein 1975). Wang et al. showed that the mechanical properties like strength and work of fracture decreased with age in human femora taken from patients of ages 19–89 years (Wang et al. 2002). Simultaneously, an increase in nonenzymatically induced cross-links was also observed. The shrinkage properties of collagen were also found to correlate with bone toughness, and decreased with age (Zioupos et al. 1999). Changes in collagen structure have been implicated in increased fracture risk, osteoporosis and other metabolic bone diseases (Langdahl et al. 1998, Mann et al. 2001, Bernad et al. 2002, Misof et al. 1997a, b, Oxlund et al. 1995, Banse et al. 2002, Bailey et al. 1992, Kowitz et al. 1997, Spotila et al. 1991)

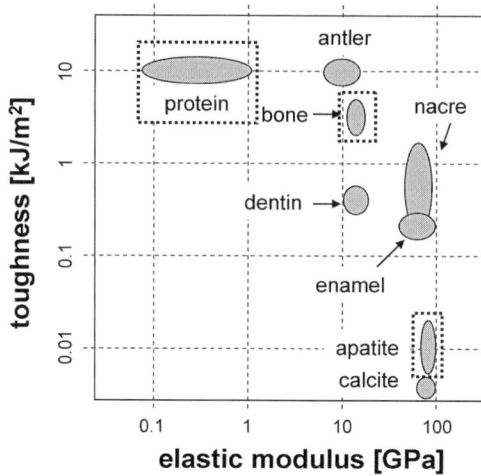

Fig. 7.6 An Ashby plot of the complementary mechanical properties of toughness (vertical axis) and stiffness (horizontal axis) (Fratzl et al. 2004a, b). Note that mineralized tissues like bone achieve both a high stiffness as well as a high toughness

Models for the deformation of bone matrix have a long history, from the work of Katz (1980) who attempted to model the tissue level of bone as consisting of ductile fibers (osteons) embedded in a more highly mineralized collagen matrix (interstitial bone). The mineral particles were proposed to play a role in "straight-jacketing" and pre-stressing the collagen fibrils (Mccutchen 1975), which would also be a method to effectively stiffen the collagen fibril itself, due to its S-shaped stress/strain curve. H. D. Wagner and coworkers modeled the anisotropic stiffness of mineralized fibrils by considering a parallel stacking of mineral platelets in a collagen matrix at the fibril level (Akiva et al. 1998). Scanning electron microscopic observations of the plywood arrangement in bone lamellae (Weiner et al. 1999) were used to determine the angular distribution of such orthotropic-mineralized fibrils in compact bone. To account for the high strength (breaking stress) of bone as well as its high stiffness, a model of the collagen fibril where mineral platelets are arranged in a staggered manner inside a collagen fibril was proposed from theoretical considerations (Jager and Fratzl 2000). However, experimental investigation of these deformation mechanisms has made significant progress only in the last 5 years or so (Gupta et al. 2005, 2006a,b, Nalla et al. 2003, Fantner et al. 2005, Tai et al. 2006), due to the application of high-resolution structural and mechanical characterization techniques.

In situ tensile testing of fibrolamellar bovine bone showed that the mineralized fibrils, in tension, exhibit a strain which is less than, and on average half of, the total tissue strain in bone tissue (Gupta et al. 2005). When extended to the internal structure of the collagen fibrils, by measuring the strain in the embedded mineral particles (Gupta et al. 2006a, b), the strain was found to decrease in a ratio of 12:5:2 from the macro- to the microscale (Fig. 7.7). This hierarchical deformation arises directly from the hierarchical structure of bone tissue, and the collagen plays a crucial role. Indeed, the intra-fibrillar deformation is thus shown to be similar to the

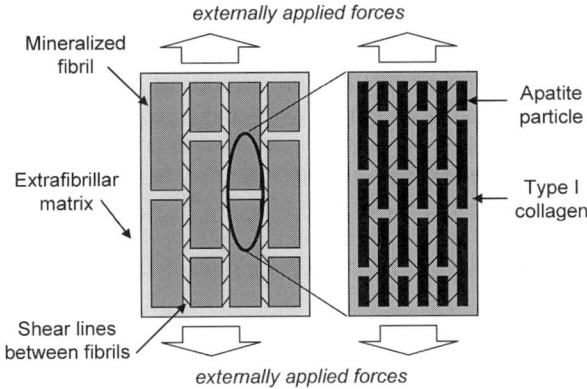

Fig. 7.7 Cooperative deformation in mineralized bone collagen. Inside the fibrils, the collagen molecules are separated by anisotropic mineral apatite particles, and strain is taken up in shear in the softer collagen matrix. The mineralized fibril itself behaves as a stiff element as in Figure 7.5, with an amorphous, partially mineralized interfibrillar matrix between the fibrils

"staggered model" (Jager and Fratzl 2000), since the reduced strain in the mineral particles relative to the overall fibril (5:2) means that the remaining three-fifths of the strain is being transferred by shearing of the ductile collagen matrix. The degree of hydration of the collagen matrix influences the amount of load on the mineral particles: dry collagen strains less and hence the brittle mineral phase bears relatively more load (Fig. 7.8). Highlighting the importance of the organic matrix in the strength and stiffness of bone (Fratzl et al. 2004a, b, Landis 1995), this result suggests that in diseases like osteogenesis imperfecta, the alteration of the collagen

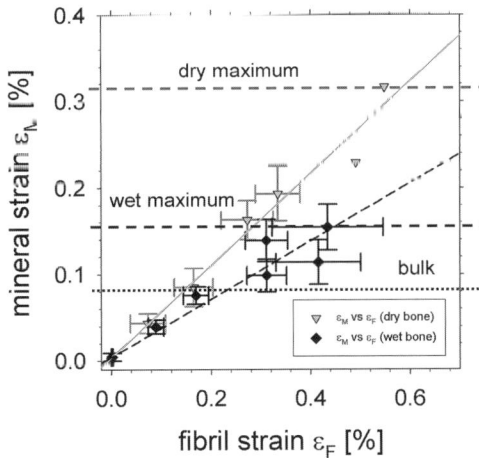

Fig. 7.8 Strain in the brittle mineral phase versus strain in mineralized fibrils, for wet and dry calcified collagen tissue (Gupta et al. 2006a, b). The remainder of the strain inside the fibrils is taken up by shearing in the collagen matrix. Note the significantly higher values of mineral (*brittle phase*) strain in the dry state. Horizontal dashed lines show the (*low*) fracture strains of bulk apatite (Ruys et al. 1995)

matrix could directly affect the load borne by the more brittle mineral phase, and make it easier to break (Gupta et al. 2006a, b). Compressive loading of bone reveals stress concentrations in the mineral phase (Almer and Stock 2005, Borsato and Sasaki 1997), and it has been proposed that local compressive loading by nanoindentation sets up a high heterogeneous field of deformation inside fibrils, which could be responsible for the large energy dissipative capabilities of bone (Tai et al. 2006).

The origin of the lower fibril strain relative to the overall tissue strain may be related to the observation of an organic "glue" between the fibrils in bone (Fantner et al. 2005, 2006) as well as in other mineralized tissues like nacre (Smith et al. 1999). High-resolution scanning electron microscopy of fractured bone surfaces, as well as atomic force microscopy single molecule pulling experiments, has suggested that an amorphous, noncollagenous protein or polysaccharide exists between the mineralized fibrils, whose initially crumpled or rolled up structure unwinds gradually under the application of external forces, exposing "hidden length" and dissipating energy (Fantner et al. 2005). Model calculations and simulations show that the energy scale of these bonds is of the order of 1 eV (Fantner et al. 2006). Independently, thermal activation analysis of the macroscopic plastic properties of bone has experimentally shown that the energy scale associated with bone deformation is $\sim$ 1 eV and the fundamental deformation step takes place in a very small volume, less than 1 nm^3 (Gupta et al. 2007). These results supported the idea of ionic-mediated breaking of bonds between polyelectrolyte molecules (the noncollagenous component) in the extrafibrillar matrix as an important basic step in the fracture of bone. Hence, the initial picture which emerges is that while the collagenous fibrils are important for the stiffness of bone, the mechanics and structure of a small fraction (1–5%) of noncollagenous proteins is even more crucial to the fracture of bone (Fig. 7.9).

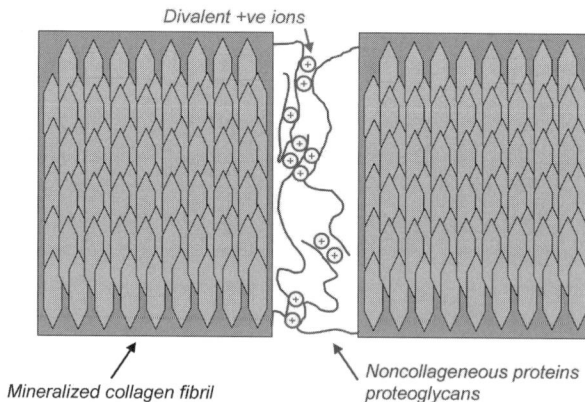

Fig. 7.9 Fracture of mineralized collagen fibrils may occur via the breaking of negatively charged polyelectrolyte molecules (proteoglycans or noncollagenous proteins) between the fibrils. Evidence from thermal activation analysis suggests that the microscopic "activation volume" is of the order of 1 nm^3, and the enthalpy of the order of 1 eV (Gupta et al. 2007)

Due to the remarkable capacity of the mineralized fibrils to aggregate into bio-logically diverse motifs depending on their functional requirements (Weiner and Wagner 1998), the in vivo deformation of fibrils in bone depend to a large extent on the organ and tissue location in which they are found. In dentine, for example, the fibrils are expected to be oriented perpendicular to the stress direction and resist compressive loads. In the osteons, which are cylindrical aggregates of bone lamellae around blood vessels (Weiner et al. 1999, Giraudguille 1988), the twisted plywood motif results in a spiraling of the fiber orientation around the central blood vessel, analogous to a spring construction (Wagermaier et al. 2006). Such a construction may have biophysical importance in absorbing mechanical energy and protecting the sensitive inner blood vessel from cracks in the interstitial bone (Gupta et al. 2006a, b). It has been suggested that the fibrils in bone lamellae adapt their orientation relative to the central Haversian canal depending on the nature of the in vivo forces to which they are subject at that location (Barbos et al. 1984, Goldman et al. 2003).

7.5 Conclusion

The conclusion from this survey of our current understanding of the nanoscale deformation mechanisms in collagen-containing tissues emphasizes several com-mon themes that are observed:

In fibrous composites, shearing in a soft interfibrillar matrix can play a crucial role in transmitting the load between fibers. Long thin fibrils enable even relatively weak, nonspecific bonds between the matrix and the fibrils to give rise to an effec-tively large adhesion. However, the actual situation is more complex, with some of the proteoglycans in tendon bound specifically to the fibril surface and the others perhaps only indirectly interacting with the fibril via steric interactions with the bound proteoglycans.

Hierarchical structuring leads to smaller and smaller units of strain being passed down to the lower levels, as seen in the studies of tissue, fibril and collagen helix deformation in tendon and the analogous study in mineralized bone collagen.

Alterations in the nature of the organic matrix, whether by changing interfib-rillar cross-links as in the beta-aminoproprionitrile study or by altering the state of hydration of bone collagen, can significantly alter the microscopic load balance in the tissue. In addition to changing the mechanosensory environment locally for cells to respond to, this microstructural alteration manifests itself directly in the macroscopic mechanical properties like stiffness, breaking stress, yield stress and failure strain.

Local alterations in small components of the tissue, for example in the organic "glue" between the fibrils in bone, could play a significant role in changing the mechanical properties of the entire organ.

The application of in situ techniques to the study of deformation mechanisms in collagen-containing tissues has revealed a wealth of quantitative information on the behavior of molecules, fibrils and their aggregates under external mechanical stress.

Extension of the techniques to include the wide range of tissues (e.g., cartilage and ligaments) not included in the studies so far would undoubtedly lead to a wide range of further discoveries on the correlation of microscopic form to biological function.

Acknowledgments In addition to the numerous workers cited in this chapter, I would like to specially acknowledge Wolfgang Wagermaier, Peter Fratzl and Hazel R. C. Screen for valuable discussions on calcified and uncalcified collagen mechanics at the fibrillar level.

References

Akiva U, Wagner H D and Weiner S (1998) Modelling the three-dimensional elastic constants of parallel-fibred and lamellar bone. Journal of Materials Science 33:1497–1509

Almer J D and Stock S R (2005) Internal strains and stresses measured in cortical bone via high-energy X-ray diffraction. Journal of Structural Biology 152:14–27

Bailey A J (2001) Molecular mechanisms of ageing in connective tissues. Mechanisms of Ageing and Development 122:735–755

Bailey A J, Wotton S F, Sims T J and Thompson P W (1992) Posttranslational modifications in the collagen of human osteoporotic femoral-head. Biochemical and Biophysical Research Communications 185:801–805

Banse X, Sims T J and Bailey A J (2002) Mechanical properties of adult vertebral cancellous bone: correlation with collagen intermolecular cross-links. Journal of Bone & Mineral Research 17:1621–1628

Barbos M P, Bianco P, Ascenzi A and Boyde A (1984) Collagen orientation in compact-bone.2. distribution of lamellae in the whole of the human femoral-shaft with reference to its mechanical-properties. Metabolic Bone Disease & Related Research 5:309–315

Barenberg S A, Filisko F E and Geil P H (1978) Ultrastructural Deformation of Collagen. Connective Tissue Research 6:25–35

Bernad M, Martinez M E, Escalona M, Gonzalez M L, Gonzalez C, Garces M V, Del Campo M T, Martin Mola E, Madero R and Carreno L (2002) Polymorphism in the type I collagen (COLIA1) gene and risk of fractures in postmenopausal women. Bone 30:223–228

Borsato K S and Sasaki N (1997) Measurement of partition of stress between mineral and collagen phases in bone using X-ray diffraction techniques. Journal of Biomechanics 30:955–957

Cribb A M and Scott J E (1995) Tendon response to tensile-stress - an ultrastructural investigation of collagen – proteoglycan interactions in stressed tendon. Journal of Anatomy 187:423–428

Dale W C and Baer E (1974) Fiber-buckling in composite systems – model for ultrastructure of uncalcified collagen tissues. Journal of Materials Science 9:369–382

Fantner, Hassenkam T, Kindt J H, Weaver J C, Birkedal H, Pechenik L, Cutroni J A, Cidade G A G, Stucky G D, Morse D E and Hansma P K (2005) Sacrificial bonds and hidden length dissipate energy as mineralized fibrils separate during bone fracture. Nature Materials 4:612–616

Fantner G E, Oroudjev E, Schitter G, Golde L S, Thurner P, Finch M M, Turner P, Gutsmann T, Morse D E, Hansma H and Hansma P K (2006) Sacrificial bonds and hidden length: Unraveling molecular mesostructures in tough materials. Biophysical Journal 90:1411–1418

Folkhard W, Geercken W, Knorzer E, Mosler E, Nemetschekgansler H, Nemetschek T and Koch M H J (1987a) Structural dynamic of native tendon collagen. Journal of Molecular Biology 193:405–407

Folkhard W, Mosler E, Geercken W, Knorzer E, Nemetschekgansler H, Nemetschek T and Koch M H J (1987b) Quantitative-analysis of the molecular sliding mechanism in native tendon collagen – time-resolved dynamic studies using synchrotron radiation. International Journal of Biological Macromolecules 9:169–175

Fratzl P, Burgert I and Gupta H S (2004a) On the role of interface polymers for the mechanics of natural polymeric composites. Physical Chemistry Chemical Physics 6:5575–5579

Fratzl P, Fratzl-Zelman N and Klaushofer K (1993) Collagen packing and mineralization – an x-ray-scattering investigation of turkey leg tendon. Biophysical Journal 64:260–266

Fratzl P, Gupta H S, Paschalis E P and Roschger P (2004b) Structure and mechanical quality of the collagen-mineral nano-composite in bone. Journal of Materials Chemistry 14:2115–2123

Fratzl P, Misof K, Zizak I, Rapp G, Amenitsch H and Bernstorff S (1998) Fibrillar structure and mechanical properties of collagen. Journal of Structural Biology 122:119–122

Giraudguille M M (1988) Twisted plywood architecture of collagen fibrils in human compact-bone osteons. Calcified Tissue International 42:167–180

Goldman H M, Bromage T G, Thomas C D L and Clement J G (2003) Preferred collagen fiber orientation iin the human mid-shaft femur. Anatomical Record Part a-Discoveries in Molecular Cellular and Evolutionary Biology 272A:434–445

Gupta H S, Fratzl P, Kerschnitzki M, Benecke G, Wagermaier W and Kirchner H O K (2007) Evidence for an elementary process in bone plasticity with an activation enthalpy of 1 eV. Journal of the Royal Society Interface 4:277–282

Gupta H S, Seto J, Wagermaier W, Zaslansky P, Boesecke P and Fratzl P (2006a) Cooperative deformation of mineral and collagen in bone at the nanoscale. Proceedings of the National Academy of Sciences of the United States of America 103:17741–17746

Gupta H S, Stachewicz U, Wagermaier W, Roschger P, Wagner H D and Fratzl P (2006a) Mechanical modulation at the lamellar level in osteonal bone. Journal of Materials Research 21: 1913–1921

Gupta H S, Wagermaier W, Zickler G A, Aroush D R B, Funari S S, Roschger P, Wagner H D and Fratzl P (2005) Nanoscale deformation mechanisms in bone. Nano Letters 5:2108–2111

Hassenkam T, Fantner G E, Cutroni J A, Weaver J C, Morse D E and Hansma P K (2004) High-resolution AFM imaging of intact and fractured trabecular bone. Bone 35:4–10

Hosemann R, Bonart R and Nemetschek T (1974) Inhomogeneous stretching process of collagen. Colloid and Polymer Science 252:912–919

Jager I and Fratzl P (2000) Mineralized collagen fibrils: A mechanical model with a staggered arrangement of mineral particles. Biophysical Journal 79:1737–1746

Kastelic J, Galeski A and Baer E (1978) Multicomposite structure of tendon. Connective Tissue Research 6:11–23

Katz J L (1980) Anisotropy of Youngs modulus of bone. Nature 283:106–107

Knorzer E, Folkhard W, Geercken W, Boschert C, Koch M H J, Hilbert B, Krahl H, Mosler E, Nemetschekgansler H and Nemetschek T (1986) New aspects of the etiology of tendon-rupture – an analysis of time-resolved dynamic-mechanical measurements using synchrotron radiation. Archives of Orthopaedic and Trauma Surgery 105:113–120

Kowitz J, Knippel M, Schuhr T and Mach J (1997) Alteration in the extent of collagen 1 hydroxylation, isolated from femoral heads of women with a femoral neck fracture caused by osteoporosis. Calcified Tissue International 60:501–5

Landis W J (1995) The strength of a calcified tissue depends in part on the molecular structure and organization of its constituent mineral crystals in their organic matrix. Bone 16:533–44

Landis W J, Hodgens K J, Arena J, Song M J and McEwen B F (1996) Structural relations between collagen and mineral in bone as determined by high voltage electron microscopic tomography. Microscopy Research and Technique 33:192–202

Landis W J, Hodgens K J, Song M J, Arena J, Kiyonaga S, Marko M, Owen C and McEwen B F (1996) Mineralization of collagen may occur on fibril surfaces: evidence from conventional and high-voltage electron microscopy and three-dimensional imaging. J Struct Biol 117:24

Landis W J, Song M J, Leith A, Mcewen L and Mcewen B F (1993) Mineral and organic matrix interaction in normally calcifying tendon visualized in 3 dimensions by high-voltage electron-microscopic tomography and graphic image-reconstruction. Journal of Structural Biology 110:39–54

Langdahl B L, Ralston S H, Grant S F and Eriksen E F (1998) An Sp1 binding site polymorphism in the COLIA1 gene predicts osteoporotic fractures in both men and women. Journal of Bone & Mineral Research 13:1384–1389

Li L P, Buschmann M D and Shirazi-Adl A (2000) A fibril reinforced nonhomogeneous poroelastic model for articular cartilage: inhomogeneous response in unconfined compression. Journal of Biomechanics 33:1533–1541

Li L P, Soulhat J, Buschmann M D and Shirazi-Adl A (1999) Nonlinear analysis of cartilage in unconfined ramp compression using a fibril reinforced poroelastic model. Clinical Biomechanics 14:673–682

Mann V, Hobson E E, Li B, Stewart T L, Grant S F, Robins S P, Aspden R M and Ralston S H (2001) A COL1A1 Sp1 binding site polymorphism predisposes to osteoporotic fracture by affecting bone density and quality. Journal of Clinical Investigation 107:899–907

Mccutchen C W (1975) Do Mineral Crystals Stiffen Bone by Straitjacketing Its Collagen. Journal of Theoretical Biology 51:51–58

Misof K, Landis W J, Klaushofer K and Fratzl P (1997a) Collagen from the osteogenesis imperfecta mouse model (oim) shows reduced resistance against tensile stress. Journal of Clinical Investigation 100:40–45

Misof K, Rapp G and Fratzl P (1997b) A new molecular model for collagen elasticity based on synchrotron x-ray scattering evidence. Biophysical Journal 72:1376–1381

Mosler E, Folkhard W, Knorzer E, Nemetschekgansler H, Koch M H J and Nemetschek T (1985a) Localization of stress-induced molecular-rearrangements in collagen. Colloid and Polymer Science 263:87–88

Mosler E, Folkhard W, Knorzer E, Nemetschekgansler H, Nemetschek T and Koch M H J (1985b) Stress-induced molecular rearrangement in tendon collagen. Journal of Molecular Biology 182:589–596

Nalla R K, Kinney J H and Ritchie R O (2003) Mechanistic fracture criteria for the failure of human cortical bone. Nature Materials 2:164–168

Orgel J P R O, Irving T C, Miller A and Wess T J (2006) Microfibrillar structure of type I collagen in situ. Proceedings of the National Academy of Sciences of the United States of America 103:9001–9005

Oxlund H, Barckman M, Ortoft G and Andreassen T T (1995) Reduced concentrations of collagen cross-links are associated with reduced strength of bone. Bone 17:365S–371S

Purslow P P, Wess T J and Hukins D W L (1998) Collagen orientation and molecular spacing during creep and stress-relaxation in soft connective tissues. Journal of Experimental Biology 201:135–142

Puxkandl R, Zizak I, Paris O, Keckes J, Tesch W, Bernstorff S, Purslow P and Fratzl P (2002) Viscoelastic properties of collagen: synchrotron radiation investigations and structural model. Philosophical Transactions of the Royal Society of London Series B-Biological Sciences 357:191–197

Raspanti M, Alessandrini A, Ottani V and Ruggeri A (1997) Direct visualization of collagen-bound proteoglycans by tapping-mode atomic force microscopy. Journal of Structural Biology 119:118–122

Redaelli A, Vesentini S, Soncini M, Vena P, Mantero S and Montevecchi F M (2003) Possible role of decorin glycosaminoglycans in fibril to fibril force transfer in relative mature tendons – a computational study from molecular to microstructural level. Journal of Biomechanics 36:1555–1569

Reilly D T and Burstein A H (1975) The elastic and ultimate properties of compact bone tissue. Journal of Biomechanics 8:393–405

Ruys A J, Wei M, Sorrell C C, Dickson M R, Brandwood A and Milthorpe B K (1995) Sintering effects on the strength of hydroxyapatite. Biomaterials 16:409–415

Sasaki N and Odajima S (1996a) Elongation mechanism of collagen fibrils and force-strain relations of tendon at each level of structural hierarchy. Journal of Biomechanics 29: 1131–1136

Sasaki N and Odajima S (1996b) Stress-strain curve and Young's modulus of a collagen molecule as determined by the X-ray diffraction technique. Journal of Biomechanics 29:655–658

Scott J E (1991) Proteoglycan – Collagen interactions in connective tissues – ultrastructural, biochemical, functional and evolutionary aspects. International Journal of Biological Macromolecules 13:157–161

Scott J E (1992) Supramolecular organization of extracellular-matrix glycosaminoglycans, invitro and in the tissues. FASEB Journal 6:2639–2645

Scott J E (2003) Elasticity in extracellular matrix 'shape modules' of tendon, cartilage, etc. A sliding proteoglycan-filament model. Journal of Physiology-London 553:335–343

Screen H R C, Bader D L, Lee D A and Shelton J C (2004a) Local strain measurement within tendon. Strain 40:157–163

Screen H R C, Chhaya V H, Greenwald S E, Bader D L, Lee D A and Shelton J C (2006) The influence of swelling and matrix degradation on the microstructural integrity of tendon. Acta Biomaterialia 2:505–513

Screen H R C, Lee D A, Bader D L and Shelton J C (2003) Development of a technique to determine strains in tendons using the cell nuclei. Biorheology 40:361–368

Screen H R C, Lee D A, Bader D L and Shelton J C (2004a) An investigation into the effects of the hierarchical structure of tendon fascicles on micromechanical properties. Proceedings of the Institution of Mechanical Engineers Part H-Journal of Engineering in Medicine 218:109–119

Screen H R C, Shelton J C, Bader D L and Lee D A (2005a) Cyclic tensile strain upregulates collagen synthesis in isolated tendon fascicles. Biochemical and Biophysical Research Communications 336:424–429

Screen H R C, Shelton J C, Bader D L and Lee D A (2005b) Cyclic tensile strain upregulates collagen production in isolated tendon fascicles. International Journal of Experimental Pathology 86:A8-a9

Screen H R C, Shelton J C, Chhaya V H, Kayser M V, Bader D L and Lee D A (2005c) The influence of noncollagenous matrix components on the micromechanical environment of tendon fascicles. Annals of Biomedical Engineering 33:1090–1099

Smith B L, Schaffer T E, Viani M, Thompson J B, Frederick N A, Kindt J, Belcher A, Stucky G D, Morse D E and Hansma P K (1999) Molecular mechanistic origin of the toughness of natural adhesives, fibres and composites. Nature 399:761–763

Spotila L D, Constantinou C D, Sereda L, Ganguly A, Riggs B L and Prockop D J (1991) Mutation in a Gene for Type-I procollagen (Col1a2) in a woman with postmenopausal osteoporosis – evidence for phenotypic and genotypic overlap with mild osteogenesis imperfecta. Proceedings of the National Academy of Sciences of the United States of America 88:5423–5427

Tai K, Ulm F J and Ortiz C (2006) Nanogranular origins of the strength of bone. Nano Letters 6:2520–2525

Wagermaier W, Gupta H S, Gourrier A, Burghammer M, Roschger P and Fratzl P (2006) Spiral twisting of fiber orientation inside bone lamellae. Biointerphases 1:1 5

Wainwright S A, Biggs W D, Currey J D and Gosline J M (1982) Mechanical design in organisms. Princeton University Press, Princeton, New Jersey

Wang X, Shen X, Li X and Agrawal C M (2002) Age-related changes in the collagen network and toughness of bone. Bone 31:1–9

Weiner S, Traub W and Wagner H D (1999) Lamellar bone: Structure-function relations. Journal of Structural Biology 126:241–255

Weiner S and Wagner H D (1998) The material bone: Structure mechanical function relations. Annual Review of Materials Science 28:271–298

Wess T J, Hammersley A P, Wess L and Miller A (1998) A consensus model for molecular packing of type I collagen. Journal of Structural Biology 122:92–100

Yanagishita M (1993) Function of proteoglycans in the extracellular-matrix. Acta Pathologica Japonica 43:283–293

Zioupos P, Currey J D and Hamer A J (1999) The role of collagen in the declining mechanical properties of aging human cortical bone. Journal of Biomedical Materials Research 45:108–116

Chapter 8
Hierarchical Nanomechanics of Collagen Fibrils: Atomistic and Molecular Modeling

M.J. Buehler

Abstract This chapter describes hierarchical multi-scale modeling of collagenous tissues, with a particular focus on the mechanical properties. Studies focus on elastic behavior, plastic behavior and fracture. Starting at the atomistic scale, we review development and application of a hierarchical multi-scale model that is capable of describing the dynamical behavior of a large number of tropocollagen molecules, reaching length scales of several micrometers and time scales of tens of microseconds. Particular emphasis is on elucidating the deformation mechanisms that operate at various scales, the scale-dependent properties, the effect of specific hierarchical features and length scales (cross-link densities, intermolecular adhesion, etc.) as well as on the effect of addition of mineral platelets during formation of nascent bone. This chapter contains a review of numerical techniques associated with modeling of chemically complex and hierarchical biological tissue, including first principles-based reactive force fields, empirical force fields, large-scale parallelization and visualization methods. A set of scaling relationships are summarized that enable one to predict deformation mechanisms and properties based on atomistic, molecular and other hierarchical features. The results are presented in deformation maps that summarize deformation modes, strength, dissipative properties and elastic behavior for various conditions, providing structure property relationships for collagenous tissue. This chapter is concluded with a discussion of how insight of nanomechanical behavior at the smallest scales relates with the physiological role of collagen. The significance of universal structural patterns such as the staggered collagen fibril architecture versus specific structures in different collagen tissues is reviewed in light of the question of universality versus diversity of structural components.

8.1 Introduction

Proteins are the fundamental building blocks of a vast array of biological materials that are involved in critical functions of life, many of which are based on highly characteristic nanostructured arrangements of protein components that include tropocollagen (here sometimes also abbreviated as "TC") molecules, alpha-helices or beta-sheets. Bone, providing structure to our body, or spider silk, used for prey

P. Fratzl (ed.), *Collagen: Structure and Mechanics*,
© Springer Science+Business Media, LLC 2008

procurement, are examples of materials that have incredible elasticity, strength and robustness unmatched by many synthetic materials, mainly attributed to its structural formation with molecular precision.

Collagen, the most abundant protein on earth, is a fibrous structural protein with superior mechanical properties, and provides an intriguing example of a hierarchical biological nanomaterial (Borel and Monboisse 1993; Hellmich and Ulm 2002; Puxkandl et al. 2002; Bhattacharjee and Bansal 2005; Bozec et al. 2005; Bozec and Horton 2005; Fratzl and Weinkamer 2007). The hierarchical structure of collagen is summarized in Fig. 8.1. Collagen plays an important role in many biological tissues, including tendon, bone, teeth, cartilage or in the eye's cornea focusing on small-scale structural features. Severe mechanical tensile loading of collagen is significant under many physiological conditions, as in joints and in bone. Further, significant mechanical deformation of collagenous tissues may occur during injuries.

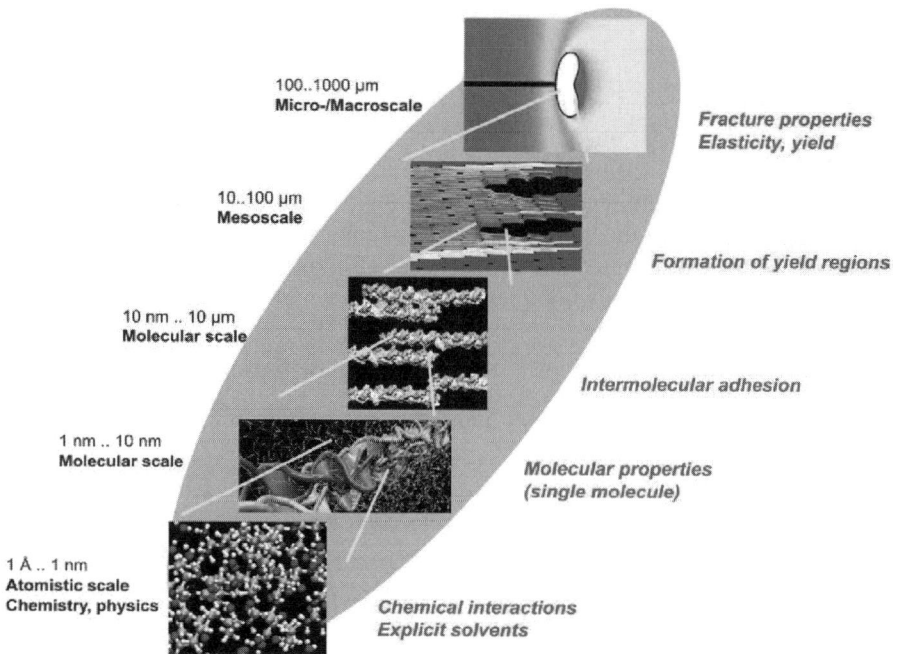

Fig. 8.1 Overview of different material scales, from nano to macro, here exemplified for collagenous tissue (Ramachandran and Kartha 1955; Bhattacharjee and Bansal 2005; Hulmes et al. 1995; Puxkandl et al. 2002; An et al. 2004; Fratzl et al. 2004; Buehler 2006a,b). The macroscopic mechanical material behavior is controlled by the interplay of properties throughout various scales, in particular molecular interaction at the mesoscale. In order to understand deformation and fracture mechanisms, it is crucial to elucidate atomistic and molecular mechanisms at each scale

The goal of this chapter is to review the elastic, plastic and fracture behavior of collagen fibrils, linking atomistic-scale studies with the mesoscopic level of collagen fibrils. We particularly focus on the large deformation behavior of collagen-based tissues, which is particularly important under physiological conditions and during injuries. The studies reviewed here explain the limiting factors in strength of collagen fibrils, as well as the origins of its toughness. These investigations complement experimental efforts focused on the deformation mechanics of collagen fibril at nanoscale, including characterization of changes of D-spacing and fibril orientation (Hulmes et al. 1995), analyses that featured X-ray diffraction and synchrotron radiation experiments (Puxkandl et al. 2002).

8.1.1 Deformation and Fracture: An Introduction

When materials are deformed, they display a small regime in which deformation is reversible or elastic (Broberg 1990; Anderson 1991). Once the forces on the material reach a critical level, deformation becomes irreversible and remains even after the load is removed. This is referred to as the plastic regime (Courtney 1990). Plastic deformation is typically followed by fracture, when the material breaks and fails. Many materials, including metals, ceramics, polymers and biological tissue, show this generic behavior. However, the details of the response to mechanical load depend on the atomic and molecular makeup of the material; from nano to macro (for a review on this topic, please see Buehler and Ackbarow 2007) (see Fig. 8.2).

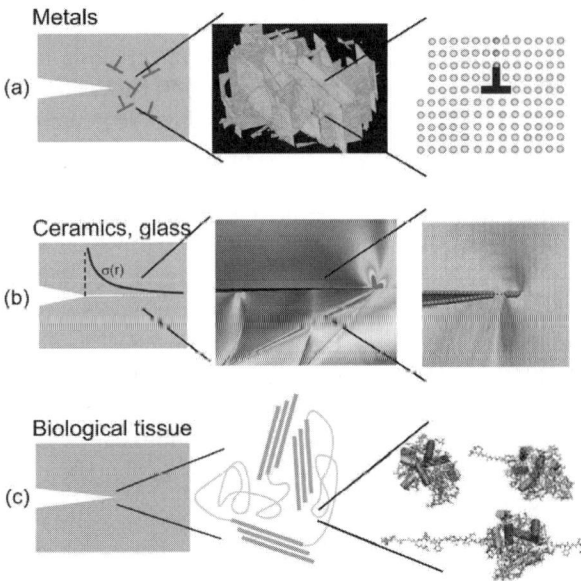

Fig. 8.2 Overview of the deformation and fracture behavior of different classes of materials, including (**a**) ductile materials (Hirth and Lothe 1982), (**b**) brittle materials (Broberg 1990) and (**c**) BPMs (Buehler and Ackbarow 2007). Each subplot shows a multi-scale view of associated deformation mechanisms. In ductile materials, deformation is mediated by creation of dislocation networks; each dislocation represents localized shear of an atomic lattice. In brittle materials, fracture occurs by spreading of cracks, which is mediated by continuous breaking of atomic bonds. In BPMs, a complex interplay of different protein structures controls the mechanical response. At the ultra-scale, unfolding of individual protein molecules by rupture of hydrogen bonds (HBs) represents the most fundamental deformation mechanism

For example, ductile metals such as copper or nickel can rather easily undergo large permanent (or "plastic") deformation without breaking (Hirth and Lothe 1982). On the other hand, brittle materials like glass cannot easily be deformed, but instead fracture rapidly once the applied load exceeds a threshold value (Broberg 1990). Contrarily, biological protein materials (BPMs) such as the cell's cytoskeleton or collagen networks in tendon or bone represent intriguing protein networks that can dynamically adapt to load application by self-organization and self-arrangement; developing stronger filaments when needed and disposing of those that do not contribute to the strength, making the material utilization overall more efficient and robust against failure.

Advancing the understanding of the origin of deformation and fracture has fascinated generations of material scientists. Currently, a major challenge is the elucidation of mechanisms in increasingly complex materials – materials that consist of multiple components or multiple hierarchies, or those whose atomic nanostructures and microstructures contain a concurrent interplay of a variety of chemical bonds. An important concept in understanding the deformation and failure properties of materials are the underlying fundamental atomic mechanisms, as illustrated in Fig. 8.2 for ductile materials, brittle materials and BPMs (Buehler and Ackbarow 2007) (see caption of Fig. 8.2 for an overview of the various deformation modes). While the basic deformation mechanisms of crystalline solids are relatively well understood, analogous mechanisms have only recently been discovered in BPMs. Permanent plastic deformation in these materials is mediated by intermolecular slip, unfolding of proteins (Rief et al. 1997; Lu et al. 1998; Buehler and Ackbarow 2007), breaking of intermolecular cross-links or stretching of convoluted protein chains, as it has for instance been demonstrated in multi-scale studies for collagenous materials and bone.

Due to the multi-scale hierarchical structure of these materials, different deformation mechanisms may occur at each scale, while the inter- and intra-hierarchal interactions might be of competing or of reinforcing character. The most fundamental deformation mode is often, however, breaking of weak bonds, for instance the rupture of individual H-bonds (HBs). Even though they are 100–1,000 times weaker than covalent bonds, HBs are the most important type of chemical bonds that hold together proteins, assemblies of proteins and control their adhesion behavior. An important open question thereby is how weak interactions can be utilized to make up macroscopically strong materials. Up until now, few systematic classifications, nor fundamental theories exist for the different deformation and fracture mechanisms in BPM.

8.1.2 Collagen Structure – From Atoms to Tissue

Collagen consists of tropocollagen molecules that have lengths of $L \approx 300$nm with approximately 1.5 nm in diameter, leading to an aspect ratio of close to 200 (Ramachandran and Kartha 1955; Bhattacharjee and Bansal 2005). Staggered arrays of tropocollagen molecules form fibrils, which arrange to form collagen fibers. A schematic of the main hierarchical features of collagen is shown in Fig. 8.3.

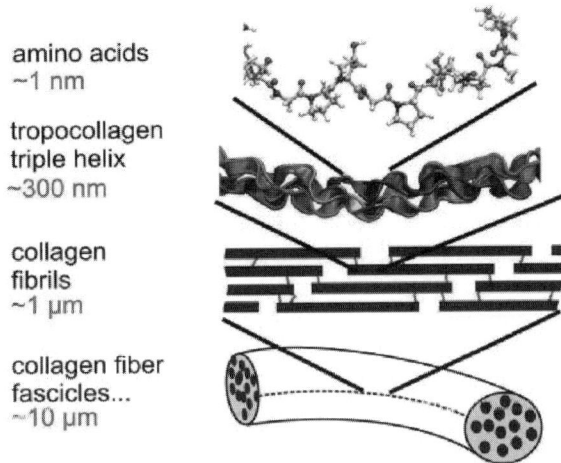

Fig. 8.3 Schematic view of some of the hierarchical features of collagen, ranging from the amino acid sequence level at nanoscale up to the scale of collagen fibers with lengths on the order of 10 μm (Figure adapted from Buehler 2006a,b). The present study is focused on the mechanical properties of collagen fibrils, consisting of a staggered array of TC molecules. The red lines in the graph indicate intermolecular cross-links that are primarily developed at the ends of tropocollagen molecules

amino acids
~1 nm

tropocollagen
triple helix
~300 nm

collagen
fibrils
~1 μm

collagen fiber
fascicles...
~10 μm

Each tropocollagen molecule consists of a spatial arrangement of three polypeptides. These three molecules or polypeptides are arranged in a helical structure, stabilized primarily by H-bonding between different residues. Every third residue in each of these molecules is a GLY amino acid, and about one-fourth of the tropocollagen molecule consists of proline (PRO) and hydroxyproline (HYP). The structure of collagen has been known since classical works focusing on theoretical understanding of how tropocollagen molecules are stabilized (Ramachandran and Kartha 1955; Bhattacharjee and Bansal 2005). Recently, various types of tropocollagen molecules have been crystallized and analyzed using X-ray diffraction techniques to determine their precise atomic configuration (Kramer et al. 2000). TEM experiments have also been used to study the structure of collagen in various environments, including in bone, in particular focusing on larger length scale features and its three-dimensional arrangement (Weiner and Wagner 1998; Currey 2002; Ritchie et al. 2004; Nalla et al. 2005).

Collagen is the most fundamental building block of bone, providing additional evidence for the great significance of collagen. Bone has evolved to provide structural support to organisms, and therefore, its mechanical properties are of great physiological relevance. A total of seven hierarchical levels are found in bone. The smallest scale hierarchical features of bone include the protein phase composed of tropocollagen molecules, collagen fibrils (CFs) as well as mineralized collagen fibrils (MCFs). Tropocollagen molecules assemble into collagen fibrils in a hydrated environment, which mineralizes by formation of hydroxyapatite (HA) crystals in the gap regions that exist due to the staggered geometry (Laudis et al. 2002; Currey 2002; Weiner and Wagner 1998).

8.1.3 Outline of This Chapter

The analysis reported in this chapter is focused on molecular and supermolecular deformation mechanisms as well as prevalent molecular length scales in collagenous tissues. The results help to explain the particular molecular architecture as observed in tendon, bone and the eye's cornea, and provide models that predict how molecular properties influence the deformation and fracture mechanics of tissues. This chapter consists of seven sections that are dedicated to a review of atomistic and molecular simulation techniques, geared toward improving our understanding of the mechanical behavior of collagenous tissue, at the atomistic, molecular and supermolecular scales. It is noted that while this chapter includes a review of the broader field, the focus is on results from our group.

In Section 8.2, we discuss the numerical foundation and theoretical framework of the analysis techniques reviewed in this chapter, in particular atomistic simulation approaches. The subsequent sections systematically discuss various scales of collagen, ranging up to the scale of mineralized collagen fibrils. Section 8.3 is dedicated to atomistic simulations of individual tropocollagen molecules. This section also includes an analysis of the interaction between two tropocollagen molecules. In Section 8.4, the mechanical behavior of collagen fibrils is discussed, including studies of effects of the molecular length and cross-link densities. In Section 8.5, we review studies of mineralized collagen fibrils, forming the fundamental building block of bone. Section 8.6 is dedicated to a broader discussion on structure–function relationships of hierarchical biological materials. We conclude in Section 8.7 with a discussion.

8.2 Numerical Simulation Techniques and Theoretical Framework

In order to develop a fundamental and quantitative understanding of collagen mechanics, theoretical models encompassing the mesoscopic scales between the atomistic and the macroscopic levels, considering atomistic and chemical interactions during deformation, are vital. This represents an alternative strategy capable of predicting the properties of collagen tissue from bottom up.

In order to achieve this goal, a parameter-free atomistic-based model of the mechanical properties of collagen fibrils, based solely on atomistic simulation input data, can be used (Buehler 2006a, b, Buehler 2008).

Materials failure processes begin with the erratic motion of individual atoms around flaws or defects within the material that evolve into formation of macroscopic fractures as chemical bonds rupture rapidly, eventually compromising the integrity of the entire structure. Thus the behavior of chemical bonds under large stretch – on a small scale – controls how structures respond to mechanical load and fail on much larger material scales, as has for instance been demonstrated by our group for model materials and silicon (Buehler et al. 2003; Buehler et al. 2006; Buehler and Gao 2006; Buehler et al. 2007).

Fracture mechanisms in brittle and ductile materials are representative examples for an intrinsic multi-scale problem that cannot be understood by considering one scale alone (see Figs. 8.1 and 8.2). Experimentation, simulation and development of theories therefore must consider a complex interplay of mechanisms at several scales. In particular in hierarchical BPMs, development of a rigorous understanding of deformation depends critically on the elucidation of the deformation mechanism at each scale and on how these mechanisms interact dynamically, across the scales. These examples illustrate the importance of developing fundamental, atomistic or molecular scale models of the behavior of collagenous tissues.

8.2.1 Multi-scale Modeling of Deformation and Failure

Multi-scale modeling is a particularly useful approach to gain insight into complex deformation and fracture phenomena. In order to allow the best resolution at any length and time scale, a set of computational methods is integrated seamlessly, which enables one to bridge scales from nano to macro (see Fig. 8.4).

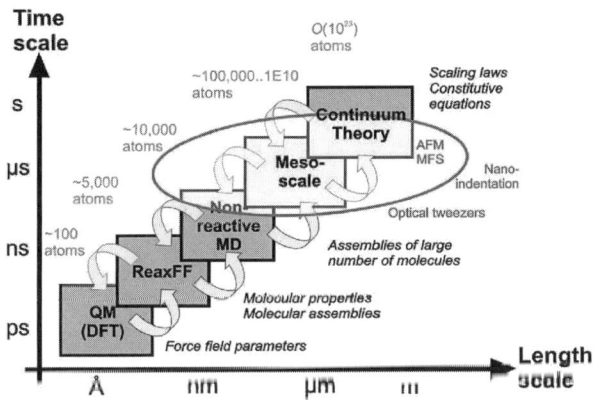

Fig. 8.4 Schematic that illustrates the concept of hierarchical multi-scale modeling (Figure adapted from (Buehler and Ackbarow 2007)). Hierarchical coupling of different computational tools can be used to traverse throughout a wide range of length and time scales. Such methods enable to provide a fundamental insight into deformation and fracture phenomena, across various time and length scales. Handshaking between different methods enables one to transport information from one scale to another. Eventually, results of atomistic, molecular or mesoscale simulation may feed into constitutive equations or continuum models. While continuum mechanical theories have been very successful for crystalline materials, BPMs require statistical theories, e.g., the Extended Bell Model (Ackbarow et al. 2007; Ackbarow and Buehler 2007b). Experimental techniques such as atomic force microscope (AFM), molecular force spectroscopy (MFS), nanoindentation or optical tweezers now overlap into atomistic and molecular approaches, enabling direct comparison of experiment and simulation (Lim et al. 2006)

Atomistic simulation, or molecular dynamics (MD) (Allen and Tildesley 1989), provides one with a fundamental view on materials deformation – describing the patterns of fracture, yield, diffusion and other mechanisms at resolutions that cannot yet be reached by experiments. After careful validation of these computational models with experiments, atomistic and multi-scale modeling has predictive power. Predictive multi-scale simulation could play an important role in science, engineering and materials design in the coming decades.

8.2.2 Basics of Atomistic Modeling

The basic concept behind atomistic simulation via MD is to calculate the dynamical trajectory of each atom in the material, by considering their atomic interaction potentials, by solving each atom's equation of motion according to $F = ma$. Numerical integration of this equation by considering proper interatomic potentials enables one to simulate a large ensemble of atoms that represents a larger material volume, albeit typically limited to several nanoseconds of time scale. The availability of such potentials for a specific material is often a limiting factor in applicability of this method.

Classical molecular dynamics generates the trajectories of a large number of particles, interacting with a specific interatomic potential, leading to positions $r_i(t)$, velocities $v_i(t)$ and accelerations $a_i(t)$. It can be considered an alternative to methods like Monte Carlo, with the difference that MD actually provides full dynamical information – and deterministic trajectories. The total energy of the system (E) is written as the sum of kinetic energy (K) and potential energy (U),

$$E = K + U, \tag{8.1}$$

where the kinetic energy is

$$K = \frac{1}{2}m \sum_{j=1}^{N} v_j^2, \tag{8.2}$$

and the potential energy is a function of the atomic coordinates r_j,

$$U = U(r_j), \tag{8.3}$$

with a properly defined potential energy surface $U(r)$, where $r = \{r_i\}$ describes the set of all atomic coordinates in the system. The numerical problem to be solved is a system of coupled second-order nonlinear differential equations:

$$m \frac{d^2 r_j}{dt^2} = -\nabla_{r_j} U(r) \quad j = 1 \ldots N, \tag{8.4}$$

which can only be solved numerically for more than two particles, $N > 2$. Typically, MD is based on updating schemes that yield new positions from the old positions, velocities and the current accelerations of particles:

$$r_i(t_0 + \Delta t) = -r_i(t_0 - \Delta t) + 2r_i(t_0)\Delta t + a_i(t_0)(\Delta t)^2 + \cdots . \tag{8.5}$$

The forces and accelerations are related by $a_i = f_i/m$. The forces are obtained from the potential energy surface – sometimes also called force field – as

$$F = m\frac{d^2 r_j}{dt^2} = -\nabla_{r_j} U(r) \quad j = 1\ldots N. \tag{8.6}$$

This technique can not only be used for particles that are atoms, but also be applied for particles that represent groups of atoms, such as in bead models. Provided interatomic potentials are available, MD is capable of directly simulating a variety of materials phenomena, for instance the response of an atomic lattice to applied loading under the presence of a crack-like defect, or the unfolding mechanisms of proteins.

One of the strengths of atomistic methods is its very fundamental viewpoint of materials phenomena. The only physical law that is put into the simulations is Newton's law and a definition of how atoms interact with each other. Despite this very simple basis, very complex phenomena can be simulated. Unlike many continuum mechanics approaches, atomistic techniques require no a priori assumption on the defect dynamics. A drawback of atomistic simulations is the difficulty of analyzing results and the large computational resources necessary to perform the simulations. Once the atomic interactions are chosen, the complete material behavior is determined. Different interatomic potentials are used in the studies of collagen at different scales; specific methods and theories will be introduced in the subsequent sections.

8.2.3 Large-Scale Parallelized Computing

Large-scale molecular dynamics simulations often require a significant amount of computing resources. Classical molecular dynamics can be quite efficiently implemented on modern supercomputers using parallelized computing strategies. Such supercomputers are composed of hundreds of individual computers (see, e.g. www.top500.org).

We now expect petaflop computers by the middle or end of the current decade. Based on the concept of concurrent computing, modern parallel computers are made out of hundreds or thousands of small computers working simultaneously on different parts of the same problem. Information between these small computers is shared by communicating, which is achieved by message passing procedures, enabled via software libraries such as the "Message Passing Interface" (MPI) (Gropp et al. 1999; Kadau et al. 2004).

Implemented based on spatial domain decomposition, parallel MD reaches linear scaling, that is the total execution time scales linear with the number of particles $\sim N$, and scales inversely proportional with the number of processors used to solve the numerical problem, $\sim 1/P$ (where P is the number of processors).

With a parallel computer whose number of processors increases with the number of cells (the number of particles per cell does not change), the computational burden remains constant. To achieve this, the computational space is divided into cells such that in searching for neighbors interacting with a given particle, only the cell in which it is located and the next-nearest neighbors have to considered. This scheme allows to treat huge systems with several billion particles.

8.2.4 Analysis and Visualization

A versatile, powerful and widely used visualization tool is the visual molecular dynamics (VMD) program (Humphrey et al. 1996). This software enables one to render complex molecular geometries using particular coloring schemes. It also enables us to highlight important structural features of proteins by using a simple graphical representation, such as alpha-helices, or the protein's backbone. The simple graphical representation is often referred to as cartoon model. In this particular cartoon model, the triple-helical structure of the tropocollagen motif is clearly visible. This is shown in Fig. 8.5.

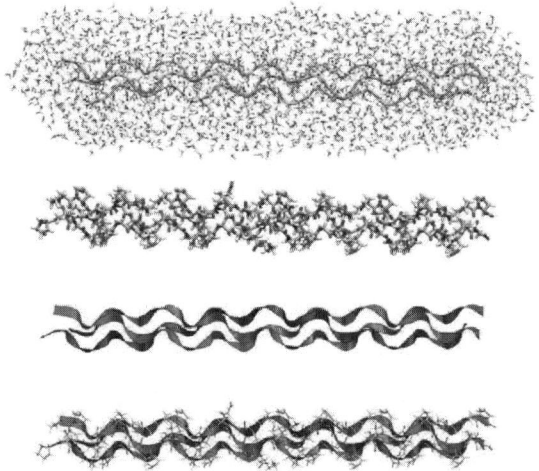

Fig. 8.5 Different representation of a single tropocollagen molecule, realized by the visualization program VMD (Humphrey et al. 1996). The upper plot shows the TC molecule solvated in water, the second plot from the top shows a tropocollagen molecule without the water molecules and the lower parts show a cartoon representation of the molecular geometry. In each subplot, the same molecular structure is visualized, illustrating the potential of visualization methods

An important quantity in the analysis of mechanical properties is the Cauchy stress tensor. For most studies discussed here we use the virial stress to calculate the Cauchy stress tensor directly from atomistic data (Tsai 1979; Zimmerman et al. 2004). The atomistic data is averaged over all particles (spatial average) and over several snapshots (temporal average). The virial stress is calculated by considering the volume of the computational sample, including the free volume in the molecular structure. For details regarding the calculation of the virial stress tensor we refer the reader to the literature (Tsai 1979; Zimmerman et al. 2004).

Energy dissipation per unit volume is calculated by integrating over the stress–strain curve until the fibril has fractured (strain ε_F), according to

$$E_{\text{diss},V} = \int_{\tilde{\varepsilon}=0}^{\varepsilon_F} \sigma(\tilde{\varepsilon})\mathrm{d}\tilde{\varepsilon}. \tag{8.7}$$

8.2.5 Complementary Experimental Methods

Recent advances in experimental techniques further facilitate analyses of ultra-small-scale material behavior. For instance, techniques such as nanoindentation, optical tweezers or atomic force microscopy (AFM) can provide valuable insight to analyze the molecular mechanisms in a variety of materials, including metals, ceramics and proteins. A selection of experimental techniques is summarized in Fig. 8.4, illustrating the overlap with multi-scale simulation methods.

An important experimental technique in conjunction with atomistic modeling of protein materials is X-ray diffraction; results of such experiments provide the initial atomistic and molecular structure, the starting point for all atomistic simulations. The structure of many proteins, elucidated using such experiments, has been deposited in the Protein Data Bank (PDB).

8.2.6 Summary

In this section, we have summarized the key aspects of molecular modeling, including basic molecular dynamics, a brief discussion on supercomputing as well as a brief review of complementary experimental methods.

8.3 Deformation and Fracture of Single Tropocollagen Molecules

This section is focused on the nanomechanical properties of single tropocollagen molecules, as originally reported in (Buehler 2006a, b; Buehler and Wong 2007). This approach in describing collagenous tissues represents a bottom-up approach, focusing on the finest, atomistic scales of detail governed by quantum mechanics (QM) as starting point, reaching up to large, macroscopic scales, using hierarchical multi-scale modeling. The first step in achieving this goal is the careful study of the properties of a single molecule.

There are several reports of experimental studies focused on the mechanics of single tropocollagen molecules (Waite et al. 1998; Arnoux et al. 2002; Sun et al. 2002; An et al. 2004). However, despite its relatively simple structure (for a recently crystallized model protein please see Kramer et al. 2000), single tropocollagen molecules have rarely been studied using molecular dynamics (MD) studies. In one of the few reports found in the literature, Lorenzo and coworkers (Lorenzo and Caffarena 2005) have reported investigations of the mechanical properties of collagen fibers, using MD studies, focusing on their Young's modulus. Other studies focused on the stability of collagen molecules (Israelowitz et al. 2005) and other structural investigations (Mooney et al. 2001; Mooney and Klein 2002; Mooney et al. 2002), or the effect of point mutations on the stability (Israelowitz et al. 2005). Some researchers modeled collagen at the continuum scale, using techniques such as the Finite Element Method (Bischoff et al. 2000).

Questions of particular interest include: How does a tropocollagen molecule respond to mechanical stretching force, in particular at large stretches? How does it fracture? How can these properties be linked to the folded structure? How do ultra-long collagen molecules with realistic lengths of several hundred nanometers behave in solution, under mechanical stretch? Such insight is important to understand the role of individual tropocollagen molecules in the context of tissue mechanics.

After a brief review of our computational technique in Section 8.3.1, we report atomistic modeling of the mechanics of single collagen fibers under different types of loading in Section 8.3.2. In Section 8.3.3 we provide a discussion of the results.

Table 8.1 provides an overview of important mathematical symbols used throughout this chapter.

8.3.1 Atomistic Model

Definition of the atomic interactions by force fields is at the heart of MD methods, as it defines the complete materials behavior. The basis for our investigations is a combination of the classical CHARMM force field (MacKerell et al. 2000) and the ReaxFF reactive force field (Duin et al. 2001; Strachan et al. 2005). The CHARMM

Table 8.1 Description of the main parameters and material or molecular properties used in the manuscript. Units are provided for some of the variables

F_{tens}	Tensile strength of a bimolecular fibril (geometry see Fig. 8.23)
A_{C}	Cross-sectional area of a TC molecule
L_0	Length of an individual TC molecule
L	Length of a TC molecule in a collagen fibril
L_{C}	Contact length between different TC molecules (e.g., in a bimolecular assembly or in a collagen fibril)
α	Overlap parameter in an assembly of TC molecules, note that $\alpha = L_{\text{C}}/L$
τ_{shear}	Shear strength between two TC molecules (units: force/length)
σ_{R}	Critical molecular tensile stress to nucleate slip pulse
E	Young's modulus, e.g., of an individual TC molecule or a collagen fibril
σ_{tens}	Tensile stress in a TC molecule (note that $\sigma_{\text{tens}} = F_{\text{tens}}/A_{\text{c}}$)
F_{max}	Maximum tensile force a single TC molecule can sustain
F_{F}	Maximum tensile force a BM collagen fibril can sustain
χ_{S}	Critical molecular length scale beyond which slip pulse nucleation occurs
χ_{R}	Critical molecular length scale beyond which fracture occurs
L_{χ}	Critical molecular length scale at which maximum strength is reached
γ	Energetic barrier to nucleation of a slip pulse (units: energy per length^2)
E_{diss}	Energy dissipation during deformation (units: energy)
$E_{\text{diss,V}}$	Energy density dissipation during deformation (units: energy per volume)

model is a widely used model to describe the behavior of proteins and related materials and structures.

However, for extreme mechanical loading and large deformation close to the breaking point, such classical approaches fail and new methods are required that take into consideration the behavior of chemical bonds at large deformation. We employ a new generation of reactive force fields to account for these chemical effects in protein mechanics.

8.3.1.1 Classical CHARMM Force Field

The classical force field CHARMM, implemented in the MD program NAMD (Nelson et al. 1996; MacKerell et al. 1998), is used for deformation studies of tropocollagen molecules and assemblies of tropocollagen molecules. The CHARMM force field (Anderson 2005) is widely applied in the protein and biophysics community and provides a basic description of proteins. It is based on harmonic and anharmonic terms describing covalent interactions, in addition to long-range contributions describing van der Waals (vdW) interactions, ionic (Coulomb) interactions, as well as hydrogen bonding. Since the bonds between atoms are modeled by harmonic springs or its variations, bonds between atoms cannot be broken and new bonds cannot be formed. Also, the charges are fixed and cannot change, and the equilibrium angles do not change depending on stretch. We have added an extension to the standard CHARMM force field to include a description of the hydroxyproline residue ("HYP" in the PDB file), which is not one of the 20 natural amino acids, following the procedure suggested in Anderson (2005).

8.3.1.2 Reactive Force Field: A New Bridge to Integrate Chemistry and Mechanics

Reactive force fields represent an important milestone in overcoming the limitations of classical force fields in not being able to describe chemical reactions. For mechanical properties, this translates into the properties of molecules at large strain, a phenomenon referred to as hyperelasticity. Several flavors of reactive potentials have been proposed in recent years (Stuart et al. 2000; Duin et al. 2001; Brenner et al. 2002). Reactive potentials can overcome the limitations of empirical force fields and enable large-scale simulations of thousands of atoms with quantum mechanics accuracy. The reactive potentials, originally only developed for hydrocarbons (Duin et al. 2003; Strachan et al. 2003; van Duin et al. 2004; Chenoweth et al. 2005; Cheung et al. 2005; Han et al. 2005; Nielson et al. 2005; Strachan et al. 2005; Buehler et al. 2006; Buehler 2007; Buehler et al. 2007), have been extended recently to cover a wide range of materials, including metals, semiconductors and organic chemistry in biological systems such as proteins. Here we employ a particular flavor of the ReaxFF potentials as suggested in Datta et al. (2005), with slight modifications to include additional QM data suitable for protein modeling.

Reactive potentials are based on a more sophisticated formulation than most nonreactive potentials. A bond length/bond order relationship is used to obtain smooth transition from non-bonded to single-, double- and triple-bonded systems.

All connectivity-dependent interactions (that means, valence and torsion angles) are formulated to be bond order dependent. This ensures that their energy contributions disappear upon bond dissociation so that no energy discontinuities appear during reactions. The reactive potential also features non-bonded interactions (shielded van der Waals and shielded Coulomb).

The reactive formulation uses a geometry-dependent charge calculation (QEq) scheme similar to Goddard's QEq (Rappé and Goddard 1991) that accounts for polarization effects and modeling of charge flow. This is a critical advance leading to a new bridge between QM and empirical force fields. All interactions feature a finite cutoff. Figure 8.6 shows a comparison between reactive and nonreactive formulations, illustrating that both descriptions agree for small deviations from the equilibrium, but disagree significantly for large strains.

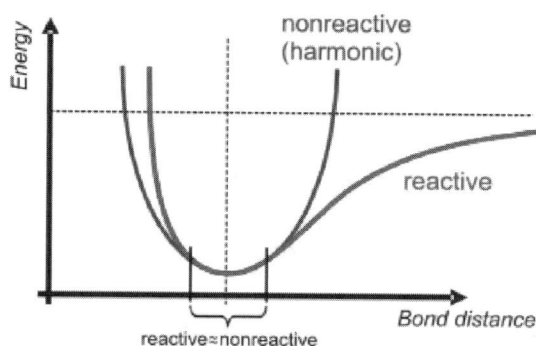

Fig. 8.6 Schematic illustration of the differences between a reactive and nonreactive force field. The schematic illustrates that the nonreactive model is only valid for small deformation from the equilibrium bond configuration and cannot describe dissociation of the chemical bond. The reactive description overcomes these limitations (for further information, see for instance Buehler 2007)

8.3.1.3 Preprocessing

Preprocessing of the simulations is done using the CMDF framework (Buehler and Dodson et al. 2006), a computational Python-based simulation environment capable of seamless integration of various file formats, computational engines, including molecular and crystal building tools.

Python scripts are used to analyze and post-process the simulation data, as required for example to compute statistical averages of force–displacement curves.

8.3.1.4 Atomistic Simulation Procedure

We build the atomistic based on the crystal unit cells according to X-ray diffraction data obtained by experiment; a short tropocollagen segment is solvated in a water skin. These structures are taken directly from the Protein Data Bank (PDB). We use the crystal structure PDB ID 1QSU, with 1.75 Å resolution, as reported by Kramer and coworkers (Kramer et al. 2000). The 1QSU is a triple-helical collagen-like molecule with the sequence (Pro-Hyp-Gly)$_4$-Glu-Lys-Gly-(Pro-Hyp-Gly)$_5$ (structure is also shown in Fig. 8.5 in various visualization modes).

The charges of each atom are assigned according to the CHARMM rules. Hydrogen atoms are added according to pH 7. The CHARMM input files (structure and topology files) are then used to perform NAMD calculations. For ReaxFF calculations, no atom typing is necessary (only element types are assigned), and charges are determined dynamically during the simulation. Hydrogen atoms are added using the NAMD/CHARMM procedure, according to the same conditions as outlined above.

Before finite temperature, dynamical calculations are performed; we carry out an energy minimization, making sure that convergence is achieved, thus relieving any potential overlap in vdW interactions after adding hydrogen atoms. In the second step, we anneal the molecule after heating it up to a temperature $T = 300$ K. The heat up rate is $\Delta T = 25$ K every 25 steps, and we keep the temperature fixed after the final temperature $T = 300$ K is achieved (then we apply a temperature control in an NVT ensemble). We also ensure that the energy remains constant after the annealing procedure. The relaxed initial length of each molecule (consisting of 30 residues in each of the three chains) is $L_0 = 84$ Å.

Depending on the details of the loading case, we then apply mechanical forces using varied types of constraints and investigate the response of the molecule due to the applied loading. Typically, we obtain force-versus-displacement data, which are then used to extract mechanical quantities such as stress and strain, using continuum mechanical concepts by drawing analogies between the molecular level and continuum mechanical theories. Steered MD is based on the concept of adding restraint force to groups of atoms by extending the Hamiltonian by an additional restraint potential of the form $1/2\,k_{SMD}(r - r_\lambda)^2$. The SMD method mimics a AFM nanomechanics experiment, as illustrated in Fig. 8.7.

Unless indicated otherwise, we use a steered molecular dynamics (SMD) scheme with spring constant $k_{SMD} = 10$ kcal/mol/Å^2. It was shown in previous studies that this is a reasonable choice leading to independence of the measured molecular mechanical properties from the choice of the SMD spring constant (Lorenzo and Caffarena 2005).

8.3.1.5 Computational Experiments of Stretching Short Tropocollagen Segments

In the following section, we present a suite of studies with different mechanical loading. The different loading cases studied in this chapter are summarized in Fig. 8.8. The different loading conditions and the objective of the specific calculation are:

- Tensile/compressive testing (obtain Young's modulus/buckling load), Fig. 8.8(a)
- Bending (obtain bending stiffness and persistence length), Fig. 8.8(b)
- Shearing two tropocollagen molecules (obtain fiber–fiber interactions), Fig. 8.8(c)

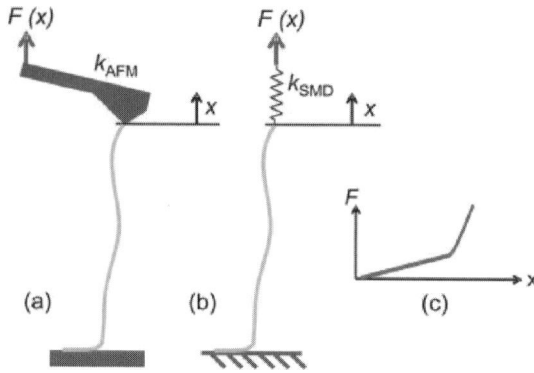

Fig. 8.7 Single molecule pulling experiments, carried out on a single protein molecule (Buehler and Ackbarow 2007). Subplot (**a**) depicts an experimental setup based on AFM and subplot (**b**) depicts a steered molecular dynamics (SMD) analogue. In the SMD approach, the end of the molecule is slowly pulled at a pulling velocity v vector. This leads to a slowly increasing force $F = k(v \cdot t - x)$ ($k = 10\,\text{kcal/mol/Å}^2$), where t is the time and x is the current position vector of the end of the molecule ($F(x)$, schematically shown in subplot (**c**)). Both approaches, AFM and SMD, lead to force–displacement information. In addition to the $F(x)$ curve, SMD provides detailed information about associated atomistic deformation mechanisms. Due to the time scale limitations of MD to several nanoseconds, there is typically a large difference between simulation and experiment with respect to pulling rates. Whereas MD simulations are limited to pulling rates of ≈ 0.01 m/s, experimental rates are six to eight magnitudes smaller than those. This requires additional consideration in order to interpret MD results in light of experimental findings

8.3.2 Tensile and Compressive Loading

8.3.2.1 Small Deformation

First we discuss tensile testing of the fibers using the nonreactive CHARMM force field. After careful equilibrium of the structure of the collagen molecule, we apply a force at one end, while we keep the other end of the molecule fixed. The loading case is shown in Fig. 8.8(a). By slowly increasing the load applied to the collagen

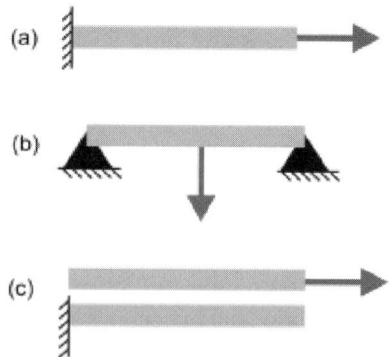

Fig. 8.8 Overview of various load cases studied, as reported in Buehler (2006a,b). Our load cases include (**a**) tensile loading in axial direction of the molecule (also including compressive test), (**b**) bending test and (**c**) shear test

molecule, while measuring the displacement d, we compute force–displacement curves. The force-versus-displacement curve $F(d)$ can be used to determine a stress versus strain curve, by proper normalization:

$$\sigma(d) = \frac{F(d)}{A_C}, \tag{8.8}$$

where A_C denotes an equivalent area of the cross-section of a collagen molecule, and $A_C = \pi \cdot R^2 \approx 214.34\,\text{Å}^2$, assuming that $R = 8.26\,\text{Å}$ (obtained from studies of an assembly of two tropocollagen molecule as described below). Note that the stress is typically dependent on the stretch d. The local (in terms of strain) Young's modulus $E(d)$ is given by

$$E(d) = \frac{d_0}{A_C} \frac{\partial F(d)}{\partial d}, \tag{8.9}$$

where d_0 is the initial, undeformed length of the collagen fiber and $d_0 = 84\,\text{Å}$. Note that Young's modulus is independent of the length of the molecule. The definition in Eq. (8.9) is a consequence of the fact that the stretching force is expressed as a function of stretch d rather than strain ($\sigma = E\varepsilon$).

Figure 8.9 shows force-versus-displacement plots for tensile loading, for three different loading rates. The loading rates in the three cases are $\dot{r}_\lambda = 0.0001$, 0.0002 and 0.001 Å/step, respectively. Young's modulus is determined as the tangential slope corresponding to 10% tensile strain. Young's modulus is obtained to be $E_{\text{tens}} = 6.99$, 8.71 and 18.82 GPa, respectively, for increasing loading rates as specified above. We observe an increase in stiffness for higher loading rates. These results indicate that collagen fibers show a rate-dependent elastic response.

The estimate for Young's modulus is somewhat in agreement with results reported in Lorenzo and Caffarena (2005), presenting a value for Young's modulus

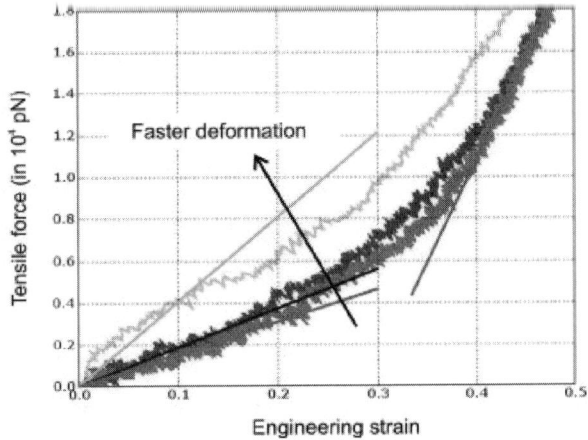

Fig. 8.9 Force versus strain, pulling of a single tropocollagen molecule, for three different pulling velocities, as reported in Buehler (2006a). The results indicate that faster pulling velocities lead to a stiffening effect (the loading rates in the three cases are $\dot{r}_\lambda = 0.0001$, 0.0002 and 0.001 Å/step, respectively)

of around 4.8 ± 1.0 GPa. The reason for the difference to our results could be the different force field used, different boundary conditions or different strain rates.

Figure 8.10 depicts snapshots of the tropocollagen molecule under increasing stretch. At large strains, the helical structure is lost and the three polypeptides appear as individual strands, then defining its elasticity by the behavior of covalent bonds. This is also confirmed in the force–strain plot (Fig. 8.9). The "local" Young's modulus associated with large strains is given by $E_{\text{tens}}^{\text{large}} = 46.7$ GPa.

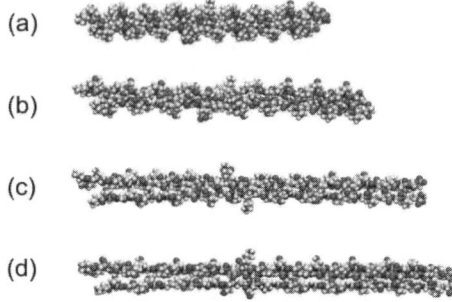

Fig. 8.10 Stretching of a single tropocollagen molecule, using a nonreactive CHARMM force field (water molecules are not shown for clarity) (Buehler 2006). The helical structure unfolds with increasing strain and vanishes at large deformation and the three strands become independent and the covalent bonds between atoms govern the elasticity. This loss in tertiary helical structure is represented by a change in tangent elasticity, as seen in Fig. 8.9, at strains beyond 35%

Under compressive loading, it is observed that the tropocollagen molecule can only sustain a relatively small load before buckling. Figure 8.11 shows load versus displacement curves, and Fig. 8.12 depicts several snapshots as the collagen molecule is subjected to compressive loading. The maximum force level sustained before buckling is $F_{\text{max,compr}} \approx 1,050$ pN, reached at about 5% compressive strain. Young's modulus under compressive loading, calculated for very small strains

Fig. 8.11 Compressive loading of the single tropocollagen molecule, at a loading rate of $\dot{r}_{\lambda} = 0.0002$ Å/step (Buehler 2006a). The behavior under compression is significantly different than under tension, also revealing an asymmetry of the elastic properties at small strains. The tropocollagen molecule starts to buckle at about 5% compressive strain

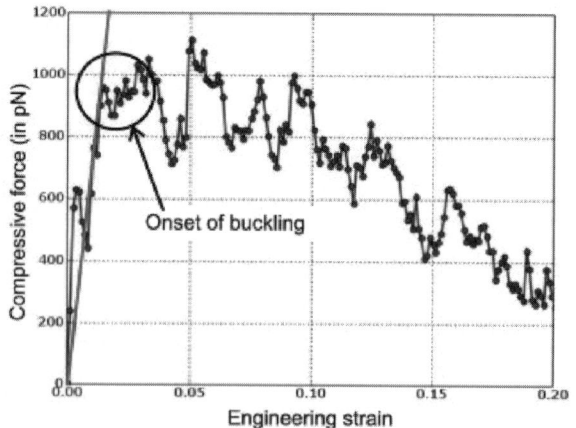

up to 1.25%, is given by $E_{compr} \approx 29.86\,GPa$. This indicates that the elastic tensile/compressive behavior is asymmetric around the equilibrium position, with significantly higher compressive modulus than tensile modulus. However, the maximum load that can be sustained under pure compression is much lower than under tension due to buckling.

Fig. 8.12 Tropocollagen molecule under compressive loading (water molecules are not shown for clarity) (Buehler 2006a). The results suggest that upon a relatively small load, the tropocollagen molecule starts to buckle and goes into a bending mode

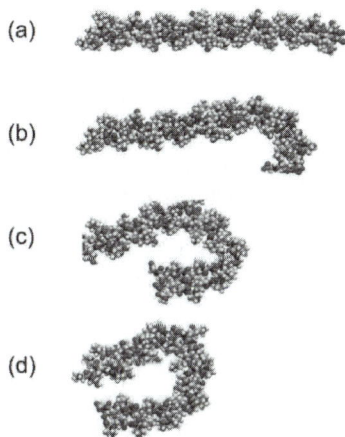

(a)

(b)

(c)

(d)

8.3.2.2 Large Deformation and Fracture Mechanics of Individual Tropocollagen Molecules

Now we focus on the mechanical behavior of tropocollagen molecules using a reactive force field. The dominating forces in tropocollagen molecules occur typically in the axial direction of the molecule, so this type of loading is most critical for the mechanical integrity of tissues.

Questions that we would like to investigate include: Under which conditions do classical, nonreactive force fields break down, and what are the limitations of these methods? Do the results agree for small deformation? How are the mechanical properties different once mechanical deformation is large and formation and breaking of bonds are allowed? We will pull the tropocollagen molecule until fracture occurs and study the details of the fracture mechanisms. A central question we would like to address is, to which strain and deformation levels – or, equivalently, applied force level – can we rely on the assumption of nonreactive force fields. A solid understanding of the large deformation, nonlinear and fracture properties of tropocollagen molecules becomes important when the mesoscale model is introduced at the next hierarchical level.

Figure 8.13 depicts snapshots of results obtained from ReaxFF simulations of tensile deformation of tropocollagen molecules. We observe that the shape of the fiber changes from a straight shape to an S-like shape as the load increases. This change of shape leads to an increasingly large radius of curvature in localized regions of the fiber (toward the left and right ends). These regions with high curvature represent regions of higher tensile stress. In agreement with this notion, we observe that these regions lead to onset of failure. A possible explanation for the transformation into the S-like shape may be the different energy expression in the reactive potential. The change of bond behavior at large deformation is not linear, but instead, it is nonlinear, eventually leading to rupture of covalent bonds.

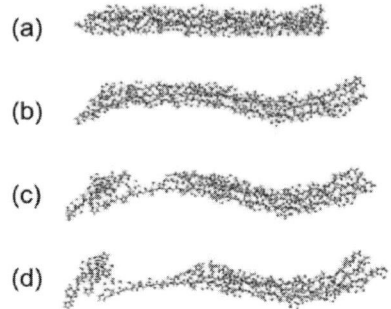

Fig. 8.13 Fracture mechanics of a single tropocollagen molecule, as reported in Buehler (2006a). We observe a transformation from the initial straight shape to an S-shaped structure at large strain, leading to fracture of an individual polypeptide. This transformation is found consistently for a variety of loading conditions

Figure 8.14 shows plots of force versus strain, comparing the CHARMM model (curve (a)) with the ReaxFF model (curve (b)). The strength of the tropocollagen molecule is determined to be 2.35×10^4 pN, reached at approximately 50% tensile strain. We find that the CHARMM description and the ReaxFF model agree for

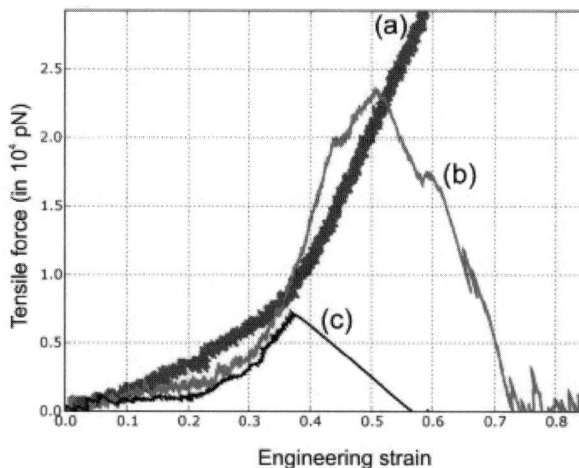

Fig. 8.14 Force versus strain, pulling of a single tropocollagen molecule, as reported in Buehler (2006a). An initial regime of flat, almost linear elastic extension is followed by onset of nonlinear, stiffening behavior at larger strains beyond approximately 30–35% strain. This behavior can be explained by the fact that the tertiary, helical structure of the tropocollagen molecule begins to disappear and the elasticity of each covalent bond in the single strand governs the elastic response. This represents a significant distinction to the results obtained with reactive force fields, suggesting failure due to strain/force localization

small strains up to approximately 10% strain. The tensile stress for onset of permanent deformation (at around 43% strain) occurs at approximately 9.3 GPa, and the fracture stress is determined to be 11.2 GPa.

We have repeated the tensile deformation simulation for a single polypeptide out of the three making up the entire tropocollagen molecule. The strength of an individual polypeptide molecule is determined to be 0.713×10^4 pN, reached at about 37% tensile strain. This is significantly lower than that of the tropocollagen molecule (see Fig. 8.14, curve (c)).

8.3.3 Bending a Single Tropocollagen Molecule

We perform a computational experiment to describe the bending of a single tropocollagen molecule by clamping it at the boundaries and applying a force in the center of the molecule, as shown in Fig. 8.8(b). This is equivalent to the three-point bending test widely used in engineering mechanics. From the force–displacement data obtained by atomistic modeling, the bending stiffness EI is given by

$$EI = \frac{dL^3}{48 F_{\text{appl}}},\qquad(8.10)$$

where F_{appl} is the applied force and d is the bending displacement. Figure 8.15 depicts load versus displacement curves for various deformation speeds. Figure 8.16 depicts the resulting bending stiffness as a function of loading speed.

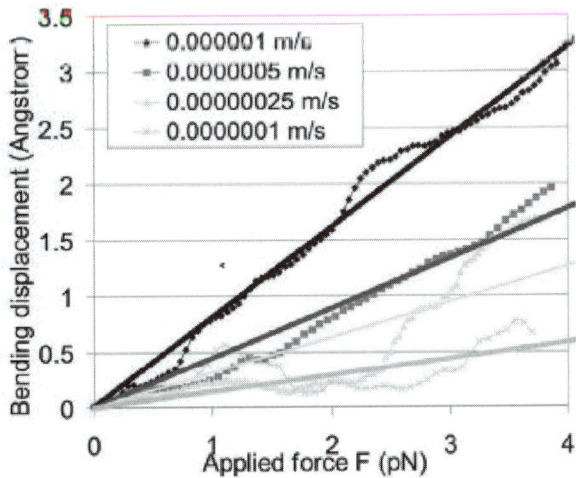

Fig. 8.15 Bending displacement over bending force for the three-point bending test of a single tropocollagen molecule, as reported in Buehler and Wong (2007). Results are shown for various loading rates. The linear curves are linear fits to the MD simulation results

Fig. 8.16 Dependence of
bending stiffness EI of a
single TC molecule on the
deformation rate (based on
results shown in Fig. 8.15)
(Buehler and Wong 2007).
The results indicate that EI
decreases linearly with
decreasing loading rate. A
linear fit to the data enables
us to extrapolate to smaller
deformation speeds

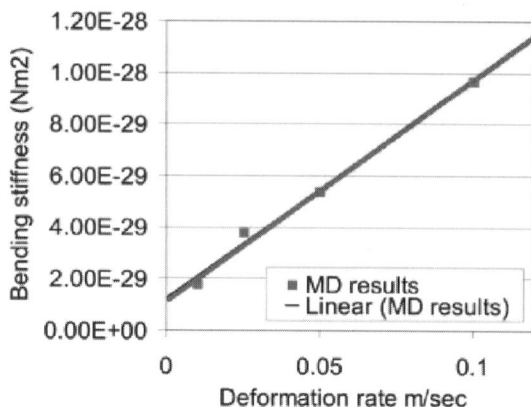

8.3.4 Shearing Two Tropocollagen Molecules

We continue with shearing experiments of an assembly of two tropocollagen
molecules. The objective is to gain insight into the mechanisms and type of interactions between two tropocollagen molecules in aqueous solution. We start with a
geometry as depicted schematically in Fig. 8.8(c). We first equilibrate the system
without application of any mechanical shear load. We find that the equilibrium
distance between two molecules depends on the presence of solvent; being reduced
if solvent is present. With solvent present, we find an equilibrium distance between
two tropocollagen molecules of $r_{EQ} \approx 16.52\,\text{Å}$. All studies reported here are
carried out with water molecules present. It is noted that this equilibrium distance
leads to a molecular radius of $8.26\,\text{Å}$.

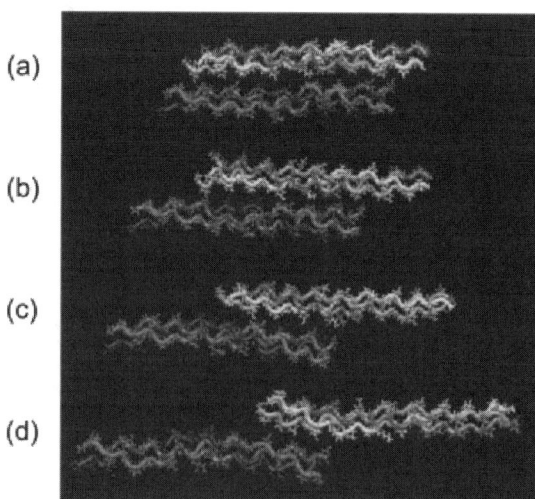

Fig. 8.17 Snapshots of
shearing of two tropocollagen
molecules (Buehler 2006a).
The subplots (**a**)–(**d**) show
the behavior as the shear
strain is increased

Fig. 8.18 Atomistic modeling of shear experiments between two TC molecules (shear load applied using the SMD method) (Buehler 2006a). The plot shows the shear resistance as a function of loading rate, obtained by fully atomistic modeling using the CHARMM potential

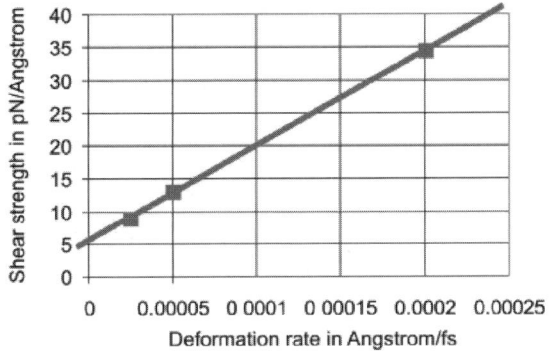

Figure 8.17 depicts snapshots of the system as it undergoes shear deformation. Figure 8.18 depicts the maximum shear force versus pulling velocity as obtained during the shearing experiment.

We perform calculations with three different loading rates, $\dot{r}_\lambda = 0.0002$, 0.00005 and 0.000025 Å/step. We find that the resulting values are strain rate dependent. The maximum force decreases with decrease in loading rate, assuming values $F_{max,shear} \approx 2,900$ pN, $F_{max,shear} \approx 1,100$ pN and $F_{max,shear} \approx 750$ pN, corresponding to the strain rates provided above. Using a linear extrapolation to vanishing loading rate, we estimate a maximum shear force of $F_{max,shear} \approx 466$ pN, corresponding to adhesion strength $\tau_{max} = 5.55$ pN/Å (see Fig. 8.18). It is noted that the units of τ_{shear} are force/length (this is not a "stress" but rather adhesion force per unit length).

8.3.5 Development of a Mesoscopic, Molecular Model

The atomistic modeling results, carried out at the level of individual polypeptide chains, tropocollagen molecules and assemblies of those, helped to develop a better qualitative and quantitative understanding of the competing mechanisms and forces during deformation of collagen, at a microscopic level. This information is now used to develop a mesoscopic model, in which beads connected by different types of springs represent collagen molecules, whereas all parameters are completely derived from atomistic calculations. The motivation for these studies is the desire to model larger length and time scales. This approach is similar to training empirical potentials from quantum mechanical data using the force-matching approach, as done successfully earlier for metallic systems (Ercolessi and Adams 1994). The geometrical approach of coarse graining is visualized in Fig. 8.19.

The reduction of degrees of freedom in the mesoscale model enables one to study very long tropocollagen molecules with lengths on the order of several hundred nanometers, as well as bundles of tropocollagen molecules. This approach enables reaching a "material scale" and makes the overall mechanics of the material accessible to atomic and molecular scale modeling.

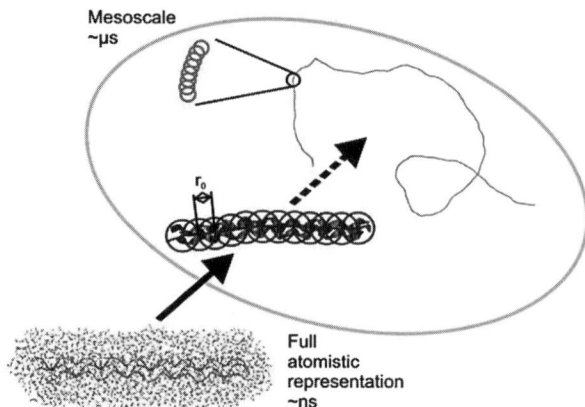

Fig. 8.19 Schematic showing the development of the coarse-grained molecular model from a full atomistic description, illustrating the procedure utilized in various studies reviewed in this chapter. The full atomistic representation of the triple-helical TC structure is replaced by a collection of beads. The mesoscale model enables the treatment of ultra-long TC molecules at time and length scales not in reach by full atomistic models

8.3.5.1 Model Development: Training from Pure Atomistic Results Using Energy and Force Matching

The goal is to develop the simplest model possible to perform large-scale studies of the mechanics of collagen molecules, eventually leading to understanding of the behavior of assemblies of such fibers. We assume that we can write the total energy of the system as

$$U = U_T + U_B + U_{weak}. \tag{8.11}$$

The bending energy is given by

$$\phi_B(\varphi) = \frac{1}{2}k_B (\varphi - \varphi_0)^2, \tag{8.12}$$

with k_B as the spring constant relating to the bending stiffness. The resistance to tensile load is characterized by

$$\phi_T(r) = \frac{1}{2}k_T (r - r_0)^2, \tag{8.13}$$

where k_T refers to the resistance of the molecule to deform under tensile load. To account for the nonlinear stress–strain behavior of a single molecule under tensile loading, we replace the harmonic approximation with a bilinear model (Buehler 2006a,b; Buehler and Wong 2007). The force between two particles is

$$F_T(r) = -\frac{\partial \phi_T(r)}{\partial r},$$ (8.14)

where

$$\frac{\partial \phi_T(r)}{\partial r}(r) = H(r_{\text{break}} - r) \begin{cases} k_T^{(0)}(r - r_0) & \text{if } r < r_1 \\ k_T^{(1)}(r - \tilde{r}_1) & \text{if } r \geq r_1 \end{cases}.$$ (8.15)

In Eq. (8.15), $H(r - r_{\text{break}})$ is the Heaviside function $H(a)$, which is defined to be zero for $a < 0$, and one for $a \geq 0$, and $k_T^{(0)}$ as well as $k_T^{(1)}$ for the small and large deformation spring constants. The parameter $\tilde{r}_1 = r_1 - k_T^{(0)}/k_T^{(1)}(r_1 - r_0)$ is determined from force continuity conditions. The function U_T is given by integrating $F_T(r)$ over the radial distance.

In addition, we assume weak, dispersive interactions between either different parts of each molecule or different molecules, defined by a Lennard–Jones (LJ) function

$$\phi_{\text{weak}}(r) = 4\varepsilon \left(\left[\frac{\sigma}{r} \right]^{12} - \left[\frac{\sigma}{r} \right]^6 \right),$$ (8.16)

with σ as the distance parameter and ε describing the energy well depth at equilibrium.

Note that the total energy contribution of each part is given by the sum over all pair-wise and triple (angular) interactions in the system

$$U_1 = \sum_{\text{pairs}} \psi_1(r) \quad \text{and} \quad U_B = \sum_{\text{angles}} \phi_B(\varphi).$$ (8.17)

8.3.5.2 Equilibrium Distances of Beads and Corresponding Masses

The mass of each bead is determined by assuming a homogeneous distribution of mass in the molecular model. The total mass of the tropocollagen molecule used in our studies is given by 8,152.2 amu. We divide the total length of the tropocollagen molecule used in the MD studies into $N_{\text{MD}} = 6$ pieces, each bead containing five amino acid residues. Each bead then has a weight of 1,358.7 amu. Since the total length of the molecule is $L_0 = 84$ Å, the beads are separated by a distance $r_0 = 14$ Å (for a finer discretization of beads, say 7 Å, the mass will be half of this value). The beads represent different sequences in tropocollagen that when added together make the entire sequence.

8.3.5.3 Dispersive and Non-bonding Interactions

The LJ parameters are determined from the calculations of shearing two collagen fibers. In all these considerations, we assume that a pair-wise interaction between different tropocollagen molecules is sufficient and that there are no multi-body contributions. Based on these assumptions, we model the interactions between different molecules using a LJ 12:6 potential. The distance parameter σ is given by

$$\sigma = \frac{D}{\sqrt[6]{2}} \approx 14.72 \,\text{Å}, \tag{8.18}$$

where D is the equilibrium distance as measured in the MD simulations, $D = 16.52 \,\text{Å}$.

The shear strength can be used to extract the LJ parameters for the weak, dispersive interactions between two fibers. Note that this interaction includes the effect of solvation water and other bondings (e.g., H-bonds, etc.).

The maximum force in a LJ potential, assuming a single LJ "bond" is given by

$$F_{\text{max,LJ}} = \frac{\Lambda \varepsilon}{\sigma}, \tag{8.19}$$

while noting that $\Lambda \approx 2.3964$ for the LJ 12:6 potential. The parameter σ is already determined, leaving only ε to be trained using a force-matching approach.

The parameter ε can be obtained by requiring a force balance at the point of rupture:

$$F_{\text{max,LJ}} N_{\text{MD}} = \tau_{\text{max}} \quad L = F_{\text{max}}. \tag{8.20}$$

This expression can be used to determine ε as

$$\varepsilon = \frac{F_{\text{max}} \sigma}{\Lambda N_{\text{MD}}}. \tag{8.21}$$

From atomistic modeling we calculate F_{max} allowing to estimate numerical values for ε. We find that $\varepsilon \approx 11.06 \,\text{kcal/mol}$ predicted from Eq. (8.21). Based on the extrapolation of shear force $F_{\text{max,shear}} \approx 466 \,\text{pN}$ corresponding to vanishing strain rate we finally arrive at a value $\varepsilon \approx 6.87 \,\text{kcal/mol}$.

Presence of intermolecular cross-links effectively leads to an increased intermolecular adhesion in the region where cross-links are formed, as discussed in Robins and Bailey (1973), Lodish et al. (1999) and Bailey (2001). To model the effect of cross-links, the adhesion parameter ε_{LJ} is modified to account for the stronger interaction between molecules. Variation of the parameter ε_{LJ} along the molecular axis enables one to account for specific spatial distributions of cross-links. Experimental analyses of the molecular geometry suggest that intermolecular aldol cross-links between lysine or hydroxylysine residues (Lodish et al. 1999; Alberts et al. 2002) primarily develop at the ends of tropocollagen molecules (Grandbois et al. 1999; Lantz et al. 2001). The aldol cross-link is a C–C bond that forms between side chains of residues of two tropocollagen molecules.

The presence of cross-links is modeled by increased adhesion at the ends of each molecule, in segments of 60 Å to the left and right end of each tropocollagen molecule. According to this idea, the LJ potential parameter ε_{LJ} is increased by a factor $\beta \geq 1$ compared with the rest of the molecule in regions where cross-links are formed, and therefore

$$\varepsilon_{LJ,XL} = \beta \, \varepsilon_{LJ}. \tag{8.22}$$

For a choice $\beta = 12.5$, the additional shear force exerted at the end of the molecule corresponds to ≈ 4.2 nN, which is on the order of the bond strength of covalent cross-link bonds (Buehler 2006). The parameter $\beta = 12.5$ therefore corresponds to the case when approximately one cross-link is present at each end of a tropocollagen molecule, leading to a cross-link density of $2.2 \times 10^{24}/m^3$ (the cross-link density is defined as the number of cross-links per unit volume). Similarly, doubling the value $\beta = 25$ corresponds to two covalent cross-links.

8.3.5.4 Tensile Spring Parameter

The tensile spring constant is determined from various calculations of stretch versus deformation, while being constrained to the regime of small loads and consequently small displacements. The spring constant k_T is then defined as

$$k_T = \frac{N_{MD} F_{appl}}{\Delta d} = \frac{A_C}{L_0} E, \tag{8.23}$$

with $\Delta d = L - L_0$ being the displacement of the atomistic model due to applied force F_{appl}. Based on the low-strain rate tensile testing data discussed earlier, we find that $k_T^{(0)} \sim 15.41$ kcal/mol/Å^2. Similar considerations can be used to determine a value for $k_T^{(1)}$, thereby considering the large deformation elastic behavior.

The parameters r_1 and r_{break} (unit: length) are related to the critical strains at which the tangent slopes in the stress–strain curve changes (denoted by ε_1, which is approximately 30%), and from the breaking strain (denoted by ε_{break}, which is approximately 50%),

$$r_1 = (1 + \varepsilon_1)r_0. \tag{8.24}$$

$$r_{break} = (1 + \varepsilon_{break})r_0. \tag{8.25}$$

8.3.5.5 Bending Spring Parameter

Using an argument of energy conservation between the atomistic and the mesoscale model, we arrive at an expression for the bending stiffness parameter k_B

$$k_{\mathrm{B}} = \frac{3}{2}\frac{EI}{r_0}. \tag{8.26}$$

We find that $k_{\mathrm{B}} \approx 14.98\,\mathrm{kcal/mol/rad^2}$ as reported in Buehler and Wong (2007). These expressions are only valid for small deformations.

8.3.6 Validation of Mesoscale Model in Tensile Deformation

Figure 8.20 shows the force–displacement curve of a tensile stretching experiment of tropocollagen molecules, comparing results obtained with the CHARMM method, ReaxFF and the mesoscale model. The results confirm that the mesoscale model indeed approximates the results of reactive MD.

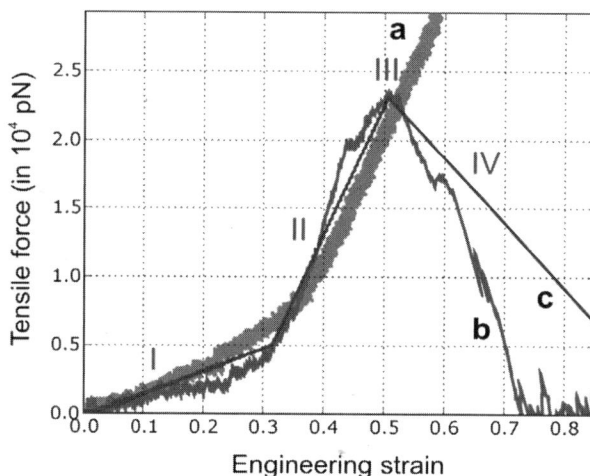

Fig. 8.20 Modeling of a tensile stretching experiment of a TC molecule, comparing atomistic and mesoscale models for validation (using the SMD method) (results as reported in Buehler and Wong (2007)). The plot shows the force versus displacement of a TC molecule, using the CHARMM force field (curve "a") and the reactive ReaxFF force field (curves "b" and "c"). Curve "b" shows the force–displacement response of a single TC molecule, displaying three regimes. Regime I is characterized by uncoiling of the TC molecule, regime II is associated with a larger modulus due to stretching of covalent bonds and in regime III we observe fracture of the molecule, followed by a rapid decay of the force in regime IV. Curve "c" shows the results obtained using the reactive mesoscale model, illustrating the agreement between the different methods

8.3.7 Stretching an Ultra-long Tropocollagen Molecule: Mesoscale Modeling

The mesoscale model now enables direct comparison with experimental results, validating the major predictions of our model. The present section reviews molecular simulation results reported in Buehler and Wong (2007) (for details regarding the modeling procedure see this reference).

Figure 8.21 shows a stretching experiment obtained by using the mesoscale model, as reported in Buehler and Wong (2007). Loading of the tropocollagen molecule starts from a coiled entangled configuration of the molecule with end-to-end distance of approximately 100 nm. During the initial regime (I), the molecule loses its entangled structure, while the applied forces remain relatively low. Once the contour length is reached ($x = L$), (II) uncurling of the triple helix, (III) stretching of covalent bonds in the individual polypeptides and (IV) rupture of the tropocollagen molecule occurs, followed by a sharp drop of the forces to zero. The qualitative behavior of tropocollagen molecules under stretch is similar to recent experimental studies carried out on collagen fibrils that reach forces on the order of μN, also showing a significant hyperelastic stiffening effect (Buehler and Wong 2007).

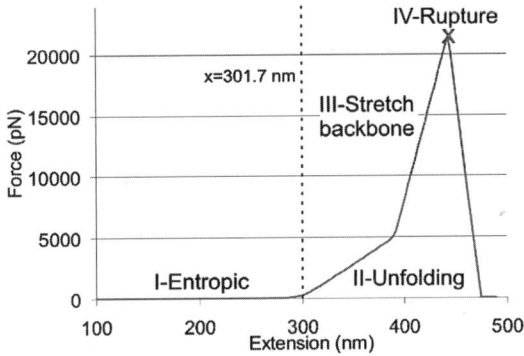

Fig. 8.21 Force–displacement ($F(x)$) curves of stretching a single TC molecule, $L = 301.7$ nm, at 300 K, as reported in Buehler and Wong (2007). The plot depicts the force–displacement curve over the entire deformation range, covering four stages: (I) uncoiling of the entangled configuration, (II) uncurling of the triple helix, (III) stretching of covalent bonds in the individual polypeptides and (IV) rupture of the TC molecule. The dashed line indicates the contour length of the molecule

Figure 8.22 shows a zoom into the small-force regime, providing a quantitative comparison with optical tweezers experiments (Sun et al. 2002; An et al. 2004; Sun et al. 2004). The plot reveals very good agreement, even though the deformation rate in MD is still much higher.

8.3.8 Discussion and Conclusion

We have reported atomistic modeling to calculate the elastic, plastic and fracture properties of tropocollagen molecules. Using full atomistic calculations, we presented a suite of calculations of different mechanical loading types to gain insight into the deformation behavior of tropocollagen molecules. Our results suggest that

Fig. 8.22 This plot depicts a subset of the results depicted in the previous figure, focusing on the small force, entropic response ($F < 14$ pN) (original results reported in Buehler and Wong (2007)). This plot also depicts experimental results obtained for TC molecules with similar contour lengths (Sun et al. 2002; An et al. 2004; Sun et al. 2004), as well as the prediction of the WLC model with persistence length approximately 16 nm

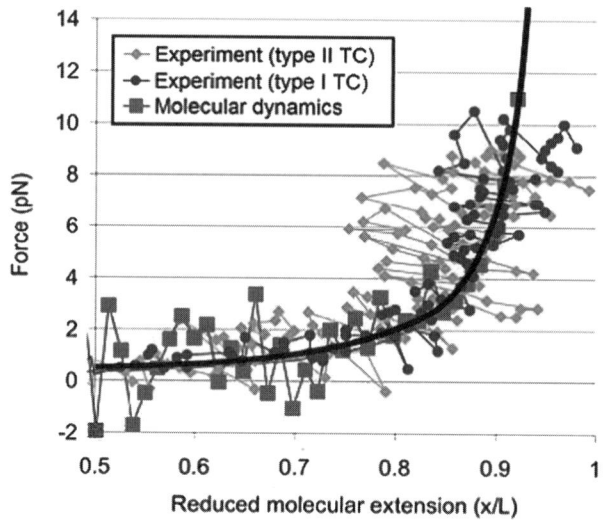

it is critical to include a correct description of the bond behavior and breaking processes at large bond stretch, information stemming from the quantum chemical details of bonding. A critical outcome of these studies is the observation that tropocollagen molecules undergo a transition from straight molecules to an S-shaped structure at increasingly large tensile stretch. As a consequence, we find that rupture of a single molecule does not occur homogeneously and thus at random locations, but instead, a local stress concentration develops leading to rupture of the molecule. Such information about the fracture behavior of collagen may be essential to understand the role of collagen components in biological materials. For example, the mechanics of collagen fibers at large stretch may play a critical role in the mechanical properties of bone during crack propagation, and elucidation of its mechanical response in particular at large strains is of critical importance during crack bridging in bone-like hard tissues (Ritchie et al. 2004).

We believe that reactive modeling that takes into account the complexity of chemical bonding may be critical to understand the fracture and deformation behavior of many other biological and protein-based materials. Further, we find a strong rate dependence of the mechanical properties, including Young's modulus. This is in agreement with the fact that collagen is known to be a viscoelastic material and suggests that this behavior may at least partly originate from processes and mechanisms at the molecular scale.

We find that the properties of collagen are scale dependent. For example, the fracture strength of individual polypeptides is different from the fracture strength of a tropocollagen molecule (see, e.g. Figure 8.14). The results further provide estimates of the fracture and deformation strength for different types of loading, enabling a comparison of different relative strengths.

The mesoscale model was used to predict the force-extension curve of a long tropocollagen molecule, including a direct and quantitative comparison with experimental results (as shown in Fig. 8.22).

8.4 Deformation and Fracture of Collagen Fibrils

Now we move up in the hierarchical scale to study the mechanics of assemblies of many tropocollagen molecules into a collagen fibril (see Fig. 8.3). Particular focus of this section is on studies of effects of the molecular length and cross-link densities. The present section reviews the molecular simulation results reported in Buehler (2006b) and Buehler (2008) (for details regarding the modeling procedure see these references).

8.4.1 Model Geometry and Molecular Simulation Approach

A two-dimensional plane stress model of collagen fibrils with periodic boundary conditions in the in-plane direction orthogonal to the pulling orientation is considered here, with a periodic array of 2×5 tropocollagen molecules (total number of tropocollagen molecules is 10). The collagen fibrils show the characteristic staggered arrangement as observed in experiment. The plane stress condition is used to mimic the fact that the system is not periodic in the out-of-plane direction. Fully three-dimensional models are computationally very expensive. However, the model could treat such cases as well since there appears to be no intrinsic limitation of a two-dimensional case. No additional constraints other than the molecular interactions are applied to the system.

The simulations are carried out in two steps, (i) relaxation, followed by (ii) loading. Relaxation is achieved by slowly heating up the system, then annealing the structure at constant temperature, followed by energy minimization. Finite temperature calculations enable the structure to reassemble more easily, whereas energy minimization ensures finding the energetically optimal configuration of the molecules. If the initial relaxation is not carried out, pulling may be applied to a structure that is not in equilibrium and yield may be observed that is actually not due to the applied load but due to rearrangements toward the equilibrium structure. After relaxation, the structure displays the characteristics of collagen fibrils in agreement with experiment. To model tensile deformation of collagen fibrils, displacement boundary conditions are implemented by continuously displacing a set of particles in the boundary regions (in a region 40 Å to the left and right of the end of the collagen fibril).

The simulations are carried out by constantly minimizing the potential energy as the external strain is applied, where a displacement rate of 0.4 m/s is used for all simulations. Such rather high-strain rates are a consequence of the time scale limitation of the molecular model; total time spans of several microseconds are the most that can be simulated since the time step has to be on the order of several femtoseconds.

As indicated above, the virial stress is used to calculate the stress tensor (Tsai 1979) for analyses of the stress–strain behavior. The yield stress σ_Y is defined as the stress at which permanent deformation of the collagen fibril begins. This is characterized either by intermolecular shear or by molecular fracture, leading to permanent deformation. The yield strain ε_Y is defined as the critical strain at which these mechanisms begin. The fracture stress σ_F is defined as the largest stress in the stress–strain curve, corresponding to the maximum load the collagen fibril can sustain. The fracture strain ε_F is the corresponding strain at which the largest stress occurs.

The strain is defined as $\varepsilon = (x - x_0)/x_0$, where x_0 is the initial, undeformed length of the collagen fibril, and x is the current, deformed length. It is noted that the extension ratio or stretch λ is related to the strain via $\lambda = 1 + \varepsilon$.

8.4.2 Size-Dependent Properties: Effects of Molecular Length

Here we focus on atomistic and molecular modeling of the mechanical properties of collagen under large stretch, leading to permanent deformation or fracture. We show that the key to understanding the mechanics of collagen is to consider the interplay among the mechanics of individual tropocollagen molecules, the intermolecular chemical interactions and the mesoscopic properties arising from hundreds of molecules arranged in fibrils. We explore the mechanics of collagen by considering different nanostructural designs and pay specific attention to the details of molecular and intermolecular properties and its impact on the mechanical properties.

Under macroscopic tensile loading of collagen fibrils, the forces are distributed predominantly as tensile load carried by individual and as shear forces between different tropocollagen molecules (Buehler 2006a, b). Energetic effects rather than entropic contributions govern the elastic and fracture properties of collagen fibrils and fibers. The fracture strength of individual tropocollagen molecules is largely controlled by covalent polypeptide chemistry. The shear strength between two tropocollagen molecules is controlled by weak dispersive and hydrogen bond interactions and by some intermolecular covalent cross-links.

8.4.2.1 Theoretical Considerations: Homogeneous Shear

We first consider a simplistic model of a collagen fibril by focusing on a staggered assembly of two tropocollagen molecules (Buehler 2006b), as illustrated in Fig. 8.23. The shear resistance between two tropocollagen molecules, denoted as τ_{shear}, leads to a contact length dependent force

Fig. 8.23 Simplistic representation of the staggered collagen fibril geometry as a simple bimolecular assembly. The lower part of the figure shows the distribution of intermolecular shear forces and tensile forces within each TC molecule

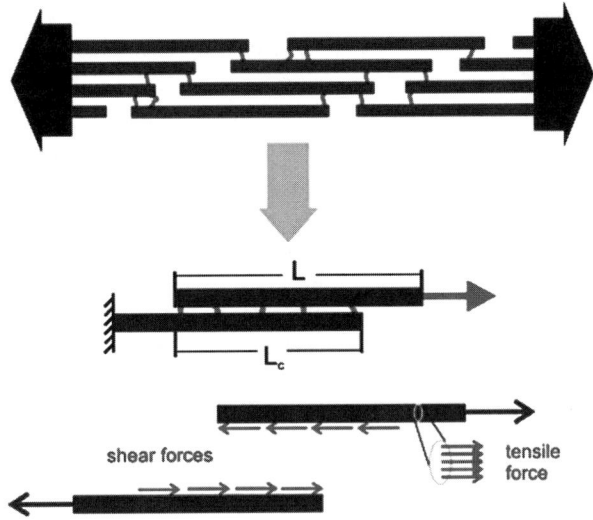

$$F_{\text{tens}} = \tau_{\text{shear}} L_C = \alpha \tau_{\text{shear}} L, \qquad (8.27)$$

where L_C is the contact length and F_{tens} is the applied force in the axial molecular direction, which can alternatively be expressed as tensile stress $\sigma_{\text{tens}} = F_{\text{tens}}/A_C$ by considering the molecular cross-sectional area A_C. The parameter α describes the fraction of contact length relative to the molecular length, $\alpha = L_C/L$. Due to the staggered geometry, the shear resistance increases linearly with L, thus $F_{\text{tens}} \sim \tau_{\text{shear}} L$. This model holds only if shear deformation between the molecules is homogeneous along the axial direction.

8.4.2.2 Theoretical Considerations: Nucleation of Slip Pulses

An alternative to homogeneous intermolecular shear is propagation of slip pulses due to localized breaking of intermolecular "bonds". This analysis is based on a one-dimensional model of fracture initially proposed by Hellan (Griffith 1920; Hellan 1984). The model describes a one-dimensional strip of material attached on a substrate, which is under tensile loading in the axial direction. At a critical load, the energy released per advancement length of the adhesion front is equal to the energy required to break the bonding between the material strip and the substrate, leading to initiation of failure front. The failure front – corresponding to a dynamic crack tip – propagates at a fraction of the sound velocity, eventually displacing the material permanently in the direction of the applied load. We now apply this model to intermolecular deformation in collagen fibrils. The energy release rate is given by

$$G_0 = \frac{\sigma_R^2}{2E}, \qquad (8.28)$$

where E is Young's modulus of the tropocollagen molecule and σ_R the applied stress. With γ as the energy necessary to nucleate this defect, at the onset of nucleation the condition

$$G_0 = \gamma \tag{8.29}$$

needs to be satisfied (similar to the Griffith condition (Griffith 1920)). The detachment front corresponds to the front of decohesion. Bonds behind the fracture front reform, thus forming a "slip pulse". The slip pulse is a region with increased tensile strain in the tropocollagen molecule, which is several nanometers wide.

The existence of slip pulses is not a consequence of the discretization of the mesoscale model. Instead, this theoretical framework is developed based on continuum mechanics, assuming a homogeneous distribution of adhesive interactions along the molecular surface. In the spirit of Griffith's energy argument describing the onset of fracture, nucleation of slip pulses is controlled by the applied tensile stress σ_R, where

$$\sigma_R = \sqrt{2E\gamma}, \tag{8.30}$$

where E is Young's modulus of an individual tropocollagen molecule and γ relates to the energy required to nucleate a slip pulse.

When $\sigma_{\text{tens}} < \sigma_R$, deformation is controlled by homogeneous shear between tropocollagen molecules. However, when $\sigma_{\text{tens}} \geq \sigma_R$ intermolecular slip pulses are nucleated. This leads to a critical molecular length

$$\chi_S = \frac{\sqrt{2\gamma E}}{\tau_{\text{shear}}\alpha} A_C. \tag{8.31}$$

For fibrils in which $L < \chi_S$, the predominant deformation mode is homogeneous shear. When $L > \chi_S$, propagation of slip pulses dominates. The strength of the fibril is then independent of L (Eq. (8.31)), approaching $\tau_{\text{shear}}\alpha\chi_S$. This concept is similar to the flaw tolerance length scale proposed for mineral platelets in bone (Gao et al. 2003).

The length scale χ_S depends on the material parameters and interaction between molecules. If γ assumes very large values – for instance due to high cross-linking density, or due to the effects of solvents (e.g., low water concentration) – the tensile forces in each tropocollagen molecule (Eq. (8.27), or $F_{\text{tens}} \sim L$) reach the tensile strength of tropocollagen molecules, denoted by F_{max}, before homogeneous shear or slip pulses are nucleated (F_{max} is a material constant that ultimately depends on the molecular structure of the tropocollagen molecule).

Considering $F_{\text{tens}} = F_{\text{max}}$ leads to a second critical molecular length scale,

$$\chi_R = \frac{F_{\text{max}}}{\tau_{\text{shear}}\alpha}. \tag{8.32}$$

This molecular length χ_R characterizes when the transition from molecular shear to brittle-like rupture of individual tropocollagen molecules occurs. The response of collagen fibrils to mechanical load changes from shear or glide between tropocollagen molecules, to molecular fracture as L increases. For $L > \chi_R$, tropocollagen molecules break during deformation, whereas for $L \leq \chi_R$ deformation is characterized by homogeneous intermolecular shear.

The integrity of a complete collagen fibril is controlled by the strength of the weakest link. Thus, the interplay of the critical length scales χ_S/χ_R controls the deformation mechanism.

When $\chi_S/\chi_R < 1$, slip pulse nucleation governs at large molecular lengths, whereas when $\chi_S/\chi_R > 1$, fracture of individual tropocollagen molecules occurs. For $L/L_\chi < 1$ homogeneous intermolecular slip dominates deformation. In both cases, the strength does not increase by making L larger. The maximum strength of the fibril is reached at $L \approx L_\chi = \min(\chi_R, \chi_S)$. This is true for any arbitrary length L of a tropocollagen molecule. Homogeneous shear deformation dominates below the critical molecular length L_χ. For molecules with $L > L_\chi$, either slip pulses or fracture sets in, depending on which of the two length scales χ_S or χ_R is smaller. For short tropocollagen molecules, the strength of collagen fibrils tends to be small and depends on L_C. When $L \approx L_\chi$, the maximal tensile strength of fibrils is reached.

Further, choosing $L \approx L_\chi$ leads to maximized energy dissipation during deformation. The work necessary to separate two fibers in contact along a length L_C under macroscopic tensile deformation is

$$E_{\text{diss}} = \int_{l=0}^{l=L_C} l \tau_{\text{shear}}\, dl = \frac{1}{2} L_C^2 \tau_{\text{shear}}. \tag{8.33}$$

Equation (8.33) predicts an increase of the dissipated energy with increasing molecule length, therefore favoring long molecules. If $\chi_R < \chi_S$, the critical length L_χ constitutes an upper bound for L_C, since molecules rupture before shear deformation sets in. After bond rupture and formation of shorter molecules, E_{diss} decreases significantly, suggesting that $L > L_\chi$ is not favored. Energy dissipation is at a maximum for $L \approx L_\chi$. If $\chi_S < \chi_R$, the dissipated energy can be approximated by (assuming $L_C > \chi_S$)

$$E_{\text{diss}} \approx \left(\frac{1}{2}\alpha^2 \chi_S^2 \tau_{\text{shear}} + (L_C - \alpha\chi_S) \cdot F_{\text{max}} \right), \tag{8.34}$$

suggesting that after a quadratic increase for small molecular lengths, the dissipated energy increases linear with L_C.

8.4.2.3 Molecular Modeling of Mechanical Properties of Collagen Fibrils

We now model the deformation behavior of a more realistic fibril geometry as shown in Fig. 8.23 ("fibril", upper part), studying the change in mechanical properties due to variations in molecule length L.

Due to the staggered design of collagen fibrils with an axial displacement of about 25% of the molecular length, the contact length between tropocollagen molecules in a fibril is proportional to L. The length scales suggested in the previous section therefore have major implications on the deformation mechanics of collagen fibrils.

We consider fully hydrated cross-link-free collagen fibrils serving as a model for cross-link-deficient collagen. Figure 8.24 shows the stress versus strain response of a collagen fibril for different molecular lengths L. The results suggest that the onset of plastic deformation, the maximum strength and large-strain mechanics of collagen fibrils depends on the molecular length.

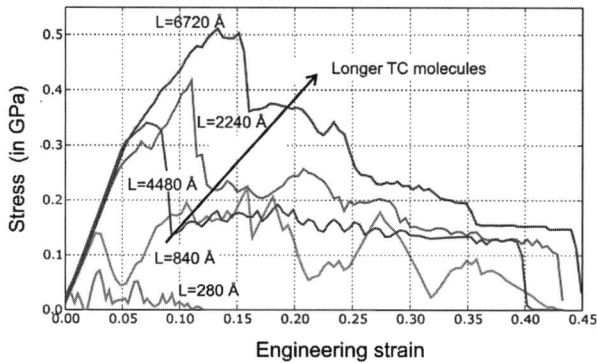

Fig. 8.24 Stress versus strain of a collagen fibril, for different molecular lengths (model for cross-link-deficient collagen, as no covalent cross-links are present in the collagen fibril) (results as reported in (Buehler 2006)). The results suggest that the longer the molecular length, the stronger the fibril. The maximum elastic strength achieved by collagen fibrils approaches approximately 0.3 GPa, with largest stress around 0.5 GPa. The onset of intermolecular shear can be recognized by the deviation of the stress–strain behavior from a linear-elastic relationship

Figure 8.25 shows the normalized elastic strength of the fibril as a function of molecular length L. The results suggest an increase up to about 200 nm, then reaching a plateau value of around 0.3 GPa (results normalized by this value). The elastic uniaxial strains of collagen fibrils reach up to approximately 5%. The maximum stress reaches up to 0.5 GPa, during plastic deformation.

The molecular length at which the saturation occurs corresponds to a change in deformation mechanism, from homogeneous shear ($L \rightarrow 0$) to nucleation of slip pulses ($L \rightarrow \infty$). The corresponding molecular length provides an estimate for the critical molecular length scale $\chi_S \approx 200$ nm.

We note that $\chi_R \approx 436$ nm, as described in the previous section (it is a material property of the reference system). Therefore, the ratio $\chi_S/\chi_R < 1$, suggesting a

Fig. 8.25 This plot shows the critical stress at the onset of plastic shear between TC molecules (Buehler 2006b). An initial regime of linear increase of strength with molecular length is followed by a regime of finite strength, at a plateau value

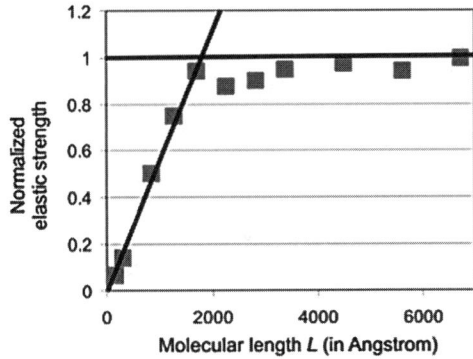

competition between slip pulses and homogeneous shear as the molecular length is varied. This suggests that cross-link-deficient collagen may predominantly undergo intermolecular shear deformation.

Figure 8.26 depicts the energy dissipated during deformation, per unit volume. We observe continuous increases with molecule length L, reaching a maximum at a critical molecular length L_χ, then a slight decrease. Energy dissipation increases further at ultra large molecular lengths beyond 400 nm, due to longer shear paths during slip pulse propagation. The modest increase in energy dissipation for ultra-long molecules may be an inefficient solution, since assembling such ultra-long molecules into regular fibrils is challenging.

Fig. 8.26 The plot depicts the dissipated energy during deformation, per unit volume, in a collagen fibril, as a function of molecular length, normalized by the maximum value (Buehler 2006b). An initial steep increase is followed by a plateau regime, with a local maximum around 220 nm. The smooth curve is a fit of a third-order expansion to the simulation data

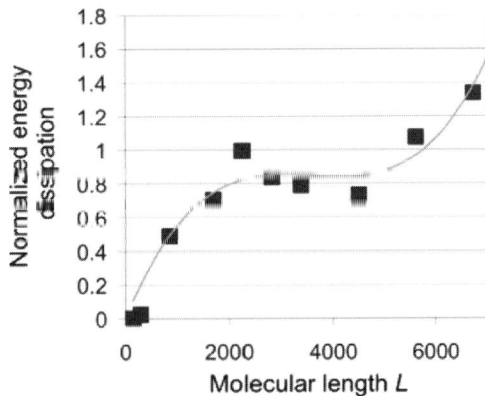

8.4.2.4 Discussion

The results suggest that the length of tropocollagen molecules plays a significant role in determining the deformation mechanics, possibly explaining some of the universal structural features of collagen found in Nature.

The two length scales χ_S and χ_R provide a quantitative description of the three different deformation mechanisms in collagen fibrils: (i) intermolecular shear, (ii) slip pulse propagation and (iii) fracture of individual tropocollagen molecules.

The governing deformation mechanism is controlled by the ratio χ_S / χ_R: Whether molecular fracture ($\chi_S / \chi_R > 1$) or slip pulses ($\chi_S / \chi_R < 1$) dominate deformation, the strength of the fibril approaches a maximum that cannot be overcome by increasing L. When $L_\chi = \min(\chi_R, \chi_S)$, tensile forces due to shear are in balance with either the fracture strength of tropocollagen molecules ($\chi_S / \chi_R > 1$) or with the critical load to nucleate slip pulses ($\chi_S / \chi_R < 1$). In either case, the maximum strength of the fibril is reached when $L \approx L_\chi$, including maximum energy dissipation.

When the length of collagen molecules is close to the critical length scale L_χ, two objectives are satisfied: (i) Under large deformation, tropocollagen molecules reach their maximum strength without leading to brittle fracture and (ii) energy dissipation during deformation is maximized. This concept may explain the typical staggered geometry of collagen fibrils found in experiment, with extremely long molecules – leading to large energy dissipation during deformation.

The mechanisms of deformation and their dependence on the molecular design are summarized in a deformation map, shown in Fig. 8.27.

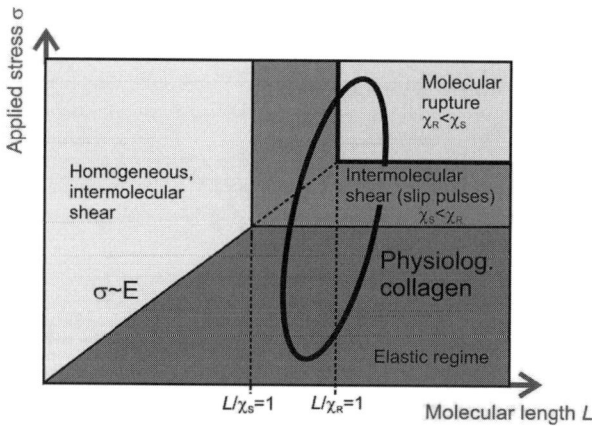

Fig. 8.27 Deformation map of collagen fibrils, as reported in Buehler (2006b). The mechanical response is controlled by two length scales χ_S and χ_R. Intermolecular shear governs deformation for small molecular lengths, leading to relatively small strength of the collagen fibril. For large molecular lengths, either intermolecular slip pulses ($\chi_S / \chi_R < 1$) or rupture of individual TC molecules ($\chi_S / \chi_R > 1$) dominate. This regime refers to the case of strong intermolecular interactions (e.g., increased cross-link densities). Physiological collagen typically features long molecules, with variations in molecular interaction, so that either intermolecular shear (e.g., slip pulses) or molecular fracture are predicted to dominate

Slip pulses are nucleated by localized larger shear stresses at the end of the tropocollagen molecules. Thus, cross-links at these locations provide a molecular-scale mechanism to prevent slip pulse nucleation, as this leads to an increase of the energy required to nucleate slip pulses, thus to a larger value of γ. This results in an increase of χ_S, due to the scaling law:

$$\chi_S \sim \sqrt{\gamma}. \tag{8.35}$$

As a consequence, the ratio χ_S/χ_R increases, making collagen fibrils stronger. Remarkably, this nanoscale distribution of cross-links agrees with the natural collagen design seen in experiment, often showing cross-links at the ends of the tropocollagen molecules (Bailey 2001) (reminiscent of crack bridging (Nalla et al. 2003a,b; Ritchie et al. 2004)).

Cross-links provide additional strength to the fibrils, in agreement with experiment (Bailey 2001). However, extremely large cross-link densities lead to negative effects as the material is not capable to dissipate much energy during deformation – leading to a brittle collagen that is strong, but not tough. Such behavior is observed in dehydrated collagen, or in aged collagen featuring higher cross-link density (Bailey 2001). In contrast, decreased cross-linking as it occurs in the Ehlers–Danlos V disease (Lichtens Jr et al. 1973; Glorieux 2005) leads to significantly reduced tensile strength of collagen, as $\chi_S/\chi_R < 1$. The ratio L/L_χ decreases, resulting in skin and joint hyperextensibility due to extremely weak collagen tissue, incapable of dissipating significant energy.

Our model can be used to study different design scenarios. A design with many cross-links and short molecules would lead to a very brittle collagen, even in the hydrated state. Such behavior would be highly disadvantageous under physiological conditions. In contrast, long molecules provide robust material behavior with significant dissipation of energy (Fig. 8.26). Some experiments (Hulmes et al. 1995; Sasaki and Odajima 1996; Puxkandl et al. 2002; Sun et al. 2004; Bhattacharjee and Bansal 2005; Bozec et al. 2005) support the notion that cross-link-deficient collagen shows wide yield regions and large plastic deformation.

Both elastic strength and energy dissipation approach a finite value for large molecular lengths, making it inefficient to create collagen fibrils with tropocollagen molecules much longer than L_χ, which is on the order of a few hundred nanometers (Fig. 8.25). This length scale agrees somewhat with experimental results of tropocollagen molecules with lengths around 300 nm.

Large deformation is a critical physiological condition for collagen-rich tissue. The risk of catastrophic brittle-like failure needs to be minimized in order to sustain optimal biological function. The nanoscale ultrastructure of collagen may be designed to provide robust material behavior under large deformation by choosing long tropocollagen molecules. Robustness is achieved by the design for maximum strength and maximized energy dissipation by shear-like mechanisms. The requirement for maximum energy dissipation plays a crucial role in determining the optimal molecular length L_χ. The layered design of collagen fibrils plays a critical role in enabling long deformation paths with large dissipative stresses. This is reminiscent of the "sacrificial bond" concept (Hansma et al. 2007).

The properties of collagen are scale dependent (Sasaki and Odajima 1996). The fracture strength of an individual tropocollagen molecule (11.2 GPa) differs from the fracture strength of a collagen fibril (0.5 GPa). Similarly, Young's modulus of an individual tropocollagen molecule is approximately 7 GPa, while Young's modulus of a collagen fibril is smaller, approaching 5 GPa (for $L \approx 224$ nm). This is in qualitative agreement with experiment.

Quantitative theories of the mechanics of collagen have many applications, ranging from the development of new biopolymers to studies in tissue engineering, where collagen is used as a scaffolding material. In addition to optimization for mechanical properties, other design objectives such as biological function, chemical properties or functional constraints may be responsible for the structure of collagen. However, the physiological significance of large mechanical deformation of collagen fibers suggests that mechanical properties could indeed be an important design objective.

8.4.3 Effect of Cross-Link Densities

To understand the influence of cross-links on the deformation mechanics of collagen fibrils, a series of computational experiments of pulling individual collagen fibrils with increasing density of cross-links are carried out (Buehler 2008). All results are compared with a control system of a cross-link-free collagen fibril. Systematic increases of the density of cross-links enable one to observe the difference in mechanical behavior. Particular attention is paid to the small- and large deformation behavior and the effect of intermolecular cross-links on the mechanical properties and deformation mechanisms.

In particular, studies are carried out that focus on the changes in the elastic and fracture behavior of the collagen fibril as the parameters are varied. An analysis of the molecular mechanisms allows one to develop a mechanistic understanding of the deformation behavior of collagen fibrils.

Figure 8.28 visualizes how the molecular model describes the presence of intermolecular cross-links.

Fig. 8.28 Schematic showing how the presence of cross-links is modeled by increased adhesion at the ends of each molecule, in segments of 60 Å to the left and right of each tropocollagen molecule. Implementing a variation of the amplification of the adhesion strength constitutes a simplistic model for varying cross-link densities. A parameter β is introduced that describes the increase of adhesion at the ends of each TC molecule, so that $\tau = \beta \tau XL$ (τ is the adhesion force/length between two TC molecules). The parameter $\beta = 15$ corresponds to the case when approximately one cross-link is present at each end of a tropocollagen molecule

8.4.3.1 Tensile Deformation: Stress–Strain Curves

Figure 8.29 depicts the stress–strain curve for various cross-link densities, expressed in terms of the parameter β. For small values of cross-link densities ($\beta < 10$), the fibril starts to yield at strain in the range of 5–10% and shows rather long dissipative deformation paths, leading to fracture at strains between 50 and 100%.

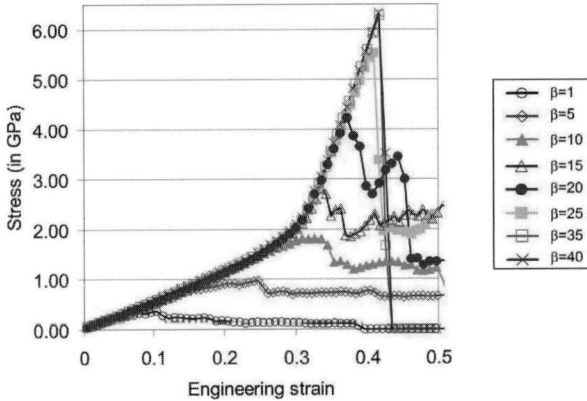

Fig. 8.29 Stress versus strain of a collagen fibril, for different cross-link densities, as reported in Buehler (2008). The results clearly show that larger cross-link densities lead to larger yield strains, larger yield stresses as well as larger fracture stresses. For larger cross-link densities, the second elastic regime (seen as much steeper, second slope) is activated. As the cross-link density increases, the collagen fibril shows a more "brittle-like" deformation behavior. For values of $\beta > 25$, the deformation mechanisms is characterized by molecular fracture, and as a consequence, the maximum fracture stress of the collagen fibril does not increase with increasing cross-link densities. This cross-link density corresponds to the case when two cross links per molecule are present

It is found that larger cross-link densities lead to larger yield strains, larger yield stresses as well as larger fracture stresses. At a critical cross-link density corresponding to one cross-link per molecule ($\beta \approx 15$), the second, steeper elastic regime is activated. This strong increase in tangent modulus corresponds to stretching of the protein backbone. This molecular deformation mode dominates after the uncoiling of the tropocollagen molecule under breaking of H-bonds (Sasaki and Odajima 1996). The results clearly confirm the significance of the presence and density of cross-links on the deformation behavior.

Large elastic tensile strains of up to 50% are possible since each tropocollagen molecule itself can sustain strains of up to 50% tensile deformation (this is shown in Fig. 8.4, curve for a single tropocollagen molecule). In the collagen fibril, such large strains at the molecular scale are only possible if strong links exist which prevent molecular slip and therefore enable transfer of large loads to the individual tropocollagen molecules. As shown in this chapter, developing cross-links between molecules is a possible means of achieving this situation.

8.4.3.2 Comparison: Single Tropocollagen Molecule and Collagen Fibril

Figure 8.30 shows a comparison between the stress–strain curves of a collagen fibril ($\beta = 25$) and a single tropocollagen molecule, for tensile strains below 40%. Both structures are completely in the elastic regime (the tropocollagen molecule fractures at approximately 50% tensile strain and the collagen fibril starts to yield at slightly above 40% strain). These results show that the stresses in the single tropocollagen molecule are larger than in the collagen fibril.

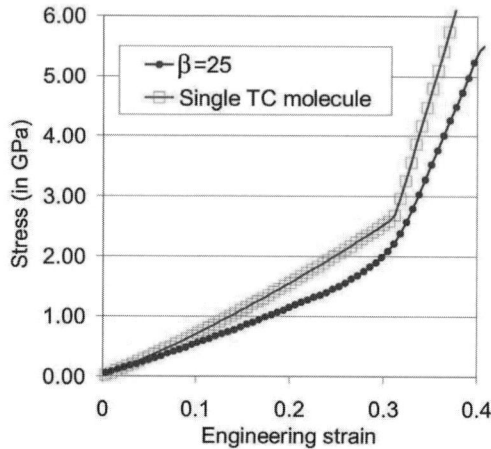

Fig. 8.30 Stress versus strain, comparing a collagen fibril ($\beta = 25$) with a single TC molecule (Buehler 2008). Both structures are completely in the elastic regime (the TC molecule fractures at approximately 50% tensile strain and the collagen fibril starts to yield at slightly above 40% strain). This plot shows that the stresses in the single TC molecule are larger than in the collagen fibril and that the tangent modulus is larger throughout deformation. This agrees well with experimental results (Sasaki and Odajima 1996)

Figure 8.31 plots the tangent modulus of the stress–strain curve depicted in Fig. 8.30. The results clearly indicate that the tangent modulus of the single tropocollagen molecule is larger than that of a collagen fibril. The results suggest that the modulus of a single tropocollagen molecule is approximately 40% larger throughout deformation. This agrees well with experimental results, suggesting an increase of the stiffness from fibril to molecule close to 40% (Sasaki and Odajima 1996).

Even though cross-links are stiffer than the tropocollagen molecule itself, the overall density of cross-links is rather small so that the stiffening effect is negligible. The origin of the softening is the combination of rather weak intermolecular interactions with the single molecule elasticity along most of the axial length of the tropocollagen molecules. This leads to an effective softening of the fibrillar structure even when cross-links are present. This may change for extremely large cross-link densities, for example when cross-links form along the entire axial dimension of the chain.

Fig. 8.31 Tangent modulus versus strain, comparing a single TC molecule and a collagen fibril (cross-link parameter $\beta = 25$) (Buehler 2008). The results show that the tangent modulus of the single TC molecule is approximately 40% larger, except for the transition region during which the modulus of the fibril is larger (between 20 and 30% fibril strain)

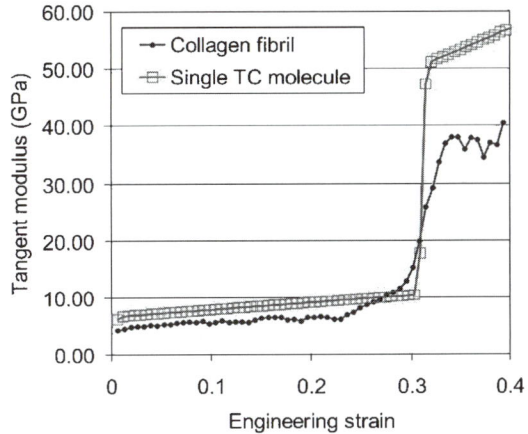

8.4.3.3 Yield Stress and Fracture Stress Analysis

Figure 8.32 depicts the yield stress of a collagen fibril as a function of the cross-link density (curve "relative strength"). The plot shows that for larger cross-link densities, the material becomes stronger. However, when $\beta > 25$, the yield stress and fracture stress do not depend on the cross-link density any more as the yield stress reaches a plateau value. The plateau can be explained by a change in molecular deformation mechanism from predominantly intermolecular shear (for $\beta < 25$) to molecular fracture (for $\beta > 25$). Whereas the strength of the fibril is controlled by intermolecular adhesion for $\beta < 25$, the strength is dominated by the molecular fracture properties. This observation confirms a change in mechanisms as suggested in an earlier study (Buehler 2006b).

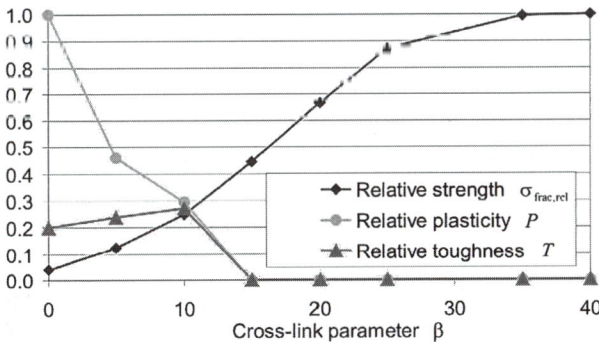

Fig. 8.32 Relative strength $\sigma_{frac,rel}$, relative amount of plasticity P as well as relative toughness of a collagen fibril T as a function of the cross-link parameter β. For large cross-link densities $\beta > 15$, the material behavior becomes increasingly brittle and the failure strength (or yield strength, equivalently) saturates, as failure is controlled by rupture of individual TC molecules. The variation of the toughness T suggests that a maximum relative toughness is reached for a cross-link parameter of approximately $\beta \approx 10$

As the cross-link density increases, the collagen fibril becomes more "brittle-like". The increasingly brittle character is clearly illustrated by the ratio of fracture stress versus yield stress. For smaller values of $\beta < 15$, the stress–strain curves show a stiffening effect after onset of yield, similar to work-hardening as known in metal plasticity. However, this stress increase decreases with increasing cross-link density. The data show that as the cross-link parameter exceeds 15, the material becomes "brittle-like", characterized by immediate drop of the stress after onset of yield without dissipative deformation.

An analysis of the stress–strain behavior provides further insight into the elastic and plastic deformation modes. The analysis of the stress–strain curves for varying cross-link densities corroborates the notion that for increasing cross-link densities, the material becomes increasingly "brittle-like".

Figure 8.32 further depicts the relative strength ($\sigma_{\text{frac,rel}} = \sigma_{\text{frac}}/\sigma_{\text{frac,max}}$), relative amount of plasticity (calculated with yield strain $\varepsilon_{\text{yield}}$ and yield stress $\varepsilon_{\text{frac}}$, $P = (\varepsilon_{\text{frac}}/\varepsilon_{\text{yield}} - 1)/(\max(\varepsilon_{\text{frac}}/\varepsilon_{\text{yield}}) - 1)$), as well as the relative toughness of a collagen fibril ($T = \sqrt{P \cdot \sigma_{\text{frac,rel}}}$) as a function of the cross-link parameter β. Hereby the "toughness" of a material is defined as the property of being both strong and requiring a lot of energy to break. The results suggest that an optimal relative toughness is reached for a cross-link parameter of $\beta \approx 10$. This value density corresponds to an approximate spacing of cross-links in the molecular axial direction of ≈ 420 nm.

8.4.3.4 Comparison with Experimental Results

This section is dedicated to a brief discussion of our computational results in light of recent experimental reports of stretching experiments of tropocollagen molecules and individual collagen fibrils.

Table 8.2 provides an overview of moduli obtained for single tropocollagen molecules. The comparison shows that our predictions for the moduli are close to experimental results, albeit they fall into the higher end of the range of values reported.

Table 8.3 summarizes results for elastic moduli of collagen fibrils from various sources. Unlike as for the single molecule case the agreement between experiment and simulation is not as good. A few important observations are discussed in more detail. Recently, MEMS devices were used to carry out tensile studies of single collagen fibrils (Eppell et al. 2006). The authors obtained a small-deformation modulus of approximately 0.4 GPa and a large deformation modulus of 12 GPa. The absolute values of the small-strain moduli are approximately 10 times smaller than in our simulation results.

One possible explanation for this disagreement could be entropic effects that may make the fibril softer in particular in the small-deformation regime. Such entropic effects are not considered in the present study, since all molecules are completely stretched out to their contour length at the beginning of the simulation and thus enter the energetic stretching regime instantaneously.

Table 8.2 Comparison of Young's modulus of single tropocollagen molecules, experiment and computation

Study/case and approach	Young's modulus
Single molecule stretching (Lorenzo and Caffarena 2005) – atomistic modeling	4.8 ± 1 GPa
Single molecule stretching (Buehler 2006a) – reactive atomistic modeling	≈ 7 GPa
Single molecule stretching (Vesentini, Fitie et al. 2005) – atomistic modeling	2.4 GPa
X-ray diffraction (Sasaki and Odajima 1996)	≈ 3 GPa
Brillouin light scattering (Harley, James et al. 1977)	9 GPa
Brillouin light scattering (Cusack and Miller 1979)	5.1 GPa
Estimate based on persistence length (Hofmann, Voss et al. 1984)	3 GPa
Estimate based on persistence length (Sun, Luo et al. 2004)	0.35–12 GPa

Table 8.3 Comparison of Young's modulus of collagen fibrils, experiment and computation

Study, case and approach	Young's modulus
MEMS stretching of collagen fibrils (Ballarini et al. (Eppell, Smith et al. 2006))	≈ 0.4–0.5 GPa (small-strain modulus) ≈ 12 GPa (large-strain modulus)
X-ray diffraction (Gupta, Messmer et al. 2004)	1 GPa
AFM testing (van der Rijt, van der Werf et al. 2006)	2.7 GPa (ambient conditions) 0.2–0.8 GPa (aqueous media)
Molecular multi-scale modeling (Buehler 2006a,b; Buehler 2008)	4.36 GPa (small-strain modulus) ≈ 38 GPa (large-strain modulus)

Another possible reason may be the large deformation rates used in atomistic modeling, which often lead to overestimation of forces during mechanical deformation. Considering smaller deformation rates, for instance, may lead to smaller values for Young's modulus, as typically unfolding forces are larger for larger deformation rates (e.g., based on concepts related to Bell theory (Ackbarow et al. 2007; Buehler and Ackbarow 2007). Since the molecular model used in this study is based solely on atomistic input data, overestimation of the modulus value from the MD simulations will be transported throughout the multi-scale modeling scheme. This may explain why the values reported in the present work are close to the upper end of the range of experimental measurements.

However, the ratio of large-strain modulus to small-strain modulus is on the same order of magnitude, being between 24 and 30 in experiment and approximately 8.4 in simulation. The transition from small to large deformation modulus occurs at strains of approximately 30%, which is found in both experiment and simulation.

It has been suggested in Sasaki and Odajima (1996) and Borsato and Sasaki (1997) that the tensile strength may be greater than 1 GPa, which is corroborated by our results that suggest strengths ranging from 300 MPa (cross-link-deficient fibrils) to 6 GPa (highly cross-linked collagen fibrils). These values agree with the strengths predicted in our simulation (see Fig. 8.6, for example). On the other hand, other results (Sasaki and Odajima 1996; Borsato and Sasaki 1997) show much lower failure stresses on the order of several MPa. Possible explanation for this discrepancy could be molecular defects, high loading rates or different geometries in those experiments that do not resemble the perfect patterns as considered in our study.

8.4.3.5 Discussion

Molecular modeling has been employed to predict the small and large deformation mechanics of collagen fibrils, as a function of varying cross-link densities. The results suggest that the cross-link density governs the large deformation and in particular the yield or fracture mechanics. However, it influences the small-deformation mechanics only marginally (see, e.g., Fig. 8.29).

The model predicts that collagen fibrils are capable of undergoing extremely large deformation without fracturing; how much of this is elastic or dissipative depends on the cross-link densities. It is found that two prominent molecular mechanisms of permanent deformation dominate: molecular glide and molecular rupture.

Formation of covalent cross-links are essential to reach the elastically stiffer, second regime in the stress–strain curve of collagen, which corresponds to backbone stretching in the tropocollagen molecule. This phenomenon can be understood based on the mechanisms and the effect of the presence of cross-links: The increased traction at the end of the molecule allows for larger molecular strains to be reached. The larger strains give rise to larger overall yield and fracture stress. However, collagen fibrils become more "brittle-like" under these conditions as their ability to undergo dissipative, plastic deformation is reduced. These findings confirm some of the key hypotheses put forward in Bailey (2001), including effect of cross-links in making the material appear more "brittle", observed deformation mechanics and reduction of elastic modulus. This is confirmed by several analyses shown in Fig. 8.32, for instance.

The results improve the understanding of how molecular changes during aging contribute to modifications of tissue properties. Aging of organisms is primarily controlled by changes in the protein structure of elastin and collagen, when increased cross-linking between molecules develops due to non-enzymatic processes. These changes in the molecular architecture may lead to diseases that are induced by the modification of the mechanical properties of tissues. The analysis confirms that cross-linking indeed leads to stiffening and increasing "brittleness" of collagen-based tissues. It is noted that the results shown in Fig. 8.29 are in good qualitative agreement with the results of stress–strain responses of collagen during aging. Both the present model and experiment predict a stronger and less dissipative behavior with the development of additional cross-links.

It is found that the material properties of collagen are scale dependent. A softening of the modulus is observed when tropocollagen molecules are assembled into a collagen fibril. The modeling suggests a reduction of modulus on the order of 40%, which is close to experimental results (Sasaki and Odajima 1996) of similar comparisons between the mechanics of collagen fibrils and tropocollagen molecules (see Fig. 8.30). This can also be found by taking a simple average value of all values for tropocollagen molecules reported in the literature (5.1 GPa, average of Table 8.2) divided by the average value of moduli for the collagen fibril (2.8 GPa, average of Table 8.3), which suggests an increase of modulus by approximately 80%.

The results show several features of the stress–strain behavior also found in experiment (Hulmes et al. 1995; Puxkandl et al. 2002; Eppell et al. 2006), notably the two regimes of moduli with a strong progressive stiffening with increasing strains. However, the magnitude of the stress is different, as MD modeling predicts larger stresses and larger moduli than seen in experiment.

A limitation of the present study is that spatial inhomogeneities of cross-link distributions are not considered. In principle, this can be implemented straightforwardly. Also, changes of molecular properties along the molecular length have not been considered, an important characteristic feature of many collagen-based tissues. This aspect is particularly significant to account for entropic effects that stem from more floppy labile regions of the tropocollagen molecules (Miles and Bailey 2001). These important aspects will be addressed in future work.

An improved understanding of the nanomechanics of collagen may help in the development of biomimetic materials, or for improved scaffolding materials for tissue engineering applications (Kim et al. 1999). Diseases such as Ehlers–Danlos (Lichtens Jr et al. 1973), osteogenesis imperfecta, scurvy or the Caffey disease (Glorieux 2005) are caused by defects in the molecular structure of collagen altering the intermolecular and molecular properties due to genetic mutations, modifying the mechanical behavior of collagen fibrils.

8.5 Nanomechanics of Mineralized Collagen Fibrils: Molecular Mechanics of Nascent Bone

One of the most intriguing protein materials found in Nature is bone, a material composed out of assemblies of tropocollagen molecules and tiny hydroxyapatite crystals, forming an extremely tough, yet lightweight material (Weiner and Wagner 1998; Currey 2002, 2005). Bone has evolved to provide structural support to organisms, and therefore, its mechanical properties are of great physiological relevance. Since collagen is the most fundamental building block of bone, here we review some insight into bone's smallest scale during bone formation (here referred to as nascent bone), mostly based on the study reported in Buehler (2007).

Mineralized collagen fibrils are highly conserved nanostructural building blocks of bone. By a combination of molecular dynamics simulation and theoretical analysis it is shown that the characteristic nanostructure of mineralized collagen fibrils is vital for its high strength and its ability to sustain large deformation, as relevant to the physiological role of bone, creating a strong and tough material. An analysis of the molecular mechanisms of protein and mineral phases under large deformation of mineralized collagen fibrils reveals a fibrillar toughening mechanism that leads to a manifold increase of energy dissipation compared to fibrils without mineral phase. This fibrillar toughening mechanism increases the resistance to fracture by forming large local yield regions around crack-like defects, a mechanism that protects the integrity of the entire structure by allowing for localized failure.

As a consequence, mineralized collagen fibrils are able to tolerate micro-cracks on the order of several hundred micrometers size without causing any macroscopic failure of the tissue, which may be essential to enable bone remodeling. The analysis proves that adding nanoscopic small platelets to collagen fibrils increases their Young's modulus, yield strength as well as their fracture strength. It was found that mineralized collagen fibrils have a Young's modulus of 6.23 GPa (versus 4.59 GPa for the collagen fibril), yield at a tensile strain of 6.7% (versus 5% for the collagen fibril) and feature a fracture stress of 0.6 GPa (versus 0.3 GPa for the collagen fibril).

The work reviewed here (Buehler 2007; additional details regarding the numerical procedure can be found therein) is limited to the scale of mineralized fibrils, with the objective to provide insight into the most fundamental scales of bone and its deformation mechanics under tensile loading.

8.5.1 Introduction

Figure 8.1 depicts the geometry of the nanostructure of bone, showing several hierarchical features from atomic to microscale. The smallest scale hierarchical features of bone include the protein phase composed of tropocollagen molecules, collagen fibrils (CFs) as well as mineralized collagen fibrils (MCFs) (see Fig. 8.33). Tropocollagen molecules assemble into collagen fibrils in a hydrated environment, which mineralize by formation of hydroxyapatite (HA) crystals in the gap regions

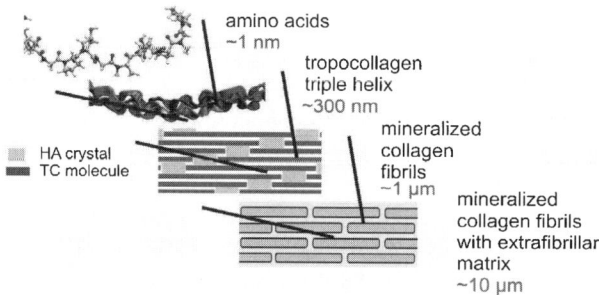

Fig. 8.33 Geometry of the nanostructure of bone, showing several hierarchical features from atomic to microscale. Simple schematic of the hierarchical design of mineralized collagen fibrils, forming the most basic building block of bone (Weiner and Wagner 1998; Currey 2002). Three polypeptide strands arrange to form a triple-helical tropocollagen molecule. Tropocollagen molecules assemble into collagen fibrils in a hydrated environment, which mineralize by formation of hydroxyapatite (HA) crystals in the gap regions that exist due to the staggered geometry. Mineralized collagen fibrils combine with the extrafibrillar matrix to fibril arrays, which form fibril array patterns (Weiner and Wagner 1998; Currey 2002; Gupta et al. 2005). Typically, a total of seven hierarchical levels are found in bone. The present work is limited to the scale of mineralized fibrils, with the objective to provide insight into the most fundamental scales of bone and its deformation mechanics under tensile loading

that exist due to the staggered geometry. MCFs arrange together with an extrafibrillar matrix (EFM) to form the next hierarchical layer of bone. While the structures at scales larger than MCFs vary for different bone types, mineralized collagen fibrils are highly conserved, nanostructural primary building blocks of bone that are found universally (Thompson et al. 2001; Fratzl et al. 2004; Aizenberg et al. 2005; Fantner et al. 2005; Gupta et al. 2005; Gao 2006; Gupta et al. 2006; Tai et al. 2006). Each MCF consists of tropocollagen molecules with approximately 300 nm length, arranged in a characteristic staggered pattern. Gap regions in this arrangement are filled with tiny hydroxyapatite (HA) crystals. The present work is limited to the scale of mineralized fibrils, with the objective to provide insight into the most fundamental scale of bone and its deformation mechanics under tensile loading.

The mechanical properties of bone have received significant attention. Particular effort has been devoted to understanding the mechanisms that make bone tough. Whereas some experimental evidence suggests that the sub-micrometer structure is critical for the mechanical properties of macroscopic bone, other results indicate that macroscopic mechanisms such as crack bridging or micro-cracking contribute to the toughness of bone (Nalla et al. 2003a,b; Ritchie et al. 2004; Nalla et al. 2005; Ritchie et al. 2006). Concepts such as sacrificial bonds and hidden length (Thompson et al. 2001; Fantner et al. 2005; Hansma et al. 2007) suggest toughening mechanisms that occur between different mineralized collagen fibrils.

However, due to the structural complexity of bone, the analysis and quantification of the deformation mechanisms at the ultra-scale of individual mineralized collagen fibrils (MCFs) remains an area that is not well understood. Limited knowledge exists whether, and if yes, how molecular scale mechanisms within single MCFs contribute to the toughness and stiffness of bone, as well as for its ability to repair itself. The effect of precipitating mineral crystals during bone formation remains unknown.

Most theoretical and computational analyses of bone have been carried out at continuum scales, neglecting the particular complexities of molecular interactions and chemistry. To date, there exists no molecular model of the nanostructure of bone that enables a rigorous linking between molecular and tissue scales. It is emphasized that previous atomistic and molecular models of collagen fibrils (described in the previous sections) have not included any mineral phase (Buehler 2006b).

In the studies reported here, a simple molecular model of MCFs is utilized that provides a fundamental description of its nanomechanical properties. The small and large deformation mechanics of a pure collagen fibril and a mineralized collagen fibril are systematically compared (see Fig. 8.34 for a comparison of the two model systems). Both structures are subject to identical tensile loading in the direction of the molecular axis of the tropocollagen molecules. Comparing the deformation mechanisms, stress–strain behavior and energy dissipation reveals insight into the effect of mineralization. It is found that mineralization leads to an increase in stiffness, yield stress, fracture stress and energy dissipation. The study reveals how a highly dissipative, yet strong material can be formed out of a soft polymeric collagen phase and hard, brittle HA by arranging molecules and crystals at characteristic nanostructured length scales.

Fig. 8.34 Overview of the
two structures considered
here; subplot (a) shows a CF
and subplot (b) shows a MCF
(original study reported in
Buehler 2007). The structures
are loading in uniaxial
tension along the axis of the
TC molecules

8.5.2 Molecular Model

The computational model is developed with the aim to elucidate generic behavior
and deformation mechanisms of MCFs.

The model is a simple 2D system of a mineralized collagen fibril, based on
the model of pure collagen fibrils discussed above, here extended to describe an
additional mineral phase. The CF consists of a staggered array of tropocollagen
molecules. The gap zones in the CF are filled with a single crystal that has a planar
size of approximately 28×1.4 nm, filling the entire open space, resembling the
presence of the HA phase, leading to a MCF.

It is noted that as bone is formed, mineral crystals exceed the size of the gap
region and penetrate into the collagen phase. The present study does not include
these effects and is thus limited to the early stages of bone formation ("nascent
bone"). The term "hydroxyapatite" (HA) is used to refer to the mineral phase in
bone, although this component is also referred to as "dahllite" or carbonated apatite.

The study reviewed in this section is the first molecular-scale model of the nanostructure of bone. The experimental paper (Gupta et al. 2004) considers a geometry
that is closest to the one considered in this chapter. However, to the best of our
knowledge no other molecular scale model of the nanostructure of bone has been
reported thus far.

The studies are carried out using a reactive mesoscopic model describing
tropocollagen molecules as a collection of beads interacting according to interparticle multi-body potentials; the pure collagen model described in Buehler (2006b)
is extended here to describe the HA phase and HA–TC interactions. The equations
of motion are solved according to a classical molecular dynamics (MD) scheme
implemented in the LAMMPS simulation code (Plimpton 1995).

The total potential energy of the model is

$$U = U_T + U_B + U_{TC} + U_{HA} + U_{HA-TC}. \tag{8.36}$$

The difference to the model for pure collagen fibrils is extended here by two terms,
U_{HA} (potential function for HA phase) and U_{HA-TC} (interaction potential HA–TC).
The terms U_T and U_B are only applied within the tropocollagen molecules, as
described above.

Intermolecular interactions between tropocollagen particles, HA particles and between HA and tropocollagen particles are described by a Lennard–Jones 12:6 (LJ) potential

$$\phi_{\text{TC/HA/HA}-\text{TC}}(r) = 4\varepsilon \left(\left[\frac{\sigma}{r} \right]^{12} - \left[\frac{\sigma}{r} \right]^{6} \right), \tag{8.37}$$

with σ as the distance and ε as energy parameter, defined separately for different materials.

Model parameters for tropocollagen properties and interactions are identical as in the pure collagen model. The parameter $\varepsilon_{\text{MF}} = 26.72 \, \text{kcal/mol}$, with $\sigma_{\text{MF}} = 3.118 \, \text{Å}$, which leads to a Young's modulus of $E_{\text{HA}} \approx 135 \, \text{GPa}$. These parameters are determined by fitting against the experimentally determined elastic modulus of HA. This 2D LJ model leads to extremely brittle material behavior (Buehler et al. 2004), thus providing a good model for the physical and mechanical properties of the HA phase.

Interactions between the HA crystal and tropocollagen molecules are described by a LJ potential with $\sigma_{\text{MF}-\text{TC}} = 7 \, \text{Å}$. The adhesion strength in this potential is chosen to be $\varepsilon_{\text{MF}-\text{TC}} = 25 \, \text{kcal/mol}$ for all HA–TC interactions, while the beginning and end of each tropocollagen molecule interacts with $\varepsilon_{\text{MF}-\text{TC}} = 15 \, \text{kcal/mol}$. The distinction of interaction mimics weaker adhesion between HA and tropocollagen at the head of each tropocollagen molecule due to smaller contact area. The choice of these parameters corresponds to interface surface energies of $\gamma_{\text{MF}-\text{TC}} \approx 0.375 \, \text{J/m}^2$ and $\gamma_{\text{MF}-\text{TC}} \approx 0.225 \, \text{J/m}^2$.

Classical MD is used to solve the equations of motion by performing a continuous energy minimization of the system as the loading is increased, with a time step of 55 fs. After energy minimization and relaxation of the initial structure, loading is applied by displacing a thin layer of particles at the ends of the system with a strain rate 7.558×10^{-8} per integration step. Periodic boundary conditions are applied in the direction orthogonal to pulling mimicking an infinitely large fibril, subject to uniaxial tensile loading.

It is emphasized that this simple model of the molecular and physical behavior of the nanocomposite is designed to deliberately avoid modeling the atomistic details of bonding within the HA crystal or across the HA–TC interface. However, it enables one to model the inhomogeneous stress and strain fields as well as the fracture behavior. Thus, the model system enables some first fundamental insight into the nanomechanics of mineralized fibrils.

8.5.3 Computational Results: Elastic, Plastic Regime and Fracture

Figure 8.35 plots the stress–strain response of a pure collagen fibril (CF) compared with that of a MCF, under tensile loading, for tensile strains up to 50%. The stress–strain response for CF and MCF is qualitatively and quantitatively different, indicating that precipitation of HA crystals during bone formation significantly alters the material response. The MCF features a larger strength and much increased energy dissipation under deformation. Plastic deformation starts at approximately 6.7% tensile strain for the mineralized fibril, whereas it occurs at approximately 5% tissue strain in the case of a pure CF. Further, the MCF shows significant softening at larger strains, with a characteristic sawtooth-shaped stress–strain curve due to repeated intermolecular slip. The mineralized fibril features a higher stiffness than the pure collagen fibril.

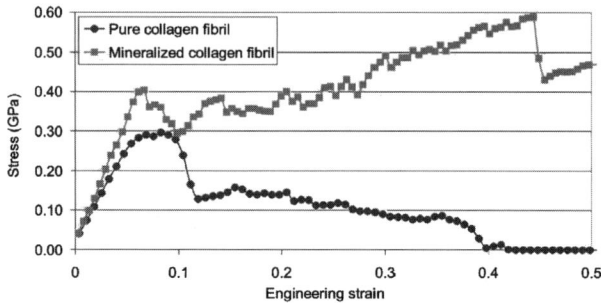

Fig. 8.35 Stress–strain response of a mineralized collagen fibril (MCF) and a nonmineralized, pure collagen fibril (CF) (Buehler 2007). The plot shows the stress–strain curve, for the entire deformation up to tensile tissue strains of 50%. It is apparent that the MCF features a larger strength and much increased energy dissipation under deformation. Plastic deformation starts at approximately 6.7% tensile strain for the mineralized fibril, whereas it occurs at approximately 5% tissue strain in the case of a pure CF. The MCF shows significant softening at larger strains, with a characteristic sawtooth-shaped stress–strain curve due to repeated intermolecular slip. The mineralized fibril is stiffer than the pure collagen fibril

Figure 8.36 shows snapshots of the molecular geometry under increasing tensile load, clearly showing the deformation mechanism of intermolecular slip. Figure 8.36(a) shows snapshots of the deformation mechanisms of pure CF. Fibrillar yield is characterized by intermolecular slip (see red circle highlighting a local area of repeated molecular slip). Slip leads to formation of regions with lower material density. Figure 8.36(b) displays snapshots of the deformation mechanisms of MCFs. Slip initiates at the interface between HA particles and tropocollagen molecules. Slip reduces the density, leading to formation of nanoscale voids.

The details of the differences between the MCF and CF and associated deformation mechanisms will be discussed in the following sections.

8.5.3.1 Elastic and Plastic Deformation

Up to the onset of yield, the mechanical response of the MCF is elastic (within the range of normal physiological function). In this regime, the increase of tissue strain leads to continuous increase of the strain in each tropocollagen molecule. Both MCF and CF display a linear-elastic regime for small deformation. The MCF is 36% stiffer than the CF (Young's modulus of a CF is 4.59 GPa versus 6.23 GPa for a MCF). The presence of HA crystals further changes the onset of plastic deformation,

(a)   (b)

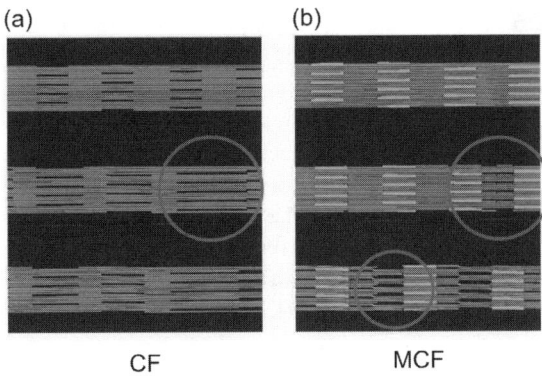

CF   MCF

Fig. 8.36 Molecular geometry of plastic deformation (Buehler 2007). Subplot (**a**): Snapshots of the deformation mechanisms, pure CF, for increasing strain. Fibrillar yield is characterized by intermolecular slip (see circle highlighting a local area of repeated molecular slip). Slip leads to formation of regions with lower material density. Subplot (**b**): Snapshots of the deformation mechanisms, MCF, for increasing strain. Slip initiates at the interface between HA particles and TC molecules. Slip reduces the density, leading to formation of nanoscale voids

characterized by a sudden drop in the stress–strain response. Whereas the CF begins to yield at approximately 5%, the MCF yields at 6.7%. This represents a 34% increase in yield strain. These results are summarized in Table 8.4. The data shown in Table 8.4 was generated based on the stress–strain curve shown in Fig. 8.35.

After onset of yield, the stress does not drop to zero rather quickly, but instead remains at levels of 0.4 GPa, with a slight increase with strain, approaching 0.6 GPa. After onset of yield, the MCF becomes softer, that is, less force is required for identical extension. The reduction in slope is due to the fact the strain in some of the tropocollagen molecules does not increase with tissue strain, since an increasing

Table 8.4 Quantitative comparison of the deformation and fracture properties of CFs and MCFs (tensile strains up to 50%), as reported in (Buehler 2007)

Property	CF	MCF	Ratio value MCF/CF
Young's modulus (small deformation)	4.59 GPa	6.23 GPa	1.36
Yield strain	5%	6.7%	1.34
Maximum stress	0.3 GPa	0.6 GPa	2
Failure mode	Molecular slip	Molecular slip and slip along HA–TC interface	N/A
Energy dissipation	3.83 GJ/m^3	19.48 GJ/m^3	≈ 5
Ratio of TC strain versus tissue strain	Approaches 87% at yield	Approaches 100% at yield	1.15
Ratio of HA strain versus tissue strain	N/A	11%	N/A
Size of fracture process zone ξ_{cr}	$\approx 150\mu m$	$\approx 200\mu m$	1.33

number of bonds to HA crystals and other tropocollagen molecules are broken. When fracture occurs, all molecular bonds inside the MCF are broken and the strains inside each component drops to zero.

Both CF and MCF yield by intermolecular slip. Repeated glide between tropocollagen molecules and between HA particles and tropocollagen molecules initiating by slip at the HA–TC interface enables a large regime of dissipative deformation after beginning of yield. In the case of the MCF, larger stresses can be maintained after initiation of slip due to additional resistance to slip at the interface between the tropocollagen molecules and HA particles. Mineralization of the CF leads to increase in strength by a factor of two. Most importantly, mineralization leads to a fivefold increase in energy dissipation.

8.5.3.2 Molecular Mechanisms of Deformation and Toughening

An analysis of the strain field within tropocollagen molecules and HA platelets reveals that the observations discussed in the previous section can be explained by molecular nanomechanical mechanisms, since mineralization significantly changes the strain distribution.

In pure CF, the tissue strain (applied strain) is always larger than the strain within tropocollagen molecules, reaching approximately 87% immediately before yield begins.

In MCF, the tissue strain and tropocollagen strain remain much closer during deformation, approaching similar strain levels at the onset of plastic deformation. This is due to the good adhesion between HA platelets and tropocollagen molecules, which hinders initiation of intermolecular slip. The HA phase carries up to 11% of the tissue strain. Such large tensile strains correspond to a stress of several GPa. Evidence for the molecular failure mechanisms of intermolecular slip is also found in experiment, as for instance shown in Gupta et al. (2004).

8.5.3.3 Comparison with Experimental Results

The most direct comparison of our molecular simulation results can be done with a recent experimental study reported in Gupta et al. (2004). We briefly summarize the main findings. By carrying out tensile tests of MCFs obtained from mineralized turkey leg tendon, it was shown that the stiffness increases continuously with increasing mineral content. It was shown that different mineralization stages correspond to stiffness values from 500 MPa (low mineral content) to 3 GPa (high mineral content). Further, the experiments revealed that the stress–strain behavior shows a characteristic softening behavior: An initial, rather stiff regime persists up to strains of approximately 3%, which is followed by a significant softening. The observed stress–strain response is reminiscent of a bilinear softening stress–strain behavior.

The mechanical behavior calculated based on the molecular model of MCF reported in this chapter agrees with several observations made in experiment (Gupta et al. 2004). For example, the MCF yield strain is somewhat close to experimental results (3% in experiment and 6.7% in simulation) (Gupta et al. 2004).

Further, the reduced slope at large strains agrees qualitatively with experiment (Gupta et al. 2004).

The finding that Young's modulus increases is in qualitative accordance with experiment (Gupta et al. 2004) comparing mineralized and non-mineralized tendon CFs. Experimental results suggest a continuous increase in Young's modulus under mineralization, ranging up to a factor of 3 for high mineral content (Borsato and Sasaki 1997; Gupta et al. 2006).

Further, the finding that component strains are smaller than tissue strains are consistent with experiment in bone (Screen et al. 2004) and tendon (Taylor et al. 2007), albeit these studies were carried out at larger scales. The results prove that this is also true at the smallest hierarchical scale of bone.

8.5.3.4 Local Yield Protects the Integrity of the Entire Structure

The fracture process zone describes the geometric extension of the region around a crack-like flaw that undergoes plastic deformation when the specimen is loaded. For brittle materials, the fracture process zone is extremely small, limited to a few atomic distances. In ductile materials, the fracture process zone can become very large, approaching the specimen dimensions.

The size of the plastic, dissipative zone for a crack oriented orthogonal to the alignment direction of tropocollagen molecules can be approximated as

$$\xi_{cr} \approx \frac{2\gamma E}{\sigma_{max}^2},\qquad(8.38)$$

where σ_{max} is the maximum fracture stress, γ is the energy necessary to create a new surface and E is Young's modulus. Equation (8.38) shows that the size of the fracture process zone is proportional to the fracture surface energy. Thus increases in the dissipative work required to create two new surfaces lead to much larger plastic zones.

Based on the parameters extracted from molecular simulation (numerical values for E and σ_{max} are given in Table 8.4; $\gamma = 11,460 \, J/m^2$) $\xi_{cr} \approx 400 \, \mu m$ for MCFs.

This length scale has another important implication: For any defect smaller than ξ_{cr}, fracture will not be controlled by the presence of this flaw. The material is insensitive to the presence of crack-like flaws below this characteristic defect dimension.

Notably, this length scale is on the same order of magnitude as small microcracks typically found in bone, with characteristic dimensions of several hundred micrometer diameter (Turner 2006; Taylor et al. 2007). It may also play a significant role in bone remodeling. Bone is remodeled in so-called basic multi-cellular units – BMUs – a combination of osteoclasts and osteoblasts forming small cavities inside the tissue. It has been shown that BMUs represent defects with dimensions of approximately $200 \, \mu m$, thus on similar orders of magnitude as ξ_{cr} (Buehler et al. 2006). Thus the particular properties of MCF could be a vital component in allowing the presence of BMUs inside the tissue without compromising its strength.

Further, by limiting the dimensions of individual MCF in the hierarchical structure, failure will occur homogeneously within each MCF, with plastic strains distributed over the entire geometry.

This analysis provides insight into how the particular MCF structure contributes to toughness by comparing a pure HA crystal. The surface energy of pure HA ranges from 0.3 to $1.6\,J/m^2$ (density functional theory calculations (Zhu and Wu 2004), leading to a rather small fracture process zone on the order several nanometers. Thus any larger crack-like defect will lead to catastrophic failure. The estimate for γ of a MCF obtained from MD studies of a fibril is several orders of magnitudes higher. Even though the modulus is much reduced in comparing a pure HA crystal with the MCF, the significant increase in γ outruns the reduction in modulus. Further, the fact that σ_{max} is smaller further leads to increase in the length scale.

In comparison with a CF, the MCF has a larger fracture process zone due to an increase in γ as well as an increase in E. The effect of these two parameters outruns the effect of a smaller σ_{max}.

8.5.4 Discussion

The work overcomes the limitations of the existing models of bone by explicitly considering tropocollagen molecules interacting with HA phases, providing a physics-based material description that enables one to make direct links between molecular structure, topology and fracture behavior. It is found that the nanostructural arrangement of the MCF is key to its mechanical properties, notably by allowing molecular slip as a major toughening mechanism.

In the following sections, implications of the findings for the understanding of bone formation and bone mechanics are discussed.

8.5.4.1 Hierarchical Toughening Mechanisms

Past research has revealed that toughening occurs at different scales. Our studies and results from investigations at other length scales suggest that each level in the hierarchy of bone may be designed to provide optimal toughness, thus being capable of taking advantage of nanoscale molecular and crystal properties, at larger scales.

The behavior discussed in this chapter is qualitatively similar to that suggested by the sacrificial bond model (Thompson et al. 2001). However, the mechanism described here operates at a smaller length scale in bone's structural hierarchy and has a different nanostructural origin; it is closely linked to the particular staggered molecular structure of the collagen fibrils and does not involve presence of metal ions. It is found that at the level of individual MCF, intermolecular slip is a major mechanism of dissipation (see, for instance in Fig. 8.36).

To enable this dissipation mechanism, the adhesion energy between HA crystals and tropocollagen molecules must be in a critical regime. This regime is characterized by the following condition: It must allow strengthening by making it more difficult to initiate molecular slip, but it must be small enough so that covalent bonds inside the tropocollagen molecules are not broken. Interface energies on the order of magnitude that allow for these deformation mechanisms correspond to ionic interactions across the TC–HA interface. Indeed, ionic interactions have recently been suggested based on NMR studies of a TC–HA interface in physiological bone (Jaeger et al. 2005; Wilson et al. 2006).

Pure vdW or H-bond interactions would lead to adhesion energies of approximately $0.01 \, \mathrm{J/m^2}$. This would be insufficient to make MCFs stronger or increase its toughness, thus rendering the presence of minerals in the gap regions insignificant. In the other extreme case, increasing γ_{MF-TC} to values corresponding to covalent bonds ($\gamma_{MF-TC} > 1 \, \mathrm{J/m^2}$), the deformation mechanics changes so that plastic yield does not set in until tropocollagen molecule rupture occurs, leading to a shutdown of the toughening mechanism.

The large aspect ratio of the mineral platelets leads to large shear forces between the tropocollagen molecule and the HA crystal, since $F_{\mathrm{shear}} \sim A_C$ (the variable $A_C \sim L_C$ is the contact area between tropocollagen and HA).

The analysis of the strain distribution inside the MCF shows that the stress in HA platelets approaches several GPa. However, macroscopic HA crystals break at 0.1% tensile strain and stresses as low as 65 MPa. It was shown in earlier molecular simulation studies that by reducing the size of a HA crystal to dimensions below 30 nm, the strength of the crystal approaches the theoretical value, even under presence of cracks or other defects (Buehler et al. 2006). Under flaw-tolerant conditions, the material does not sense the existence of defects and is thus capable of reaching its theoretical strength. Thus the flaw-tolerance concept could be a possible explanation for the fact that mineral platelets can sustain large stresses that approach 1 GPa, without fracturing.

8.5.4.2 Molecular Design Scenarios

Our model enables one to develop different design scenarios. As reported in a previous study, high cross-link densities in a pure CF without HA phase make the material stronger, but lead to a brittle polymer with low toughness and low stiffness. Such behavior is undesirable for the physiological role of bone.

As shown in the analyses reported in this chapter, adding very stiff ceramic platelets inside the collagen fibril represents a strategy to insert high densities of covalent chemistry in order to make the material stiffer and stronger without compromising toughness. The addition of mineral platelets allows the material to yield under large load in order to protect the entire structure. The molecular role of HA platelets in MCFs thus appears to be related to the increase of the strength by providing a larger energy barrier against intermolecular slip. At the same time, presence of HA platelets increases the dissipative nature of large-strain deformation of MCFs. Also, the molecular arrangement of MCFs allows to achieve a good weight–strength efficiency, since the dominating protein phase is lighter than the HA phase.

As discussed in a previous study (Buehler et al. 2006b), the length of tropocollagen molecules controls the mechanical behavior of CFs, and we expect a similar behavior for MCFs. Short tropocollagen molecules lead to reduced strength and MCFs may become rather brittle. Long tropocollagen molecules are vital to yield large toughness, as they provide a means to enable long deformation paths with large slipping inside the material. The physiological significance of toughness may explain why extremely long tropocollagen molecules in MCFs is a highly conserved molecular feature. However, if tropocollagen molecules become too long, utilization of the intermolecular "glue" becomes inefficient. As shown in an earlier study (see also the discussion above), molecular lengths at approximately 200 nm provide an optimal basis for CFs. Further analysis of the dependence of mechanical properties of MCFs on the tropocollagen molecule length is left to future studies.

The molecular toughening mechanism described in this work unifies controversial attempts of explaining sources of toughness of bone, as it illustrates that both crack tip mechanisms and flaw tolerance concepts play a key role in the mechanical response of bone under extreme load.

In addition to mineralizing CFs, structural features at larger length scales of bone, the dependence of material properties on time (e.g., via osteoblast and osteoclast cells) and extrafibrillar matrix properties are important for the macroscopic mechanical properties of bone (Taylor et al. 2007). However, our results clearly show the significance of the nanoscale TC–HA patterning as a toughening mechanism at the nanoscale and microscale.

8.5.4.3 Bone Formation and Tissue Growth

Our analysis shows that the particular properties of MCF allow to tolerate cracks at dimensions of several hundred micrometers; this may be critical to enable operation of basic molecular units (BMUs) in repair of bone, which require the presence of small cavities inside the tissue (Taylor et al. 2007).

The mechanical properties of a scaffolding material can influence the growth rate and quality of the bone tissue (Alsberg et al. 2006), providing evidence that not only chemical growth factors, but also the nanomechanical and micromechanical material properties play a role in tissue development.

Further, it has been shown that the presence of a stiff matrix directs stem cell differentiation toward osteoblasts (Engler et al. 2006). Thus the increase in stiffness due to mineralization – as shown in Fig. 8.35 and Table 8.4 – could be a critical aspect during formation of nascent bone.

8.5.5 Conclusion

The studies reveal that the mechanical properties of CF change significantly after mineralization. Whereas pure tropocollagen fibrils are soft and the HA minerals are stiff and extremely fragile, the stiffness of mineralized fibrils assumes intermediate values, but with much increased energy dissipation during deformation (Table 8.4 summarizes the main effects of mineralization). Important structural features in MCFs and their effects on the mechanical behavior are :

- Presence of HA crystals, to provide additional resistance against plastic deformation, to increase Young's modulus and the fracture strength
- Adhesion forces between HA and tropocollagen remain weak enough, to allow for slip under large load instead of inducing fracture inside the tropocollagen molecules, but strong enough to provide significant strengthening
- Characteristic nanoscopic dimensions, to utilize the intermolecular adhesion forces most efficiently
- Further, as illustrated in a previous study (Buehler 2006b), presence of long tropocollagen molecules in order to provide the basis for long deformation paths for high energy dissipation

Modifications of the mechanical properties of CF under mineralization control the fracture properties of MCFs. Our analysis reveals that the particular constitutive behavior of MCFs induces a crack tip mechanism known as plastic shielding, effectively increasing the toughness of the tissue. The concept behind this mechanism is to sacrifice a small part of the structure in order to rescue the integrity of the entire structure. Presence of large yield regions on the order of several hundred micrometers leads to a more equal stress distribution under loading and enables operation of BMUs. In contrast, cracks in pure HA crystals lead to potentially dangerous large stress concentrations around flaws. The concept of equal stress distribution is known as a driving force for topology and shape evolution of natural structures such as bone and trees. The results suggest that this appears to be a universal principle that also holds at nanoscale.

Development of the yield region represents a toughening mechanism whose origin is intimately linked with molecular geometry and mechanisms, underlining the significance of nanostructure for bone properties. The results provide molecular scale explanations of experiments that show an increase in yield stress, maximum fracture stress, as well as failure strength with increasing mineral content.

8.6 Structure–Property Relationships in Biological Protein Materials

Historically, the classes of materials has been used to classify stages of civilizations, ranging from stone age more than 300,000 years ago to the bronze age, and possibly the silicon age in the late twentieth and early twenty-first century. However, a systematic analysis of materials in the context of linking chemical and physical concepts with engineering applications has not been achieved until very recently. For instance, 50 years ago, E. Orowan, M. Polanyi and G.I. Taylor have discovered dislocations, a concept proposed theoretically in 1905 by V. Volterra. It was discovered that dislocations represent the fundamental mechanism of plastic deformation of metals (Taylor 1934; Hirth and Lothe 1982). Remarkably, it was not until dislocations and other nanoscopic and microscopic mechanisms were understood theoretically that major breakthroughs have been possible that utilize this knowledge, to enable building high performance and reliable

airplanes, cars, space shuttles and more recently, nanodevices, through synthesis of
ultra-strong and heat-resistant materials, for instance.

Perhaps, today we stand at another cross-road: Biological materials and systems
are vital elements of life, and therefore, a rigorous understanding of the matter that
makes life "work" is in reach. This may enable us eventually to integrate concepts from
living systems into materials design, seamlessly. Optical, mechanical and electrical
properties at ultra-small material scales, their control, synthesis and analysis as well
as their theoretical description represent major scientific and engineering challenges
and opportunities. However, just like in the case of more conventional materials, these
breakthroughs will probably only be accessible provided that the fundamentals are
understood very well. Characterization of the materials found in biology within a rigorous materials science approach is aimed toward the elucidation of these fundamental
principles of assembly, deformation and fracture of these materials.

It is known from other fields in materials science that nanoscopic or microscopic structures control the macroscopic material behavior: For example, grain
size reduction or confinement leads to an increase of the strength of crystalline
metals (Nieh and Wadsworth 1991; Yip 1998; Blanckenhagen et al. 2001; Wolf
et al. 2003). Deformation maps have been proposed to characterize material properties for engineering applications (Frost and Ashby 1982). Discovering similar
insight for biological structures and materials represents an important frontier of
research. A particularly challenging question is the elucidation of the significance
and role of nanostructures for macroscopic properties, that is, carry out sensitivity
analyses that show how small-scale features influence larger scale properties.

What are the most promising strategies in order to analyze these materials?
Perhaps, an integrated approach that uses experiment and simulation concurrently
could evolve into a new paradigm of materials research. Experimental techniques
have gained unparalleled accuracy in both length and time scales (see Fig. 8.4), as
reflected in development and utilization of atomic force microscope (AFM), optical
tweezers or nanoindentation (Dao et al. 2003; Sun et al. 2004; Lim et al. 2006)
to analyze biological materials. At the same time, modeling and simulation have
evolved into predictive tools that complement experimental analyses (see Fig. 8.4)
(Goddard 2006). It is now achievable to start from smallest scales – considering
electrons and atoms – to reach all the way up to macroscopic scales of entire tissues
(Goddard 2006), by explicitly considering the characteristic structural features at
each scale. Even though there are still major challenges ahead of us, this progress
is amazing and provides one with infinite possibilities and potentials, transforming materials science as a discipline through increased integration of computational
approaches in scientific research (see Fig. 8.37).

8.6.1 Cross-Scale Interactions: Fracture Mechanisms in Collagenous Tissue

Deformation and fracture are intimately linked to the atomic microstructure of
the material. A central theme of the efforts in developing the materials science of

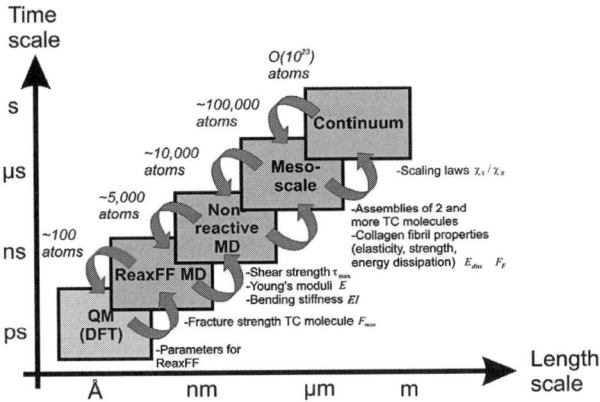

Fig. 8.37 Summary of the multi-scale scheme used in this work. First principles quantum mechanics (QM) calculations (e.g., density function theory or DFT) are carried out to train a reactive force field ReaxFF. The reactive force field is used together with nonreactive force fields to obtain properties of individual TC molecules and assemblies of two TC molecules. Parameters include max F (from ReaxFF), E, EI and shear τ (from nonreactive CHARMM). These parameters are used to develop a mesoscale model, which enables studies of ultra-long tropocollagen molecules and assemblies of those into collagen fibrils. The calculation results are coupled to the continuum scale using scaling laws. The scaling laws and associated length scales contain parameters that were obtained from mesoscale simulations (e.g., the length scale χ_s)

biological materials is to appreciate the structure–property or structure-processing-property paradigm, constituting the heart of the materials science community. This paradigm has guided materials science for many decades. For biological materials, there are many challenges that make developing these rigorous links increasingly difficult.

For example, bond energies in biological materials are often comparable to the thermal energy, as for instance in the case of hydrogen bonding, the most abundant chemical bond in biology. Biological materials show highly viscoelastic behavior, since their response to mechanical deformation is intrinsically time dependent. In many cases, biological structures contain extremely compliant filaments, in which entropic contributions to free energy are important and can even control the deformation behavior. Many material properties are also length scale dependent and can vary significantly across various length scales. Quite often, this can be quite perplexing, since measuring different volumes of material leads to different values of Young's modulus. Size effects are very strong and possibly utilized systematically to ensure physiological functioning of the material in its biological context. However, why and how these size effects are exploited within this context remains less understood. The presence of hierarchical structures calls for new paradigms in thinking about the structure–property paradigm, since corresponding concepts must include an explicit notion of the cross-scale and inter-scale interactions.

It has become evident that the atomistic scale, and in particular the notion of a chemical bond, provides a very fundamental, universal platform at which a variety of scientific disciplines can interact: chemists, through the molecular structure of proteins; physicists, through the statistical mechanics of a large number of atoms; and materials scientists through analysis of phenomena such as elasticity, optical properties, electrical properties or thermodynamics, linking structure and function.

A particularly exciting aspect of the materials science of biological materials is that it is interdisciplinary, by nature. Nature does not know of scientific disciplines, since they were invented by humans many centuries ago. Performing research in this field thus often means to overcome barriers between scientific disciplines and to develop strategies that enable us speak to each other more openly. Structures in universities and research institutions may have to be modified to facilitate such investigations.

It is vital to overcome the barrier that currently separates the scales, through development of new methods, better model systems and an advanced appreciation for a multi-scale view, in order to fully understand multi-scale or cross-scale interactions. To facilitate these developments, we must also develop a proper nomenclature to capture the various scales involved in a material. Current terminologies referring to atomistic, meso, micro and macro are insufficient to capture the subtleties of the various scales. Research should address the following questions: What are the opportunities in integrating nanoscience and nanotechnology into biological research? What will and can our impact be, in a long perspective, in understanding fundamental biology? For instance, is the nanomechanics of protein materials significant for biology, and have biologists missed out on important effects due to lack of consideration of the nanomechanics? How does Nature design materials that are environmentally friendly, lightweight and yet tough and robust and can serve multiple objectives? How is robustness achieved? How do universality and diversity integrate into biological structures?

From a theoretical viewpoint, major challenges are the development of new materials theories that include atomistic and statistical effects into an effective description, while retaining a system theoretical perspective, maybe eventually leading to a merger between system biology and materials science.

Similar to dislocation mechanics for metal plasticity, what is the theoretical framework for biological materials and structures? It is possible that statistical theories may evolve into the theoretical language of nanomechanics. Atomistic simulations of complex protein structures with explicit solvents are often prohibitive, and coarse-graining techniques are often used. However, how effective are coarse-graining techniques? Can we indeed average out over atomistic or mesoscale structures? How important are atomistic features at macroscale? What are the best numerical strategies to simulate the role of water in very small confinement? How does confined water influence the mechanics of natural and biological materials?

Progress in these various challenging fields will probably occur specific to problems and applications, perhaps in those which have most impact in medical or economic fields. Eventually, we must generalize our insight into the formulation of a *holistic theory* that extends the current nomenclature, theory and experimental thinking. These efforts will provide the scientific and engineering fundamentals to develop and maintain the infrastructures to enable and evolve modern civilization. Materials – and materials science – will surely play a seminal role in these developments.

8.6.2 The Significance of Hierarchical Features

A major trait of BPMs is the occurrence of hierarchies and the abundance of weak interactions. The presence of hierarchies in biological materials may be vital to take advantage of molecular and sub-molecular features, often characterized by weak interactions, and multiply their properties so that they become visible at larger scales. Utilization of weak interactions makes it possible to produce strong materials at moderate temperatures and thus with limited energy use.

Another distinction between traditional and biological materials is the geometrical occurrence of defects. While defects are often distributed randomly over the volume in crystalline materials, biological materials consist of an ordered structure that reaches down to the nanoscale. Defects are placed with atomistic or molecular precision and may play a major role in the material behavior observed at larger scales. These results suggest that analogies can be drawn between biological and synthetic materials.

In addition to the long-term impact in biology, bioengineering and medicine, this research may eventually contribute to our understanding of how different scales interact with one another. It may also enable synthesis of novel complex structural materials, designed from nano to macro. In order to achieve these goals, major challenges must be overcome, in particular in relating molecular processes to larger scale phenomena. As illustrated in this chapter for the example of collagenous tissues, protein materials constitute exceedingly complex structures. While the behavior of individual proteins is reasonably well understood, the properties of large assemblies of proteins remain largely unknown.

8.6.3 Universality Versus Diversity

An important trait of protein materials is that they display highly specific hierarchical structures, from nano to macro. Some of these features are commonly found in different species and tissues, that is, they are highly conserved. Examples include alpha-helices, beta-sheets or collagen fibrils that represent universal building blocks forming the basis for diverse range of protein materials. In contrast, other features are highly specific to species or tissue types, such as tendon fascicles or beta-sheet nanocrystals in spider silk (Ackbarow and Buehler 2007a).

Universal features in protein materials are most common at nanoscale, as can be seen in examples such as the beta-strand motif, alpha-helices or collagen fibrils. Structural diversity becomes more prominent at larger scales. Biological materials often feature a coexistence of universality and diversity. Universality is linked to robustness, diversity is linked to optimality. The collagen fibril motif – one of the most abundant protein structures found in biology – may be highly conserved since it provides a very robust means to achieve highly dissipative materials.

For instance, collagen consists of triple-helical tropocollagen molecules that have lengths of 300 nm with 1.5 nm in diameter. Staggered arrays of tropocollagen molecules form fibrils, which arrange to form collagen fibers. Whereas the fibrillar, staggered structure is a universal feature of collagenous tissue, structures at length scales beyond those of fibrils vary drastically for different species or different tissue. For instance, fibrils form fascicles in tendon, but mineralize to a hybrid composite in bone. In the eye's cornea, they align in a regular highly ordered orthogonal pattern. Universality persists up to the fibrillar level but vanishes at larger scales, when structural diversity dominates. Figure 8.38 depicts a schematic of the variation of structural features across the scales.

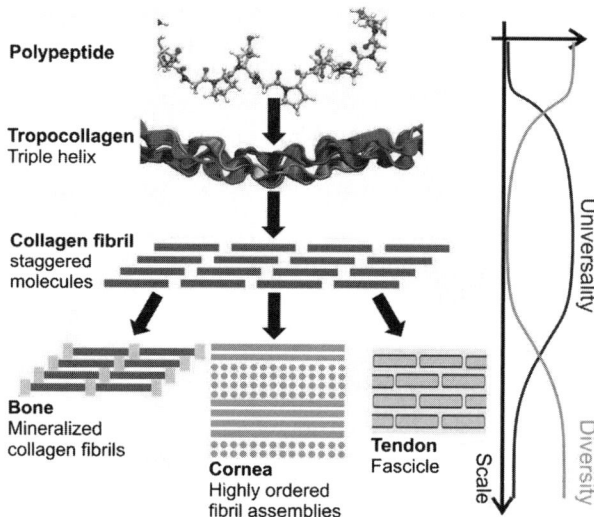

Fig. 8.38 The nanostructure and microstructure of various collagen tissues, in light of the universality–diversity paradigm as discussed in Ackbarow and Buehler (2007a). Beyond the fibril scale, structural features vary significantly, here shown for bone, cornea and tendon

Notably, instead of creating a multitude of distinct secondary protein structures, Nature creates complexity through hierarchies and internal degrees of freedom that arise from the lower scale. Through applying hierarchies, Nature keeps the opportunity to adapt systems without significantly changing their structure. Formation of hierarchical structures enables one to overcome the physical limitations of a scale-specific design space. By simultaneously adapting a multitude of structures at different length scales, it is possible to create materials whose properties by far exceed those of each constituent or those that could be reached at a single scale alone. Understanding the fundamentals of this trait of biological materials could lay the foundation for a new class of biomimetic materials in which precisely controlled hierarchical features are exploited to tailor their properties.

8.7 Discussion and Conclusion

Deformation and fracture are fundamental phenomena with major implications on the stability and reliability of machines, buildings and biological systems. All deformation processes begin with erratic motion of individual atoms around flaws or defects that quickly evolve into formation of macroscopic fractures as chemical bonds rupture rapidly, eventually compromising the integrity of the entire structure. However, most existing theories of fracture treat matter as a continuum, neglecting the existence of atoms or nanoscopic features. Clearly, such a description is questionable. An atomistic approach as discussed in this chapter provides unparalleled insight into the complex atomic-scale deformation processes, linking nano to macro, without relying on empirical input.

The study reported here illustrates that molecular multi-scale modeling of collagen can be used to predict the elastic and fracture properties of hierarchical protein materials, marvelous examples of structural designs that balance a multitude of tasks, representing some of the most sustainable material solutions that integrate structure and function across the scales.

Breaking the material into its building blocks enables one to perform system atic studies of how microscopic design features influence the mechanical behavior at larger scales. The studies elucidate intriguing material concepts that balance strength, energy dissipation and robustness by selecting nanopatterned, hierarchical features.

Over the last century, engineers have developed understanding of how to create complex man-made structures out of a diverse range of constituents, at various scales (machines, buildings, airplanes, nuclear reactors and many others). Increased development and research funding into these areas of research will lead to breakthroughs not only on the fundamental sciences, but also in technological applications. Research in the area of mechanics of biological materials will extend our ability to carry out structural engineering, as used for buildings or bridges today, to the ultimate scale – nanoscale, and may be a vital component of the realization of nanotechnology.

A better understanding of the mechanics of biological and natural materials, integrated within complex technological systems, will make it possible to combine living and non-living environments to develop sustainable technologies. New materials technologies such as protein-based materials produced by recombinant DNA techniques represent new frontiers in materials design and synthesis (Langer and Tirrell 2004; Zhao and Zhang 2007). These questions have high impact on the understanding and design of environmentally friendly technologies and may enhance the quality of life of millions of people, through advances in the medical sciences as well as through improvements of the living environment. A currently pressing question is the development of new technologies to address the energy problem. Advances may be possible by utilization of bacteria to produce and process fuel from crops or by enabling the synthesis of materials at reduced processing temperature.

Nanoscience and nanotechnology enable us to make structures at the ultimate scale (self-assembly, recombinant DNA, utilization of motor proteins for nanomachines and many others). This will perhaps lead to novel complex structural materials, designed from nano to macro. The theoretical progress in understanding hierarchical biological materials will facilitate to use an extended physical space, through the use of multiple hierarchies, in an efficient and controlled manner, that is, lead to a bottom-up structural design on the sub-macroscopic scale, instead of trial-and-error approaches. For example, the extended design space might serve as a means to realize new physical realities that are not accessible to a single scale, such as material synthesis at moderate temperatures, or fault-tolerant hierarchical assembly pathways (Cui et al. 2007; Hule and Pochan 2007; Winey 2007), which enable biological systems to overcome the limitations to particular chemical bonds (soft) and chemical elements (organic) present under natural conditions. The increased understanding of the hierarchical design laws might further enable the development and application of new organic and organic–inorganic multi-featured composites (such as assemblies of carbon nanotubes and proteins or polymer–protein composites (Petka et al. 1998; Langer and Tirrell 2004; Smeenk et al. 2005)), which will mainly consist of chemical elements that appear in our environment in an almost unlimited amount (C, H, N, O, S). These materials might consequently help to solve humans' energy and resource problems (e.g., fossil resources, iron, etc.) and allow us to manufacture nanomaterials, which will be produced in the future by techniques like recombinant DNA (Mershin et al. 2005; Zhao and Zhang 2006, 2007) or peptide self-assembly, techniques where the borders between materials, structures and machines vanish.

Applications of these new materials and structures are new biomaterials, new polymers, new composites, engineered spider silk, new scaffolding tissues, improved understanding of cell–ECM interactions, cell mechanics, hierarchical structures and self-assembly. In addition to the long-term impact in biology, bioengineering and medicine, this research may eventually contribute to our theoretical understanding of how structural features at different scales interact with one another. In light of the "extended physical design space" discussed above, this may transform engineering approaches not only for materials applications, but also in manufacturing, transportation or designs of networks.

Acknowledgments This research was supported by the Army Research Office (ARO), grant number W911NF-06-1-0291 (program officer Dr. Bruce LaMattina), the Solomon Buchsbaum AT&T Research Fund, as well as a National Science Foundation CAREER Award (CMMI-0642545, program officer Dr. Jimmy Hsia).

References

Ackbarow, T. and M. J. Buehler (2007a). "Hierarchical coexistence of universality and diversity controls robustness and multi-functionality in protein materials" *Nature Precedings* http://hdl.nature.com/10101/npre.2007.826.1.

Ackbarow, T. and M. J. Buehler (2007). "Superelasticity, energy dissipation and strain hardening of vimentin coiled-coil intermediate filaments: Atomistic and continuum studies." *Journal of Materials Science* 42(21): 8771–8787. DOI 10.1007/s10853-007-1719-2.

Ackbarow, T., X. Chen, et al. (2007). "Hierarchies, multiple energy barriers and robustness govern the fracture mechanics of alpha-helical and beta-sheet protein domains." *Proceedings of the National Academy of Sciences of the USA* 104: 16410–16415.

Aizenberg, J., J. C. Weaver, et al. (2005). "Skeleton of Euplectella sp.: Structural hierarchy from the nanoscale to the macroscale." *Science* 309(5732): 275–278.

Alberts, B., A. Johnson, et al. (2002). *Molecular Biology of the Cell*, Taylor & Francis, London.

Allen, M. P. and D. J. Tildesley (1989). *Computer Simulation of Liquids*, Oxford University Press, Oxford.

Alsberg, E., H. J. Kong, et al. (2003). "Regulating bone formation via controlled scaffold degradation." *Journal of Dental Research* 82(11): 903–908.

An, K. N., Y. L. Sun, et al. (2004). "Flexibility of type I collagen and mechanical property of connective tissue." *Biorheology* 41(3–4): 239–246.

Anderson, T. L. (1991). *Fracture Mechanics: Fundamentals and Applications*, CRC Press, Boca Raton.

Anderson, D. (2005). Collagen Self-Assembly: A Complementary Experimental and Theoretical Perspective. Toronto, Canada, University of Toronto. PhD.

Arnoux, P. J., J. Bonnoit, et al. (2002). "Numerical damage models using a structural approach: Application in bones and ligaments." *European Physical Journal-Applied Physics* 17(1): 65–73.

Bailey, A. J. (2001). "Molecular mechanisms of ageing in connective tissues." *Mechanisms of Ageing and Development* 122(7): 735–755.

Bailey, N. P. and J. P. Sethna (2003). "Macroscopic measure of the cohesive length scale: Fracture of notched single-crystal silicon." *Physical Review B* 68(20).

Bell, G. I. (1978). "Models for specific adhesion of cells to cells." *Science* 200(4342): 618–627.

Bhattacharjee, A. and M. Bansal (2005). "Collagen structure: The Madras triple helix and the current scenario." *IUBMB Life* 57(3): 161–172.

Bischoff, J. E., E. M. Arruda, et al. (2000). "Finite element modeling of human skin using an isotropic, nonlinear elastic constitutive model." *Journal of Biomechanics* 33(6): 645–652.

Blanckenhagen, B. v., P. Gumbsch, et al. (2001). "Dislocation sources in discrete dislocation simulations of thin film plasticity and the Hall-Petch relation." *Modelling and Simulation in Materials Science and Engineering* 9: 157–169.

Borel, J. P. and J. C. Monboisse (1993). "Collagens – Why such a complicated structure." *Comptes Rendus Des Seances De La Societe De Biologie Et De Ses Filiales* 187(2): 124–142.

Borsato, K. S. and N. Sasaki (1997). "Measurement of partition of stress between mineral and collagen phases in bone using X-ray diffraction techniques." *Journal of Biomechanics* 30(9): 955–957.

Bozec, L. and M. Horton (2005). "Topography and mechanical properties of single molecules of type I collagen using atomic force microscopy." *Biophysical Journal* 88(6): 4223–4231.

Bozec, L. et al. (2005). "Atomic force microscopy of collagen structure in bone and dentine revealed by osteoclastic resorption. *Ultramicroscopy* 105: 79–89.

Brenner, D. W., O. A. Shenderova, et al. (2002). "A second-generation reactive empirical bond order (REBO) potential energy expression for hydrocarbons." *Journal Of Physics-Condensed Matter* 14(4): 783–802.

Broberg, K. B. (1990). *Cracks and Fracture*, Academic Press, New York.

Buehler, M. J. (2006a). "Atomistic and continuum modeling of mechanical properties of collagen: Elasticity, fracture and self-assembly." *Journal of Materials Research* 21(8): 1947–1961.

Buehler, M. J. (2006b). "Nature designs tough collagen: Explaining the nanostructure of collagen fibrils." *Proceedings of the National Academy of Sciences of the USA* 103(33): 12285–12290.

Buehler, M. J. (2007a). "Hierarchical chemo-nanomechanics of stretching protein molecules: Entropic elasticity, protein unfolding and molecular fracture." *Journal of Mechanics of Materials and Structures* 2(6): 1019–1057.

Buehler, M. J. (2007b). "Molecular nanomechanics of nascent bone: fibrillar toughening by mineralization." *Nanotechnology* 18: 295102.

Buehler, M. J. (2007). "Nano- and micromechanical properties of hierarchical biological materials and tissues" *Journal of Materials Science* 42(21): 8765–8770. DOI 10.1007/s10853-007-1952-8.

Buehler, M. J. (2008). "Nanomechanics of collagen fibrils under varying cross-link densities: Atomistic and continuum studies." *Journal of the Mechanical Behavior of Biomedical Materials* 1(1): doi:10.1016/j.jmbbm.2007.04.001

Buehler, M. J., F. F. Abraham, et al. (2003). "Hyperelasticity governs dynamic fracture at a critical length scale." *Nature* 426: 141–146.

Buehler, M. J., F. F. Abraham, et al. (2004). "Stress and energy flow field near a rapidly propagating mode I crack." *Springer Lecture Notes in Computational Science and Engineering* ISBN 3-540-21180-2: 143–156.

Buehler, M. J. and T. Ackbarow (2007). "Fracture mechanics of protein materials." *Materials Today* 10(9): 46–58.

Buehler, M. J., J. Dodson, et al. (2006). "The Computational Materials Design Facility (CMDF): A powerful framework for multiparadigm multi-scale simulations." *Materials Research Society Symposium Proceedings* 894: LL3.8.

Buehler, M. J., A. C. T. v. Duin, et al. (2006). "Multi-paradigm modeling of dynamical crack propagation in silicon using the ReaxFF reactive force field." *Physical Review Letters* 96(9): 095505.

Buehler, M. J. and H. Gao (2006). "Dynamical fracture instabilities due to local hyperelasticity at crack tips" *Nature* 439: 307–310.

Buehler, M. J., H. Tang, et al. (2007). "Threshold crack speed controls dynamical fracture of silicon single crystals." *Physical Review Letters*. 99: 165502.

Buehler, M. J. and S. Y. Wong (2007). "Entropic elasticity controls nanomechanics of single tropocollagen molecules." *Biophysical Journal* 93(1): 37–43.

Buehler, M. J., H. Yao, et al. (2006). "Cracking and adhesion at small scales: atomistic and continuum studies of flaw tolerant nanostructures." *Modelling and Simulation in Materials Science and Engineering* 14: 799–816.

Bustamante, C., J. F. Marko, et al. (1994). "Entropic elasticity of lambda-phage DNA." *Science* 265(5178): 1599–1600.

Chenoweth, K., S. Cheung, et al. (2005). "Simulations on the thermal decomposition of a poly(dimethylsiloxane) polymer using the ReaxFF reactive force field." *Journal of the American Chemical Society* 127(19): 7192–7202.

Cheung, S., W. Q. Deng, et al. (2005). "ReaxFF(MgH) reactive force field for magnesium hydride systems." *Journal of Physical Chemistry A* 109(5): 851–859.

Courtney, T. H. (1990). *Mechanical Behavior of Materials*. New York, NY, USA, McGraw-Hill.

Cressey, B. A. and G. Cressey (2003). "A model for the composite nanostructure of bone suggested by high-resolution transmission electron microscopy." *Mineralogical Magazine* 67(6): 1171–1182.

Cui, X. Q., C. M. Li, et al. (2007). "Biocatalytic generation of ppy-enzyme-CNT nanocomposite: From network assembly to film growth." *Journal of Physical Chemistry C* 111(5): 2025–2031.

Currey, J. D. (2002). *Bones: Structure and Mechanics*. Princeton, NJ, Princeton University Press.

Currey, J. D. (2005). "Materials science – Hierarchies in biomineral structures." *Science* 309(5732): 253–254.

Cusack, S. and A. Miller (1979). "Determination of the elastic-constants of collagen by Brillouin light-scattering." *Journal of Molecular Biology* 135(1): 39–51.

Cuy, J. L., A. B. Mann, et al. (2002). "Nanoindentation mapping of the mechanical properties of human molar tooth enamel." *Archives of Oral Biology* 47(4): 281–291.

Dao, M., C. T. Lim, et al. (2003). "Mechanics of the human red blood cell deformed by optical tweezers." *Journal of the Mechanics and Physics of Solids* 51(11–12): 2259–2280.

Dao, M., C. T. Lim, et al. (2005). "Mechanics of the human red blood cell deformed by optical tweezers (vol 51, pg 2259, 2003)." *Journal of the Mechanics and Physics of Solids* 53(2): 493–494.

Datta, D., A. C. T. v. Duin, et al. (2005). "Extending ReaxFF to biomacromolecules." *Unpublished.*

Duin, A. C. T. v., S. Dasgupta, et al. (2001). "ReaxFF: A reactive force field for hydrocarbons." *Journal of Physical Chemistry A* 105: 9396–9409.

Duin, A. C. T. v., A. Strachan, et al. (2003). "ReaxFF SiO: Reactive force field for silicon and silicon oxide systems." *Journal of Physical Chemistry A* 107: 3803–3811.

Engler, A. J., S. Sen, et al. (2006). "Matrix elasticity directs stem cell lineage specification." *Cell* 126(4): 677–689.

Eppell, S. J., B. N. Smith, et al. (2006). "Nano measurements with micro-devices: mechanical properties of hydrated collagen fibrils." *Journal of the Royal Society Interface* 3(6): 117–121.

Ercolessi, F. and J. B. Adams (1994). "Interatomic potentials from 1st principle-calculations – the force matching method." *Europhysics Letter* 28(8): 583–588.

Fantner, G. E., T. Hassenkam, et al. (2005). "Sacrificial bonds and hidden length dissipate energy as mineralized fibrils separate during bone fracture." *Nature Materials* 4(8): 612–616.

Fratzl, P., H. S. Gupta, et al. (2004). "Structure and mechanical quality of the collagen-mineral nano-composite in bone." *Journal of Materials Chemistry* 14(14): 2115–2123.

Fratzl, P. and R. Weinkamer (2007). "Nature's hierarchical materials." *Progress in Materials Science* 52: 1263–1334.

Freeman, J. W. and F. H. Silver (2004). "Elastic energy storage in unmineralized and mineralized extracellular matrices (ECMs): A comparison between molecular modeling and experimental measurements." *Journal of Theoretical Biology* 229(3): 371–381.

Freund, L. B. (1990). *Dynamic Fracture Mechanics,* Cambridge University Press, Cambridge, ISBN 0-521-30330-3.

Frost, H. J. and M. F. Ashby (1982). *Deformation-mechanism Maps,* Pergamon Press, Oxford.

Gao, H. J. (2006). "Application of fracture mechanics concepts to hierarchical biomechanics of bone and bone-like materials." *International Journal of Fracture* 138(1–4): 101–137.

Gao, H., B. Ji, et al. (2003). "Materials become insensitive to flaws at nanoscale: Lessons from nature." *Proceedings of the National Academy Sciences of the USA* 100(10): 5597–5600.

Glorieux, F. H. (2005). "Caffey disease: An unlikely collagenopathy." *Journal of Clinical Investigation* 115(5): 1142–1144.

Goddard, W. A. (2006). A Perspective of Materials Modeling *Handbook of Materials Modeling.* S. Yip, Springer.

Grandbois, M., M. Beyer, et al. (1999). "How strong is a covalent bond?" *Science* 283(5408): 1727–1730.

Griffith, A. A. (1920). "The phenomenon of rupture and flows in solids." *Philosophical Transactions of the Royal Society of London, Series A* 221: 163–198.

Gropp, W., W. Lusk, et al. (1999). *Using MPI,* MIT Press, Cambridge.

Gupta, H. S., P. Messmer, et al. (2004). "Synchrotron diffraction study of deformation mechanisms in mineralized tendon." *Physical Review Letters* 93(15).

Gupta, H. S., J. Seto, et al. (2006). "Cooperative deformation of mineral and collagen in bone at the nanoscale." *Proceedings of the National Academy Sciences of the USA* 103: 17741–17746.

Gupta, H. S., W. Wagermaier, et al. (2005). "Nanoscale deformation mechanisms in bone." *Nano Letters* 5(10): 2108–2111.

Han, S. S., A. C. T. van Duin, et al. (2005). "Optimization and application of lithium parameters for the reactive force field, ReaxFF." *Journal of Physical Chemistry A* 109(20): 4575–4582.

Hansma, P. K., P. J. Turner, et al. (2007). "Optimized adhesives for strong, lightweight, damage-resistant, nanocomposite materials: New insights from natural materials." *Nanotechnology* 18(4).

Harley, R., D. James, et al. (1977). "Phonons and elastic-moduli of collagen and muscle." *Nature* 267(5608): 285–287.

Hellan, K. (1984). *Introduction to Fracture Mechanics,* McGraw-Hill, Inc., New York.

Hellmich, C. and F. J. Ulm (2002). "Are mineralized tissues open crystal foams reinforced by crosslinked collagen? – some energy arguments." *Journal of Biomechanics* 35(9): 1199–1212.

Hirth, J. P. and J. Lothe (1982). *Theory of Dislocations*, Wiley-Interscience, New York.

Hofmann, H., T. Voss, et al. (1984). "Localization of flexible sites in thread-like molecules from electron-micrographs – comparison of interstitial, basement-membrane and intima collagens." *Journal of Molecular Biology* 172(3): 325–343.

Holland, J. H. (1995). *Hidden Order – How Adaptation Builds Complexity*. Reading, MA, Helix Books. http://www.top500.org/ TOP 500 Supercomputer Sites.

Hule, R., A., Pochan, D., J., (2007). "Polymer nanocomposites for biomedical application." *MRS Bulletin* 32(4): 5.

Hulmes, D. J. S., T. J. Wess, et al. (1995). "Radial packing, order, and disorder in collagen fibrils." *Biophysical Journal* 68(5): 1661–1670.

Humphrey, W., A. Dalke, et al. (1996). "VMD: Visual molecular dynamics." *Journal of Molecular Graphics* 14(1): 33.

Israelowitz, M., S. W. H. Rizvi, et al. (2005). "Computational modeling of type I collagen fibers to determine the extracellular matrix structure of connective tissues." *Protein Engineering Design & Selection* 18(7): 329–335.

Jaeger, C., N. S. Groom, et al. (2005). "Investigation of the nature of the protein-mineral interface in bone by solid-state NMR." *Chemistry of Materials* 17(12): 3059–3061.

Jager, I. and P. Fratzl (2000). "Mineralized collagen fibrils: A mechanical model with a staggered arrangement of mineral particles." *Biophysical Journal* 79(4): 1737–1746.

Kadau, K., T. C. Germann, et al. (2004). "Large-scale molecular-dynamics simulation of 19 billion particles." *International Journal of Modern Physics C* 15: 193.

Kim, B. S., J. Nikolovski, et al. (1999). "Cyclic mechanical strain regulates the development of engineered smooth muscle tissue." *Nature Biotechnology* 17(10): 979–983.

Kramer, R. Z., M. G. Venugopal, et al. (2000). "Staggered molecular packing in crystals of a collagen-like peptide with a single charged pair." *Journal of Molecular Biology* 301(5): 1191–1205.

Lakes, R. (1993). "Materials with structural hierarchy." *Nature* 361(6412): 511–515.

Laudis, V., B. L. H. Kraus, et al. (2002). "Vascular-mineral spatial correlation in the calcifying turkey leg tendon." *Connective Tissue Research* 43(4): 595–605.

Langer, R. and D. A. Tirrell (2004). "Designing materials for biology and medicine." *Nature* 428(6982): 487–492.

Lantz, M. A., H. J. Hug, et al. (2001). "Quantitative measurement of short-range chemical bonding forces." *Science* 291(5513): 2580–2583.

Layton, B. E., S. M. Sullivan, et al. (2005). "Nanomanipulation and aggregation limitations of self-assembling structural proteins." *Microelectronics Journal* 36(7): 644–649.

Lees, S. (1987). "Possible effect between the molecular packing of collagen and the composition of bony tissues." *International Journal Of Biological Macromolecules* 9(6): 321–326.

Lees, S. (2003). "Mineralization of type I collagen." *Biophysical Journal* 85(1): 204–207.

Lichtens Jr, G. R. Martin, et al. (1973). "Defect in conversion of procollagen to collagen in a form of Ehlers-Danlos syndrome." *Science* 182(4109): 298–300.

Lim, C. T., E. H. Zhou, et al. (2006). "Experimental techniques for single cell and single molecule biomechanics." *Materials Science & Engineering C-Biomimetic and Supramolecular Systems* 26(8): 1278–1288.

Lodish, H. B., Arnold; Zipursky, S. Lawrence; Matsudaira, Paul; Baltimore, David; Darnell, James E. (1999). *Molecular Cell Biology*. W H Freeman & Co, New York.

Lorenzo, A. C. and E. R. Caffarena (2005). "Elastic properties, Young's modulus determination and structural stability of the tropocollagen molecule: a computational study by steered molecular dynamics." *Journal of Biomechanics* 38(7): 1527–1533.

Lotz, J. C., T. N. Gerhart, et al. (1990). "Mechanical-properties of trabecular bone from the proximal femur – a quantitative Ct study." *Journal of Computer Assisted Tomography* 14(1): 107–114.

Louis, O., F. Boulpaep, et al. (1995). "Cortical mineral-content of the radius assessed by peripheral qct predicts compressive strength on biomechanical testing." *Bone* 16(3): 375–379.

Lu, H., B. Isralewitz, et al. (1998). "Unfolding of titin immunoglobulin domains by steered molecular dynamics simulation." *Biophysical Journal* 75(2): 662–671.

MacKerell, A. D., D. Bashford, et al. (1998). "All-atom empirical potential for molecular modeling and dynamics studies of proteins." *Journal of Physical Chemistry B* 102(18): 3586–3616.

Mershin, A., B. Cook, et al. (2005). "A classic assembly of nanobiomaterials." *Nature Biotechnology* 23(11): 1379–1380.

Miles, C. A. and A. J. Bailey (2001). "Thermally labile domains in the collagen molecule." *Micron* 32(3): 325–332.

Mooney, S. D., C. C. Huang, et al. (2001). "Computed free energy differences between point mutations in a collagen-like peptide." *Biopolymers* 58(3): 347–353.

Mooney, S. D. and T. E. Klein (2002). "Structural models of osteogenesis imperfecta-associated variants in the COL1A1 gene." *Molecular & Cellular Proteomics* 1(11): 868–875.

Mooney, S. D., P. A. Kollman, et al. (2002). "Conformational preferences of substituted prolines in the collagen triple helix." *Biopolymers* 64(2): 63–71.

Nalla, R. K., J. H. Kinney, et al. (2003a). "Effect of orientation on the in vitro fracture toughness of dentin: the role of toughening mechanisms." *Biomaterials* 24(22): 3955–3968.

Nalla, R. K., J. H. Kinney, et al. (2003b). "Mechanistic fracture criteria for the failure of human cortical bone." *Nature Materials* 2(3): 164–168.

Nalla, R. K., J. J. Kruzic, et al. (2005). "Mechanistic aspects of fracture and R-curve behavior in human cortical bone." *Biomaterials* 26(2): 217–231.

Nalla, R. K., J. S. Stolken, et al. (2005). "Fracture in human cortical bone: local fracture criteria and toughening mechanisms." *Journal of Biomechanics* 38(7): 1517–1525.

Nelson, M. T., W. Humphrey, et al. (1996). "NAMD: A parallel, object oriented molecular dynamics program." *International Journal Of Supercomputer Applications And High Performance Computing* 10(4): 251–268.

Nieh, T. G. and J. Wadsworth (1991). "Hall-Petch relation in nanocrystalline solids." *Scripta Metallurgica* 25(4).

Nielson, K. D., A. C. T. v. Duin, et al. (2005). "Development of the ReaxFF reactive force field for describing transition metal catalyzed reactions, with application to the initial stages of the catalytic formation of carbon nanotubes." *Journal of Physical Chemistry A* 109: 49.

Orgel, J. P. R. O., T. C. Irving, et al. (1995). "Microfibrillar structure of type I collagen in situ." *Proceedings of the National Academy Sciences of the USA* 103(24): 9001–9005.

Persikov, A. V., J. A. M. Ramshaw, et al. (2005). "Electrostatic interactions involving lysine make major contributions to collagen triple-helix stability." *Biochemistry* 44(5): 1414–1422.

Peterlik, H., P. Roschger, et al. (2006). "From brittle to ductile fracture of bone." *Nature Materials* 5(1): 52–55.

Petka, W. A., J. L. Harden, et al. (1998). "Reversible hydrogels from self-assembling artificial proteins." *Science* 281(5375): 389–392.

Phillips, J. C., R. Braun, et al. (2005). "Scalable molecular dynamics with NAMD." *Journal of Computational Chemistry* 26(16): 1781–1802.

Plimpton, S. (1995). "Fast parallel algorithms for short-range molecular-dynamics." *Journal of Computational Physics* 117: 1–19.

Prater, C. B., H. J. Butt, et al. (1990). "Atomic force microscopy." *Nature* 345(6278): 839–840.

Puxkandl, R., I. Zizak, et al. (2002). "Viscoelastic properties of collagen: Synchrotron radiation investigations and structural model." *Philosophical Transactions of the Royal Society of London Series B-Biological Sciences* 357(1418): 191–197.

Ramachandran, G. N., Kartha, G. (1955). "Structure of collagen." *Nature* 176: 593–595.

Rappé, A. K. and W. A. Goddard (1991). "Charge equilibration for molecular-dynamics simulations." *Journal of Physical Chemistry* 95(8): 3358–3363.

Rief, M., M. Gautel, et al. (1997). "Reversible unfolding of individual titin immunoglobulin domains by AFM." *Science* 276(5315): 1109–1112.

Ritchie, R. O., J. J. Kruzic, et al. (2004). "Characteristic dimensions and the micro-mechanisms of fracture and fatigue in 'nano' and 'bio' materials." *International Journal of Fracture* 128(1–4): 1–15.

Ritchie, R. O., R. K. Nalla, et al. (2006). "Fracture and ageing in bone: Toughness and structural characterization." *Strain* 42(4): 225–232.

Robins, S. P. and A. J. Bailey (1973). "The chemistry of the collagen cross-links." *Biochemical Journal*. 135: 657–665.

Sasaki, N. and S. Odajima (1996). "Elongation mechanism of collagen fibrils and force-strain relations of tendon at each level of structural hierarchy." *Journal of Biomechanics* 29(9): 1131–1136.

Screen, H. R. C., D. L. Bader, et al. (2004). "Local strain measurement within tendon." *Strain* 40(4): 157–163.

Smeenk, J. M., M. B. J. Otten, et al. (2005). "Controlled assembly of macromolecular beta-sheet fibrils." *Angewandte Chemie-International Edition* 44(13): 1968–1971.

Smith, B. L., T. E. Schaffer, et al. (1999). "Molecular mechanistic origin of the toughness of natural adhesives, fibres and composites." *Nature* 399(6738): 761–763.

Strachan, A., E. M. Kober, et al. (2005). "Thermal decomposition of RDX from reactive molecular dynamics." *Journal of Chemical Physics* 122(5): 054502.

Strachan, A., A. C. T. van Duin, et al. (2003). "Shock waves in high-energy materials: The initial chemical events in nitramine RDX." *Physical Review Letters* 91(9): 098301.

Stuart, S. J., A. B. Tutein, et al. (2000). "A reactive potential for hydrocarbons with intermolecular interactions." *Journal of Chemical Physics* 112(14): 6472-6486.

Sun, Y. L., Z. P. Luo, et al. (2002). "Direct quantification of the flexibility of type I collagen monomer." *Biochemical and Biophysical Research Communications* 295(2): 382–386.

Sun, Y. L., Z. P. Luo, et al. (2004). "Stretching type II collagen with optical tweezers." *Journal of Biomechanics* 37(11): 1665–1669.

Tai, K., F. J. Ulm, et al. (2006). "Nanogranular origins of the strength of bone." *Nano Letters* 11: 2520–2525

Taylor, G. I. (1934). "Mechanism of plastic deformation in crystals." *Proceedings of the Royal Society A* 145: 362.

Taylor, D., J.G. Hazenberg, et al. (2007). "Living with cracks: Damage and repair in human bone." *Nature Materials* 6(4): 263–266.

Thompson, J. B., J. H. Kindt, et al. (2001). "Bone indentation recovery time correlates with bond reforming time." *Nature* 414(6865): 773–776.

Tsai, D. H. (1979). "Virial theorem and stress calculation in molecular-dynamics." *Journal of Chemical Physics* 70(3): 1375–1382.

Turner, C. H. (2006). Bone strength: Current concepts. *Skeletal Development and Remodeling in Health, Disease, and Aging*. 1068: 429–446.

van der Rijt, J. A. J., K. O. van der Werf, et al. (2006). "Micromechanical testing of individual collagen fibrils." *Macromolecular Bioscience* 6(9): 697–702.

van Duin, A. C. T., K. Nielson, et al. (2004). "Application of ReaxFF reactive force fields to transition metal catalyzed nanotube formation." *Abstracts of Papers of the American Chemical Society* 227: U1031–U1031.

Vesentini, S., C. F. C. Fitie, et al. (2005). "Molecular assessment of the elastic properties of collagen-like homotrimer sequences." *Biomechanics and Modeling in Mechanobiology* 3(4): 224–234.

Wachter, N. J., G. D. Krischak, et al. (2002). "Correlation of bone mineral density with strength and microstructural parameters of cortical bone in vitro." *Bone* 31(1): 90–95.

Waite, J. H., X. X. Qin, et al. (1998). "The peculiar collagens of mussel byssus." *Matrix Biology* 17(2): 93–106.

Weiner, S. and H. D. Wagner (1998). "The material bone: Structure mechanical function relations." *Annual Review of Materials Science* 28: 271–298.

Wilson, E. E., A. Awonusi, et al. (2006). "Three structural roles for water in bone observed by solid-state NMR." *Biophysical Journal* 90(10): 3722–3731.

Winey, K. I., Vaia R.A., (2007). "Polymer nanocomposites." *MRS Bulletin* 32(4): 5.

Wolf, D., V. Yamakov, et al. (2003). "Deformation mechanism and inverse Hall-Petch behavior in nanocrystalline materials." *Zeitschrift Fur Metallkunde* 94: 1052–1061.

Yip, S. (1998). "The strongest size." *Nature* 391: 532–533.

Zervakis, M., V. Gkoumplias, et al. (2005). "Analysis of fibrous proteins from electron microscopy images." *Medical Engineering & Physics* 27(8): 655–667.

Zhao, X. J. and S. G. Zhang (2006). "Molecular designer self-assembling peptides." *Chemical Society Reviews* 35(11): 1105–1110.

Zhao, X. J. and S. G. Zhang (2007). "Designer self-assembling peptide materials." *Macromolecular Bioscience* 7(1): 13–22.

Zhu, W. H. and P. Wu (2004). "Surface energetics of hydroxyapatite: a DFT study." *Chemical Physics Letters* 396(1–3): 38–42.

Zimmerman, J. A., E. B. Webb, et al. (2004). "Calculation of stress in atomistic simulation." *Modelling and Simulation in Materials Science and Engineering* 12: S319–S332.

Chapter 9
Mechanical Adaptation and Tissue Remodeling

M. Kjær and S.P. Magnusson

Abstract The adaptive response of connective tissue to mechanical loading includes an increased synthesis and turnover of matrix proteins, including the collagen. Collagen formation and degradation increases with acute loading of tendon and skeletal muscle, in vivo. This increased activity is associated with local and systemic release of growth factors (e.g., IGF-1, TGF-beta, IL-6) that is temporally coupled with a rise in procollagen expression. Chronic loading of tissue, such as with physical training, will lead to increased collagen turnover and a net collagen synthesis which are together associated with a modification of the mechanical properties, including a reduction in tendon stress. Altogether this likely yields a more load-resistant tissue. The adaptation time to chronic loading is longer in tendon tissue compared to contractile elements of skeletal muscle or heart, and it is only with very prolonged loading that significant changes in gross dimensions of the tendon can be observed. Current observations support the notion that mechanical loading leads to collagen-rich tissue adaptation, and that this requires an intimate interplay between mechanical signaling and biochemical changes in the matrix, such that chemical changes can be converted into adaptations in morphology, structure and material properties.

9.1 Introduction

Collagen type I rich structures such as tendons, ligaments, bone or intramuscular connective tissue have been studied for decades, and the biomechanical properties of these tissues are well appreciated. However, at least for tendon and ligaments the structures have been considered to be relatively inert (Kjær, 2004). The methods available to study these collagen type I rich tissues have in the past been somewhat limited with regard to in vivo techniques, and very often findings were based solely on studies of cadaver tissue properties, or relied on cell culture work. Whereas these approaches have provided very valuable insight, they did not allow for an integrated approach toward understanding how collagen-rich tissue responds to changes in mechanical loading in living humans. However, over the last 5–10 years methodological developments and renewed interest for metabolic, circulatory and tissue protein turnover in collagen tissue such as tendon – in conjunction with

P. Fratzl (ed.), *Collagen: Structure and Mechanics*,
© Springer Science+Business Media, LLC 2008

a simultaneous determination of morphology and biomechanical tissue properties – have lead to a greater insight into the way that mechanical loading influences both the cells and the matrix of tendon. It has been demonstrated that in response to mechanical loading the human tendon increases its blood flow, metabolic activity and substrate uptake (e.g., glucose) by three- to fourfold acutely associated with exercise (for references please see Kjær, 2004).

9.2 Collagen Adaptation to Loading – Biochemical Approaches

9.2.1 Dynamics of Collagen Metabolism in Human Tendon and Skeletal Muscle with Mechanical Loading

Matrix adaptation to loading involves several proteins, but by far the most mechanically important load-bearing structure in tendon tissue is collagen type I. In theory the magnitude of collagen synthesis can be determined in various indirect or direct ways. A determination of increased procollagen mRNA in tissue evidently indicates an upregulation of transcriptional activity, but does not guarantee for any mature collagen formation. Furthermore, activity in the intracellular biosynthesis pathway for collagen can be estimated by determination of enzymes like prolyl-4-hydroxylase (P-4-H), galactosylhydroxy-lysyl-glucotransferase (GGT) or lysyl hydroxylase (LHy) and increased collagen synthesis has been estimated from such methods in animal experiments (Kovanen, 1989). Collagen formation in tendon is shown to rise with acute and chronic loading, as determined by the use of the microdialysis technique, where catheters are put into the region of interest (e.g., around or through a tendon) and the interstitial concentration of the procollagen propeptides that are cleaved of in the maturation from procollagen to collagen (PICP or PINP) is determined of both previously loaded and unloaded tendon (Langberg et al. 1999, 2001). Using the microdialysis technique locally around a tendon provides the possibility to determine local changes of PICP or PINP in regions that otherwise would contribute only marginally to any change in these parameters globally (i.e., changes in concentrations of PINP or PICP in the circulating blood stream) (Langberg et al. 2000).

In collagen-containing tissues like tendon and bone, other methods like the use of stable isotopes for incorporation of representative tracers into tissue protein have been tried, and given that representative samples of the tissue can be obtained, the direct protein synthesis can be determined (Babraj et al. 2005; Miller et al. 2005). Tendon tissue sampling can be performed by percutaneous tendon biopsies, and has been used in protocols where protein turnover, mRNA transcription and collagen fibril diameter have been determined in both young and elderly, patients and healthy subjects (Miller et al. 2005). The use of stable isotope techniques to study incorporation of labeled amino acids into tissue in order to study the kinetics of collagen has been tried in tendon, ligament and bone (Babraj et al. 2005). Briefly, the principle of the direct incorporation technique using the precursor–product approach applicable on tendon tissue is to label the amino acid, proline, e.g., L-^{13}C1-Proline or L-^{15}N-Proline. Proline is abundant in collagen, the major structural protein in tendinous tissue, and is incorporated directly into new collagen proteins. Newly synthesized procollagens are posttranslationally hydroxylated at the proline residues (forming hydroxyproline) before being assembled into the triplehelical structure. Thus, the collagen-specific hydroxyproline is labeled. Measuring the enrichment of hydroxyproline from a tendon sample gives a very specific synthesis measure of collagen protein in the tendon tissue (Babraj et al., 2005). Use of this method, and taking tendon biopsies, demonstrates that acute exercise increases the fractional synthesis rate of collagen in the patella tendon from approximately 0.05%/h to around 0.10%/h within 24 h after exercise, showing a significant rise already after 6 h post exercise (Miller et al. 2005). This corresponds to a collagen synthesis that on a 24-h level increases from around 1% at rest to 2–3% after exercise. The collagen synthesis rate remained elevated for at least 3 days after acute exercise (Heinemeier et al. 2003). In the patella tendon, the collagen synthesis response to 36 km of running was positively related to the intensity (expressed as running speed) of the exercise (Langberg, unpublished data). This is interesting as overuse symptoms in the patella tendon is closely related to high-intensity activities like jumping, suggesting a gradual interplay between physiological adaptation and pathological development in relation to connective tissue like tendon.

Similar to collagen synthesis, there is indication that protein degradation is activated after exercise, in that local levels of matrix metalloproteinases in tendon tissue are increased after acute exercise (Koskinen et al. 2004). Although no good determination of collagen degradation has been provided so far, it indirectly points at an increased collagen turnover with exercise. It is difficult to state how much of this increased turnover is in fact turned into assembled load-bearing collagen, but from long-term training studies indicating an increase in tendon cross-sectional area, there is reason to believe that at least a certain percentage of the newly formed collagen will end up as insoluble collagen type I in the final tendon structure. Within skeletal muscle, the collagen synthesis increases with acute exercise by two- to threefold and in a time fashion that resembles that of myofibrillar protein synthesis, at least suggesting a coordinated response with regard to structural proteins both in contractile elements and in the extracellular matrix of human skeletal muscle.

Prolonged, repeated mechanical loading chronically elevates the collagen synthesis in tendon almost three- to fourfold (Langberg et al. 2001, 2007), and more recently it has been shown that inactivity for only 10 days resulted in a decrease in protein synthesis of both collagen and myofibrillar protein of around 20–25% (DeBoer et al. 2007). Interestingly, neither of these studies resulted in dramatic morphological changes of the tendon when determined by ultrasound. Clearly, in none of the studies an accurate determination of protein degradation could be performed, and maybe more importantly, it has been shown that determination of collagen synthesis per se does not always correlate with true incorporation of new load-bearing mature collagen into existing structures (Laurent, 1987). Finally, it cannot from cross-measurements of a tendon diameter be clearly stated whether or not the total collagen content of the tendon has been changed.

9.2.2 Regulatory Factors for Collagen Adaptation to Exercise

An important question in relation to increased collagen turnover is how tendon senses the external loading during muscular contraction and more specifically what factors are involved in this regulation. The first example of factors involved in collagen synthesis regulation comes from the observation that apparently a gender difference exists, in that females respond less than males with regard to the increase in collagen formation after exercise (Miller et al. 2006a,b). The expected increase in collagen synthesis to exercise was less pronounced in females than in males, and furthermore the basal collagen synthesis rate was also lower in females compared to male counterparts (Miller et al. 2006a,b). Further experiments in females who have varying levels of sex hormones (e.g., oral contraceptives) suggest that estradiol may contribute to a diminished collagen synthesis response in females (Hansen, unpublished observations). Interestingly, this finding of a lower collagen synthesis rate both at rest and in response to exercise in females vs males, is correlated to a demonstration of lower stress tolerance in patella tendons of females compared to males (Haraldsson et al. 2005). Overall this may contribute to our understanding with regard to the fact that women experience a larger number of soft tissue injuries in, e.g., cruciate ligaments than men, and may also be important in understanding not only how rapidly humans of both genders adapt to physical training but also how well they may resist prolonged periods of reduced activity. What exact mechanism lies behind this gender-specific response is not definitively answered, but the finding of a correlation between increased estrogen levels and a reduction in collagen synthesis in humans is supported by in vitro studies where it has been shown that estradiol receptors are present in ligaments (Sciore et al. 1998) and that estradiol per se can exert a collagen synthesis inhibiting effect in tendons and ligaments (Liu et al. 1996; Yu et al. 2001). In addition to a suggested direct inhibiting effect of estradiol upon collagen synthesis, it may also be that estradiol levels exert an indirect effect by influencing other hormonal components of the human endocrine system. As an example, more lately high levels of circulating estradiol in women have been shown to be associated with low levels of circulating insulin-like growth factor (IGF-I), a substance that may be directly coupled to the degree of collagen synthesis rise with exercise (see below).

A documented important regulating factor in collagen synthesis is the growth hormone (GH) – insulin-like growth factor 1 (IGF-1) axis, where in vitro data has shown a role for collagen formation (Døssing et al. 2005). As an example, it has been shown that IGF-I (and IGF-II) administration in rabbits will accelerate the protein synthesis in tendons (Abrahamsson, 1997), and likewise the recovery after tendon injury was accelerated when IGF-I was administered (Kurtz et al. 1999). Although GH in skeletal muscle has been shown to exert an effect upon muscle growth in GH-deficient individuals, the effect of GH supplementation upon muscle protein synthesis is absent in both young and elderly humans (Lange et al. 2002). Despite this, it seems that GH/IGF-I influences connective tissue, and administration of GH over 3 weeks has been shown to elevate circulation blood levels of procollagen propeptides (Longobardi et al. 2000). Thus it may be tempting to speculate that

GH/IGF-I has a stimulating effect upon tendon tissue. In dwarf rats, GH administration has been shown to increase the expression of both collagen type I and III in intramuscular fibroblasts (Wilson et al. 1995). Recently it has been shown that IGF-1 is present in human Achilles tendon linked directly to fibroblasts, and furthermore a detectable interstitial concentrations has been demonstrated in human tendon (Olesen et al. 2006, 2007a). Although this supports the possibility that IGF-I may play a role in human tendon, the question however remains: To what extent is IGF-1 upregulated with exercise? By analyzing RNA extracted from whole rodent tendons, it has recently been shown that functional overloading of a single muscle attached to the tendon-muscle (by ablation of other muscle synergists) leads to increased expression of IGF-I within the tendon and furthermore that short-term strength training induces the expression of both TGF-β1 and IGF-I in rat Achilles tendon (Heinemeier et al. 2007a, Olesen et al. 2007b). Quantification of specific mRNA species has also been used as a tool for studying human tendon pathology, and by analyzing RNA extracted from surgical tendon specimens several changes in expression of matrix proteins and enzymes have been identified in overused/damaged tendons compared to healthy tendons (Riley et al. 2002). Furthermore, increased expression in tendon of an IGF-1 isoform called MGF (mechanogrowth factor) was found (Heinemeier et al. 2007a). This is notable since until recently, this isoform was thought to be expressed in skeletal muscle cells only (Goldspink, 2006).

In the rodent overload study mentioned above (where synergist muscles were abladed), also an upregulation of transforming growth factor beta 1 (TGF-β1) was found after 2 days of stimulation, and it preceded the rise in collagen expression. As an indication that TGF-β1 could play a role in collagen synthesis (Heinemeier et al. 2007a, 2007b), the mRNA for connective tissue growth factor (CTGF) was increased similar to TGF-β1. CTGF is thought to mediate many of the effects of IGF-1 in relation to matrix proteins. A number of in vitro studies have demonstrated a coupling between mechanical loading of tissue and TGF-beta expression (Skutek et al. 2001), and furthermore it has been shown that mechanically induced type I collagen synthesis can be ablated by inhibiting TGF-beta activity (Lindahl et al. 2002). In human tendon the presence of TGF-β1 has been demonstrated in association with fibroblasts, and in human skeletal muscle TGF β1 has been demonstrated to be located mainly in the perimysium in relation to fibroblasts and endothelium (Heinemeier et al. 2007a). Furthermore, after exercise of human skeletal muscle TGF-β1 was upregulated by 2.5 h after exercise followed by increased collagen synthesis 6 h post exercise. Using the microdialysis technique around the Achilles tendon in humans, it was found that both local and circulating levels of TGF-beta increased in response to running, and furthermore, the time relation between TGF-beta and indicators of local collagen type I synthesis supported a role of TGF-beta in regulation of local collagen type I synthesis of tendon-related connective tissue (Heinemeier et al. 2003). In a recent study, different types of muscular contractions were used in order to study responses of collagen synthesis. Both concentric and eccentric contractions were performed in two different experimental groups. When performing eccentric contractions, the loading of both muscle and tendon was higher than with concentric contractions, whereas strain was not assumed to be different to a major degree in the two situations (Heinemeier et al. 2007a). The expression of IGF-1 and TGF-β1 in muscle increased the most during eccentric exercise, whereas the response in tendon was independent of the type of contraction (Heinemeier et al. 2007a, 2007b). Likewise, the procollagen expression was similar in concentric and eccentric exercises, indicating that stress was not a major determiner in the magnitude of collagen synthesis response in tendon, and the findings suggest a more prominent role for strain in signaling a growth factor release and subsequent increase in collagen formation. Taken together, the data support the view that TGF-beta plays a role in collagen synthesis of tendon with mechanical loading of collagen-rich tissue. In addition, it is important to note that whereas acute TGF-beta responses to mechanical loading most likely represent an important physiological response, a more prolonged and chronic elevation of TGF-beta in association with a variety of pathological situations more likely will result in an uncontrolled formation of fibrotic tissue (Leask et al. 2002).

9.2.3 Interplay Between Collagen-Rich Matrix and Contracting Skeletal Muscle

In the developing skeletal muscle it is clearly shown that there is a close coupling between myogenesis and development of the intramuscular extracellular matrix components. It is more unclear to what extent such an interplay exists in mature skeletal muscle, but given the important role that intramuscular collagen plays in force transmission, it would be relevant to assume some kind of coordinated response to loading. Experiments have shown that the myogenic stem cell in skeletal muscle, the satellite cell, is activated not only in pathological situations associated with skeletal muscle injury, but also in physiological situations associated with muscle hypertrophy development (Kadi et al. 2005), but to what extent the connective tissue within the skeletal muscle is involved in this activation of satellite cells is not known. When human muscle is subjected to eccentric loading, it will result in activation of satellite cells in the absence of major damage of the muscle cell itself (Crameri et al. 2004a). In fact, the activation of satellite cells was accompanied by increased staining intensity for the procollagen propeptide PINP and for tenacin-C, a marker for connective tissue and known to reinforce lateral adhesion of the myofiber to the surrounding endomysium (Crameri et al. 2004b). These observations suggest that activation of collagen synthesis in the intramuscular connective tissue is associated with satellite cell activation. Although this suggests a potential coupling of signaling to the myofiber that involves activation of the collagen-rich tissue within the muscle, it has to be stated that performing even more heavy eccentric exercise in association with electrical stimulation can result in a marked degree of myofiber injury and this will cause an even more pronounced activation of the satellite cells (Crameri et al. 2007). This illustrates that even if a coupling between connective tissue and the muscle cell exists in mature muscle, it certainly does not exclude a direct link between muscle cell injury and satellite cell activation. A further dissociation between muscle cells and connective tissue comes from a recent study that investigated protein synthesis in response to heavy resistance training and more prolonged low-resistance exercise, respectively. Whereas, the exercise-induced rise in protein synthesis for myofibrillar proteins was dependent upon the exercise intensity, the rise in protein synthesis for collagen of the muscle was similar in the two situations where intensity was varied but the total work output was the same (Holm et al., unpublished). This may illustrate that muscle cells are sensing the exercise intensity, whereas the collagen-rich tissue does have a similar detailed sensing of mechanical loading.

9.2.4 Role for Stem Cells in Tendon Adaptation and Healing

The ability to harvest mesenchymal cells from the bone marrow, grow them in culture and re-introduce them into an injured tendon has been widely exploited in racehorses, despite a lack of controlled studies endorsing this treatment (Smith and Webbon, 2005). In a recent controlled study however, mesenchymal cells, extracted and treated in a similar way, were applied to severed rabbit tendons, where some positive findings support the use of these stem cells in this way, at least in the early stages of tendon healing (Chong et al. 2007). The most limiting factor in stem cell research in tendon tissue, as in all tissues, is the availability of suitable markers. Many of the traditional markers, used in cell-sorting methods, stain endothelial cells, pericytes and inflammatory cells as well as stem cells, are rendered unsuitable for use in situ where these different cell types are present. In vitro studies do not suffer from this complication and recently cells derived from human hamstring tendons were observed to behave similarly to bone marrow-derived cells in vitro, suggesting the presence of a population of cells with differentiation potential in human tendon (DeMos et al. 2007). The temporal appearance of tendon-produced and circulation-derived cells during tendon healing was elegantly demonstrated with the aid of green fluorescent protein (GFP) chimeric rats (Kajikawa et al. 2007). There, circulation-derived GFP cells were observed in the tendon wound 24 h after injury, whereas tendon-derived GFP cells were not evident until 3 days after the injury, but appeared to take over from the circulation-derived cells by day 7. The labeling and tracing method employed in this study opens up a new way of investigating tendon remodeling in response to injury. In vivo cell tracking using magnetic resonance imaging (MRI) of magnetically labeled cells is the latest non-invasive method for following the movement of cells around the body. The success of this method in tracking the migration of adult stem cells in the central nervous system has recently been reviewed (Sykova and Jendelova, 2007), and one can surmise that it is merely a matter of time before this method will be applied to the area of tendon research, which, together with the other methods mentioned here, will help enrich our understanding of the involvement of stem cells in adaptation of this tissue.

9.3 Mechanical Properties of Human Tendon, In Vivo

The tendons of the human lower extremity are subjected to appreciable loads during human locomotion. In fact, the Achilles and patellar tendons can be subjected to several times the body weight during running. Unfortunately, it is also recognized that tendons are unable to adapt to certain loading conditions and therefore end up with pathology, including so-called overuse injuries and complete tendon ruptures (Maffulli et al., 1999; Kannus and Jozsa, 1991). However, if overuse injuries and ruptures are associated with predisposing morphological changes or if the tendon inadvertently exceeds a narrow safety margin remains unknown. Nevertheless, the mechanical properties of the tendon are generally thought to play some pivotal role. Because the etiology of these injuries, and exactly how tendons adapt to physical activity or lack thereof, are at best incompletely understood, the ability of clinicians to provide optimal treatment and prevent injury remain circumscribed.

9.3.1 Tendon Hypertrophy

Tendons subjected to considerable stress, including the Achilles tendon, are believed to operate as springs by storing and releasing elastic strain energy during loco-motion (Alexander & Bennet-Clark, 1977; Biewener & Roberts, 2000). From this standpoint it is advantageous to have a thin (and long) tendon. On the other hand, a thicker tendon, which would yield less strain energy, would reduce the average stress (force/area) across the tendon and thereby provide a greater safety margin. For example, if the tendon were to increase in size (hypertrophy) as a function of physical activity/exercise, the average stress would be reduced for a given applied force (e.g., some multiple of the body weight). However, the time course and to what extent human tendon adapts by changing its material and/or undergo hypertrophy in response to exercise remain an enigma. Animal data show that tendon may undergo qualitative (Buchanan & Marsh, 2001), hypertrophic changes (Woo et al., 1982; Birch et al., 1999), or both (Woo et al., 1982) in response to endurance-type exer-cise and do therefore not provide a coherent picture. In humans cross-sectional data suggest that habitual long-distance running (> 5 yrs.) is associated with a markedly greater cross-sectional area (22%) of the Achilles tendon compared to that of non-runners (Rosager et al., 2002). However, a total training stimulus of ∼9 months of running in previously untrained young subjects did not result in tendon hypertrophy of the Achilles tendon (Hansen et al., 2003). At the same time it has been shown that resistance training for 3 months induced remarkable changes in the material properties of human tendon in the absence of any tendon hypertrophy in older people (Reeves et al., 2003), while the opposite was demonstrated in young male subjects (Kongsgaard et al., 2007). It is possible that these apparent discrep-ancies may relate to the training mode (endurance vs resistance training), age and tendon type (Achilles vs patellar tendon). It is also possible that gender (hormonal milieu) may play a role (Miller et al., 2006b): Recent cross-sectional data indirectly suggest that the ability of patellar tendon to adapt in response to habitual loading such as running is attenuated in women as evidenced by the similar tendon cross-sectional area, stiffness and modulus in habitual runners and non-runners (Westh et al., 2007) (Fig. 9.1). Therefore, although it has been elegantly and repeatedly shown in a human model that there is an increased metabolic activity of human tendon tissue in response to an acute bout of loading (Bojsen-Moller et al., 2006; Langberg et al., 1999; Miller et al., 2005), it is not entirely understood to what extent this results in a larger structure or a different material.

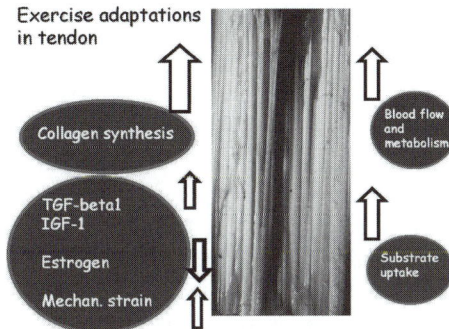

Fig. 9.1 Schematic overview of adaptations in tendon to exercise. Blood flow, metabolism and substrate uptake is elevated acutely with exercise. Collagen synthesis rises both acutely and chronically in response to exercise. Evidence is provided that IGF-1, TGF-beta1 and mechanical strain has a stimulating effect upon this process, whereas estrogen (estradiol) has an inhibiting influence upon exercise-induced collagen synthesis

9.3.2 Regional Differences in Cross-Sectional Area

When tendon stress is calculated it is commonly based on a single or average value of the tendon cross-sectional area (CSA). However, it appears that there is a large variation in the tendon CSA along the length of both the human Achilles and patellar tendon (Kongsgaard et al., 2005; Magnusson & Kjær, 2003; Rosager et al., 2002; Westh et al., 2007). In fact, both the Achilles and patellar tendon can have a CSA that differs by more than 50% along its length such that the most proximal segment is considerably smaller than the most distal tendon segment. Stress along the length of the tendon may consequently differ considerably. It is interesting to note that clinical conditions such as patellar and Achilles tendinopathies occur in the region with the greatest average stress for a given applied force, and the etiology may therefore be somehow related to the stress in the region. Furthermore, if in fact there is tendon hypertrophy in response to increased loading (see above), there is some evidence that it may occur in a region-specific manner (Kongsgaard et al., 2007; Magnusson & Kjær, 2003).

9.3.3 The Ultrasonography Method

Much of the current knowledge regarding mechanical behavior is based on data from experimental animal or cadaver tissue, while some more recent studies have studied mechanics on a microscopic level. Currently, the most widely used technique to investigate, in vivo, human tendon displacement during muscle contractions is B-mode ultrasonography, which was developed in the mid-1990s for measurements in the whole tendon–aponeurosis complex (Fukashiro et al., 1995) and late 1990s for measurements in isolated tendon (Maganaris & Paul, 1999). The non-invasive approach of studying human tendons in vivo is attractive, and many limitations have been successfully addressed, but several details have to be considered. The method accounts for only two dimensions of structural deformation (i.e., in the sagittal plane). The technique was originally intended for the displacement of intramuscular fascicular structures, and therefore the resulting deformation does not represent that of the tendon per se, but rather the total deformation of the combined tendon and aponeurosis distal to the measurement site. This can, in some muscle–tendon complexes, but not all, be circumvented by identifying the very junction between the aponeurosis and free tendon (the myotendinous junction), or alternatively by introducing a visible landmark such as a needle (Magnusson et al., 2003). For the patellar tendon the problem can be overcome by having two bony landmarks (tibia and patella) (Hansen et al., 2006). The technique is most commonly employed during "isometric" conditions in which small amounts of joint rotation or body movement may take place that can affect the displacement measurements, which need to be accounted for (Magnusson et al., 2003). A true resting length of the tendon (i.e., 0% strain) may be difficult to obtain in vivo, but may be defined as that corresponding to zero net joint moment. The current time-resolution of ultrasonography (frequency of sampling) and of video recording apparatus is a limiting factor that precludes analysis of faster contraction velocities. However, the technique still lacks an associated accurate measurement of tendon force. Notwithstanding these methodological difficulties, the technique has provided recent insights into human, in vivo, tendon behavior.

9.3.4 Human Aponeurosis Shear, In Vivo

The fact that the CSA of tendon differs along its length and that the stress seen by the tendon therefore also differs considerably for a given applied force prompts questions as to how force is transmitted in general throughout the tendon. In this regard the human Achilles tendon is of interest because of its unique anatomy and because it is often implicated clinically. The human triceps surae muscle-tendon structure is composed of three separate muscle compartments that merge via their aponeuroses into one common tendon to ultimately insert on the calcaneus. To what extent activation of the individual muscles of the triceps surae complex influences aponeurosis and tendon strain has been unknown. An in vitro study demonstrated that differences in medial and lateral forces in the Achilles tendon can be observed when single muscles of the triceps surae were subjected to force (Arndt et al., 1999). Non-uniform tendon force would theoretically result in intratendinous shear strain and cause sliding between planes of tissue layers parallel to the acting forces. Because the triceps surae includes the gastrocnemii muscles that cross both the ankle and knee joints, and the soleus muscle that crosses the ankle joint alone, the relative contribution of these muscles to the tendon force will be influenced by the degree of knee flexion (Cresswell et al., 1995).

It has been shown that during maximal isometric contractions with the plantarflexor muscles, there is a differential displacement between the soleus and gastrocnemius aponeuroses proximal to the junction with the Achilles tendon (Fig. 9.2) (Bojsen-Moller et al., 2004). When the knee joint was extended, displacement of the medial gastrocnemius aponeurosis exceeded that of the soleus aponeurosis, whereas the converse occurred when the knee joint was flexed. These differences in aponeurosis displacement created a "shear" effect with a direction that was governed by the knee joint position. Since the collagen fibers of the aponeurosis fuse distally to become free tendon structures, any difference in displacement at the level of the aponeuroses may be manifested as shear strain at the level of the tendon, although this has yet to be confirmed. The difference in displacement between

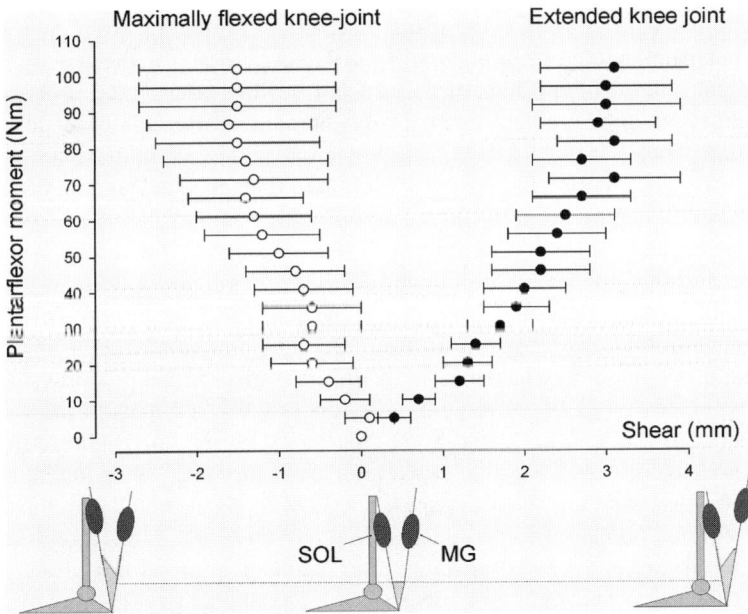

Fig. 9.2 Anterior–posterior shear between the MG and Sol aponeuroses. Shear displacement between the two aponeuroses was determined by subtracting the displacement of the soleus aponeurosis from that of the medial gastrocnemius throughout ramp contractions in the two knee joint positions. In the extended knee position, displacement of the gastrocnemius exceeded that of the soleus, whereas the opposite occurred in the flexed knee position

separate aponeuroses within the triceps surae amounted to more than 30% of the maximal observed displacement, indicating that a considerable "shear potential" exists during movement tasks. The observed differential aponeurosis displacement was likely caused by differences in force output of the medial gastrocnemius and soleus muscles.

9.3.5 Mechanical Properties of Individual Human Tendon Fascicles

Patellar tendinopathy chiefly concerns the proximal and posterior portion of the patellar tendon (Johnson et al., 1996), and this poses the question as to whether the mechanical properties differ by region within a whole tendon. Data on strain properties of the anterior and posterior portion of the human patellar tendon are sparse and conflicting. In cadaver knees of older persons it has been shown that during quadriceps loading tensile strain was uniform in the anterior and posterior region with the knee in full extension, but tensile strain increased on the anterior side and decreased on the posterior with knee flexion (Almekinders et al., 2002). But it has also been shown that quadriceps loading in flexion caused a greater strain on the posterior compared to the anterior side of the patellar tendon (Basso et al., 2002).

An alternative way to examine if the mechanical properties differ by region in the human patellar tendon is to test the isolated portions of the tendon. It was recently possible to obtain human tissue from healthy young men in association with elective anterior cruciate ligament reconstruction (Haraldsson et al., 2005). Thin collagen bundles that were ~ 35 mm in length and ~ 3.5 mm in diameter were obtained from the anterior and posterior portion of the harvested tendon. From these collagen bundles, individual strands of fascicles (ø ~300 μm) were dissected and tested for mechanical properties (Fig. 9.3). The data showed that tendon fascicles from the anterior portion of the human patellar tendon in young men displayed substantially greater peak and yield stress and tangent modulus compared to the posterior portion of the tendon (Fig. 9.3). This observation suggests that regions of the patellar tendon may have markedly different mechanical properties. It is not surprising since there may be regional differences when it comes to the structural components of the tendon. In the horse there may be region-specific differences with respect to collagen fibril diameter (Patterson-Kane et al., 1997), and it has also been shown that there may be a site-specific loss of larger size collagen fibrils in the core of ruptured human Achilles tendons (Magnusson et al., 2002). It remains to be established if the above-noted differences in mechanical properties of the patellar tendon can be attributed to factors such as fibril size and density (Parry et al., 1978), cross-links (Thompson & Czernuszka, 1995), fibril length (Redaelli et al., 2003) and components of the extracellular matrix (Kjær, 2004). Also, it is unknown to what extent training regimes with varying degrees of knee flexion may influence the region-specific material properties of the tendon, and if such exercises can reduce the site-specific patellar tendinopathy.

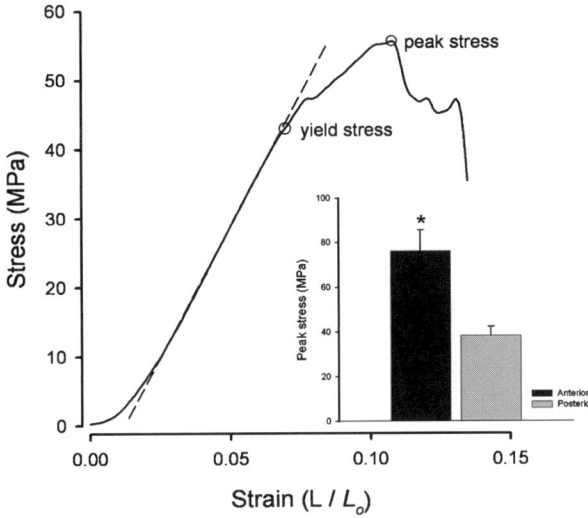

Fig. 9.3 The stress–strain relationship for a human patella fascicle. Dashed line is the linear portion of the relationship; identified yield stress and peak stress. The bar graph shows the data from the anterior and posterior portion of the patellar tendon demonstrating the region-specific greater stress in the anterior fascicles

9.3.6 Force Transmission Between Human Tendon Fascicles

There is some understanding of force transmission in skeletal muscle, but relatively little is known about force transmission in tendon. Skeletal muscle fibers do not always extend from one tendon plate to the other, which requires that contractile force can be transmitted laterally to adjacent fibers to initiate movement, which was demonstrated elegantly by Street (1983). Force is transmitted from the muscle fibers to the aponeurosis via the myotendinous junction, and can also be transmitted to parallel adjacent structures via the aponeurosis (Huijing & Jaspers, 2005). The aponeurosis, which has different mechanical properties than free tendon (Magnusson et al., 2003), may be loaded heterogeneously during skeletal muscle contraction (Finni et al., 2003), although how this affects tendon force transmission and to what extent lateral force transmission exists in human-free tendon remain unknown.

The tendon is a hierarchical structure (Kastelic et al., 1978), and to understand how force is transmitted within the whole tendon it is necessary to examine the mechanical properties at the various levels. A disparity in magnitude of strain at different hierarchical levels has been observed, and while the exact mechanism for such a hierarchy of strain distribution within the tendon structure remains unknown, it suggests that "gliding" processes may be involved (Mosler et al., 1985; Puxkandl et al., 2002). Studies on force pathways in tendons have largely focused on molecular strain and intermolecular force transmission (Mosler et al., 1985; Puxkandl et al., 2002; Sasaki & Odajima, 1996), fibril–fibroblast mechanotransduction (Arnoczky et al., 2002) and contributing factors to interfibrillar force transmission (Scott, 2003; Vesentini et al., 2005). The tendon fascicle, however, is a distinct entity within the tendon hierarchy.

The ability of the fascicle to transmit force to the adjacent parallel fascicle was recently examined in human patellar and Achilles tendon tissue obtained during surgery (Haraldsson et al. 2008). From each of the tendon specimens, two adjoining strands of collagen fascicles enclosed in an intact fascicular membrane were carefully dissected and special care was taken to ensure that the fascicular membrane between adjoining fascicles was left intact. The specimens were subjected to a series of load–displacement cycles to a given displacement ($\sim 3\%$ strain). The first load–displacement cycle was performed with both fascicles and fascicular membrane intact (Fig. 9.4). The second load–displacement cycle was performed with one fascicle transversally cut while the other fascicle and the fascicular membrane remained intact. The third load–displacement cycle was performed with both fascicles transversally cut on opposite ends while the fascicular membrane was kept intact. Theoretically, if there is no or inconsequential lateral transfer of force, a reduction in the cross-sectional area of a structure by 50% (one of two fascicles of similar size) should reduce the stiffness of the structure by half. Severing one fascicle from the patellar tendon reduced the material stiffness by $\sim 40\%$ while the same procedure reduced the material stiffness by more than 50% in the Achilles tendon, indicating that lateral force transmission between adjacent human tendon fascicles was marginal. Cycle 3 represented two adjacent fascicles that could potentially only transmit force in parallel rather than in series, and these data confirm the findings of the comparison between cycles 1 and 2 by demonstrating that the lateral force transmission between fascicles was relatively small or negligible (Fig. 9.4).

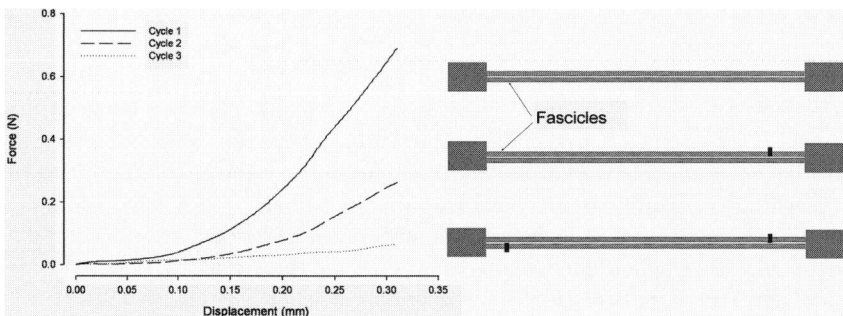

Fig. 9.4 Force deformation data of fascicles cutting experiments. Top: two adjacent fascicles and intact fascicular membrane. Middle: one fascicle transversally cut while the other fascicle and the fascicular membrane remained intact. The stiffness was reduced by ~ 40–50%. Bottom: both fascicles transversally cut on opposite ends while the fascicular membrane was kept intact, and could therefore potentially only transmit force in parallel rather than in series. In this case the lateral force transmission between fascicles was relatively small or negligible

Of course, these data do not show to what extent fascicles are functionally "independent" in whole tendon in vivo. It appears that the magnitude of strain at the various hierarchical levels differ such that strain of the whole tendon exceeds that of the strain at the molecule and fibril level (Mosler et al., 1985; Puxkandl et al., 2002; Sasaki & Odajima, 1996). Specifically, it has been suggested that during loading the triple helix itself may elongate, that the gap region between longitudinally adjoining molecules increases, that there is relative slippage between laterally adjacent molecules and that fibrils themselves may slide relative to one another. At the level of the molecule, strain will be taken up by cross-links (Puxkandl et al., 2002), while the strain between fibrils may be prevented by components of the extracellular matrix (Scott, 2003). Intuitively, this means that some of the force is "transferred" to the adjacent structure by means of lateral shear forces. The fascicle cutting experiment adds to the existing knowledge of tendon mechanical properties by demonstrating that gliding processes can freely take place between adjacent fascicles. However, the present data also suggest that the aforementioned mechanisms of lateral shear force are largely limited to within the fascicle.

References

Abrahamsson SO (1997). Similar effect of recombinant human insulin-like growth factor-I and II in cellular activities in flexor tendons of young rabbits: experimental studies in vitro. J Orthop Res 15: 256–262.

Alexander RM & Bennet-Clark HC (1977). Storage of elastic strain energy in muscle and other tissues. Nature 265: 114–117.

Almekinders LC, Vellema JH, & Weinhold PS (2002). Strain patterns in the patellar tendon and the implications for patellar tendinopathy. Knee Surg Sports Traumatol Arthrosc 10: 2–5.

Arndt AN, Bruggemann GP, Koebke J, & Segesser B (1999). Asymmetrical loading of the human triceps surae: I. Mediolateral force difference in the Achilles tendon. Foot Ankle Int 20: 445–449.

Arnoczky SP, Lavagnino M, Whallon JH, & Hoonjan A (2002). In situ cell nucleus deformation in tendons under tensile load; a morphological analysis using confocal laser microscopy. J Orthop Res 20: 29–35.

Babraj J, Cuthbertson D, Smith J, Langberg H, Miller BF, Krogsgaard M, Kjær M, & Rennie MJ (2005). Collagen synthesis in human musculoskeletal tissues and skin. Am J Physiol 99: 986–994.

Basso O, Amis AA, Race A, & Johnson DP (2002). Patellar tendon fiber strains: their differential responses to quadriceps tension. Clin Orthop 400: 246–253.

Biewener AA & Roberts TJ (2000). Muscle and tendon contributions to force, work, and elastic energy savings: a comparative perspective. Exerc Sport Sci Rev 28: 99–107.

Birch HL, McLaughlin L, Smith RK, & Goodship AE (1999). Treadmill exercise-induced tendon hypertrophy: assessment of tendons with different mechanical functions. Equine Vet J Suppl 30: 222–226.

Bojsen-Moller J, Hansen P, Aagaard P, Svantesson U, Kjær M, & Magnusson SP (2004). Differential displacement of the human soleus and medial gastrocnemius aponeuroses during isometric plantar flexor contractions in vivo. J Appl Physiol 97: 1908–1914.

Bojsen-Moller J, Kalliokoski KK, Seppanen M, Kjær M, & Magnusson SP (2006). Low-intensity tensile loading increases intratendinous glucose uptake in the Achilles tendon. J Appl Physiol 101: 196–201.

Buchanan CI & Marsh RL (2001). Effects of long-term exercise on the biomechanical properties of the Achilles tendon of guinea fowl. J Appl Physiol 90: 164–171.

Chong, AK, et al. (2007). Bone marrow-derived mesenchymal stem cells influence early tendon-healing in a rabbit Achilles tendon model. J Bone Joint Surg Am, 89: 74–81.

Crameri R, Langberg H, Jensen CH, Teisner B, Schrøder HD, & Kjær M (2004a). Activation of satellite cells in human skeletal muscle after a single bout of exercise. J Physiol 558: 333–340.

Crameri R, Langberg H, Teisner B, Magnusson P, Olesen JL, Koskinen S, Suetta C, & Kjær M (2004b). Synchronous disruption of the extracellular matrix and mechanical tenderness in skeletal muscle after a single bout of eccentric loading in humans. Matrix Biol 23: 259–264.

Crameri R, Aaagaard P, Qvortrup K, Langberg H, Olesen JL, & Kjær M (2007). Myofibre damage in human skeletal muscle: Effects of electrical stimulation vs voluntary contraction. J Physiol 583: 365–380.

Cresswell AG, Loscher WN, & Thorstensson A (1995). Influence of gastrocnemius muscle length on triceps surae torque development and electromyographic activity in man. Exp Brain Res 105: 283–290.

DeBoer J, Selby A, Atherton P, Smith K, Seynnes OR, Maganaris CN, Maffulli N, Movin T, Narici MV, & Rennie MJ (2007). The temporal response of protein synthesis, gene expression and cell signalling in human quadriceps muscle and patellar tendon to disuse. J Physiol 585: 241–251.

DeMos, M et al. (2007). Intrinsic differentiation potential of adolescent human tendon tissue: an in-vitro cell differentiation study. BMC Musculoskelet Disord, 8: 16–22.

Døssing S & Kjær M (2005). Growth hormone and connective tissue in exercise. Scand. J Med Sci Sports 15: 202–210.

Finni T, Hodgson JA, Lai AM, Edgerton VR, & Sinha S (2003). Nonuniform strain of human soleus aponeurosis-tendon complex during submaximal voluntary contractions in vivo. J Appl Physiol 95: 829–837.

Fukashiro S, Itoh M, Ichinose Y, Kawakami Y, & Fukunaga T (1995). Ultrasonography gives directly but noninvasively elastic characteristic of human tendon in vivo. Eur J Appl Physiol 71: 555–557.

Goldspink G (2006). Impairment of IGF-I gene splicing and MGF expression associated with muscle wasting. Int J Biochem Cell Biol 38: 481–489.

Hansen P, Aagaard P, Kjær M, Larsson B, & Magnusson SP (2003). The effect of habitual running on human Achilles tendon load-deformation properties and cross-sectional area. J Appl Physiol 95: 2375–2380.

Hansen P, Bojsen-Moller J, Aagaard P, Kjær M, & Magnusson SP (2006). Mechanical properties of the human patellar tendon, in vivo. Clin Biomech 21: 54–58.

Haraldsson BT, Aagaard P, Krogsgaard M, Alkjaer T, Kjær M, & Magnusson SP (2005). Region-specific mechanical properties of the human patella tendon. J Appl Physiol 98: 1006–1012.

Haraldsson BT, Aagard P, Qvortrup K, Bojscn-Moller J, Krogsgaard M, Koskinen S, Kjaer M, Magnusson SP (2008). Lateral force transmission between human tendon fascicles. Matrix Biol. 27:86–95.

Heinemeier K, Langberg H, Olesen JL, & Kjær M (2003) Role of transforming growth factor beta in relation to exercise induced type I collagen synthesis in human tendinous tissue. J Appl Physiol 95: 2390–2397.

Heinemeier KM, Olesen JL, Schjerling P, Haddad F, Langberg H, Baldwin KM, & Kjær M (2007a). Short term strength training and the expression of myostatin- and IGF-I isoforms in rat muscle and tendon: Differential effects of specific contraction types. J Appl Physiol 102: 573–581.

Heinemeier KM, Olesen JL, Haddad F, Langberg H, Kjær M, Baldwin KM, & Schjerling P (2007b). Expression of collagen and related growth factors in rat tendon and skeletal muscle in response to specific contraction types. J Physiol 582: 1303–1316.

Huijing PA & Jaspers RT (2005). Adaptation of muscle size and myofascial force transmission: a review and some new experimental results. Scand J Med Sci Sports 15: 349–380.

Johnson DP, Wakeley CJ, & Watt I (1996). Magnetic resonance imaging of patellar tendonitis. J Bone Joint Surg – Br 78: 452–457.

Kadi F, Charifi N, Denis C, Lexell J, Andersern JL, Schjerling P, Olsen S, & Kjær M (2005). The behaviour of satellite cells in response to exercise: What have we learned from human studies? Pfügers Arch 451: 319–327.

Kajikawa, Y et al. (2007). GFP chimeric models exhibited a biphasic pattern of mesenchymal cell invasion in tendon healing. J Cell Physiol 210: 684–691.

Kannus P & Jozsa L (1991). Histopathological changes preceding spontaneous rupture of a tendon. A controlled study of 891 patients. J Bone Joint Surg [Am] 73: 1507–1525.

Kastelic J, Galeski A, & Baer E (1978). The multicomposite structure of tendon. Connect Tissue Res 6: 11–23.

Kjær M (2004). Role of extracellular matrix in adaptation of tendon and skeletal muscle to mechanical loading. Physiol Rev 84: 649–698.

Kongsgaard M, Aagaard P, Kjær M, & Magnusson SP (2005). Structural Achilles tendon properties in athletes subjected to different exercise modes and in Achilles tendon rupture patients. J Appl Physiol 99: 1965–1971.

Kongsgaard M, Reitelseder S, Pedersen TG, Holm L, Aagaard P, Kjær M, & Magnusson SP (2007). Region specific patellar tendon hypertrophy in humans following resistance training. Acta Physiol 191: 111–112.

Koskinen SO, Heinemeier KM, Olesen JL, Langberg H, & Kjær M (2004). Physical exercise can influence local levels of matrix metalloproteinases and their inhibitors in tendon related connective tissue. J Appl Physiol 96: 861–864.

Kovanen V (1989). Effects of ageing and physical training on rat skeletal muscle. Acta Physiol Scand 135 suppl 577: 1–56.

Kurtz CA, Loebig TG, Anderson DD, Demeo PJ & Cambell PG (1999). Insulin-like growth factor I accelerates functional recovery from Achilles tendon injury in a rat model. Am J Sports Med 27: 363–369.

Langberg H, Skovgaard D, Bülow J, & Kjær M (2000). Time pattern of exercise-induced changes in type-I collagen turnover after prolonged endurance exercise in humans. Calcif Tissue Int 67: 41–44.

Langberg, H., Rosendal L, & Kjær M (2001). Training induced changes in peritendinous type I collagen turnover determined by microdialysis in humans. J Physiol 534: 297–302.

Langberg H, Ellingsgaard H, Madsen T, Jansson J, Magnusson SP, Aagaard P, & Kjær M (2007). Eccentric rehabilitation exercise increases peritendinous type I collagen synthesis in humans with Achilles tendinosis. Scand J Med Sci Sports 17: 61–66.

Langberg H, Skovgaard D, Petersen LJ, Bulow J, & Kjær M (1999). Type I collagen synthesis and degradation in peritendinous tissue after exercise determined by microdialysis in humans. J Physiol 521 Pt 1: 299–306.

Lange, KH, Andersen JL, Beyer N, Isaksson F, Larsson B, Rasmussen MH, Juul A, Bülow J, & Kjær M (2002). GH administration changes myosin heavy chain isoforms in skeletal muscle but does not augment muscle strength or hypertrophy, either alone or combined with resistance exercise training in healthy elderly men. J Clin Endocrinol Metab 87: 513–523.

Laurent GJ (1987). Dynamic state of collagen pathways of collagen degradation in vivo and their possible role in regulation of collagen mass. Am J Physiol 252: C1–C9.

Leask A, Holmes A, & Abraham DJ (2002). Connective tissue growth factor: a new and important player in the pathogenesis of fibrosis. Curr Rheumatol Rep 4: 136–142.

Lindahl GE, Chambers RC, Papakrivopoulou J, Dawson SJ, Jacobsen MC, Bishop JE, & Laurent GJ (2002). Activation of fibroblast procollagen alpha I(I) transcription by mechanical strain is transforming growth factor beta dependent and involves increased binding of CCAAT-binding factor (CBF/NF-y) at the proximal promotor. J Biol Chem 277: 6153–6161.

Liu SH, Shaikh R, Panossian V, Yang RS, Nelson SD, Soleiman N, Finerman GA, & Lane JM (1996). Primary immunolocalization of estrogen and progesterone target cells in the human anterior cruciate ligament. J Orthop Res 14: 526–533.

Longobardi S, Keay N, Ehrnborg C, Cittadini A, Rosen T, Dall R, Boroujerdi MA, Bassett EE, Healy ML, Pentecost C, Wallace JD, Powrie J, Jorgensen JO, & Sacca L (2000). Growth hormone (GH) effects on bone and collagen turnover in healthy adults and its potential as a marker of GH abuse in sports: a double blind, placebo-controlled study. The GH-2000 Study Group. J Clin Endocrinol Metab 85: 1505–1512.

Maffulli N, Waterston SW, Squair J, Reaper J, & Douglas AS (1999). Changing incidence of Achilles tendon rupture in Scotland: a 15-year study. Clin J Sport Med 9: 157–160.

Maganaris CN & Paul JP (1999). In vivo human tendon mechanical properties. J Physiol (Lond) 521: 307–313.

Magnusson SP, Hansen P, Aagaard P, Brond J, Dyhre-Poulsen P, Bojsen-Moller J, & Kjær M (2003). Differential strain patterns of the human gastrocnemius aponeurosis and free tendon, in vivo. Acta Physiol Scand 177: 185–195.

Magnusson SP & Kjær M (2003). Region-specific differences in Achilles tendon cross-sectional area in runners and non-runners. Eur J Appl Physiol 90: 549–553.

Magnusson SP, Qvortrup K, Larsen JO, Rosager S, Hanson P, Aagaard P, Krogsgaard M, & Kjær M (2002). Collagen fibril size and crimp morphology in ruptured and intact Achilles tendons. Matrix Biol 21: 369–377.

Miller B, Olesen JL, Hansen M, Døssing S, Crameri R, Welling RJ, Langberg H, Flyvbjerg A, Kjær M, Babraj J, Smith K, & Rennie MJ (2005). Coordinated collagen and muscle protein synthesis in human patella tendon and quadriceps muscle after exercise. J Physiol 567: 1021–1033.

Miller B, Hansen M, Olesen JL, Flyvbjerg A, Schwarz P, Babraj JA, Smith K, Rennie MJ, & Kjær M (2006a). No effect of menstrual cycle on myofibrillar and connective tissue synthesis in contracting skeletal muscle. Am J Physiol 290: E163–E168.

Miller BF, Hansen M, Olesen JL, Schwarz P, Babraj JA, Smith K, Rennie MJ, & Kjær M (2006b). Tendon collagen synthesis at rest and after exercise in women. J Appl Physiol 102: 541–546.

Mosler E, Folkhard W, Knorzer E, Nemetschek-Gansler H, Nemetschek T, & Koch MH (1985). Stress-induced molecular rearrangement in tendon collagen. J Mol Biol 182: 589–596.

Olesen JL, Heinemeier KM, Langberg H, Magnusson SP, Kjær M, & Flyvbjerg A (2006). Expression, content and localization of IGF-1 in human Achilles tendon. Conn Tissue Res 47: 200–206.

Olesen, JL, Langberg H, Heinemeier K, Flyvbjerg A, & Kjær M (2007a). Exercise-dependent IGF-I, IGFBPs, and type I collagen changes in human peritendinous connective tissue determined by microdialysis. J Appl Physiol 102:. 214–220.

Olesen JL, Heinemeier KM, Haddad F, Langberg H, Flyvbjerg A, Kjær M, & Baldwin KM (2007b). Expression of insulin-like growth factor I, insulin-like growth factor binding proteins, and collagen mRNA in mechanically loaded plantaris tendon. J Appl Physiol 101: 183–188.

Parry DA, Barnes GR, & Craig AS (1978). A comparison of the size distribution of collagen fibrils in connective tissues as a function of age and a possible relation between fibril size distribution and mechanical properties. Proc R Soc Lond B Biol Sci 203: 305–321.

Patterson-Kane JC, Wilson AM, Firth EC, Parry DA, & Goodship AE (1997). Comparison of collagen fibril populations in the superficial digital flexor tendons of exercised and nonexercised thoroughbreds. Equine Vet J 29: 121–125 [published erratum appears in Equine Vet J 1998 Mar; 30(2):176].

Puxkandl R, Zizak I, Paris O, Keckes J, Tesch W, Bernstorff S, Purslow P, & Fratzl P (2002). Viscoelastic properties of collagen: synchrotron radiation investigations and structural model. Philos Trans R Soc Lond B Biol Sci 357: 191–197.

Redaelli A, Vesentini S, Soncini M, Vena P, Mantero S, & Montevecchi FM (2003). Possible role of decorin glycosaminoglycans in fibril to fibril force transfer in relative mature tendons – a computational study from molecular to microstructural level. J Biomech 36: 1555–1569.

Reeves ND, Maganaris CN, & Narici MV (2003). Effect of strength training on human patella tendon mechanical properties of older individuals. J Physiol 548: 971–981.

Riley, GP et al. (2002) Matrix metalloproteinase activities and their relationship with collagen remodelling in tendon pathology. Matrix Biol 21: 185–195.

Rosager S, Aagaard P, Dyhre-Poulsen P, Neergaard K, Kjær M, & Magnusson SP (2002). Load-displacement properties of the human triceps surae aponeurosis and tendon in runners and non-runners. Scand J Med Sci Sports 12: 90–98.

Sasaki N & Odajima S (1996). Elongation mechanism of collagen fibrils and force-strain relations of tendon at each level of structural hierarchy. J Biomech 29: 1131–1136.

Sciore P, Frank CB, & Hart DA (1998). Identification of sex hormone receptors in human and rabbit ligaments of the knee by reverse transcription-polymerase chain reaction: evidence that receptors are present in tissue from both male and female subjects. J Orthop Res 16: 604–610.

Scott JE (2003). Elasticity in extracellular matrix 'shape modules' of tendon, cartilage, etc. A sliding proteoglycan-filament model. J Physiol 553: 335–343.

Skutek M, Van Griensven M, Zeichen J, Brauer N, & Bosch U (2001). Cyclic mechanical stretching modulates secretion pattern of growth factors in human fibroblasts. Eur J Appl Physiol 86: 48–52.

Smith, RK & Webbon PM (2005). Harnessing the stem cell for the treatment of tendon injuries: heralding a new dawn? Br J Sports Med 39: 582–584.

Street SF (1983). Lateral transmission of tension in frog myofibers: a myofibrillar network and transverse cytoskeletal connections are possible transmitters. J Cell Physiol 114: 346–364.

Sykova, E. & Jendelova P (2007). Migration, fate and in vivo imaging of adult stem cells in the CNS. Cell Death Differ 14: 1336–1342.

Thompson JI & Czernuszka JT (1995). The effect of two types of cross-linking on some mechanical properties of collagen. Biomed Mater Eng 5: 37–48.

Vesentini S, Redaelli A, & Montevecchi FM (2005). Estimation of the binding force of the collagen molecule-decorin core protein complex in collagen fibril. J Biomech 38: 433–443.

Westh E, Kongsgaard M, Bojsen-Moller J, Aagaard P, Hansen M, Kjær M, & Magnusson SP (2007). Effect of habitual exercise on the structural and mechanical properties of human tendon, in vivo, in men and women. Scand J Med Sci Sports.

Wilson VJ, Rattray M, Tomas CR, Moreland BH, & Schulster D (1995). Growth hormone increases IGF-I, collagen type I and collagen II gene expression in dwarf rat skeletal muscle. Mol Cell Endocrinol 115: 187–197.

Woo SL, Gomez MA, Woo YK, & Akeson WH (1982). Mechanical properties of tendon and ligaments. The relationship of immobilization and exercise on tissue remodeling. Biorheology 19 397–408.

Yu WD, Panossian V, Hatch JD, Liu SH & Finerman GA (2001). Combined effects of estrogen and progesterone on the anterior cruciate ligament. Clin Orthop 21: 268–281.

Chapter 10
Tendons and Ligaments: Structure, Mechanical Behavior and Biological Function

A.A. Biewener

Abstract The non-linear viscoelasticity of tendons and ligaments, for which much of their mechanical behavior reflects the properties of their collagen I fibrils, is well suited to absorbing and returning energy associated with the transmission of tensile forces across joints of the body. The high resilience of tendon means that it can serve as an effective biological spring. At the same time, the flexibility of tendons and ligaments allows them to accommodate a wide range of joint movement (or, in the case of ligaments, to restrict movement within a certain range). The high strength of tendons and ligaments also provides considerable weight savings, but this is traded off against the ability to control position and movements of the musculoskeletal system. Tendon and ligament compliance allows elastic energy to be stored and returned to offset energy fluctuations of the body's center of mass during locomotion, conserving muscle work and reducing the metabolic energy cost of locomotor movement. Tendon architecture greatly affects the storage and recovery of elastic strain energy, with long, thin tendons favoring greater strain energy/volume (and weight) of the tendon. It is likely that other elastic elements, such as muscle aponeuroses, also contribute significant energy savings. Tendon compliance may also reduce the cost of muscle contraction, by reducing a muscle's contractile velocity and length change for a given movement, as well as increasing the power output of muscle–tendon units that is key to rapid acceleration and jumping performance. This power enhancement requires a temporal decoupling of muscle work to stretch the tendon from the subsequent more rapid release of elastic strain energy from the tendon. This decoupling may be achieved by changes in inertia and mechanical advantage in vertebrates, but is facilitated by catch mechanisms in invertebrate jumpers. Although it is critical that tendons and ligaments have sufficient strength and an adequate safety factor to limit the risk of failure, tendons are likely subject to damage during their use, which favors a greater safety factor. In addition, because tendon compliance impedes position control, the thickness of many tendons suggests that having sufficient stiffness, rather than strength, is a key structural requirement. Indeed, the majority of tendons that have been studied to date appear to operate at lower stresses and strains, have larger safety factors, and are stiffer, compared with "high-stress" tendons of animals specialized for elastic energy savings.

P. Fratzl (ed.), *Collagen: Structure and Mechanics*,
© Springer Science+Business Media, LLC 2008

10.1 Introduction

Tendons and ligaments enable musculoskeletal forces to be transmitted and redirected across skeletal joints within the body. In doing so, they also facilitate provide a wide range of joint motion and considerable weight and energy savings associated with locomotor movement. Because of their high tensile strength (~100–140 MPa) and stiffness (~1.0–1.5 GPa), tendons transmit muscle forces over long lengths with minimal "in-series stretch". Nevertheless, tendon compliance is important for storing and releasing elastic energy to reduce locomotor costs, as well as for allowing muscles to contract economically at lower velocities and strains, and for minimizing the risk of damage to musculoskeletal structures.

Given that the maximum tensile stress produced by skeletal muscles is ~200–400 kPa (Biewener, 2003; Lieber, 1992), a tendon on the order of only ~1/1000th the physiological cross-sectional area of a muscle can, therefore, safely transmit the muscle's force without failing. To the extent that tendons suffer damage through repeated loading activity and require repair (Ker et al., 2000), or must be sufficiently stiff, rather than being of adequate strength, their thickness and safety factor will tend to be higher. By transmitting tensile stresses, as flexible structures tendons and ligaments also easily bend and change shape to accommodate changes in joint position and skeletal orientation.

All collagen I-based tissues exhibit non-linear elastic properties, reflected by their "J"-shaped stress–strain curves (Woo, 1982; Bennett et al., 1986; Shadwick, 1990) (Chapter 1). This means that tendons and ligaments exhibit greater compliance at low stresses than at intermediate to higher stresses, enhancing their energy absorption capacity at low stress levels. A greater compliance at the onset of loading, combined with slight viscous damping (Chapters 1 and 5), therefore, represent key properties of tendons and ligaments that reduce their susceptibility to damage, as well as the susceptibility of their skeletal attachments and the skeleton more generally. This is clearly important given that these structures must function over a large number of loading cycles during an individual's lifetime.

The "in-series" compliance of a muscle's tendon, particularly at low force levels, however, means that stretch within the tendon must be offset by shortening of the muscle's fibers to control or produce a given net movement of the overall muscle–tendon unit in relation to the skeletal joints and segments that it spans. Consequently, the relative length of a tendon in series with a muscle's fibers strongly affects the role and mechanical performance of a muscle–tendon complex as a whole. Muscle–tendon units with short muscle fibers and long tendons favor elastic energy recovery from the tendon and reduced energy for muscle force production. However, when position and control of movement are key, longer muscle fibers with a short or no tendon are favored. This allows the length changes of the muscle's fibers to be transmitted to the skeleton with minimal in-series compliance, but incurs a much larger energy cost and weight penalty than if the forces were transmitted over much of the distance via a passive, high tensile strength tendon.

The low cell density of tendons and ligaments and the low metabolic activity of their fibroblast populations (Benjamin and Ralphs, 1997; Rumian et al., 2007) means that little energy is expended to transmit mechanical forces, compared with the much greater ATP consumption of muscle fibers that actively generate force and produce movement. Further, because of their comparatively low hysteresis (Alexander, 1988; Bennett et al., 1986; Shadwick, 1990), a large fraction (~91–95%) of the elastic strain energy stored in tendons during loading can be recovered upon unloading. This makes tendons highly resilient springs that can usefully store and recover mechanical energy of the body associated with changes in kinetic and potential energy that occur during each locomotor cycle, serving to reduce muscle work and the energy cost of movement (Alexander, 1988, 2002; Roberts, 2002; Biewener, 2003). Ligaments generally are more variable in composition and organization (Rumian et al., 2007), often exhibiting less regular collagen fibril alignment, lower modulus, and greater hysteresis than tendons (Provenzano et al., 2001; Woo et al., 2006). Their more variable composition reflects the complex shear and multiaxial loading that they are often subjected to in the context of the stabilization of joints over a wide range of motion.

Because the power output of muscles is limited by their rates of force development and shortening, elastic energy stored and released by tendons also serves to enhance the power output that an animal can achieve for jumping and likely for other muscle-powered accelerations (Roberts, 2002). However, this requires a temporal redistribution of muscle work relative to elastic energy release from the tendon. During the initial phase of such movements, muscle work mainly acts to stretch the tendon storing strain energy at a relatively slow rate, rather than producing much skeletal movement. Stored elastic strain energy in the tendon can then be subsequently released at a rapid rate during unloading, as the body or limb segments are accelerated, greatly increasing the effective power output of the muscle–tendon unit as a whole.

This chapter explores these functions, as they relate to the locomotor movements of animals, and how they are affected by the structural and underlying biochemical and material properties of the tendons and ligaments.

10.2 Tendon–Ligament Force Transmission and Weight Savings

Tendons and ligaments transmit tensile forces across the joints of the skeleton, and in doing so, they redirect the path of force transmission. At the same time, they allow flexible motion of the joint(s) that they cross. In the case of ligaments, force transmission within or across the external surfaces of a joint often serves to limit the range of joint motion, helping to stabilize a joint via the ligament's passive viscoelastic properties (Woo, 1982; Provenzano et al., 2001; Woo et al., 2006). In the case of tendons, force transmission may occur over a substantial distance, crossing one or more joints, with respect to the muscle that produces the force. Over longer lengths, tendon stiffness becomes particularly critical, as it determines the amount of series elastic compliance (SEC) that the muscle fibers must accommodate as they contract and shorten to produce or control movement.

With a tensile failure strength in the range of 100–140 MPa and an elastic modulus of 1.0–1.5 GPa (Bennett et al., 1986; Alexander, 1988; Shadwick, 1990; Chapter 1), tendons provide considerable weight savings compared with force transmission via muscles on their own. Ligaments also provide weight savings and flexibility but, as noted above, are often not as stiff or as strong along their longitudinal axis. Over a given length, the weight of muscle needed to generate and transmit a given force is approximately 2800-fold greater than that required if the force were transmitted by tendon. This reflects the relative material strengths and densities of the two tissues (ratio of muscle weight to tendon weight scales $\alpha \sigma_t / \sigma_m \times \rho_m / \rho_t$; $\rho_m = 1060 \, \mathrm{kg \, m^{-3}}$ and $\rho_t = 1120 \, \mathrm{kg^{-3}}$; Alexander, 1983; Biewener, 2003). Hence, when local control of force is less important and force must be transmitted over longer lengths, selection can be expected to favor the presence of a longer tendon or ligament. An extreme example of this is the digital suspensory ligaments found in horses, antelopes and other ungulates (Alexander, 1983; Alexander and Dimery, 1985; Biewener and Bertram, 1990; Biewener, 1998), in which muscle fibers are reduced or absent, allowing forces to be transmitted solely by a ligament (Figs. 10.1 and 10.2). However, many animals, including humans, have long tendons (and ligaments) that reflect a design for weight savings, while allowing flexible movement at one or more joints. Although control of joint and limb position is reduced by forces being transmitted via long ligaments and tendons, the energy cost associated with force transmission is extremely low.

In addition to the energy savings that come from a reduction of weight, force transmission by tendons and ligaments also reduces the utilization of metabolic energy during locomotion. Unlike muscle fibers that consume considerable ATP to produce force and actively change their length, as passive structures, tendons and ligaments incur no metabolic cost on their own to operate in the context of locomotor movement. Force transmission by tendons is linked to the metabolic cost of force generation by their muscles; however, they can act to reduce the cost of muscle contraction by reducing the velocity and length changes of the muscle's fibers (Roberts, 2002). As a result, a subdivision of muscle–tendon architecture is commonly found within the limbs of most animals (Fig. 10.1A; Biewener and Roberts, 2000). More proximal muscles typically have long, parallel fibers that run much of the muscle's length and connect to the skeleton with little or no tendon. In contrast, distal muscles are frequently short-fibered and connect to the skeleton via longer tendons that may span multiple joints. In more extreme cases, as noted above, some distal muscle–tendon units have evolved into passive ligaments, with their muscle fibers having become vestigial or lost.

These varying muscle–tendon and ligament designs reflect trade-offs and functional specialization for control of force transmission and length change versus energy cost (Fig. 10.2). Long, parallel muscle fibers provide the greatest control of muscle length, joint position and ultimately limb motion. Short-fibered pennate muscles that attach to the skeleton via longer tendons require less metabolic energy to generate and transmit force, but reduce the ability for active control of length change and position. Due to tendon stretch, the series elastic compliance of the tendon often exceeds the functional strain range of the muscle's fibers

10.3 Tendon and Ligament Compliance, Resilience and Functional Stress Limits

The viscoelastic properties of tendons and ligaments provide both spring energy savings and shock-absorbing functions for the body. Because their properties are non-linear (Fig. 10.3, Chapter 1), tendons and ligaments stretch proportionately more at lower forces (stresses), as occurs during the onset of loading during the stance phase of running or, as noted previously, when gripping forces are applied to manipulate an object with the digits of the hand. This results from the greater compliance (lower slope) of tendon and ligament at low loads. At lower loads, stresses are transmitted by the collagen fibrils, which are slightly crimped when unloaded (Fratzl et al., 1998), in association with their matrix via cross-links and other molecular bonds (Chapter 4). As loading increases, stresses act more in line with and are borne more directly by the collagen fibrils, leading to an increase in the tendon's overall stiffness (increased slope of the stress–strain curve). There is evidence that relative sliding between collagen fibrils occurs over much of the physiological loading range of tendon, up to the point that the fibrillar structure is disrupted and the tendon begins to fail (Fratzl et al., 1998). It is likely that similar

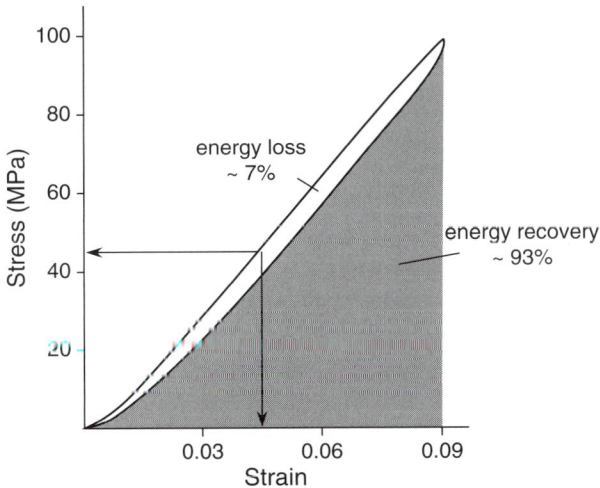

Fig. 10.3 Stress–strain relationship typical for many vertebrate tendons. Tendons (and ligaments) exhibit non-linear "J"-shaped stress–strain curves and viscoelastic behavior. Tendons are most compliant (lowest slope) at low stresses and become increasingly stiff as stress increases from 0 to 20 MPa. At greater stresses and strains (from ~0.02 to 0.09), tendons maintain a fairly uniform modulus (1.2–1.5 GPa). Tendons return about 90–95% of the stored strain energy when they are unloaded (shaded region: showing 93% storage) and dissipate about 5–10% of the energy (open, hysteresis loop, showing 7% energy loss). *Arrows* denote functional stress and strain limits of most tendons (even "high-stress" tendons specialized for elastic energy savings during locomotion). Most tendons operate within a lower stress–strain range, reflecting the need to reduce their structural compliance and minimize their risk of failure. See Chapters 1 and 5 for additional discussion of tendon stress–strain behavior and damage

molecular and biophysical structural dynamics also occur during the loading of ligaments. However, this has not been as well explored.

Direct measurements of tendon forces from animals made during locomotor
activity indicate that peak tendon stresses can range as high as 30–40 MPa during
strenuous activity (Fig. 10.4). Whereas relatively high stresses have been determined

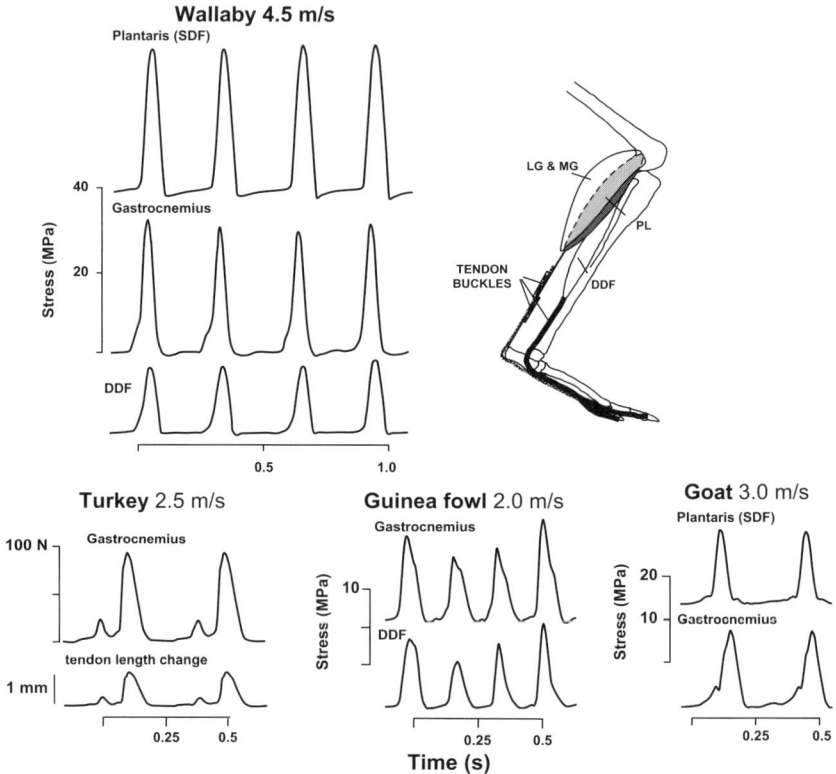

Fig. 10.4 In vivo tendon stresses recorded from leg tendons of (**A**) wallaby (Biewener and
Baudinette, 1995), showing the tendon force buckles instrumented on the tendons of the gastrocnemius (medial and lateral, MG & LG), plantaris (PL, or SDF) and the deep digital flexor (DDF),
and (**B**) a turkey (Roberts, 2002), guinea fowl (Daley and Biewener, 2003) and a goat (unpublished
data). Gray horizontal bars denote the support phase of the stride. Stresses were calculated from the
in vivo force recordings based on an average cross-sectional area of the free tendon. Recordings for
the turkey show the force transmitted by the tendon (tendon cross-sectional area was not reported)
and length changes of the tendon. The tendon's series elastic compliance was determined from an
in situ fixed length contraction of the muscle–tendon unit (Roberts, 2002). Note that tendon stresses
are scaled two-fold less for the guinea fowl and goat recordings, due to the low stresses developed
in comparison with the wallaby leg tendons. Differences in tendon stress levels reflect, in part,
differences in body mass and the speeds of the four species; however, the wallaby's "high-stress"
tendons are specialized for elastic savings in comparison with the guinea fowl and goat, whose
tendons operate at lower stresses, reflecting their more limited compliance to enhance position
control

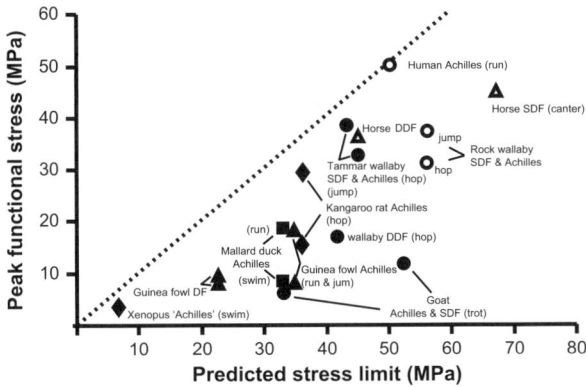

Fig. 10.5 Graph of peak functional tendon stresses determined for various animals during locomotion plotted against their predicted stress limit. This limit is estimated assuming a peak muscle stress (0.3 MPa) transmitted to the tendon (determined as $0.3 \times A_\mathrm{m}/A_\mathrm{t}$, the ratio of muscle to tendon cross-sectional area), following the "stress-in-life" estimates and definition of Ker et al. (1988, 2000). *Solid points* represent in vivo stresses determined from direct tendon force measurements. *Open points* (human, horse and rock wallaby) are calculated from joint mechanics and estimates of muscle–tendon force. In some species, values are shown for two different locomotor conditions (e.g., run versus jump, or run versus swim). *Dashed line* indicates values for which functional tendon stresses match those predicted by the muscle stress limit. In nearly all cases (except human Achilles), peak functional tendon stresses are well below those predicted to act at their muscle's stress limit. This reflects the fact that most tendons are thicker and operate with higher safety factors to reduce their series elastic compliance to reduce risk of damage and failure and/or to enhance position control. Data sources are human (Ker et al., 1987); horse (Biewener, 1998); rock wallaby (McGowan et al., 2008, in press); tammar wallaby (Biewener and Baudinette, 1995); kangaroo rat (Biewener and Blickhan, 1988); goat (McGuigan et al., unpublished); guinea fowl (Daley and Biewener, 2003); mallard duck (Biewener and Corning, 2001); *Xenopus* frog (Richards and Biewener, 2007)

for wallabies during steady hopping, considerably lower stresses have been found in other species during moderate to strenuous activity (Figs. 10.4 and 10.5). Tendon stresses 30–40 MPa correspond to peak strains in the range of 0.03–0.04, based on an average modulus of 1.0 GPa appropriate to the lower stress range of tendon (Bennett et al., 1986; Shadwick, 1990; Biewener and Baudinette, 1995). Strains have also been measured non-invasively in human Achilles tendon based on ultrasound imaging under a variety of conditions. Reported strains range from 0.025 for walking (Fukunaga et al., 2001) to 0.05 for countermovement jumps (Kawakami et al., 2002), and 0.055 for running (Lichtwark et al., 2007). For isometric maximum voluntary contractions of ankle plantar flexor muscles, free Achilles tendon strains of 0.049 (Maganaris and Paul, 2002) and 0.08 (Magnusson et al., 2003) have been measured via ultrasound imaging. In cases where tendon stresses have been estimated from mechanical analysis of external ground reaction forces and limb kinematics (Alexander and Vernon, 1975; Ker et al., 1987; Biewener, 1998), generally higher stresses (40–80 MPa) have been calculated. Measurements obtained in animals, such as wallabies, kangaroos and horses reflect species that are likely adapted to operate at high stresses (and strain levels). However, Achilles tendon strains in humans appear to be as high or higher than strains determined in these species. Given a failure strength of 100–140 MPa (Bennett et al., 1986; Alexander, 1988; Shadwick, 1990), and that tendons may suffer damage associated with regular cyclical loading and use (Ker et al., 2000), it seems likely therefore that tendons have been selected to operate at stresses that provide a safety factor to failure of at least 2–4, and in many cases (see below), much higher than this. Nevertheless, additional in vivo data for tendon strains measured under varying locomotor conditions and for humans in comparison with other animals are needed to assess how tendon material properties and structural dimensions are suited to the mechanical requirements of varying locomotor movements.

In other animals, for which direct tendon forces have been recorded (Figs. 10.4 and 10.5), maximal tendon stresses during regular locomotor activity are in the range of 5–24 MPa (peak ε: 0.005–0.024, given $E_t = 1.0$ GPa). Consequently, strains up to 0.05 appear to represent the normal functional range of most tendons; although, as noted above, strains greater than this have been measured for humans during moderate to more strenuous exercise. In comparison, measurements of in vivo ligament strains, or forces (and calculation of stresses), are generally more limited (see Woo et al., 2006 for review).

Direct force measurements of tendons indicate that in most cases (Fig. 10.5), tendons operate with stresses less than those predicted by the peak forces that their muscles can produce (Ker et al., 1988, 2000). This reflects their greater thickness and safety factor relative to the forces that their muscles can transmit. Available data, therefore, suggest that tendons most regularly experience strains of < 2–3% during low to moderate activity, and from 3–5% during more strenuous activities. It seems unlikely that tendons have been selected to operate at strains much greater than 5%, as this would place them at significant risk for damage and rupture. The higher strains reported for the human (free) Achilles tendon under maximum voluntary contraction, or during running and jumping, suggest that the material properties of human Achilles tendon may differ from those generally obtained from tests on animal tendons (Ker et al., 2000; Pollock and Shadwick, 1994). However, to date a clear demonstration of this has remained elusive, as available data for human tendons suggest fairly similar strength and stiffness (Schechtman and Bader, 1997; Maganaris and Paul, 2002; Magnusson et al., 2003; Wren et al., 2001). Clearly, under rare accidental or high-intensity loading circumstances, tendon strains exceeding 5% and resulting in rupture (~9–10%) may occur.

The low operating stresses of many tendons (Figs. 10.4 and 10.5) suggests that their geometry, in association with their material properties, has been selected to provide an appropriate stiffness for transmission of the forces developed by their muscles. Indeed, in a broad survey of mammalian tendons Ker et al. (1988, 2000) used muscle and tendon dimensions to estimate the operating stresses of tendons, based on the peak forces their muscles could produce ("predicted stress limit", Fig. 10.5), and found that tendons could be divided into "high-stress" (safety factor to rupture < 3) and "low-stress" (safety factor > 7) tendons. Similarly, in situ measurements of tendon strain, based on estimates of forces produced by five primary wrist extensors and flexors in humans (Fig. 10.6), yielded tendon strains that ranged from 1.8 to 3.8% at the peak forces the muscles were calculated to develop (Loren and Lieber, 1995). Thus, most tendons appear to operate at lower stresses and strains than the "high-stress" tendons specialized for substantial elastic savings during locomotion. The safety factors of these "low-stress" tendons are sufficiently high (8–10, or greater) to suggest that even if fatigue damage is a problem (Ker et al., 2000), limits to strength and effective elastic energy savings are not critical determinants of their design. Instead, by being thicker (and/or by having a greater elastic modulus), reduced tendon compliance allows the muscle to have greater control of length and position of those segments on which the muscle–tendon unit acts. Given the importance of tendon compliance in relation to muscle force–velocity contractile behavior, it seems likely that tendon compliance may well be adjusted to match the particular architecture and role of the muscle–tendon unit (Loren and Lieber, 1995; Ker et al., 2000; Roberts, 2002).

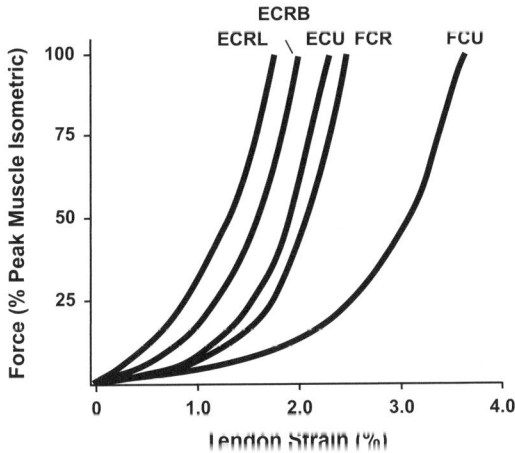

Fig. 10.6 Graph of human wrist flexor and extensor muscle tendon strains in response to in situ loads applied to the free tendon. Loads were applied up to the peak isometric force estimated for the muscle based on its physiological cross-sectional area and assuming a peak isometric muscle stress of 225 kPa (Loren and Lieber, 1995; adapted from their Fig. 1)

Recent study of distal muscle–tendon units of guinea fowl during running and obstacle negotiation (Daley and Biewener, 2006), for which relatively low-peak tendon stresses have been measured (Fig. 10.4; Daley and Biewener, 2003), support the importance of control and stability that stiffer tendons provide. Similarly, tendon stresses determined from in vivo force measurements in goats (Fig. 10.4; McGuigan et al., unpublished) also indicate comparatively low stress levels in distal hind leg tendons (6–18 MPa) over a range of speed and changes in locomotor grade. During jumping, stresses acting in the Achilles tendon of guinea fowl and goats increase to 18–20 MPa, but even this more strenuous activity provides a safety factor to rupture of $\sim$5–6. The low stresses operating in the leg tendons of goats likely reflect the need for stiffer, thicker tendons favored in animals that must negotiate mountain terrain. Rock wallabies that inhabit steep, rocky terrain also have proportionately thicker tendons and operate at lower stresses (McGowan et al., 2008, in press) than a closely related plains-dwelling species, the tammar wallaby, which is capable of substantial elastic savings during hopping (Fig. 10.5; Baudinette and Biewener, 1998).

Based on the available in vivo data, and estimates of tendon stresses from muscle force generating capacity, it is clear that many tendons are designed to transmit force at comparatively low operating stress levels and with minimal stretch. This suggests that for many tendons control of position and length (Alexander, 1988; Daley and Biewener, 2006), as well as minimizing fatigue damage accumulation (Chapter 5; Ker et al., 2000), are key determinants of their design.

For "high-stress" tendons that experience strains of up to 5% or more, and stresses in the range of 50 MPa, their resiliency assures that $>$ 90% of the energy absorbed during stretching can be usefully recovered to help support and move the animal. As a result, elastic energy savings is a key property of many distal leg tendons. In addition to its material properties, a tendon's basic architecture (area versus length) strongly affects its capacity for elastic energy recovery. For a given volume (and weight), longer thin tendons store considerably more energy than short thick tendons, reflecting the fact that strain energy per volume of a material varies Proportional to strain squared. Hence, tendons well suited for elastic energy recovery are typically long and relatively thin (Fig. 10.1), such that they operate at fairly high strains and relatively low safety factors.

10.4 Tendon Elastic Energy Savings During Locomotion

When humans and other animals run, each step involves a cycle of mechanical energy (Cavagna et al., 1977). During the first half of a runner's step, the limb absorbs potential and kinetic energy as the body's center of mass falls and slows down. During the second half of the step, mechanical energy must be provided to raise and reaccelerate the body's center of mass. Rather than all of this energy being produced by muscles doing work, much of the energy can be stored and recovered in tendons, ligaments and other elastic elements of the limbs. Elastic strain energy storage and recovery, therefore, can save considerable metabolic energy by reducing muscular work. As noted above, whereas muscles consume considerable metabolic energy when they shorten to perform mechanical work, at least at the level of the tendon, the storage and release of elastic strain energy is "free", with more than 90% of this energy returned passively during the second half of stance, as the muscles, tendons and limb are unloaded. In running humans, elastic energy stored and recovered in the Achilles tendon and plantar ligaments of the foot have been calculated to save 50% or more of the work that is required for each step (Ker et al., 1987). In kangaroos and wallabies, elastic savings has been estimated to be even higher, offsetting up to 80% of the work required per step (Alexander and Vernon, 1975; Biewener and Baudinette, 1995). In tammar wallabies, elastic energy savings in the hind leg tendons enable the animals to reduce their metabolic energy expenditure by 50% or more, and allow females to carry their pouch young for "free" (Baudinette and Biewener, 1998). Although long, thin leg tendons are especially well suited for elastic energy savings, it is likely that other collagen I-based connective tissue structures, such as ligaments and muscle aponeuroses, serve as important elastic structures that store and recover useful strain energy (Alexander, 1988; Lieber et al., 2000; Maganaris and Paul, 2002; Magnusson et al., 2003). In addition to their importance during running, elastic aponeuroses and ligaments in the trunk are likely important energy savings structures in quadrupeds that gallop and in animals that flex their trunks to swim (Alexander, 1988).

10.5 Role of Tendon Elasticity in Jumping and Acceleration

In addition to saving metabolic energy by reducing the work that muscles must perform during steady locomotion, tendon elasticity also serves an important role in providing increased power output for ballistic movements, such as jumping (Alexander, 1995, 2002; Roberts and Marsh, 2003). This is also likely important to the power output that muscle–tendon units produce during more general locomotor accelerations (Roberts, 2002). This is because the rate at which muscles can develop force and shorten constrains their power output. Further, when a muscle contracts at faster shortening velocities, the force that its fibers can produce is reduced (Hill, 1938). This has also been long known to cost more metabolic energy to produce work at a higher rate (Fenn, 1924). As a result, activities that require high-power outputs, such as jumping and acceleration, are limited by the rates that muscles can perform mechanical work (Bennet-Clark, 1975).

When a muscle shortens and does work against the stretch of its tendon, although its work rate is relatively modest, the strain energy stored in the tendon can subsequently be released at an extremely rapid rate. The rapid release of strain energy, therefore, greatly increases the effective power output that the muscle-tendon unit as a whole can achieve. Whereas maximal muscle power outputs range from 50 to 300 W/kg-muscle, jumping vertebrate animals (e.g., Peplowski and Marsh, 1997; Aerts, 1998) often achieve power outputs of up to 800–1600 W/kg-muscle (invertebrate animals achieve even more impressive power outputs (Bennet-Clark, 1975). The temporal transformation of muscle work into tendon strain energy, however, requires an uncoupling of muscle shortening relative to body movement. This can only be achieved if muscle shortening and work precedes the rapid release of energy from the tendon that powers the movement. This has been long recognized as a requirement for specialized movements, such as jumping. Whereas insects, such as fleas and locusts, employ a catch mechanism that enables muscle work to be decoupled from the subsequent rapid release of elastic energy (Bennet-Clark, 1975), larger vertebrate jumpers appear to rely on interactions of their body's inertial load with changes in muscle mechanical advantage to allow the muscle to store energy in its tendon spring before the energy is suddenly released, as the muscle's mechanical advantage shifts to allow the limb joint to extend (Roberts and Marsh, 2003).

The temporal redistribution of muscle work by tendon elastic storage and recovery, characteristic of specialized movements such as jumping, also likely applies to muscle-powered accelerations more generally (Roberts, 2002). In studies of accelerating turkeys (Roberts and Scales, 2002), the magnitude of mechanical work measured for the animal, as a whole, is low during the first half of the step and much greater during the second half of the step. This is consistent with the pattern of muscle work initially stretching the tendon, with the tendon's elastic energy subsequently being released at a greater rate during the second half of the step. The magnitude of instantaneous power output for the limb as a whole (400 W/kg-muscle, Roberts and Scales, 2002) later in stance also exceeds the muscles' capacity, indicating that power is enhanced by the more rapid release of tendon strain energy. The importance of elastic energy storage and recovery to enhance muscle-powered accelerations is also relevant to such physical activities as throwing. The "wind-up" motion of a pitcher allows muscle work of the legs and trunk to be stored in elastic elements and transferred to the arm, allowing the elastic energy to be released as the ball leaves the pitcher's hand. As in jumping, the velocity of the ball is largely determined by the temporal shift in muscle work and elastic energy storage and release from different elements of the legs, trunks and arm. The challenge in such a complex movement is in determining the temporal distribution of muscle work, in relation to the location and timing of elastic energy storage and release.

References

Aerts, P. (1998). Vertical jumping in Galago senegalensis: the quest for an obligate power amplifier. Philos. Trans. R. Soc. Lond. B 353, 1607–1620.

Alexander, R. M. (1983). Animal Mechanics, 2nd ed. London: Blackwell Scientific.

Alexander, R. M. (1988). Elastic Mechanisms in Animal Movement. Cambridge: Cambridge University Press.

Alexander, R. M. (1995). Leg design and jumping technique for humans, other vertebrates and insects. Philos. Trans. R. Soc. Lond. B 347, 235–248.

Alexander, R. M. (2002). Tendon elasticity and muscle function. Comp. Biochem. Physiol. A 133, 1001–1011.

Alexander, R. M. and Dimery, N. J. (1985). Elastic properties of the forefoot of the donkey (Equus asinus). J. Zool. Lond. 205, 511–524.

Alexander, R. M. and Vernon, A. (1975). The mechanics of hopping by kangaroos (Macropodidae). J. Zool. Lond. 177, 265–303.

Baudinette, R. V. and Biewener, A. A. (1998). Young wallabies get a free ride. Nature 395, 653–654.

Benjamin, M. and Ralphs, J. R. (1997). Tendons and ligaments—an overview. Histol. Histopathol. 12, 1135–1144.

Bennet-Clark, H. C. (1975). The energetics of the jump of the locust, Schistocerca gregaria. J. Exp. Biol. 63, 53–83.

Bennett, M. B., Ker, R. F., Dimery, N. J. and Alexander, R. M. (1986). Mechanical properties of various mammalian tendons. J. Zool. Lond. 209, 537–548.

Biewener, A. A. (1998). Muscle-tendon stresses and elastic energy storage during locomotion in the horse. Comp. Biochem. Physiol. B 120, 73–87.

Biewener, A. A. (2003). Animal Locomotion. Oxford: Oxford University Press.

Biewener, A. A. and Baudinette, R. V. (1995). In vivo muscle force and elastic energy storage during steady-speed hopping of tammar wallabies (Macropus eugenii). J. Exp. Biol. 198, 1829–1841.

Biewener, A. A. and Bertram, J. E. A. (1990). Design and optimization in skeletal support systems. In Concepts of Efficiency in Biological Systems., vol. (in press) (ed. R. W. Blake), pp. xx. Cambridge: Cambridge University Press.

Biewener, A. A. and Blickhan, R. (1988). Kangaroo rat locomotion: design for elastic energy storage or acceleration? J. Exp. Biol. 140, 243–255.

Biewener, A. A. and Corning, W. R. (2001). Dynamics of mallard (Anas platyrynchos) gastrocnemius function during swimming versus terrestrial gait. J. Exp. Biol. 204, 1745–1756.

Biewener, A. A. and Roberts, T. J. (2000). Muscle and tendon contributions to force, work, and elastic energy savings: a comparative perspective. Exerc. Sport Sci. Rev. 28, 99–107.

Cavagna, G. A., Heglund, N. C. and Taylor, C. R. (1977). Mechanical work in terrestrial locomotion: two basic mechanisms for minimizing energy expenditures. Am. J. Physiol. 233, R243–261.

Daley, M. A. and Biewener, A. A. (2003). Muscle force-length dynamics during level versus incline locomotion: a comparison of in vivo performance of two guinea fowl ankle extensors. J. Exp. Biol. 206, 2941–2958.

Daley, M. A. and Biewener, A. A. (2006). Running over rough terrain reveals limb control for intrinsic stability. PNAS 103, 15681–15686.

Fenn, W. O. (1924). The relation between the work performed and the energy liberated in muscular contraction. J. Physiol. 58:373–395.

Fratzl, P., Misof, K., Zizak, I., Rapp, G., Amenitsch, H. and Bernstorff, S. (1998). Fibrillar structure and mechanical properties of collagen. J. Struct. Biol. 122, 119–122.

Fukunaga, T., Kubo, K., Kawakami, Y., Fukashiro, S., Kanehisa, H. and Maganaris, C. N. (2001). In vivo behavior of human muscle tendon during walking. Proc. R. Soc. Lond. B 268, 229–233.

Hill, A. V. (1938). The heat of shortening and the dynamic constants of muscle. Proc. R. Soc. Lond. B 126, 136–195.

Kawakami, Y., Muraoka, T., Ito, S., Kanehisa, H. and Fukunaga, T. (2002). In vivo muscle fibre behaviour during counter-movement exercise in humans reveals a significant role for tendon elasticity. J. Physiol. 540, 635–646.

Ker, R. F., Bennett, M. B., Bibby, S. R., Kester, R. C. and Alexander, R. M. (1987). The spring in the arch of the human foot. Nature 325, 147–149

Ker, R. F., Alexander, R. M. and Bennett, M. B. (1988). Why are mammalian tendons so thick? J. Zool. Lond. 216, 309–324.

Ker, R. F., Wang, X. T. and Pike, A. V. L. (2000). Fatigue quality of mammalian tendons. J. Exp. Biol. 203, 1317–1327.

Lichtwark, G. A., Bougoulias, K. and Wilson, A. M. (2007). Muscle fascicle and series elastic element length changes along the length of the human gastrocnemius during walking and running. J. Biomech. 40, 157–164.

Lieber, R. L. (1992). Skeletal Muscle Structure and Function. Baltimore: Williams and Wilkins.

Lieber, R. L., Leonard, M. E. and Brown-Maupin, C. G. (2000). Effects of muscle contraction on the load-strain properties of frog aponeurosis and tendon. . Cells Tissues Organs 166, 48–54.

Loren, G. J. and Lieber, R. L. (1995). Tendon biomechanical properties enhance human wrist muscle specialization. J. Biomech. 28, 791–799.

Maganaris, C. N. and Paul, J. P. (2002). Tensile properties of the in vivo human gastrocnemius tendon. J. Biomech. 35, 1639–1646.

Magnusson, S. P., Hansen, P., Aagaard, P., Brond, J., Dyhre-Poulsen, P., Bojsen-Moller, J. and Kjær, M. (2003). Differential strain patterns of the human gastrocnemius aponeurosis and free tendon, in vivo. Acta. Physiol. Scand. 177, 185–195.

McGowan, C. P., Baudinette, R. V. and Biewener, A. A. (2008). Differential design for hopping in two species of wallabies. Comp. Biochem. Physiol. in press.

McGuigan, M. P., Yoo, E. and Biewener, A. A. (2007). In vivo dynamics of distal limb muscle function during level versus graded locomotion in goats. J. Appl. Physiol. unpublished.

Peplowski, M. M. and Marsh, R. L. (1997). Work and power output in the hindlimb muscles of Cuban tree frogs Osteopilus septentrionalis during jumping. J. Exp. Biol. 200, 2861–2870.

Pollock, C. M. and Shadwick, R. E. (1994). Relationship between body mass and biomechanical properties of limb tendons in adult mammals. Am. J. Physiol. 266, R1016–1021.

Provenzano, P., Lakes, R., Keenan, T. and Vanderby Jr., R. (2001). Nonlinear ligament viscoelasticity. Ann. Biomed. Eng. 29, 908–914.

Rack, P. M. H. and Ross, H. F. (1984). The tendon of flexor pollicis longus; its effects on the muscular control of force and position in the human thumb. J. Physiol. Lond. 351, 99–110.

Richards, C. T. and Biewener, A. A. (2007). Modulation of in vivo muscle power output during swimming in the African clawed frog (Xenopus laevis). J. Exp. Biol. 210, 3147–3159.

Roberts, T. J. (2002). The integrated function of muscles and tendons during locomotion. Comp. Biochem. Physiol. A 133, 1087–1099.

Roberts, T. J. and Marsh, R. L. (2003). Probing the limits to muscle-powered accelerations: lessons from jumping bullfrogs. J. Exp. Biol. 206, 2567–2580.

Roberts, T. J. and Scales, J. A. (2002). Mechanical power ouput during running accelerations in wild turkeys. J. Exp. Biol. 205, 1485–1494.

Rumian, A. P., Wallace, A. L. and Birch, H. L. (2007). Tendons and ligaments are anatomically distinct but overlap in molecular and morphological features–a comparative study in an ovine model. J. Orthop. Res. 25, 458–464.

Schechtman, H. and Bader, D. L. (1997). In vitro fatigue of human tendons. J. Biomech. 30, 829–835.

Shadwick, R. E. (1990). Elastic energy storage in tendons: mechanical differences related to function and age. J. Appl. Physiol. 68, 1033–1040.

Wilson, A. M., McGuigan, M. P. Su, A. and van den Bogert, A. J. (2001). Horses damp the spring in their step. Nature 414, 895–899.

Woo, S. L.-Y. (1982). Mechanical properties of tendons and ligaments. I. Quasistatic and nonlinear viscoelastic properties. Biorheology 19, 385–396.

Woo, S. L.-Y., Abramowitch, S. D., Kilger, R. Land Liang, R. (2006). Biomechanics of knee ligaments: injury, healing, and repair. J. Biomech. 39, 1–20.

Wren, T. A., Yerby, S. A., Beaupre, G. S. and Carter, D. R. (2001). Mechanical properties of the human Achilles tendon. Clin. Biomech. 16, 245–251.

Chapter 11
Collagen in Arterial Walls: Biomechanical Aspects

G.A. Holzapfel

Abstract This chapter is written with an emphasis on the biomechanical role of collagen in normal and diseased arterial walls, its structural quantification and its consideration in material models including phenomena such as growth and remodeling. Collagen is the ubiquitous load-bearing and reinforcing element in arterial walls and thus forms an important structural basis. The structural arrangement of collagen leads to the characteristic anisotropic behavior of the arterial wall and its respective layers. The organization of collagen fibers, and the tension within, maintains the function, integrity and strength of arteries. This chapter starts by reviewing the structure of the arterial wall and the biomechanical properties of the individual wall layers. Subsequently, structural quantifications of the collagen fabric are discussed with focus on polarized light microscopy, small-angle X-ray scattering and computer vision analysis. A basic building block for soft collagenous tissues in which the material is reinforced by one family of collagen fibers is next presented. On this basis, a structural model for arteries with an ideal alignment of collagen fibers is reviewed, and subsequently extended to consider collagen crimping and the dispersion of collagen fiber orientations. In order to capture structural modifications such as collagen reorientation, phenomenologically based microstructural and continuum models are presented which consider stress-modulated collagen remodeling. Finally, a constitutive model in which continuous remodeling of collagen is responsible for the growth of saccular cerebral aneurysms is outlined. All of the provided models have been implemented in finite element codes, and have proven to be efficient in the computational analysis of clinically relevant problems. This chapter is by no means complete, but it might help to grasp the most important biomechanical aspects of collagen in arterial tissues, and may serve as the basis for a more intense study of this fascinating topic.

11.1 Introduction

Arteries are vessels that transport blood from the heart to the tissues and organs, and supply them with nutrition and oxygen. They are prominent organs composed of soft collagenous tissues which transform the pulsatile heart output into a flow of moderate fluctuations serving as an elastic reservoir — "windkessel" (Nichols

P. Fratzl (ed.), *Collagen: Structure and Mechanics*,
© Springer Science+Business Media, LLC 2008

and O'Rourke 2005). Arterial walls are subjected to 35–40 million load cycles (heartbeats) in the course of a year. The satisfactory blood circulation depends upon the repeatability of the deformation mechanisms in the circulatory system over a lifetime. Although fatigue failure of the type engineers encounter with concrete, polymers or metals are fortunately rare in arteries, arterial dissections and even ruptures occur in a spontaneous (aneurysm) or traumatic (balloon angioplasty, blunt thoracic trauma due to road traffic accident) manner. Arterial dissections are frequently observed in clinical practice in many arterial branches.

Soft collagenous tissues have macroscopic mechanical responses that are highly nonlinear, (in)elastic and anisotropic. In arterial walls collagen is the ubiquitous load-bearing element which forms the important structural basis. Collagen tearing and collagen defects may result in disease and an alteration of the biomechanical behavior of the arterial tissue. The structural arrangement of collagen leads to the typical anisotropic mechanical behavior of arterial tissues and provides the matrix for cellular attachment.

Elastic arteries are located close to the heart, show pronounced elastic behavior and have relatively large diameters, while muscular arteries are located at the periphery and show pronounced viscoelastic behavior with hystereses (see, for example, Holzapfel et al. 2000, 2002; Gasser et al. 2006, and references therein). Several arteries exhibit morphological structures of both types. Theoretical, experimental and clinical principles of arteries may be found in the classical book by Nichols and O'Rourke (2005), while for the underlying biomechanics including physical and computational perspectives, the reader is referred to the books edited by Cowin and Humphrey (2001) and Holzapfel and Ogden (2003, 2006). The most comprehensive and updated information on cardiovascular solid mechanics, in particular on arterial wall mechanics, is documented by Humphrey (2002).

This chapter is written with the emphasis on "collagen", its biomechanical role in arterial walls, its structural quantification and its consideration in material models including phenomena such as growth and remodeling (in this chapter the words "collagen" and "collagenous" appear over 200 times). It should be noted that this text is just a snapshot of this fascinating topic. This chapter reviews in particular the structure of arterial walls in health and disease and summarizes the related biomechanical behavior with an emphasis on the layer-specific mechanical properties of human arteries. This chapter attempts to collect methods that are used to quantify the collagen structure in the individual arterial layers. The second part of the text is devoted to up-to-date artery models appropriate for implementations in finite element programs. Finally, the phenomenon of arterial remodeling is reviewed and recent constitutive models are outlined.

11.2 Structure of the Arterial Wall

Healthy arterial walls consist of three primary layers: *intima* (innermost layer of the artery), *media* (middle layer), *adventitia* (outermost layer). For a diagrammatic model of the major components of a healthy elastic artery see Fig. 11.1.

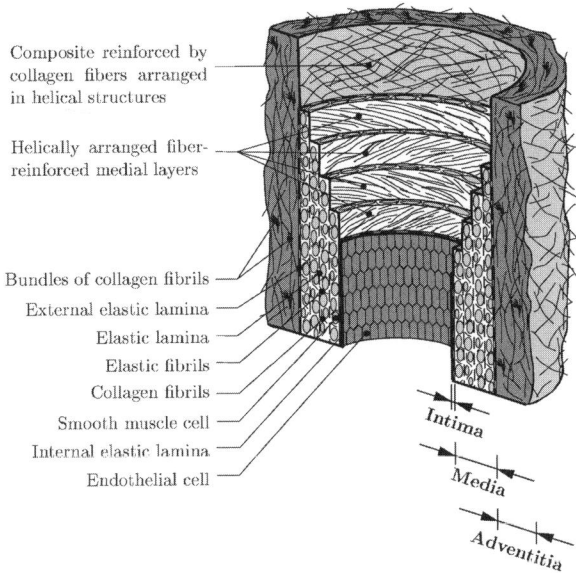

Fig. 11.1 Diagrammatic model of the major components of a healthy and young elastic artery composed of three layers: intima, media, adventitia. Reprinted with permission from Holzapfel et al. (2000)

11.2.1 Intima

In healthy and young arteries the intima is a biologically functional "membrane" that serves as an interface between the thrombogenic media and the blood. Primarily it is a single layer of endothelial cells lining the arterial wall, resting on a thin basal membrane. Histological analyses of aged arteries, however, show homogeneous diffuse intimal hyperplasia, which is the result of a non atherosclerotic process (Stary et al. 1992): intimal cells (mainly myofibroblasts) proliferate concentrically and lead to an increase of extracellular matrix containing mainly collagen fibers with thinly dispersed smooth muscle cells throughout the layer (Canham et al. 1989). The intima thickens and stiffens *(arteriosclerosis)* to restore baseline levels of stress (Glagov et al. 1993). The high content of collagen, primarily of Types I and III (von der Mark 1981; Shekhonin et al. 1985), suggests its mechanical dominance, which has recently been confirmed for non-stenotic human coronary arteries by Holzapfel et al. (2005a). The authors have identified a mean intimal thickness of 27% with respect to the total wall thickness.

In the intimal layer, the orientation of the distinct families of collagen fibers is dispersed, as shown by polarized light microscopy of stained arterial tissue (see Canham et al. 1989; Finlay et al. 1995). There is also elastin present in the intima, which is a rubber-like protein arranged in a three-dimensional network of elastic fibers (Rhodin 1980) present in nearly all vertebrates (Sage and Gray 1979).

Intimal components thicken and stiffen locally with *atherosclerosis*, which involves deposition of fatty substances, calcium, collagen fibers, cellular waste products and fibrin (Hartman 1977; Stary 2003). Biomechanical and biochemical mechanisms are then involved in the development of a lesion, known as an atherosclerotic plaque, which is composed primarily of fibrous tissue of varying density and cellularity. Atherosclerotic plaques have complex geometries and biochemical compositions leading to significant alterations in the mechanical properties of the arterial wall (Holzapfel et al. 2004b). In particular, the distribution of collagen in the different types of tissue changes significantly; see Shekhonin et al. (1985) for a detailed investigation of collagen in the regionally thickened intima, lipid streaks, fibrous plaques and the fibrous cap.

11.2.2 Media

The media consists of a complex three-dimensional network of bundles of collagen fibrils, elastin and smooth muscle cells (Clark and Glagov 1979). In the media of elastic arteries (like the aorta) collagen, elastin and smooth muscle cells are found to be organized in a varying number of medial lamellar units (for example, an average of 40 in the human abdominal aortic media, see Sommer et al. 2008), each of which is about $10 \, \mu\text{m}$ thick (Rhodin 1980). The orientation of and close interconnection between the elastic and collagen fibrils, elastic laminae and smooth muscle cells together constitute a continuous fibrous helix (Faserschraube). The helix has a small pitch so that within the media it is almost circumferentially oriented. Moreover, polarized light microscopy (Canham et al. 1989; Finlay et al. 1995) has shown that collagen, about 30% Type I and 70% Type III (von der Mark 1981; Shekhonin et al. 1985), and smooth muscle cells in the media are consistently circumferentially and coherently aligned. This structured arrangement gives the media an ability to resist high loads in the circumferential direction. There is no significant difference in the distribution of collagen in the media of normal and atherosclerotic arteries (Shekhonin et al. 1985). The elastin pattern loses its organization toward the periphery (Roach and Song 1994) so that the laminated architecture of the media is hardly present in muscular arteries. For a detailed explanation of the transmural organization of the arterial media see the electro-microscopic study of Clark and Glagov (1985).

Figure 11.2(a) shows a media of a human aorta in a kind of "frozen stress state" which straightens the elastic laminae, and the intermingled collagen fibers and smooth muscle cells straighten as well. Figure 11.2(b) shows the media embedded under stress-free conditions indicating a wavy pattern of the fibers. The laminated structure is prone to separation creating a cleavage plane parallel to the elastic lamellae (Tam et al. 1998), and hence provide the mechanism to propagate a dissection in parallel to the lumen (van Baardwijk and Roach 1987). Figure 11.2(c) shows a media which has been dissected during peeling in the axial direction, and afterward fixated and embedded in paraffin under the same loading conditions as during the peeling experiment. Figure 11.2(d) shows a magnification of the dissection tip,

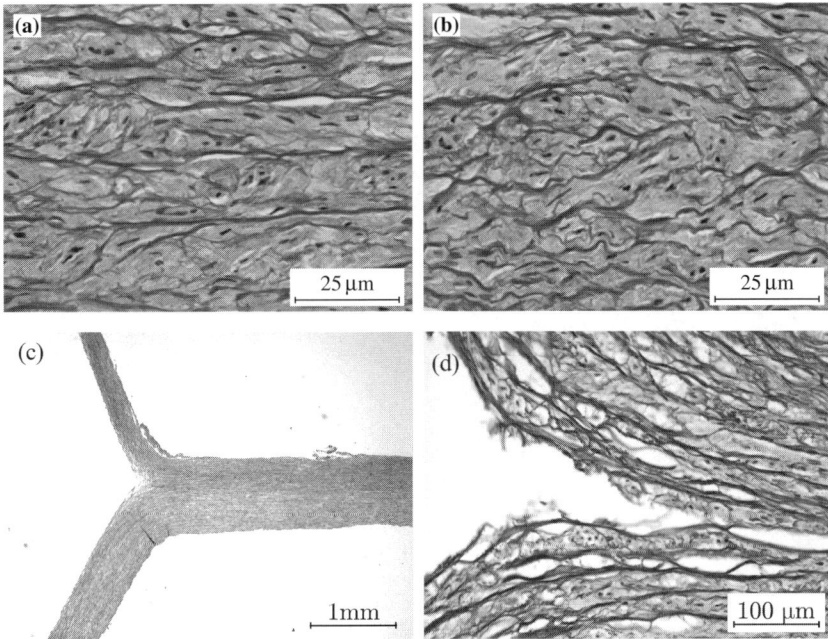

Fig. 11.2 Histological images of a representative media from a human aorta: (**a**) stretched and (**b**) unstretched samples demonstrating the microstructure of the media (original magnification 800×); (**c**) media dissected during peeling in the axial direction (original magnification 20×); (**d**) magnification of the dissection tip showing pronounced fiber bridging and a cohesive zone (original magnification 400×). Elastica van Gieson staining, 4 μm thick sections. Modified with permission from Sommer et al. (2008)

which highlights the irreversible mechanism at the microscale. In addition, the magnification shows pronounced fiber bridging and a cohesive zone.

11.2.3 Adventitia

The adventitia is surrounded continuously by loose perivascular tissue and consists mainly of fibroblasts and fibrocytes, histological ground-matrix and collagen fibers organized in thick bundles. The thickness of the adventitia depends strongly on the type (elastic or muscular) and the physiological function of the blood vessel, and its topographical site. The collagen fibers, primarily of Type I (von der Mark 1981), are arranged within the ground-matrix and form a typically fibrous tissue. Figure 11.3 shows a histological image photomicrograph of an adventitia obtained from an aged human coronary artery.

As can be seen, the assumption of homogeneity within an arterial layer, in particular for the adventitia, may result in an underestimate of the stress value for the inner part of the adventitia, because the outer parts, in which the collagenous fibers are less tightly packed, are likely to be less load bearing. In contrast to the media,

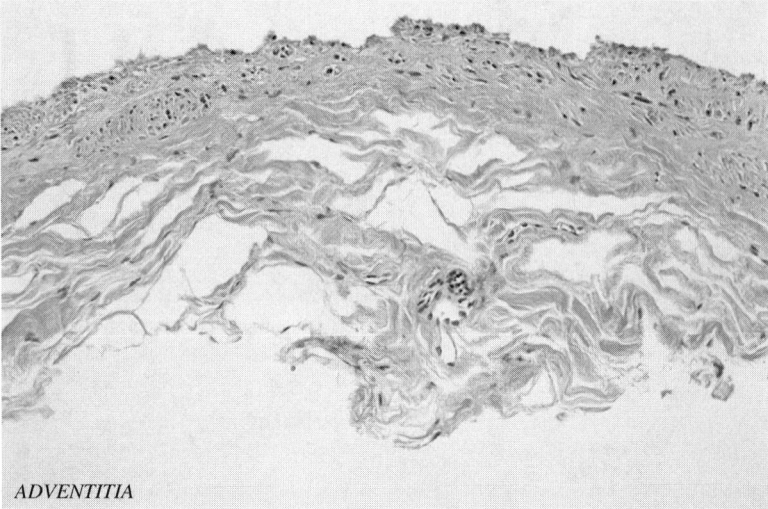

ADVENTITIA

Fig. 11.3 Photomicrograph of a 3 μm thick adventitia sample, obtained from an aged human coronary artery and stained with Hematoxylin and Eosin. Note the tendency to separate because of loose collagen fibers in the outer part of the adventitia. Original magnification 200×. Reprinted with permission from Holzapfel et al. (2005a)

in the adventitial layer the orientation of the collagen fibers is dispersed. Polarized light microscopy of the structure of the adventitia has shown that the collagen forms two helically arranged families of fibers, within which the individual collagen fibers have a large deviation from their mean orientations (Canham et al. 1989; Finlay et al. 1995). As for the media, there is no significant difference in the distribution of collagen in the adventitia of normal and atherosclerotic arteries (Shekhonin et al. 1985).

11.3 Typical Biomechanical Behavior of the Arterial Wall

Collagen is the main load-carrying element in arterial walls. It is relatively inextensible and serves as a (stiff) reinforcing structural element. The most common collagen in blood vessels is Type I. The function and integrity of arteries are maintained by the tension in collagen fibers. Collagen shrinks when heated so that healthy human and animal arteries also shrink upon heating, a phenomenon that was pointed out in the early work of Roy (1880–82) for the first time (note that, for example, rubber contracts its length under tension when its temperature is raised). For literature on the thermal denaturation of collagen, which is largely an irreversible rate process, see Wright and Humphrey (2002) and Humphrey (2003). The structural arrangement of collagen causes the arterial wall and its layers to be anisotropic. It was the early work by Patel and Fry (1969) that documented the first study on anisotropy of arterial segments in dogs.

Arteries are deformable composites that exhibit highly nonlinear stress–strain responses with a characteristic stiffening at higher pressure. Contributions of the individual tissue components (in particular elastin and collagen) to the overall non-linear response of human arterial tissue were identified in the classical work by Roach and Burton (1957). The individual mechanical role of elastin and collagen was demonstrated by selectively digesting elastin and collagen from a human external iliac artery (32 yr; male). It turns out that the initial stiffness of the arterial wall (i.e., at low pressure) is approximately equal to the slope of the tension–radius curve of the "collagen-digested" artery representing the elasticity of the elastin. In this region stress is nearly independent of strain: small tension loadings produce relatively large extensions. The final slope of the tension–radius curve (i.e., at high pressure) agreed well with the slope of the tension–radius curve of the "elastin-digested" artery representing the contribution of fully tensed collagen fibers (see Fig. 11.4). The control (untreated) artery shows the typical J-shaped mechanical response that is characteristic of many soft biological tissues. The (physiological) in vivo pressure of the untreated artery ("control" in Fig. 11.4) corresponds to the upturning region in the tension–radius curve where both elastin and collagen are load-bearing elements. The load transfer from elastin to collagen is strain dependent, i.e., at higher blood pressures more load is transferred from elastin to collagen, and the more (flexible) convoluted collagen fibrils are straightened and mechanically recruited throughout the distended wall which leads to stiffening and to the aforementioned anisotropic mechanical behavior of arteries.

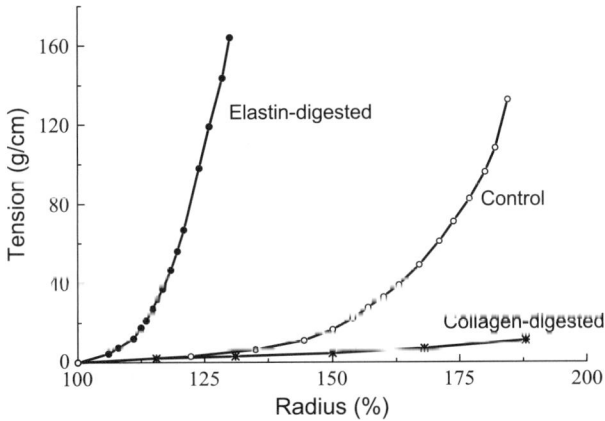

Fig. 11.4 Tension–radius responses of human iliac arteries: control (untreated artery), collagen-digested (collagen is removed from the artery by formic acid, 1 h of digestion), elastin-digested (elastin is removed by trypsin, 22 h of digestion). The elastin-digested arterial response (*filled circles*) represents the property of the collagen, while the collagen-digested response (*asterisk*) represents the property of the remaining elastin. The untreated arterial response (*open circles*) is thought to represent the elasticity of elastin in the low-loading domain and the elasticity of collagen in the higher loading domain. Modified with permission from Roach and Burton (1957)

The stress on arterial tissues is never zero. Arteries are pre-stressed; that is, in their load-free configuration they are still under tension and compression. They contain residual stresses, which are highly layer specific and axially dependent (see Holzapfel et al. 2007, with more references therein). It is important to note that residual stresses in arteries arise from certain growth (change in mass, internal structure and composition) and remodeling mechanisms of the different layers. The term remodeling means a reorganization of existing and/or synthesizing (new) constituents such as collagen fibers resulting in a different composition of the tissue. Since collagen fibers are the major load-bearing constituents in arterial walls, from the mechanical viewpoint, changes in the properties are mainly due to variations in collagen content, orientation, type of fibers and fiber thickness. Remodeling of collagen fibers is observed in various situations where the mechanical environment has changed due to, for example, hypertension, aneurysm and stenosis.

Arterial walls exhibit several types of inelastic phenomena within and above the physiological loading domain. For example, arteries show viscoelastic effects under constant load and exhibit hysteresis under cyclic loading, which is relatively *insensitive* to strain rate over several decades, although tissue response stiffens with increased strain rate (see Holzapfel et al. 2002, and references therein). In vitro tests on arterial tissues display pronounced stress softening under the first few cycles of loading, and a nearly repeatable cyclic behavior once stress softening is complete (pre-conditioning). The underlying microstructural mechanisms of pre-conditioning are still unknown, but macromolecule unfolding may play an important role. When the arterial wall exceeds its physiological range, as occurs during mechanical treatments such as balloon angioplasty, damage and failure mechanisms are activated. The associated inelastic phenomena lead to significant changes in the mechanical behavior, and residual overstretch and damage-based softening of the tissue occurs (Gasser and Holzapfel 2006, 2007; Holzapfel and Gasser 2007; Sommer et al. 2008). For example, the study of Zollikofer et al. (1984) investigates by means of electron and light microscopy the changes in normal canine arteries that follow balloon angioplasty. It turns out that repair of the dilated arterial segments occurred by formation of intima and proliferation of collagen so that after 6 months the dilated segments were characterized by persistent intimal hyperplasia and increased collagen content in the media. Balloon-induced wall overstretch seems to be mainly related to histostructural changes in the media, while the adventitia behaves nearly elastically without signs of material damage (Schulze-Bauer et al. 2002b).

In addition, chemical kinetics of smooth muscle contraction affect the mechanical properties of arteries, which can change their radii significantly due to smooth muscle cell contraction/relaxation (Canfield and Dobrin 1987). Finally, the mechanical properties of arteries change along the arterial tree and are dependent on the following: age, respective function in the organism, concentration and structural arrangement of collagen and elastin, collagen:elastin ratio, species and (vascular) risk factors.

11.3.1 Layer-Specific Mechanical Properties of Human Arteries

The three arterial layers have significantly different mechanical properties and functions. The intima, for example, is very thin in young arteries and makes an insignificant contribution to the solid mechanical properties of the arterial wall. The intima in arteries with non-atherosclerotic intimal thickening, however, has load-bearing capacity and mechanical strength when compared with the media and adventitia. It is the stiffest layer over the range of deformation (see the data of human coronary arteries documented by Holzapfel and Ogden 2006). The structured arrangement of collagen, elastin and smooth muscle cells within the medial lamellar units is responsible for the high strength and resilience of the media. From the mechanical perspective, the media is the most significant layer in a healthy artery.

Collagen in the adventitia contributes significantly to the stability and strength of the arterial wall. In unstressed tissue the collagen fibers are embedded in a wavy form in the soft ground-matrix, which causes the adventitia to be less stiff than the media in the stress-free configuration; however, at significant levels of strain the collagen fibers reach their straightened lengths and the mechanical response of the adventitia then changes to that of a stiff "jacket-like" tube that prevents the smooth muscle from acute overdistension (Schulze-Bauer et al. 2002b). Human adventitias of, for example, human coronary arteries demonstrate significant load-carrying capabilities with high ultimate tensile stresses (>1 MPa) which are on average three times higher than for medias and aged intimas. The ultimate tensile stretches are similar for all tissue layers (Sommer et al. 2008).

The in situ study of Schmid et al. (2005) used synchrotron small-angle X-ray scattering to find the relation between the stretch in the d-spacing of collagen fibers in the human aortic adventitia and the (macroscopic) stretch under uniaxial tensile test conditions. Figure 11.5 shows the correlation between the fibrillar stretch (i.e., the relative increase of the collagen d-spacing) and the macroscopic stretch of aortic strips in the axial and circumferential directions. As can be seen, it is interesting to note that the d-spacing in arterial collagen does not remain constant throughout the nonlinear stress–strain domain of the tissue. The fibrillar stretch is about one order of magnitude smaller than the macroscopic stretch over the loading domain for both samples. This supports the assumption reported in the literature that the macroscopic

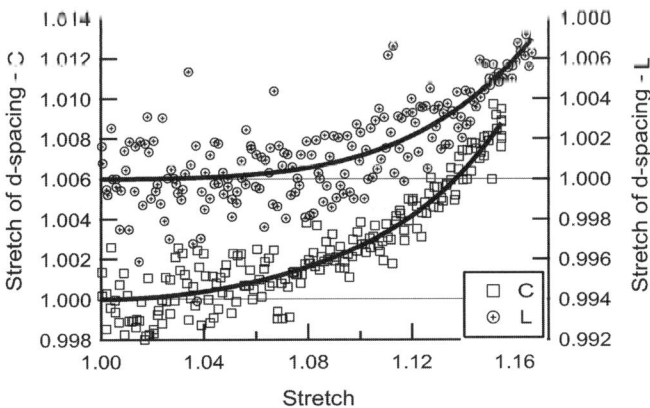

Fig. 11.5 Stretch in collagen d-spacing as a function of the macroscopic stretch for a circumferential (C) and longitudinal (L) sample of a human aortic adventitia. Reprinted with permission from Schmid et al. (2005)

stretch is only partly determined by the fibrillar stretch (Purslow et al. 1998). Consequently, elongation of the matrix material in which the collagen fibers are embedded is governing the macroscopic stretch. This is also discussed in the studies of Fratzl et al. (1998) and Sasaki and Odajima (1996) for tendon collagen, in which the proteoglycan matrix is mainly responsible for the extension.

Figure 11.6 illustrates the mean mechanical data obtained from 13 non-stenotic human left anterior descending coronary arteries (71.5 ± 7.3 yr, mean$\pm$SD). In particular, the model responses of intima, media, adventitia in the form of stress–stretch curves are illustrated, in which the different nonlinear and anisotropic characteristics of the individual tissues become clear.

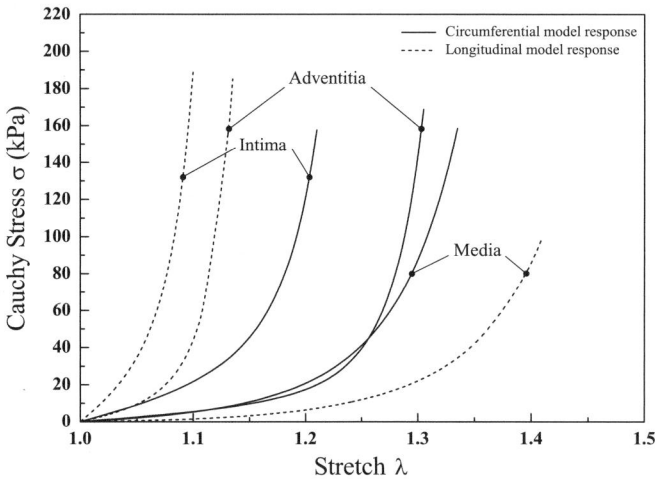

Fig. 11.6 Stress–stretch model response representing mean mechanical data of the three arterial layers in circumferential and longitudinal directions obtained from 13 non-stenotic human left anterior descending coronary arteries. Reprinted with permission from Holzapfel et al. (2005a)

The individual layers possess significantly different levels of residual deformations. On separation, for example, strips from the adventitia and media tend to shrink, while intimal strips elongate relative to the associated intact strips (Fig. 11.7). These pronounced differences in the pre-stretches (and in distensibility) of the individual layers cause the occurrence of relatively high tractions at the layer interfaces during load changes in aged arteries. These tractions, in combination with the compressive hoop stress in the unloaded intimal layer, may lead to spontaneous layer delamination and result in buckling of the intima (Holzapfel et al. 2007). In addition, the layers have individual bending behavior. For example, adventitias from aortic rings become flat, intimas open only slightly and medias spring open by more than 180°. Adventitias and intimas from axial strips remain flat, while medias bent away from the vessel axis. Hence, there exist two types of residual deformation in an arterial layer, bending *and* stretching.

For stenotic arteries the situation is more complex since the lesion is a composition of several tissue types with individual mechanical properties and consistence varying from "butter-like" (lipid) to stiff (adventitia). For example, Fig. 11.8

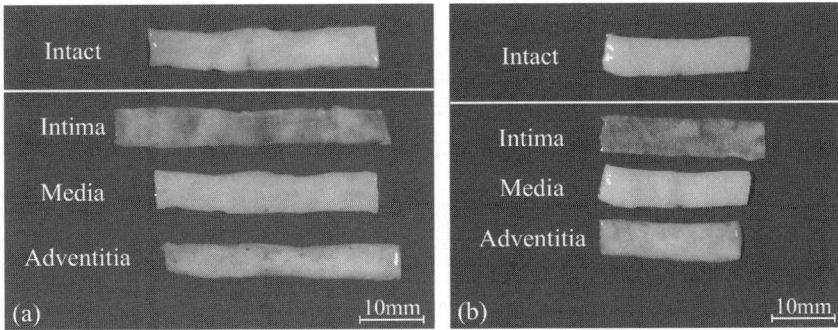

Fig. 11.7 Anatomical separation of fresh human aortic patches, oriented in (**a**) circumferential and (**b**) axial directions, into isolated patches of intima, media and adventitia. In each case the intact patch is shown at the top. Although, before separation, the separate layers had the same in-plane dimensions their dimensions differ significantly after separation. This phenomenon indicates that there are layer-specific residual stretches in the circumferential and axial directions. Reprinted with permission from Holzapfel et al. (2007)

shows a cross-sectional macroscopic view of a stenotic human external iliac artery, a transmitted light microscopic photograph with a corresponding cross-section obtained from high-resolution magnetic resonance imaging. The images show eight different tissue types: non-diseased intima I-nos, fibrous cap I-fc (fibrotic part at the luminal border), fibrotic intima at the medial border I-fm, calcification I-c, lipid pool I-lp, non-diseased media M-nos, diseased fibrotic media M-f and adventitia A. Histological analysis of the tissue component M-f shows a higher amount of collagenous tissue. Each of the tissue components so classified is mechanically relevant and contributes to the overall mechanical response. The experimental data of the individual tissue types, as characterized in Fig. 11.8, are documented by Holzapfel et al. (2004b). The calcification showed a linear response, with approximately the same stiffness as observed in the adventitia of high-stress regions. All other tissue components exhibit

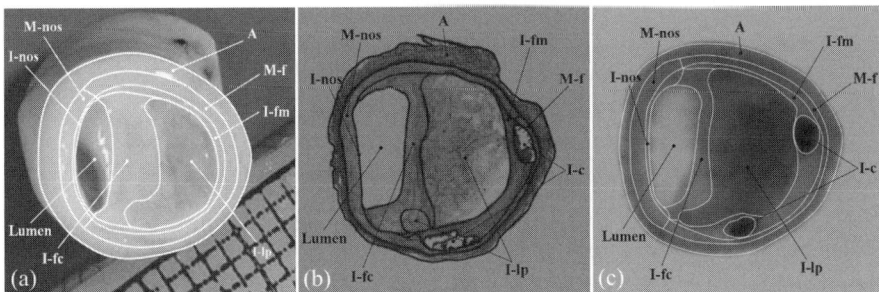

Fig. 11.8 Stenotic human external iliac artery: (**a**) segmented macroscopic view, (**b**) segmented histological section (Elastica van Gieson coloring) – transmitted light microscopic photograph, (**c**) high-resolution magnetic resonance image of the same artery, filtered and (manually) segmented. The histological section and the magnetic resonance image are taken from the same location. Reprinted with permission from Holzapfel et al. (2004b)

highly nonlinear and anisotropic properties as well as considerable interspecimen differences. The stress and stretch values at calcification fracture are smaller than for each of the other tissue components. Of all intimal tissues investigated, the lowest fracture stress occurred in the circumferential direction of the fibrous cap. On average, the adventitia demonstrated the highest, and the non-diseased media the lowest, mechanical strength.

In summary: it is necessary to model non-stenotic human arteries with non-atherosclerotic intimal thickening as composite structures composed of three solid mechanically relevant layers exhibiting different mechanical properties. Stenotic arteries are composed of various vascular tissue components with individual mechanical properties and 3D geometries, and need to be modeled as heterogeneous solids, see Holzapfel et al. (2005b) for an example of a stenotic iliac artery.

11.4 Structural Quantification of Collagen Fibers in Arterial Walls

Identifying the structural arrangement of collagen (in the form of single fibers and fiber bundles) in healthy and diseased arteries is a pressing concern in order to better understand the underlying organ functions. Organized collagen fibers are most correlated with the strength of blood vessels, and introduce strong anisotropic attributes into the tissue response. Hence, structural quantification of the collagen fiber architecture in arterial tissues (such as orientation, density, coherence) serves as a basis for the development of efficient biomechanical models for tissues. The "optimal" technique, which should be reliable, fast and inexpensive, is not yet available. In the following a brief (incomplete) overview of research in the area of structural analysis of collagen in arterial walls, i.e., polarized light microscopy, small-angle X-ray scattering and computer vision analysis, is given.

11.4.1 Polarized Light Microscopy

Collagen and muscle are birefringent, which permits the study of their 3D structural arrangement (orientation) by use of a polarizing light microscope with a four-axis Universal stage. This approach applied to arterial tissue goes back to the pioneering work of Canham et al. (1989). Polarized light microscopy (PLM) is generally recommended as a powerful method for detecting, describing and interpreting the optical (and mechanical) anisotropy of tissue specimens. For each birefringent component the Universal stage allows measurements of the azimuth angle and the angle of inclination relative to the plane of the microscope stage, which is sufficient to fully characterize a line (fiber) in space. Usually, the 3D orientation data are illustrated using Lambert equal area projections, where each data point represents alignment in space. For example, the center position on a Lambert projection relates to any fiber aligned in the circumferential direction of a cylindrical artery. A dispersion of data in the north–south direction relates to a variation toward the longitudinal direction of a cylindrical artery, while a dispersion of data in the east–west direction relates to a variation toward the radial arterial direction.

Figure 11.9 shows layer-specific results obtained from proximal and distal segments of a left anterior descending coronary artery. The histological images were prepared at a transmural pressure of 120 mmHg, which ensured preservation of arterial geometry and 3D directional organization of collagen and smooth muscle cells. Figure 11.9 reveals that collagen in the adventitia is widely dispersed in orientation, with a mean alignment in the circumferential direction. In contrast, the media has very highly ordered smooth muscle, also with a circumferential alignment (both findings are quantitatively similar between proximal and distal segments of the artery, and between left anterior descending arteries). The dominant subendothelial layer was subdivided into an outer zone (adjacent to the media) and an inner zone (adjacent to the lumen). The subendothelium consists of a multilayered fabric of collagen with larger variations in radial and longitudinal orientations.

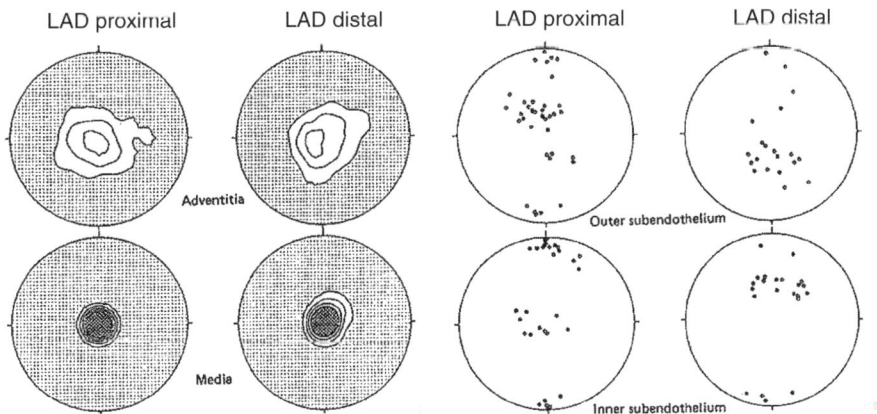

Fig. 11.9 Lambert projections showing orientations within the different layers of proximal and distal left anterior descending coronary arteries. Adventitia and media are shown in *contoured form*, and subendothelial layers in *point form*. Modified with permission from Canham et al. (1989)

The method proposed by Hilbert et al. (1996) combines PLM and morphometry to quantify (porcine aortic valve) collagen crimp morphology and to assess the related biomechanical properties. Tower et al. (2002) propose a methodology to image collagen fiber alignment during mechanical testing of both native tissue (heart valve leaflet) and tissue equivalents using quantitative polarized light microscopy (QPLM). The strength of this method is that the structural analysis can be performed dynamically *during* mechanical testing of soft biological tissues. Massoumian et al. (2003) describe a modification to a confocal microscope, which analyzes the polarization state of light emerging from the specimen so that QPLM can be performed. The system uses a novel form of rotating analyzer which, together with lock-in detection, permits images to be obtained where the image contrast corresponds to both specimen retardance and orientation. The recent work of Jouk et al. (2007) addresses the accuracies and limitations of QPLM from an experimental point of view.

11.4.2 Small-Angle X-Ray Scattering

The orientation and d-spacing of collagen fibers in arterial layers can be provided by scattering methods such as small-angle X-ray scattering (SAXS). For example, the study of Schmid et al. (2005), initiated by Schulze-Bauer et al. (2002a), documents a methodology to investigate changes in the orientation and d-spacing of collagen fibers in separated arterial layers *during* tensile testing by means of synchrotron radiation. In brief, diffraction patterns are recorded during the loading and unloading process of tissue samples undergoing uniaxial tensile tests. All SAXS images are corrected for electronic dark noise, spatial distortion and detector efficiency, and they are transformed from Cartesian (Q_x, Q_z) into polar coordinates, with the radial component Q_R and the azimuthal angle φ (Purslow et al., 1998), as suggested by Roveri et al. (1980). Figure 11.10(a) shows diffraction patterns which were recorded by the Austrian SAXS beamline at the 2 GeV electron storage ring ELETTRA, Trieste, Italy (Amenitsch et al. 1998; Bernstorff et al. 1998) The specimens were exposed to an 8 keV X-ray beam. In Fig. 11.10(a) the strongest meridional peaks of collagen are indicated by order. In the unloaded sample, the collagen fibers seem

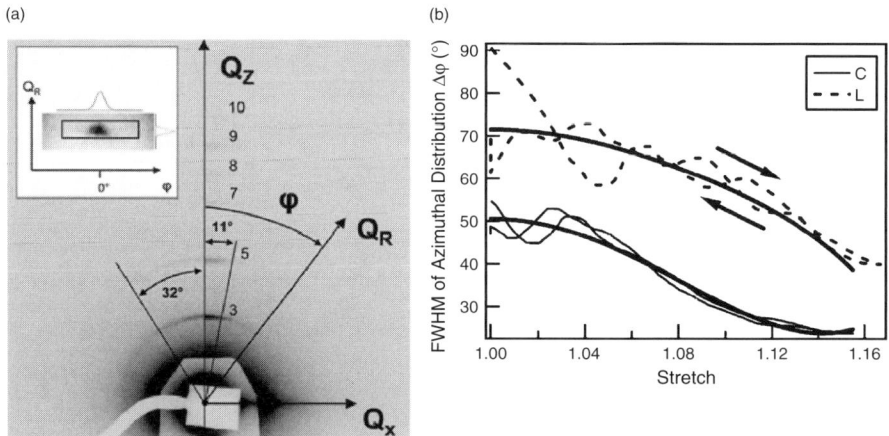

Fig. 11.10 (a) Typical X-ray diffraction patterns of an adventitia sample obtained from the circumferential direction of a human aorta (left half of image: before loading; right half of image: at a stretched state under uniaxial tensile test conditions with applied load in the Q_z-direction). The strongest meridional peaks of collagen are indicated by order. The inset shows the third-order reflection of the stretched sample. The boxed region indicates the area used for radial and azimuthal integrations. The peak shapes in the Q_R and φ-directions are almost Gaussian. (b) Full width at half-maximum (FWHM) of the azimuthal intensity distribution $\Delta\varphi$ of a circumferential (C) and a longitudinal (L) adventitia sample obtained from a human aorta as a function of stretch for one loading and unloading cycle. Modified with permission from Schmid et al. (2005)

to be rather homogenously distributed (diffraction maxima are ring-shaped), as can be seen in the broad arch of the third-order reflection (left image in Fig. 11.10(a)). An increase in the tensile load causes meridional peak intensities (small arch in the right image of Fig. 11.10(a)) indicating successive alignment of collagen fibrils toward the load axis (Q_z-direction). The half-width at half-maximum (HWHM) of the azimuthal intensity distribution $\Delta\varphi$ is given explicitly for both load cases in the corresponding diffraction patterns (32° for the unloaded sample and 11° for the loaded sample). Peak data are analyzed both in the azimuthal and the radial directions. A boxed region around the peak is integrated in the desired directions (see the inset in Fig. 11.10(a)). The diffraction patterns display an intensity distribution which is typical for wet connective tissue samples containing collagen fibers of Types I and III. This particular sample exhibited a 654 Å periodicity (d-spacing) that results from long assemblies of quarter-staggered molecules into micro-fibrils (this structure is the building block for collagen fibers). The highest intensities are observed for the adventitia, followed by the intima of aged human arteries. The media showed predominately a diffuse scatter and only a weak first-order maximum.

Figure 11.10(b) shows the full width at half-maximum (FWHM) of the azimuthal intensity distribution $\Delta\varphi$, i.e., the degree of the collagen fiber orientation in the load direction, as a function of the stretch applied on circumferential and longitudinal adventitia tissue samples. The FWHM of the azimuthal intensity distribution $\Delta\varphi$ decreases continuously with the applied stretch as the collagen fibers reorient themselves into the load direction driven by the extension of the matrix material (see also Roveri et al. 1980). The degree of collagen fiber orientation can directly be related to the stiffness of the arterial tissue. An increased tissue stiffness is caused by an increased alignment of collagen fibers in the load direction (see also Bigi et al. 1981).

Recently, Liao et al. (2007) used SAXS to investigate the relation between the mechanical properties of the mitral valve anterior leaflet (MVAL) and the collagen fibril kinematics under biaxial loading. The authors conclude that (i) the MVAL collagen fibrils do not exhibit intrinsic viscoelastic behavior, (ii) tissue relaxation results from the removal of stress from the fibrils, possibly by a slipping mecha nism modulated by non-collagenous components and (iii) the lack of creep but the occurrence of stress relaxation suggests a "load-locking" behavior under maintained loading conditions. In summary, despite the relatively complex setup of SAXS and the long acquisition time to characterize a small region of tissue, SAXS diffraction data particularly from the adventitia and intima, in combination with tensile testing, provide valuable information for related micro- and nano-structural constitutive modeling.

Small-angle light scattering (SALS) is a similar method to that of SAXS. It has the ability to characterize features on a micrometer scale (rather than the nanometer scale as for SAXS) and is thus suitable for the analysis of collagen fibers. SALS is able to provide a structural analysis superior to that of optical microscopy. Ferdman and Yannas (1993) have shown by experiment that light scattering can complement the analysis of tissue architecture (orientation and diameter of collagen fibers, degree of alignment). The analysis is typically performed with the light microscope.

SALS is a widely used technique for obtaining structural information such as the fiber orientation with a maximal spatial resolution of about 500 μm, an angular resolution of approximately 1° and a spatial resolution of ±254 μm (Billiar and Sacks 1997; Sacks et al. 1997).

11.4.3 Computer Vision Analysis

The use of computer vision techniques for the extraction of structural data guarantees objective (as opposed to subjective human observation) data acquisition. In addition, computer vision allows an exhaustive, automatic analysis of tissue samples large enough to obtain statistically significant data.

According to a comparative study by Elbischger et al. (2005), transmitted light microscopy (TLM), combined with paraffin-embedded tissue samples stained with Elastica van Gieson, turns out to be the best choice among destructive methods (in that study the recent work by Dahl et al. (2007), which proposed a sensitive ultra-structural method to measure collagen alignment on engineered and native arteries by means of transmission electron microscopy, was not considered). TLM images feature a high contrast of different tissue structures when compared with imaging techniques such as magnetic resonance imaging, computer tomography or micro-tomography. Images contain color information so that artifacts can be detected, segmented and excluded from image analysis. In TLM, all the structural information is visible in a single image, and the TLM setup is inexpensive.

The work of Elbischger et al. (2004) proposes a concept to detect and segment artifacts, to determine collagen fiber orientations and to segment regions of homogeneously oriented fibers by means of computer vision techniques. Adventitia samples were obtained from a human iliac artery and histological images were produced with Elastica van Gieson staining (see Fig. 11.11(a) for a representative microscopic image). A 3 CCD color video camera built into the optical system of a TLM was used to produce images with 760×570 pixels. The achieved resolution was 0.25 μm which allows the determination of individual collagen fibers.

Here only the determination of collagen fiber orientation, which is based on a ridge and valley analysis according to López et al. (1999), is briefly discussed. If collagen fibers appear bright in a gray value illustration, their profiles in the orthogonal direction show a dark–bright–dark transition, which forms a ridge. If collagen fibers appear dark or are located near each other and thus cause a small gap in between, their profiles show a bright–dark–bright transition, which forms a valley. Hence, two images are generated, one for ridges (positive eigenvalues) and one for valleys (negative eigenvalues). The directional information (orthogonal direction to the eigenvector) is recorded for each pixel in the two images. All possible directions in the 2D space are mapped into the angular space ranging from $-\pi/2$ to $+\pi/2$, whereby the zero value corresponds to the vertical axis. To guarantee fibers with a one pixel thickness, a morphological thinning operation is performed (see, for example, Sonka et al. 1999, Chapter 11). The analysis was performed for a particular region of interest, as indicated in Fig. 11.11(a) and illustrated in Fig. 11.11(b). For each ridge

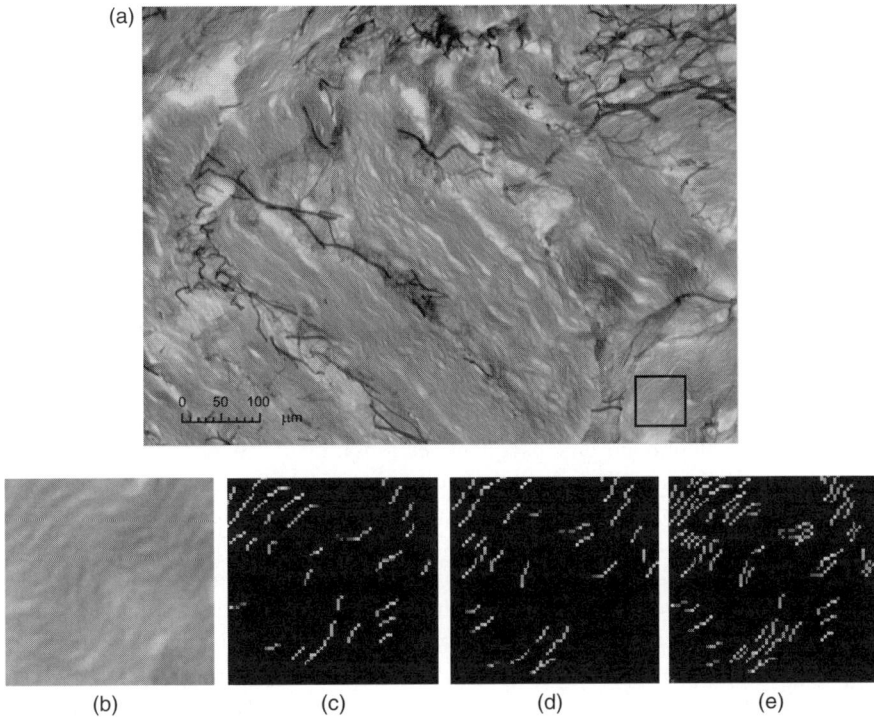

Fig. 11.11 (a) Representative microscopic image of an adventitia sample obtained from a human iliac artery with Elastica van Gieson staining showing a particular region of interest taken for the analysis of fiber orientation (*indicated by a box*); (b) region of interest; (c) image of ridges; (d) image of valleys; (e) combination of ridges and valleys. In images (b)–(e) each fiber has a distinct label, the background (no detected fiber) is represented by label 0. Modified with permission from Elbischger et al. (2004)

(valley) pixel a cross-check operation is performed with its closest neighbor valley (ridge) pixel. If the directional differences are lower than a predefined threshold, the pixel is assumed to be a real ridge (valley) pixel, otherwise it is removed. The resulting ridge and valley images are shown in the Figs. 11.11(c) and (d), respectively. Since the intersection of detected ridge and valley pixels is zero, it is possible to combine the images into one (label) image, as seen in Fig. 11.11(e). Therein, each fiber is associated with a distinct label and the background (no detected fiber) is represented by the label 0. A systematic application of the computer vision concept, to be performed by using a stack of registered histological cross-section images, is required to generate a statistically relevant structural quantification of arterial collagen fibers in 3D.

Holzapfel et al. (2002) proposed a fully automatic technique for identifying the distribution (concentration) of cell nuclei in isolated arterial tissues by performing statistical analyses on related histological images. These nuclei coincide with the preferred fiber orientations in the tissues. Manual measurements of the collagen fiber directions in the intima and the adventitia, and of the oblate nuclei of smooth muscle cells in the media, from histological images of a human aorta are documented in Holzapfel (2006). Mean fiber angles and standard deviations were determined numerically from the data by assuming a normal distribution and a symmetric arrangement with respect to the circumferential direction of the aorta.

11.5 Models for the Elastic Response of Arterial Walls

This section starts with a basic building block for a structural model which is formulated within the framework of fiber-reinforced composites. It serves as the foundation for the mechanical analysis of several types of biological materials including: arteries in health and disease (see, for example, Holzapfel and Ogden 2003, 2006, and references mentioned in subsequent sections), heart valves (Prot et al. 2008), corneas (Alastrué et al. 2006; Pandolfi and Manganiello 2006) lens capsules (David et al. 2007), ligaments (Peña et al. 2007a), temporomandibular joint disc cartilage (Peña et al. 2007b), intervertebral discs (Eberlein et al. 2001), just to name a few. The basic building block for a structural model not only may be used for biological tissues, but has also served the basis for the description of mechanical responses of, for example, engineered materials such as textile composites (Nam and Thinh 2006; Milani et al. 2007) and anisotropic hyperelastic solids in general (Lu et al. 2007).

Constitutive models for the passive *elastic* response of arterial walls with an ideal alignment of collagen fibers are summarized, and finally the model is extended by considering fiber dispersion. The model has also been extended to capture finite viscoelastic deformation and remaining deformation, not discussed here. See Holzapfel and Gasser (2001) and Gasser and Holzapfel (2002) for details. See the papers of Gasser and Holzapfel (2006, 2007) for modeling the propagation of arterial dissection. Background on the nonlinear solid mechanics required for reading this chapter can be found in the textbook of, for example, Holzapfel (2000).

11.5.1 The Basic Building Block for a Structural Model

A constitutive model for soft collagenous tissues in which the material is reinforced by one family of collagen fibers with a single preferred direction is reviewed as a basis. The stiffness of this type of composite material in the fiber direction is typically much larger than in the directions orthogonal to the fibers. It is the simplest representation of material anisotropy called transversely isotropic with respect to this preferred direction. The material response along directions orthogonal to this preferred direction is isotropic.

In order to describe the hyperelastic stress response of soft collagenous tissues, the existence of a Helmholtz free-energy function Ψ defined per unit reference volume is postulated. The decoupled form $\Psi = U(J) + \overline{\Psi}$ is assumed, where J is the volume ratio, i.e., the determinant of the deformation gradient. The given strictly convex function U, taking on its unique minimum at $J = 1$, is responsible for the volumetric elastic response of the material, while the given convex function $\overline{\Psi}$ is responsible for the isochoric elastic response. Since most of the soft biological tissues behave like incompressible materials ($J = 1$, no change in volume during deformation), U is treated as a (purely mathematically motivated) penalty function enforcing the incompressibility constraint. A possible choice is $\kappa(J-1)^2/2$, where κ is the bulk modulus, which serves as a user-specified (positive) penalty parameter. It is independent of the deformation and chosen through numerical experiments. Clearly, with increasing κ the violation of the constraint is reduced. If the restriction on the value $\kappa \rightarrow \infty$ is taken, the constraint condition is exactly enforced, and then $\Psi = U(J) + \overline{\Psi}$ represents a functional for an incompressible material with $J = 1$. In the remaining part of the chapter the emphasis is put on the form of $\overline{\Psi}$.

It is suggested to use an additive split of the isochoric strain-energy function $\overline{\Psi}$ into a part $\overline{\Psi}_g$ associated with the non-collagenous ground-matrix (indicated by subscript g) and a part $\overline{\Psi}_f$ associated with the embedded families of collagen fibers (indicated by subscript f) (Holzapfel and Weizsäcker 1998). Hence, for a representative tissue the (two-term) potential may be written as

$$\overline{\Psi}(\overline{\mathbf{C}}, \mathbf{M}) = \overline{\Psi}_g(\overline{\mathbf{C}}) + \overline{\Psi}_f(\overline{\mathbf{C}}, \mathbf{M}), \tag{11.1}$$

where the family of collagenous fibers is characterized by the (reference) direction vector $\mathbf{M}$, with $|\mathbf{M}| = 1$, and $\overline{\mathbf{C}}$ is the modified right Cauchy–Green tensor (Holzapfel 2000). The structure tensor $\mathbf{A}_1$ is now included, defined as the tensor product

$$\mathbf{A}_1 = \mathbf{M} \otimes \mathbf{M}. \tag{11.2}$$

The integrity basis for the two symmetric second-order tensors $\overline{\mathbf{C}}$ and $\mathbf{A}_1$, then consists of the five (modified) invariants $\bar{I}_1, \ldots, \bar{I}_5$. Since $\bar{I}_3 = 1$ is constant, Eq. (11.1) may be expressed in the reduced form

$$\overline{\Psi}(\overline{\mathbf{C}}, \mathbf{A}_1) = \overline{\Psi}_g(\bar{I}_1, \bar{I}_2) + \overline{\Psi}_f(\bar{I}_1, \bar{I}_2, \bar{I}_4, \bar{I}_5). \tag{11.3}$$

Note that the (modified) invariant $\bar{I}_4 = \overline{\mathbf{C}} : \mathbf{A}_1$ is the square of the stretch of the one family of collagen fibers in the directions $\mathbf{M}$ and it, therefore, has a clear physical interpretation. For simplicity, in order to minimize the number of material parameters, the reduced form of (11.3) may be considered according to

$$\overline{\Psi}(\bar{I}_1, \bar{I}_4) = \overline{\Psi}_g(\bar{I}_1) + \overline{\Psi}_f(\bar{I}_4). \tag{11.4}$$

The anisotropy then arises only through $\bar{I}_4$.

Finally, the two contributions $\overline{\Psi}_g$ and $\overline{\Psi}_f$ to the function $\overline{\Psi}$ must be particularized so as to fit the material parameters to the experimentally observed tissue response. For the low loading domain the (wavy) collagen fibers of soft collagenous tissues are not active (they do not store strain energy). Results from the study by

Gundiah et al. (2007) show that the (classical) neo-Hookean model is a satisfactory description for arterial elastin, which is a constituent of the ground-matrix considered as (solid) mechanically relevant. Hence, for $\overline{\Psi}_g$ the following function may be taken

$$\overline{\Psi}_g(\bar{I}_1) = \frac{c}{2}(\bar{I}_1 - 3), \tag{11.5}$$

where $c > 0$ is a stress-like material parameter. The strong stiffening effect of the tissue observed at higher loadings is almost entirely due to collagen fibers and motivates the use of an exponential function for the description of the strain energy stored in the collagen fibers. Thus (Holzapfel et al. 2000),

$$\overline{\Psi}_f(\bar{I}_4) = \frac{k_1}{k_2} \left\{ \exp[k_2(\bar{I}_4 - 1)^2] - 1 \right\}, \tag{11.6}$$

where $k_1 > 0$ is a stress-like material parameter and $k_2 > 0$ is a dimensionless parameter. An appropriate choice of k_1 and k_2 enables the histologically based assumption that the collagen fibers do not influence the mechanical response of the artery in the low loading domain to be modeled (see the discussion in Section 11.3 and Fig. 11.4 therein). Due to the wavy structure it is generally assumed that collagen is not able to support any compression. These fibers would buckle under the smallest compressive load. It is therefore assumed that the fibers contribute to the strain energy in extension and do not contribute in compression. Hence in the model, Eq. (11.4), the anisotropic term should only contribute when the fibers are extended, that is when $\bar{I}_4 > 1$. If, for example, $\bar{I}_4$ is less than or equal to 1, then the response of the tissue is purely isotropic. This modeling assumption is not only physically based, but is also essential for reasons of stability, see Holzapfel et al. (2004a) for the related analysis.

11.5.2 A Structural Model for Arterial Layers

Since arteries are composed of (thick-walled) layers, each of these layers is modeled with a separate strain-energy function. From the engineering point of view each layer may be considered as a composite reinforced by *two* families of collagen fibers which are arranged in symmetrical spirals.

It is assumed that each layer has a similar mechanical response, and therefore the same form of strain-energy function (but a different set of material parameters) is used for each layer. Hence, the appropriate extension of Eq. (11.1) can be written as

$$\overline{\Psi}(\overline{\mathbf{C}}, \mathbf{M}, \mathbf{M}') = \overline{\Psi}_g(\overline{\mathbf{C}}) + \overline{\Psi}_f(\overline{\mathbf{C}}, \mathbf{M}, \mathbf{M}'), \tag{11.7}$$

where the families of collagen fibers are characterized by the two (reference) direction vectors $\mathbf{M}, \mathbf{M}'$, with $|\mathbf{M}| = |\mathbf{M}'| = 1$.

In addition to $\mathbf{A}_1$ there is now a structure tensor $\mathbf{A}_2 = \mathbf{M}' \otimes \mathbf{M}'$, with the four additional (modified) invariants $\bar{I}_6, \ldots, \bar{I}_9$. Since $\bar{I}_9 = (\mathbf{M} \cdot \mathbf{M}')^2$ is constant (it does not depend on the deformation) the reduced form

$$\overline{\Psi}(\bar{I}_1, \bar{I}_4, \bar{I}_6) = \overline{\Psi}_g(\bar{I}_1) + \overline{\Psi}_f(\bar{I}_4, \bar{I}_6) \tag{11.8}$$

may be considered (Holzapfel et al. 2000), which is the analogue of Eq. (11.4). Note that $\bar{I}_6 = \overline{\mathbf{C}} : \mathbf{A}_2$ is the square of the stretch of collagen fibers in the direction of $\mathbf{M}'$. Hence, the anisotropy arises only through $\bar{I}_4$ and $\bar{I}_6$, but this form is sufficiently general to capture the typical features of arterial responses.

One possible particularization of $\overline{\Psi}_g$ that determines the isotropic response in each arterial layer is the (classical) neo-Hookean model. The strong stiffening effect of each layer observed at high pressures may be described by

$$\overline{\Psi}_f(\bar{I}_4, \bar{I}_6) = \frac{k_1}{2k_2} \sum_{i=4,6} \left\{ \exp[k_2(\bar{I}_i - 1)^2] - 1 \right\}, \tag{11.9}$$

with the parameters $k_1 > 0$, $k_2 > 0$. It is assumed that Eq. (11.9) contributes when either $\bar{I}_4 > 1$ or $\bar{I}_6 > 1$, or both. For example, if $\bar{I}_4 \leq 1$ and $\bar{I}_6 > 1$, then only $\bar{I}_6$ contributes to $\overline{\Psi}_f$, and therefore to $\overline{\Psi}$.

For a healthy and young human artery the intima is not of (solid) mechanical interest, and therefore the focus is on modeling the two remaining layers, i.e., the media and the adventitia. It is then appropriate to model the artery as a two-layer thick-walled tube with residual strains. By Eqs. (11.5) and (11.9) the free-energy functions, Eq. (11.7), may be written for the considered two-layer problem as (Holzapfel et al. 2000)

$$\overline{\Psi}_M = \frac{c_M}{2}(\bar{I}_1 - 3) + \frac{k_{1M}}{2k_{2M}} \sum_{i=4,6} \left\{ \exp[k_{2M}(\bar{I}_{iM} - 1)^2] - 1 \right\}, \tag{11.10}$$

$$\overline{\Psi}_A = \frac{c_A}{2}(\bar{I}_1 - 3) + \frac{k_{1A}}{2k_{2A}} \sum_{i=4,6} \left\{ \exp[k_{2A}(\bar{I}_{iA} - 1)^2] - 1 \right\}, \tag{11.11}$$

for the media and adventitia, respectively. The constants c_M and c_A are associated with the non-collagenous ground-matrix of the material, which describes the isotropic part of the overall response of the tissue. The constants k_{1M}, k_{2M} for the media and k_{1A}, k_{2A} for the adventitia are associated with the contribution of collagen to the overall response. The material parameters are constants and do not depend on the geometry, opening angle or fiber angle.

The (modified) invariants, associated with the media M and the adventitia A, are defined by

$$\bar{I}_{4j} = \overline{\mathbf{C}} : \mathbf{A}_{1j}, \qquad \bar{I}_{6j} = \overline{\mathbf{C}} : \mathbf{A}_{2j}, \qquad j = M, A. \tag{11.12}$$

The tensors $\mathbf{A}_{1j}$, $\mathbf{A}_{2j}$ characterize the structure of the media and adventitia and are given by

$$\mathbf{A}_{1j} = \mathbf{M}_j \otimes \mathbf{M}_j, \qquad \mathbf{A}_{2j} = \mathbf{M}'_j \otimes \mathbf{M}'_j, \qquad j = \mathrm{M, A.} \qquad (11.13)$$

In a cylindrical polar coordinate system, with the (unit) basis vectors $\mathbf{E}_R$, $\mathbf{E}_\Theta$, $\mathbf{E}_Z$ along the respective radial, circumferential and longitudinal directions of the artery, the components of the direction vectors $\mathbf{M}_j$ and $\mathbf{M}'_j$ take on the forms

$$[\mathbf{M}_j] = \begin{bmatrix} 0 \\ \cos\varphi_j \\ \sin\varphi_j \end{bmatrix}, \qquad [\mathbf{M}'_j] = \begin{bmatrix} 0 \\ \cos\varphi_j \\ -\sin\varphi_j \end{bmatrix}, \qquad j = \mathrm{M, A,} \qquad (11.14)$$

and φ_j, $j = \mathrm{M, A}$, are the (mean) angles between the collagen fibers (arranged in symmetrical spirals) and the circumferential direction of the media and the adventitia, respectively. Note that without loss of generality, the (small) radial components of collagen fiber directions are neglected. Through Eqs. (11.12)–(11.14) there exists now a unique relationship between the (modified) invariants $\bar{I}_{4j}$, $\bar{I}_{6j}$ and the collagen fiber angles φ_j, $j = \mathrm{M, A}$.

The three-dimensional formulation of the convex potentials, Eqs. (11.10) and (11.11), allows the characteristic anisotropic behavior of healthy arteries under combined bending, inflation, axial extension and torsion loading to be predicted. For material data of a carotid artery from a rabbit (experiment 71 in Fung et al. 1979 and Chuong and Fung 1983), see Table 11.1. The reader is referred to Section 5.2 in the paper of Holzapfel et al. (2000) for a detailed analysis using an axisymmetric geometry of an artery. The model is not, however, restricted to a particular geometry such as axisymmetry, and is accessible to approximation techniques such as the finite element method (Holzapfel et al. 2002).

Table 11.1 Material parameter data for a rabbit carotid artery for use in Eqs. (11.10) and (11.11) (experiment 71 in Fung et al. 1979 and Chuong and Fung 1983). Data are from Holzapfel et al. (2000)

Arterial layer	c (kPa)	k_1 (kPa)	k_2 (–)	φ (°)
Media	3.0	2.3632	0.8393	29
Adventitia	0.3	0.562	0.7112	62.0

11.5.3 Arterial Models Considering Fiber Dispersion

Collagen fiber orientations in soft biological tissues such as arteries exhibit a variation in dispersion, as discussed above. Hence, in several cases it is necessary to consider the dispersion of the collagen fiber orientation to better capture the stress–strain response of the tissue. Two approaches, which are based on the previously introduced basic building block for a constitutive model, are reviewed here. Both models incorporate one additional scalar parameter that allows the characterization of a state between isotropic distribution (equally distributed collagen fibers) and ideal alignment of collagen fibers.

The first approach is concerned with an extension of the model (11.9) to incorporate isotropic behavior as a special case. This can simply be achieved by multiplying the anisotropic function $\exp[c_2(\bar{I}_i - 1)^2]$ with the isotropic function $\exp[c_1(\bar{I}_1 - 3)^2]$, where c_1, c_2 denote two parameters. Hence, a potential $\overline{\Psi}_f$ that depends not only on $\bar{I}_4, \bar{I}_6$ but also on $\bar{I}_1$ may be postulated, and Eq. (11.9) may be replaced by (Holzapfel et al. 2005a,b)

$$\overline{\Psi}_f(\bar{I}_1, \bar{I}_4, \bar{I}_6) = \frac{k_1}{2k_2} \sum_{i=4,6} \left[\exp\{k_2[(1-\rho)(\bar{I}_1-3)^2 + \rho(\bar{I}_i-1)^2]\} - 1 \right],$$

(11.15)

where $k_2 > 0, k_1 > 0$ are dimensionless and stress-like parameters to be determined from mechanical tests of the tissue. The parameter $\rho \in [0, 1]$ may be seen as a weighting factor. Note that for the limit $\rho = 1$ the constitutive equation (11.9) is obtained (100% weight on the ideal alignment of collagen fibers and 0% weight on the isotropic distribution), while for the limit $\rho = 0$ the constitutive equation

$$\overline{\Psi}_f(\bar{I}_1) = \frac{k_1}{2k_2} \left\{ \exp[k_2(\bar{I}_1 - 3)^2] - 1 \right\}$$

(11.16)

is obtained, which is similar to that proposed by Demiray (1972) and applied within the context of finite elements by Delfino et al. (1997) (100% weight on the isotropic distribution and 0% weight on the ideal alignment of collagen fibers). The only difference with respect to the work of Demiray (1972) is that Eq. (11.16) involves the term $(\bar{I}_1 - 3)^2$ instead of $(\bar{I}_1 - 3)$. Hence, ρ may also be seen as a kind of "switch" parameter between isotropy and anisotropy describing the "degree of anisotropy." In some cases the small modification of model (11.15) with respect to Eq. (11.9) leads to a better fit to experimental data, in particular, to the mechanical response of the adventitia. By assuming that the collagen fibers are arranged in symmetrical spirals according to (11.14), the dependance of the two (modified) invariants $\bar{I}_i, i = 4, 6$, in (11.15) in terms of the collagen fiber angle φ according to Eqs. (11.12), (11.13) may again be established. Hence, for each arterial layer the material model $\overline{\Psi} = \overline{\Psi}_g + \overline{\Psi}_f$ with the particularizations, Eqs. (11.5) and (11.15), requires the five parameters $c, k_1, k_2, \varphi, \rho$.

Figure 11.12 shows a comparative study between the circumferential and longitudinal stress–stretch responses of samples obtained from the three arterial layers of a non-stenotic human left anterior descending (LAD) coronary artery (72 yr; male). In particular, a comparison between experimental data, indicated by circles and squares, with numerical results obtained from the strain-energy functions (11.5) and (11.15), indicated by solid curves, is provided. Mean values of the five parameters for the individual tissue layers of 13 non-stenotic LAD arteries are summarized in Table 11.2, while Fig. 11.6 shows the stress–stretch model results by using the

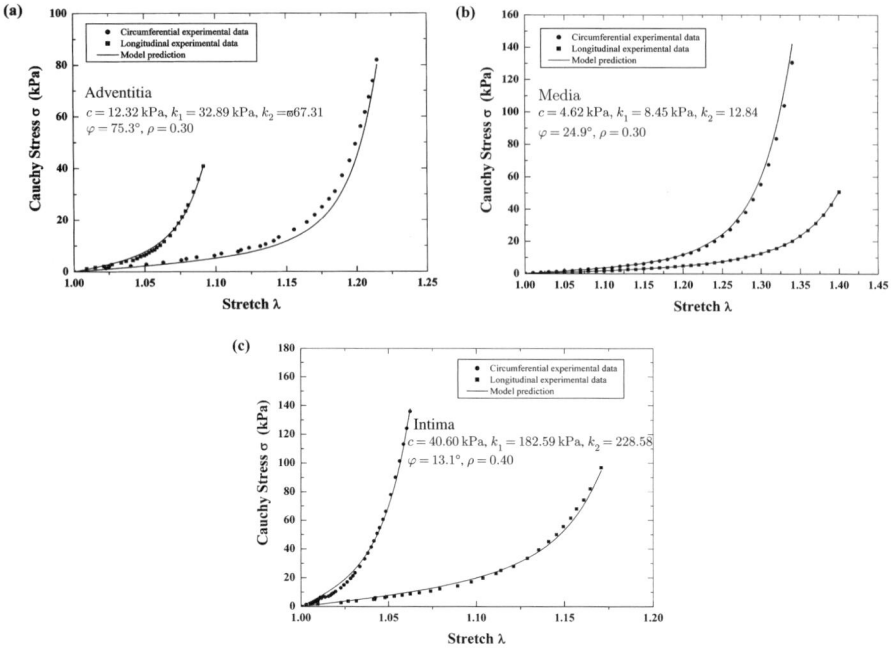

Fig. 11.12 Circumferential and longitudinal stress–stretch responses of samples obtained from the three layers of a non-stenotic human left anterior descending coronary artery (72 yr; male); (**a**) adventitia, (**b**) media, (**c**) intima. Experimental data — indicated by circles and squares — are compared with numerical results obtained from the strain-energy functions (11.5) and (11.15), indicated by solid curves. The layer-specific parameters $c, k_1, k_2, \varphi, \rho$ are provided. Reprinted with permission from Holzapfel et al. (2005a)

mean parameters of Table 11.2. Because the structural orientation of the individual layers was not investigated for that particular study, φ was treated as a phenomenological variable.

A modified form of the constitutive model, Eq. (11.15), which accounts for the (initial) crimping of collagen fibers, was recently proposed by Rodríguez et al. (2008) and applied to compute wall stresses in abdominal aortic aneurysms. Basically, in the modified form the term $(\bar{I}_i - 1)^2$ of Eq. (11.15) is replaced by $(\bar{I}_i - \bar{I}_i^0)^2$, $i = 4, 6$, so that

Table 11.2 Mean of parameters $c, k_1, k_2, \varphi, \rho$ according to Eqs. (11.5) and (11.15) for the intima, media and adventitia obtained from 13 non-stenotic human left anterior descending coronary arteries. Data are from Holzapfel et al. (2005a)

Arterial layer	c (kPa)	k_1 (kPa)	k_2 (–)	φ (°)	ρ (–)
Intima	55.80	263.66	170.88	60.3	0.51
Media	2.54	21.60	8.21	20.61	0.25
Adventitia	15.12	38.57	85.03	67.00	0.55

$$\overline{\Psi}_f(\bar{I}_1, \bar{I}_4, \bar{I}_6) = \frac{k_1}{2k_2} \sum_{i=4,6} \left[\exp\{k_2[(1-\rho)(\bar{I}_1 - 3)^2 + \rho(\bar{I}_i - \bar{I}_i^0)^2]\} - 1 \right],$$

(11.17)

where $\bar{I}_4^0 > 1$ and $\bar{I}_6^0 > 1$ are (dimensionless) parameters regarded as the crimping of the two families of collagen fibers. The anisotropic terms only contribute when either $\bar{I}_4 > \bar{I}_4^0$ or $\bar{I}_6 > \bar{I}_6^0$, or both. For example, if $\bar{I}_4 \leq \bar{I}_4^0$ and $\bar{I}_6 > \bar{I}_6^0$, then only $\bar{I}_6$ contributes to $\overline{\Psi}$. Figure 11.13 illustrates a fit of the modified model to biaxial test data of aneurysmal tissue, as documented by Vande Geest et al. (2006). The model accurately captures the strong stiffening of the tissue for different loading proto-cols ($T_{\theta\theta}$, T_{LL} and $S_{\theta\theta}$, S_{LL} denote engineering stresses and second Piola–Kirchhoff stresses in circumferential and longitudinal directions, respectively). The values of the material parameters have been obtained by imposing constraints on the load ratio used in the biaxial tests, and are summarized in Table 11.3. For the parameter fitting the mechanical response was assumed to be orthotropic and since structural orientation was not investigated, φ was treated as a phenomenological variable.

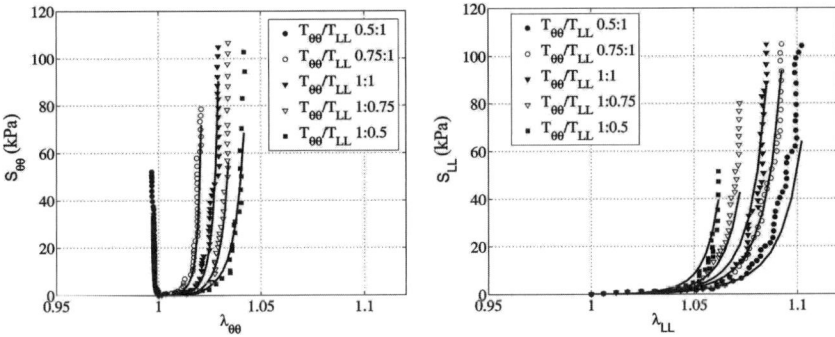

Fig. 11.13 Representative stress–stretch data for an abdominal aortic aneurysm as provided by Vande Geest et al. (2006) (*different symbols* indicate different loading protocols), and the related model response (*solid curves*) according to the model (11.5) and (11.17). Reprinted with permission from Rodríguez et al. (2008)

A second approach to consider dispersion of the collagen fiber orientation is documented in Gasser et al. (2006). Therein a hyperelastic model for arterial lay-ers is proposed by extending the model from Eq. (11.9). It incorporates one addi-tional scalar structure parameter κ characterizing the dispersed collagen orientation, so that

Table 11.3 Parameters according to the model (11.5) and (11.17) for biaxial test data of aneurys-mal tissue. Data are from Rodríguez et al. (2008)

c	k_1	k_2	φ	ρ	$\bar{I}_4^0 = \bar{I}_6^0$
(kPa)	(kPa)	(–)	(°)	(–)	(–)
0.24	244.90	1576.20	5.0	0.14	1.038

$$\overline{\Psi}_{\mathrm{f}}(\bar{I}_1, \bar{I}_4, \bar{I}_6) = \frac{k_1}{2k_2} \sum_{i=4,6} \left\{ \exp[k_2(\bar{I}_i^\star - 1)^2] - 1 \right\}, \qquad \bar{I}_i^\star = \kappa \bar{I}_1 + (1 - 3\kappa)\bar{I}_i,$$

$$(11.18)$$

where $k_1 > 0$ is a material parameter with the dimension of stress, and $k_2 > 0$ is a dimensionless material parameter, while $\bar{I}_i = \overline{\mathbf{C}} : \mathbf{A}_i, i = 4, 6$.

In Eq. (11.18) the right Cauchy–Green-like quantities $\bar{I}_4^\star, \bar{I}_6^\star$ are introduced. They are deformation measures in the direction of the mean orientations $\mathbf{M}$ and $\mathbf{M}'$ of the two fiber families, respectively. Since collagen fibers cannot support any compression it is, therefore, assumed that the anisotropic part $(1 - 3\kappa)\bar{I}_i$ contributes to $\overline{\Psi}_{\mathrm{f}}$ only if the deformation in the directions of $\mathbf{M}$, $\mathbf{M}'$ is positive, i.e., $\bar{I}_i > 1$. Note that the above (small) modification of the (modified) fourth and sixth invariants is able to represent the dispersion of the collagen fiber orientation through an additional parameter, the dispersion (structure) parameter $\kappa \in [0, 1/3]$. For $\kappa = 0$, the expression $\bar{I}_i^\star = \bar{I}_i$, and the model (11.9) according to Holzapfel et al. (2000) is simply obtained. In that case the assumption $\bar{I}_i > 1$ is also sufficient for convexity of Eq. (11.18) (Holzapfel et al. 2004a). For $\kappa = 1/3$ the expression $\bar{I}_4^\star = \bar{I}_1/3$ is obtained; this corresponds to an isotropic distribution very similar to that in Demiray (1972). The potentials resulting from Eq. (11.15), expressed through parameter ρ, and Eq. (11.18), expressed through κ, are similar in the sense that they are able to describe a certain state of collagen fiber distribution that ranges from isotropic to ideal fiber alignment.

Finally it is worthwhile to briefly discuss the interpretation of the dispersion (structural) parameter κ. It is assumed that a single fiber family is distributed with rotational symmetry about the mean referential (preferred) direction $\mathbf{M}$ (or $\mathbf{M}'$) so that the family contributes a transversely isotropic character to the overall response of the material. It can then be shown that the parameter κ takes on the integral form (Gasser et al. 2006)

$$\kappa = \frac{1}{4} \int_0^\pi \rho^\star(\Theta) \sin^3 \Theta d\Theta, \qquad (11.19)$$

where $\rho^\star$ is a density function, which characterizes the distribution of fibers in the reference configuration, and $\Theta \in [0, \pi]$ is an Eulerian angle. By assuming that the embedded fiber family follows, for example, a π-periodic *von Mises* distribution, the resulting density function $\rho^\star$ takes on the form

$$\rho^\star(\Theta) = 4\sqrt{\frac{b}{2\pi}} \frac{\exp[b(\cos(2\Theta) + 1)]}{\mathrm{erfi}(\sqrt{2b})}, \qquad \mathrm{erf}(x) = \frac{2}{\sqrt{\pi}} \int_0^x \exp(-t^2)\mathrm{d}t,$$

$$(11.20)$$

where $b > 0$ is a concentration parameter associated with the *von Mises* distribution, and $\mathrm{erfi}(x) = -i\,\mathrm{erf}(x)$ denotes the imaginary error function.

Figure 11.14 illustrates the relationship between the density function $\rho^\star$ and the Eulerian angle Θ according to Eq. (11.20). Hence, (ideal) alignment of collagen fibers occurs for $\kappa = 0$, characterized by the Dirac delta function, while the isotropic distribution of the collagen fibers ($\kappa = 1/3$) is represented by a horizontal line at $\rho^\star = 1$. In between these limits the density function is "bell-shaped".

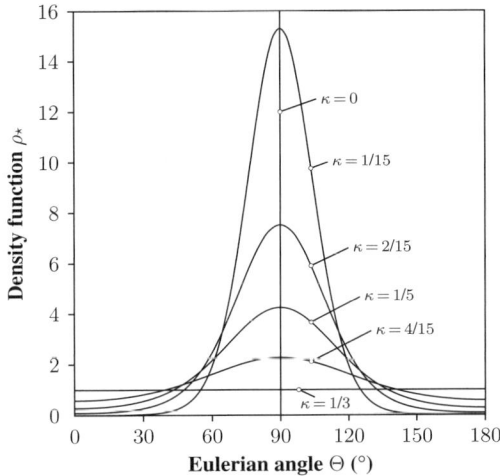

Fig. 11.14 Two-dimensional graphical representation of the (transversely isotropic) *von Mises* distribution of the collagen fibers. Reprinted with permission from Gasser et al. (2006)

Hence, the parameter κ, as provided in Eq. (11.19), represents the fiber distribution in an integral sense. It describes the "degree of anisotropy", governed by the concentration parameter b in Eq. (11.20), which needs to be determined from histology of the soft collagenous tissue in question. The methodology to identify κ by means of polarized light microscopy and preliminary results for a tissue obtained from an abdominal aortic aneurysm are documented by Landuyt (2006).

11.6 Collagen Fiber Remodeling in Arterial Walls

The structure and functionality of arterial walls are optimized according to the mechanobiological environment.[1] Arterial tissue adapts to physiological and pathophysiological stimuli through rearrangement of the microstructure (remodeling), changes in mass and geometry (growth) and shape changes (morphogenesis) (see, for example, Taber 1995; Cowin 2000) in order to keep adequate perfusion according to the metabolic demand of the tissue. In this section the emphasis is put on arterial remodeling, which is concerned with a modification of the microstructure of

---

[1] Mechanobiology is defined as the study of biological reaction of cells (cellular and extracellular activities) in response to changes in their mechanical environment. Typical phenomena are growth, remodeling, adaptation and repair.

the arterial wall caused by chronically altered mechanical forces and, consequently, by the phenotypical modulation of smooth muscle cells and endothelial cells. One possible structural modification is collagen reorientation.

Vascular cells sense and respond to changes in mechanical forces that occur in vivo via numerous receptors, and transmit and modulate compression and tension via focal adhesion sites, integrins, cellular junctions and the extracellular matrix (Lehoux et al. 2006). A great deal of interdisciplinary research effort is devoted to this (mechanical) signaling pathway (mechanotransduction) because it may enable the identification of therapeutic targets and the development of new pharmacological strategies. Arterial remodeling underlies the trend to restore basal levels of tensile and shear stresses (Tronc et al. 1996). Thereby, fibroblasts play a crucial role by synthesizing collagen molecules within minutes and secreting them in less than an hour. Fibroblasts contract the extracellular matrix and new matrix collagen is deposited in a pre-strained condition (Huang et al. 1993). Collagen deposition and degradation is a continuous process in arterial walls. The importance of mechanically induced remodeling of collagen fibers was reviewed in connection with tissue-engineered blood vessels by Nerem and Seliktar (2001).

Taber and Humphrey (2001) point out that the properties of vascular tissues vary across arterial walls. Since collagen fibers are considered to be the major load-bearing constituent in arterial walls, changes in the material (and structural) properties of arterial tissues are strongly correlated with variations in collagen content, orientation, fiber type and dimension. Leung et al. (1976, 1977) were amongst the first to verify experimentally that mechanical forces direct medial cell biosynthesis and modulate structural adaptations to hemodynamic changes. Based on in vitro studies of smooth muscle cells, they reported that aortic medial cells which were attached to elastic membranes and subjected to cyclic stretching consistently synthesized collagen of Types I and III much more rapidly than did cells growing on stationary membranes. Hence, numerical studies on remodeling require powerful constitutive descriptors that incorporate the collagen fabric.

Driessen et al. (2003), for example, considered a stretch-based theory to model changes in collagen fiber content and orientation. These authors assumed that fibers reorient with respect to the principal stretches and that collagen volume fraction is increased with the mean fiber stretch. This approach is able to predict the double-helix structure of collagen fibers within arterial walls (Driessen et al. 2004). Taber and Humphrey (2001), however, related the stresses that develop in a tissue with the anisotropic growth and the alignment of its underlined morphology, and Hariton et al. (2007a) proposed a remodeling approach for the collagen fibers in arterial walls, which is stress modulated. Key to the remodeling approach of Hariton et al. (2007a) are the assumptions that (i) the local principal tensile stresses are the mechanical factor in the formation of the collagen fiber network (collagen morphology is stress-driven), (ii) the collagen fibers are aligned in the plane spanned by the two directions of the largest principal stresses, (iii) the collagen fibers are symmetrically aligned relative to the direction of the maximal principal stress and (iv) remodeling depends solely on external loading which occurs separately from processes such as growth, resorption and production of new fibers.

More specifically, a boundary-value problem (BVP) is solved using an external load and the anisotropic constitutive function (11.9). From this function the 3D stress state and the three principal Cauchy stresses $\sigma_1 \geq \sigma_2 \geq \sigma_3$ in a local coordinate system, and the corresponding principal directions $\mathbf{e}_i$, $i = 1, 2, 3$, can be computed. Then, Hariton et al. (2007a) proposed that the angle of collagen alignment γ depends on the ratio between the magnitudes of the two principal stresses according to

$$\tan \gamma = \mathcal{M} \left(\frac{\sigma_2}{\sigma_1} \right), \tag{11.21}$$

where $\mathcal{M}$ is a monotonically increasing modulation function of the ratio σ_2/σ_1 (possibly nonmechanical factors (i.e., biochemical) may be accounted for in Eq. (11.21)).

Note that the angle γ in Eq. (11.21) is defined in the deformed state (current configuration) and that the two families of collagen fibers are characterized by the two (current) direction vectors $\mathbf{m}$ and $\mathbf{m}'$. Thus,

$$\mathbf{m} = \cos \gamma \, \mathbf{e}_1 + \sin \gamma \, \mathbf{e}_2, \qquad \mathbf{m}' = \cos \gamma \, \mathbf{e}_1 - \sin \gamma \, \mathbf{e}_2. \tag{11.22}$$

However, according to the used formulation of the constitutive relation the fiber directions must be defined in the reference configuration. Thus, the spatial vectors $\mathbf{m}$ and $\mathbf{m}'$ need to be mapped to the reference configuration according to (Holzapfel 2000)

$$\mathbf{M} = \frac{\mathbf{F}^{-1}\mathbf{m}}{|\mathbf{F}^{-1}\mathbf{m}|}, \qquad \mathbf{M}' = \frac{\mathbf{F}^{-1}\mathbf{m}'}{|\mathbf{F}^{-1}\mathbf{m}'|}, \tag{11.23}$$

where $\mathbf{M}$ and $\mathbf{M}'$ are the unit vectors, as introduced in Section 11.5. The (mean) angle φ between the collagen fibers and the circumferential direction of the artery in the reference configuration is then determined via the relation

$$\cos(2\varphi) = \mathbf{M} \cdot \mathbf{M}'. \tag{11.24}$$

The iterative algorithm for the stress-modulated collagen fiber remodeling from Hariton et al. (2007a) is as follows: the computation starts with $\mathbf{M}$, $\mathbf{M}'$ as two randomly chosen unit vectors describing the collagen fiber directions, and the BVP is solved with a given external loading. The principal stresses and directions are determined, and the (modified) directions $\mathbf{M}$, $\mathbf{M}'$ for the collagen fibers are deduced from Eqs. (11.21)–(11.24). The BVP for the artery is then resolved with the modified directions for the collagen fibers. This fiber reorientation process is repeated until the determined stress field between two consecutive iterations are sufficiently close, and thus the final morphology of the collagen fabric is achieved. The remodeling strategy works within a standard finite element framework with the remodeling equations being evaluated at the integration points.

Figure 11.15(a) illustrates the results of the described collagen remodeling algorithm for the apical ridge of a human carotid bifurcation (Hariton et al. 2007b). For this analysis the modulation function $\mathcal{M}$ in Eq. (11.21) was taken to be linear so that $\tan \gamma \equiv \sigma_2/\sigma_1$. The analysis was performed with the constitutive models (11.5) and (11.9); for explicit material parameters, the geometry and a detailed discussion of the results see Hariton et al. (2007b). As can be seen from Fig. 11.15(a), the two families of collagen fibers at the apex are oriented almost along the direction of the apical ridge and resemble a tendon-like structure, which is in agreement with the physiological collagen structure; see the uniform size and co-aligned collagen fibers of apex adventitia in the transmission electron photomicrograph depicted in Fig. 11.15(b). Basically, the fibers have only one preferred direction. Further discussions regarding the apex morphology of cerebral bifurcations may be found in Finlay et al. (1998) and Rowe et al. (2003).

(a)   (b)

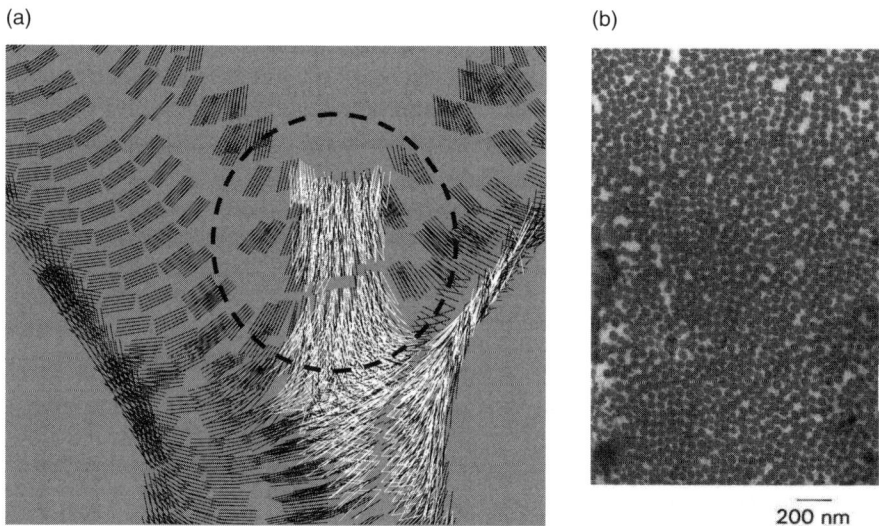

200 nm

Fig. 11.15 (**a**) Principal stress directions (segments of *black lines*) and collagen fiber morphology (segments of *white curves*). The dashed circle shows a region at the vicinity of the apex where collagen fibers are oriented almost along the direction of the apical ridge resembling a tendon-like structure. Reprinted with permission from Hariton et al. (2007b); (**b**) Transmission electron photomicrograph of collagen fibrils of an apex adventitia ($\times 40\,000$). Fibrils in the apex region are of uniform size and are co-aligned. Reprinted with permission from Finlay et al. (1998)

On the basis of the remodeling approach of Hariton et al. (2007a), the continuum model proposed by Kuhl and Holzapfel (2008) uses phenomenologically based microstructural considerations with the goal of capturing particular remodeling phenomena. Homogenization from the molecular microscale to the macroscale of the arterial tissue is performed using the concept of chain network models. On the microscopic level, collagen molecules are assumed to be represented through the worm-like chain model. The relation between applied force F and the end-to-end length $r > 0$ is given by (Bustamante et al. 1994; Marko and Siggia 1995)

$$F = \frac{k_B T}{4 l_p} \left[\frac{4r}{L_0} + \frac{1}{(1 - r/L_0)^2} - 1 \right],$$ (11.25)

where $k_B = 1.38 \times 10^{-23}$ Nm/K is the Boltzmann constant, T is the temperature (in K), $L_0 > r$ is the contour length of the filament and l_p is the persistence length. Studies performed by means of optical tweezers indicate that under physiological conditions Type I collagen molecules have a persistence length of ~ 14.5 nm which is about 5% of their contour length (~ 309 nm), see Sun et al. (2002). The extra-cellular matrix level is modeled as a representative volume element characterized by an isotropic strain energy (i.e., elastin, proteoglycans, etc.), with an anisotropic contribution due to the individual fibers, and a so-called repulsive energy in order to characterize the behavior of the arterial tissue in the reference configuration. For example, the repulsive energy may compensate collagen fiber stresses caused by non-vanishing initial end-to-end lengths characterized by r values. The driving mechanism on the tissue level is based on the idea that a representative volume element deforms according to the principal stretch space (Boyce and Arruda 2000) and that the collagen fiber directions in arterial tissues have a specific relationship with the directions of the two maximum principal stresses, as demonstrated in Eq. (11.21) (Hariton et al. 2007a). The (scalar-valued) dimensions of the representative volume element then change gradually in response to mechanical loading, while, for simplicity, collagen type and amount of collagen were assumed to be constant throughout the remodeling process.

Figure 11.16 illustrates a representative example to demonstrate mechanically induced remodeling of collagen fibers using a tube-like artery subject to 10% axial stretch in combination with internal (blood) pressure. The initial dimensions are 8, 1, 3 units, for length, inner and outer radii, respectively. Along the height 12 tri-linear finite elements, across the thickness 8 elements and in the circumferential direction 16 elements are used. Besides the Boltzmann constant and the absolute temperature, two Lamé constants characterizing the extracellular matrix, two micromechanically motivated parameters (i.e., the contour length and the persistence length that accounts for the initial stiffness), the number of chains per unit volume accounting for the degree of anisotropy, the end-to-end length of the collagen chains in the reference configuration and a relaxation parameter are used in order to characterize the exponential stiffening of the collagen morphology. The total load is applied incrementally, and then fixed to allow for remodeling toward a final state of biological equilibrium. It is important to note that the computation starts with an initially random fiber orientation. Figure 11.16 depicts the final collagen fiber orientations at biological equilibrium for the three different radial locations representing the intima, the media and the adventitia. The iteration process provided a variation

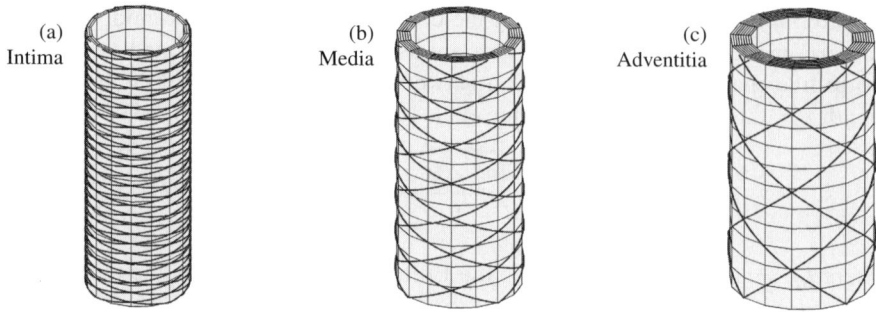

Fig. 11.16 Tube-like artery subject to axial stretch in combination with internal (blood) pressure. Final stage of the remodeling process with collagen fiber orientations at biological equilibrium projected on (**a**) the intima, (**b**) the media and (**c**) the adventitia. Reprinted with permission from Kuhl and Holzapfel (2008). The predicted collagen fiber variation agrees well with that obtained from layer-specific histological images of a human aorta: intima $= 18.8°$; media $= 37.8°$; adventitia $= 58.9°$ (see Fig. 4 in Holzapfel 2006). The angle is measured between the collagen fiber orientation and the circumferential direction of the artery

of the collagen fiber angle across the arterial wall suggesting that the transmural pitch of the (double-)helix increases from the inner to the outer wall. The same (final) collagen fiber morphology is obtained from any arbitrarily chosen initial fiber orientation. The numerically predicted architecture of collagen fibers is consistent with the mean alignment angles of collagen fibers obtained from histological images of the three layers of a human aorta, as documented in Fig. 4 by Holzapfel (2006), where intimal, medial and adventitial samples showed mean collagen fiber angles of 18.8°, 37.8° and 58.9°, respectively (see also the experimental results in Holzapfel et al. 2002). In addition, the variation of the collagen fiber alignment angle through the wall thickness, as can be seen in Fig. 11.16, agrees well with the numerical studies by Driessen et al. (2004) (Fig. 6 therein) and Hariton et al. (2007a) (Fig. 4 therein).

Kroon and Holzapfel (2007) proposed a model in which continuous remodeling of collagen is responsible for the growth of saccular cerebral aneurysms considered to be membranous structures. The cerebral aneurysm wall consists mainly of Type I collagen (Whittaker et al. 1988; Austin et al. 1993), fibroblasts and amorphous material (Espinosa et al. 1984). Hence, the only load-bearing constituent in cerebral aneurysms is the collagen fabric which is highly organized. The wall is layered with a thickness of $\approx 15–20\,\mu$m for each layer, and some layers have a wide range of azimuthal collagen fiber orientation (Canham et al. 1991, 1999). The development of saccular cerebral aneurysms is thought to be caused by a degradation of the internal elastic lamina (Miskolczi et al. 1998) due to disturbed hemodynamics. This in turn induces degenerative changes of the media (Kim et al. 1992) including apoptosis of smooth muscle cells (Kondo et al. 1998; Pentimalli et al. 2004). As the degradation process advances more and more load is carried by the adventitia, and the vessel wall develops the characteristic balloon-like shape in response. During the development of cerebral aneurysms collagen production is increased. This effect is mainly attributed to an enhanced activity of fibroblasts (Whittaker et al. 1988; Halloran and Baxter 1995).

The cerebral aneurysm wall model by Kroon and Holzapfel (2007) assumes a composition of n layers of collagen, where in each layer the newly produced collagen fibers and the fibroblasts are perfectly aligned along one direction, defined with respect to a 2D reference configuration. The collagen production rate is assumed to depend on the stretching and proliferation of fibroblasts so that

$$\dot{m}_i(t) = \beta_0 C_i^\alpha, \qquad C_i = \mathbf{C} : \mathbf{A}_1(\varphi_i), \qquad (11.26)$$

where $\dot{m}_i(t)$ is the collagen mass production rate per unit reference volume in layer i at time t, β_0 is the production rate per unit volume in the reference configuration (Baek et al. 2006) pertaining to the healthy adventitia (not necessarily stress free) and C_i is the projection of the right Cauchy–Green tensor $\mathbf{C}$ in the direction of the collagen fibers characterized by the fiber angle φ_i, and the exponent α is used to govern the influence of C_i on $\dot{m}_i$. The structure tensor $\mathbf{A}_1$ is according to Eq. (11.2).

Assuming that collagen deposition occurs at a specific time t_{dp}, the related deformation gradient is $\mathbf{F}(t_{dp})$ (the deposition may occur at any time between $-\infty$ and t). Subsequently, the current deformation gradient at time t is decomposed according to $\mathbf{F}(t) = \mathbf{F}_{loc}(t, t_{dp})\mathbf{F}(t_{dp})$, where $\mathbf{F}_{loc}(t, t_{dp}) = \mathbf{F}(t)\mathbf{F}^{-1}(t_{dp})$ is the local (current) deformation gradient to which collagen, deposited at time t_{dp}, is exposed (Humphrey and Rajagopal 2002).

The collagen fibers are deposited by the fibroblasts in a pre-stretched condition, defined by the pre-stretch λ_{pre}. The resulting deformation C_{fib} in the individual fibers may be expressed as

$$C_{fib} = \lambda_{pre}^2 \mathbf{C}_{loc} : \mathbf{A}_1(\varphi_i), \qquad C_{fib} \geq 1, \qquad (11.27)$$

where $\mathbf{C}_{loc} = \mathbf{F}_{loc}^T \mathbf{F}_{loc}$.

Assuming that the n layers all have the same reference thickness of T/n, where T is the wall thickness, the strain energy Ψ is

$$\Psi(t) = \frac{1}{n}\sum_{i=1}^{n}\int_{-\infty}^{t} g(t, t_{dp})\dot{m}_i(t_{dp})\Psi_{fib}(t, t_{dp})dt_{dp}, \qquad \Psi_{fib} = \mu(C_{fib}-1)^3, \quad (11.28)$$

where Ψ_{fib} is the energy per unit mass stored in the collagen fibers, $\mu > 0$ is a positive material parameter associated with the stiffness of the fibers and g is a pulse function accounting for the turnover of collagen which is rapid (half-life of 3–90 days) under normal conditions (Humphrey 1999). The function g is equal to 1 for $t_{dp} \in [t - t_{lf}, t]$ and 0 otherwise, and t_{lf} is the lifetime of the collagen fibers.

The model parameters boil down to one (normalized) load parameter $\bar{p} = pR_0/(\mu T\beta_0 t_{lf})$, where p is the pressure applied quasi-statically, and R_0 is the radius

(a)   (b)

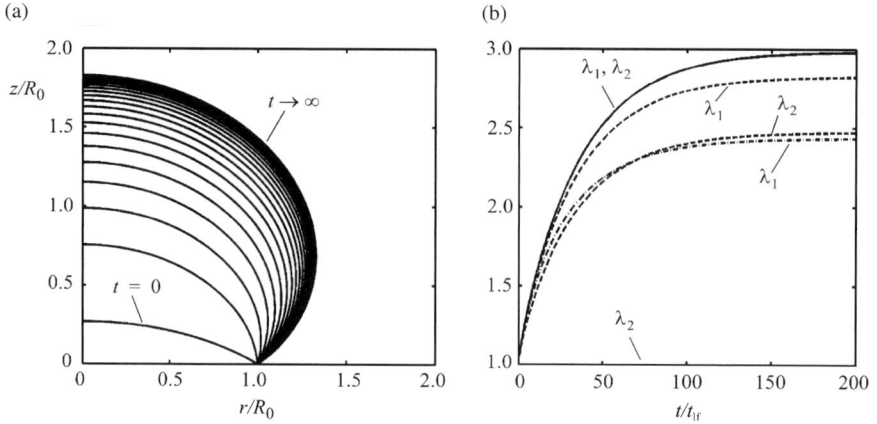

Fig. 11.17 (a) Shapes of the developing cerebral aneurysm; (b) principal stretches λ_1, λ_2 versus t/t_{lf}: highest stretch values at the fundus (two identical solid curves), lowest stretch values at the neck (dash-dotted curve for λ_1 just above 2.4, while the corresponding λ_2-stretch is always equal to unity due to the imposed boundary condition), remaining two curves for a midpoint of the aneurysm. Reprinted with permission from Kroon and Holzapfel (2007)

of a circle where the aneurysm membrane to be inflated is bounded, n is the number of membrane layers, α is the exponent in $(11.26)_2$ and λ_{pre} is the pre-stretch in Eq. (11.27). Figure 11.17(a) and (b) shows the shape of a developing axisymmetric aneurysm and the evolution of the principal stretches λ_1, λ_2 with parameters $\bar{p} = 0.06$, $n = 4$ (the fiber directions in the different layers are evenly distributed in the reference configuration, which results in an isotropic behavior), $\alpha = 2$, $\lambda_{\mathrm{pre}} = 1.02$ and with the initial conditions $\dot{m}_i(t = 0) = \beta_0$ and $C_{\mathrm{fib}}(t = 0) = \lambda_{\mathrm{pre}}^2$. As can be seen from Fig. 11.17(a), at $t > 0$ remodeling of collagen fibers starts and the aneurysm grows. The contours of the developing aneurysm are plotted every 300 time steps $\Delta t = 6t_{\mathrm{lf}}$. The deformation converges after about $t = 200t_{\mathrm{lf}}$ (see Fig. 11.17(b)), which results in a stabilized aneurysm with a height of about $1.8R_0$ (the aneurysm has reached a steady state). Figure 11.17(b) displays the evolution of the principal stretches in the aneurysm wall at the fundus, a midpoint and the neck. The principal stretches λ_1, λ_2 are highest at the fundus and are identical due to symmetry (solid curve). At the neck λ_1 decreases to a value just above 2.4 (curve with the lowest stretch values), while λ_2 is always equal to unity due to the imposed boundary condition.

The thickness change in the aneurysm wall can be quantified in terms of the third principal stretch λ_3, which is defined as the ratio between the current and the reference thickness of the membrane. Thus,

$$\lambda_3 = \frac{1}{n\lambda_1\lambda_2} \sum_{i=1}^{n} \frac{m_i}{m_0}, \tag{11.29}$$

where m_i and m_0 denote the current collagen mass content of layer i and the reference collagen mass content, respectively. Both entities pertain to the same reference volume. Relation (11.29) may be shown by means of $\sum_{i=1}^{n} m_i = Jnm_0$ together with the incompressibility condition $J = \lambda_1\lambda_2\lambda_3$. The model predicts that the thickest aneurysm wall is always at the fundus, independent of t. As the model approaches steady state the wall thickens with a factor of about 9 at the fundus. At the beginning of the collagen remodeling process the thinnest wall is located at the neck, but then at steady state is located at $0.86R_0$ from the center-point of the aneurysm.

Kroon and Holzapfel (2008) evaluated the described aneurysm growth model for a more realistic non-axisymmetric setting, a middle cerebral artery composed of two layers (media and adventitia). In particular, the influence of the structural organization of the collagen fabric on the general behavior of the growth model is investigated, with concluding remarks that the material behavior of aneurysmal tissue cannot be expected to be isotropic.

Acknowledgments I wish to acknowledge some of my co-workers, in particular D.M. Pierce, G. Sommer, B. Strametz and J. Tong, who helped finalize the manuscript. I would also like to thank Professor P. Fratzl for compiling such an interesting and important volume on the structure and mechanics of collagen and for inviting me to contribute as an author. This work was financially supported, in part, by grants from the Austrian Science Fund (START-Award Y74-TEC, grant number P11899-TEC) and The Royal Society of London (grant number No: 14467). This support is gratefully acknowledged.

References

Alastrué, V., Calvo, B., Peña, E., and Doblaré, M. (2006). Biomechanical modeling of refractive corneal surgery. *J. Biomech. Eng.* 128:150–160.

Amenitsch, H., Rappolt, M., Kriechbaum, M., Mio, H., Laggner, P., and Bernstorff, S. (1998). First performance assessment of the small-angle X-ray scattering beamline at ELETTRA. *Synchro. Rad.* 5:506–508.

Austin, G., Fisher, S., Dickson, D., Anderson, D., and Richardson, S. (1993). The significance of the extracellular matrix in intracranial aneurysms. *Ann. Clin. Lab. Sci.* 23:97–105.

Baek, S., Rajagopal, K. R., and Humphrey, J. D. (2006). A theoretical model of enlarging intracranial fusiform aneurysms. *J. Biomech. Eng.* 128:142–149.

Bernstorff, S., Amenitsch, H., and Laggner, P. (1998). High-throughput asymmetric double-crystal monochromator of the SAXS beamline at ELETTRA. *Synchro. Rad.* 5:1215–1221.

Bigi, A., Ripamonti, A., and Roveri, N. (1981). X-ray investigation of the orientation of collagen fibres in aortic media layer under distending pressure. *Int. J. Biol. Macromol.* 3:287–291.

Billiar, K. L., and Sacks, M. S. (1997). A method to quantify the fiber kinematics of planar tissues under biaxial stretch. *J. Biomech.* 30:753–756.

Boyce, M. C., and Arruda, E. M. (2000). Constitutive models of rubber elasticity: A review. *Rubber Chem. Technol.* 73:504–523.

Bustamante, C., Marko, J. F., Siggia, E. D., and Smith, S. B. (1994). Entropic elasticity of λ-phage DNA. *Science* 265:1599–1600.

Canfield, T. R., and Dobrin, P. B. (1987). Static elastic properties of blood vessels. In Skalak, R., and Chien, S., eds., *Handbook of Bioengineering*. New York: McGraw-Hill.

Canham, P. B., Finlay, H. M., Dixon, J. G., Boughner, D. R., and Chen, A. (1989). Measurements from light and polarised light microscopy of human coronary arteries fixed at distending pressure. *Cardiovasc. Res.* 23:973–982.

Canham, P. B., Finlay, H. M., Dixon, J. G., and Ferguson, S. E. (1991). Layered collagen fabric of cerebral aneurysms quantitatively assessed by the universal stage and polarized light microscopy. *Anat. Rec.* 231:579–592.

Canham, P. B., Finlay, H. M., Kiernan, J. A., and Ferguson, G. G. (1999). Layered structure of saccular aneurysms assessed by collagen birefringence. *Neurol. Res.* 21:618–626.

Chuong, C. J., and Fung, Y. C. (1983). Three-dimensional stress distribution in arteries. *J. Biomech. Eng.* 105:268–274.

Clark, J. M., and Glagov, S. (1979). Structural integration of the arterial wall. *Lab. Invest.* 40:587–602.

Clark, J. M., and Glagov, S. (1985). Transmural organization of the arterial media: The lamellar unit revisited. *Arteriosclerosis* 5:19–34.

Cowin, S. C. (2000). How is a tissue built? *J. Biomech. Eng.* 122:553–569.

Cowin, S. C., and Humphrey, J. D., eds. (2001). *Cardiovascular Soft Tissue Mechanics*. Dordrecht: Kluwer Academic Publishers.

Dahl, S. L. M., Vaughn, M. E., and Niklason, L. E. (2007). An ultrastructural analysis of collagen in tissue engineered arteries. *Ann. Biomed. Eng.* 35:1749–1755.

David, G., Pedrigi, R. M., Heistand, M. R., and Humphrey, J. D. (2007). Regional multiaxial mechanical properties of the porcine anterior lens capsule. *J. Biomech. Eng.* 129:97–104.

Delfino, A., Stergiopulos, N., Moore Jr., J. E., and Meister, J.-J. (1997). Residual strain effects on the stress field in a thick wall finite element model of the human carotid bifurcation. *J. Biomech.* 30:777–786.

Demiray, H. (1972). A note on the elasticity of soft biological tissues. *J. Biomech.* 5:309–311.

Driessen, N. J. B., Peters, G. W. M., Huyghe, J. M., Bouten, C. V. C., and Baaijens, F. P. T. (2003). Remodelling of continuously distributed collagen fibres in soft connective tissue. *J. Biomech.* 36:1151–1158.

Driessen, N. J. B., Wilson, W., Bouten, C. V. C., and Baaijens, F. P. T. (2004). A computational model for collagen fibre remodelling in the arterial wall. *J. Theor. Biol.* 226:53–64.

Eberlein, R., Holzapfel, G. A., and Schulze-Bauer, C. A. J. (2001). An anisotropic model for annulus tissue and enhanced finite element analyses of intact lumbar disc bodies. *Comput. Methods. Biomech. Biomed. Eng.* 4:209–230.

Elbischger, P. J., Bischof, H., Regitnig, P., and Holzapfel, G. A. (2004). Automatic analysis of collagen fibre orientation in the outermost layer of human arteries. *Pattern Anal. Appl.* 7:269–284.

Elbischger, P. J., Bischof, H., Holzapfel, G. A., and Regitnig, P. (2005). Computer vision analysis of collagen fiber bundles in the adventitia of human blood vessels. In Suri, J. S., Yuan, C., Wilson, D. L., and Laxminarayan, S., eds., *Plaque Imaging: Pixel to Molecular Level. Volume 113 Studies in Technology and Informatics*. IOS Press. chapter 5, 97–129.

Espinosa, F., Weir, B., and Noseworthy, T. (1984). Rupture of an experimentally induced aneurysm in a primate. *Can. J. Neurol. Sci.* 11:64–68.

Ferdman, A. G., and Yannas, I. V. (1993). Scattering of light from histologic sections: a new method for the analysis of connective tissue. *J. Invest. Dermatol.* 100:710–716.

Finlay, H. M., McCullough, L., and Canham, P. B. (1995). Three-dimensional collagen organization of human brain arteries at different transmural pressures. *J. Vasc. Res.* 32:301–312.

Finlay, H. M., Whittaker, P., and Canham, P. B. (1998). Collagen organization in the branching region of human brain arteries. *Stroke* 29:1595–1601.

Fratzl, P., Misof, K., Zizak, I., Rapp, G., Amenitsch, H., and Bernstorff, S. (1998). Fibrillar structure and mechanical properties of collagen. *J. Struct. Biol.* 122:119–122.

Fung, Y. C., Fronek, K., and Patitucci, P. (1979). Pseudoelasticity of arteries and the choice of its mathematical expression. *Am. J. Physiol.* 237:H620–H631.

Gasser, T. C., and Holzapfel, G. A. (2002). A rate-independent elastoplastic constitutive model for (biological) fiber-reinforced composites at finite strains: Continuum basis, algorithmic formulation and finite element implementation. *Comput. Mech.* 29:340–360.

Gasser, T. C., and Holzapfel, G. A. (2006). Modeling the propagation of arterial dissection. *Eur. J. Mech. A/Solids* 25:617–633.

Gasser, T. C., and Holzapfel, G. A. (2007). Modeling plaque fissuring and dissection during balloon angioplasty intervention. *Ann. Biomed. Eng.* 35:711–723.

Gasser, T. C., Ogden, R. W., and Holzapfel, G. A. (2006). Hyperelastic modelling of arterial layers with distributed collagen fibre orientations. *J. R. Soc. Interface* 3:15–35.

Glagov, S., Zarins, C. K., Masawa, N., Xu, C. P., Bassiouny, H., and Giddens, D. P. (1993). Mechanical functional role of non-atherosclerotic intimal thickening. *Front. Med. Biol. Eng.* 5:37–43.

Gundiah, N., Ratcliffe, M. B., and Pruitt, L. A. (2007). Determination of strain energy function for arterial elastin: Experiments using histology and mechanical tests. *J. Biomech.* 40:586–594.

Halloran, B. G., and Baxter, B. T. (1995). Pathogenesis of aneurysms. *Semin. Vasc. Surg.* 8:85–92.

Hariton, I., deBotton, G., Gasser, T. C., and Holzapfel, G. A. (2007a). Stress-driven collagen fiber remodeling in arterial walls. *Biomech. Model. Mechanobiol.* 6:163–175.

Hariton, I., deBotton, G., Gasser, T. C., and Holzapfel, G. A. (2007b). Stress-modulated collagen fiber remodeling in a human carotid bifurcation. *J. Theor. Biol.* 248:460–470.

Hartman, J. D. (1977). Structural changes within the media of coronary arteries related to intimal thickening. *Am. J. Pathology* 89:13–34.

Hilbert, S. L., Sword, L. C., Batchelder, K. F., Barrick, M. K., and Ferrans, V. J. (1996). Simultaneous assessment of bioprosthetic heart valve, biomechanical properties, and collagen crimp length. *J. Biomed. Mater. Res.* 31:503–509.

Holzapfel, G. A. (2000). *Nonlinear Solid Mechanics. A Continuum Approach for Engineering.* Chichester: John Wiley & Sons.

Holzapfel, G. A. (2006). Determination of material models for arterial walls from uniaxial extension tests and histological structure. *J. Theor. Biol.* 238:290–302.

Holzapfel, G. A., and Gasser, T. C. (2001). A viscoelastic model for fiber-reinforced composites at finite strains: Continuum basis, computational aspects and applications. *Comput. Methods. Appl. Mech. Eng.* 190:4379–4403.

Holzapfel, G. A., and Gasser, C. T. (2007). Computational stress–deformation analysis of arterial walls including high-pressure response. *Int. J. Cardiol.* 116:78–85.

Holzapfel, G. A., and Ogden, R. W., eds. (2003). *Biomechanics of Soft Tissue in Cardiovascular Systems.* Wien – New York: Springer-Verlag.

Holzapfel, G. A., and Ogden, R. W., eds. (2006). *Mechanics of Biological Tissue.* Heidelberg: Springer Verlag.

Holzapfel, G. A., and Weizsäcker, H. W. (1998). Biomechanical behavior of the arterial wall and its numerical characterization. *Comput. Biol. Med.* 28:377–392.

Holzapfel, G. A., Gasser, T. C., and Ogden, R. W. (2000). A new constitutive framework for arterial wall mechanics and a comparative study of material models. *J. Elast.* 61:1–48.

Holzapfel, G. A., Gasser, T. C., and Stadler, M. (2002). A structural model for the viscoelastic behavior of arterial walls: Continuum formulation and finite element analysis. *Eur. J. Mech. A/Solids* 21:441–463.

Holzapfel, G. A., Gasser, T. C., and Ogden, R. W. (2004a). Comparison of a multi-layer structural model for arterial walls with a Fung-type model, and issues of material stability. *J. Biomech. Eng.* 126:264–275.

Holzapfel, G. A., Sommer, G., and Regitnig, P. (2004b). Anisotropic mechanical properties of tissue components in human atherosclerotic plaques. *J. Biomech. Eng.* 126:657–665.

Holzapfel, G. A., Sommer, G., Gasser, C. T., and Regitnig, P. (2005a). Determination of the layer-specific mechanical properties of human coronary arteries with non-atherosclerotic intimal thickening, and related constitutive modelling. *Am. J. Physiol. Heart Circ. Physiol.* 289:H2048–2058.

Holzapfel, G. A., Stadler, M., and Gasser, T. C. (2005b). Changes in the mechanical environment of stenotic arteries during interaction with stents: Computational assessment of parametric stent design. *J. Biomech. Eng.* 127:166–180.

Holzapfel, G. A., Sommer, G., Auer, M., Regitnig, P., and Ogden, R. W. (2007). Layer-specific 3D residual deformations of human aortas with non-atherosclerotic intimal thickening. *Ann. Biomed. Eng.* 35:530–545.

Huang, D., Chang, T. R., Aggarwal, A., Lee, R. C., and Ehrlich, H. P. (1993). Mechanisms and dynamics of mechanical strengthening in ligament-equivalent fibroblast-populated collagen matrices. *Ann. Biomed. Eng.* 21:289–305.

Humphrey, J. D. (1999). Remodeling of a collagenous tissue at fixed lengths. *J. Biomech. Eng.* 121:591–597.

Humphrey, J. D. (2002). *Cardiovascular Solid Mechanics. Cells, Tissues, and Organs.* New York: Springer-Verlag.

Humphrey, J. D. (2003). Continuum thermomechanics and the clinical treatment of disease and injury. *Appl. Mech. Rev.* 56:231–260.

Humphrey, J. D., and Rajagopal, K. R. (2002). A constrained mixture model for growth and remodeling of soft tissues. *Math. Model. Methods. Appl. Sci.* 12:407–430.

Jouk, P. S., Mourad, A., Milisic, V., Michalowicz, G., Raoult, A., Caillerie, D., and Usson, Y. (2007). Analysis of the fiber architecture of the heart by quantitative polarized light microscopy. Accuracy, limitations and contribution to the study of the fiber architecture of the ventricles during fetal and neonatal life. *Eur. J. Cardiothorac. Surg.* 31:915–921.

Kim, C., Cervos-Navarro, J., Kikuchi, H., Hashimoto, N., and Hazama, F. (1992). Alterations in cerebral vessels in experimental animals and their possible relationship to the development of aneurysms. *Surg. Neurol.* 38:331–337.

Kondo, S., Hashimoto, N., Kikuchi, H., Hazama, F., Nagata, I., Kataoka, H., and Rosenblum, W. I. (1998). Apoptosis of medial smooth muscle cells in the development of saccular cerebral aneurysms in rats. *Stroke* 29:181–189.

Kroon, M., and Holzapfel, G. A. (2007). A model for saccular cerebral aneurysm growth by collagen fibre remodelling. *J. Theor. Biol.* 247:775–787.

Kroon, M., and Holzapfel, G. A. (2008a). Modelling of saccular aneurysm growth in a human middle cerebral artery. *J. Biomech. Eng.* in press.

Kuhl, E., and Holzapfel, G. A. (2008b). Continuum theory of remodeling in living structures. From collagen fibers to cardiovascular tissues. *J. Mater. Sci. Mater. Med.* in press.

Landuyt, M. (2006). Structural quantification of collagen fibers in abdominal aortic aneurysms. Master's thesis, KTH Solid Mechanics, Stockholm and Department of Civil Engineering, Gent.

Lehoux, S., Castier, Y., and Tedgui, A. (2006). Molecular mechanisms of the vascular responses to haemodynamic forces. *J. Intern. Med.* 259:381–392.

Leung, D. Y., Glagov, S., and Mathews, M. B. (1976). Cyclic stretching stimulates synthesis of matrix components by arterial smooth muscle cells in vitro. *Science* 191:475–477.

Leung, D. Y., Glagov, S., and Mathews, M. B. (1977). Elastic and collagen accumulation in rabbit ascending aorta and pulmonary trunk during postnatal growth. Correlation of cellular synthetic response with medial tension. *Circ. Res.* 41:316–323.

Liao, J., Yang, L., Grashow, J., and Sacks, M. S. (2007). The relation between collagen fibril kinematics and mechanical properties in the mitral valve anterior leaflet. *J. Biomech.* 129:78–87.

López, A. M., Lumbreras, F., Serrat, J., and Villanueva, J. J. (1999). Evaluation of methods for ridge and valley detection. *IEEE Trans. Pattern Anal. Mach. Intell.* 21:327–335.

Lu, J., Zhou, X., and Raghavan, M. L. (2007). Inverse elastostatic stress analysis in pre-deformed biological structures: Demonstration using abdominal aortic aneurysms. *J. Biomech.* 40:693–696.

Marko, J. F., and Siggia, E. D. (1995). Stretching DNA. *Macromolecules* 28:8759–8770.

Massoumian, F., Juskaitis, R., Neil, M. A., and Wilson, T. (2003). Quantitative polarized light microscopy. *J. Microsc.* 209:13–22.

Milani, A. S., Nemes, J. A., Abeyaratne, R. C., and Holzapfel, G. A. (2007). A method for the approximation of non-uniform fiber misalignment in textile composites using picture frame test. *Composites Part A Appl. Sci. Manuf.* 38:1493–1501.

Miskolczi, L., Guterman, L. R., Flaherty, J. D., and Hopkins, L. N. (1998). Saccular aneurysm induction by elastase digestion of the arterial wall: a new animal model. *Neurosurgery* 43: 595–601.

Nam, T. H., and Thinh, T. I. (2006). Large deformation analysis of inflated air-spring shell made of rubber-textile cord composite. *Struct. Eng. Mech.* 24:31–50.

Nerem, R. M., and Seliktar, D. (2001). Vascular tissue engineering. *Annu. Rev. Biomed. Eng.* 3:225–243.

Nichols, W. W., and O'Rourke, M. F. (2005). *McDonald's Blood Flow in Arteries. Theoretical, Experimental and Clinical Principles*. London: Arnold, 5th edition. chapter 4, 73–97.

Pandolfi, A., and Manganiello, F. (2006). A model for the human cornea. Constitutive formulation and numerical analysis. *Biomech. Model. Mechanobiol.* 5:237–246.

Patel, D. J., and Fry, D. L. (1969). The elastic symmetry of arterial segments in dogs. *Circ. Res.* 24:1–8.

Peña, E., Calvo, B., Martínez, M. A., and Doblaré, M. (2007a). An anisotropic visco-hyperelastic model for ligaments at finite strains. formulation and computational aspects. *Int. J. Numer. Methods. Eng.* 44:760–778.

Peña, E., Pérez del Palomar, A., Calvo, B., Martínez, M. A., and Doblaré, M. (2007b). Computational modelling of diarthrodial joints. physiological, pathological and pos-surgery simulations. *Arch. Comput. Methods. Eng.* 14:47–91.

Pentimalli, L., Modesti, A., Vignati, A., Marchese, E., Albanese, A., Di Rocco, F., Coletti, A., Di Nardo, P., Fantini, C., Tirpakova, B., and Maira, G. (2004). Role of apoptosis in intracranial aneurysm rupture. *J. Neurosurg.* 101:1018–1025.

Prot, V., Skallerud, B., and Holzapfel, G. (2008). Transversely isotropic membrane shells with application to mitral valve mechanics – constitutive modeling and finite element implementation. *Int. J. Numer. Methods. Eng.* in press.

Purslow, P. P., Wess, T. J., and Hukins, D. W. L. (1998). Collagen orientation and molecular spacing during creep and stress-relaxation in soft connective tissues. *J. Exp. Biol.* 201: 135–142.

Rhodin, J. A. G. (1980). Architecture of the vessel wall. In Bohr, D. F., Somlyo, A. D., and Sparks, H. V., eds., *Handbook of Physiology, The Cardiovascular System*, volume 2. Bethesda, Maryland: American Physiologial Society. 1–31.

Roach, M. R., and Burton, A. C. (1957). The reason for the shape of the distensibility curve of arteries. *Can. J. Biochem. Physiol.* 35:681–690.

Roach, M. R., and Song, S. H. (1994). Variations in strength of the porcine aorta as a function of location. *Clin. Invest. Med.* 17:308–318.

Rodríguez, J. F., Ruiz, C., Doblaré, M., and Holzapfel, G. A. (2008). Mechanical stresses in abdominal aortic aneurysms: influence of diameter, asymmetry and material anisotropy. *J. Biomech. Eng.* in press.

Roveri, N., Ripamonti, A., Pulga, C., Jeronimidis, G., Purslow, P. P., Volpin, D., and Gotte, L. (1980). Mechanical behaviour of aortic tissue as a function of collagen orientation. *Makromol. Chem.* 181:1999–2007.

Rowe, A. J., Finlay, H. M., and Canham, P. B. (2003). Collagen biomechanics in cerebral arteries and bifurcations assessed by polarizing microscopy. *J. Vasc. Res.* 40:406–415.

Roy, C. S. (1880–82). The elastic properties of the arterial wall. *J. Physiol.* 3:125–159.

Sacks, M. S., Smith, D. B., and Hiester, E. D. (1997). A small angle light scattering device for planar connective tissue microstructural analysis. *Ann. Biomed. Eng.* 25:678–689.

Sage, H., and Gray, W. R. (1979). Studies on the evolution of elastin–I. Phylogenetic distribution. *Comput. Biochem. Physiol. B.* 64:313–327.

Sasaki, N., and Odajima, S. (1996). Stress-strain curve and Young's modulus of a collagen molecule as determined by the X-ray diffraction technique. *J. Biomech.* 29:655–658.

Schmid, F., Sommer, G., Rappolt, M., Schulze-Bauer, C. A. J., Regitnig, P., Holzapfel, G. A., Laggner, P., and Amenitsch, H. (2005). In situ tensile testing of human aortas by time-resolved small angle X-ray scattering. *J. Synchron. Rad.* 12:727–733.

Schulze-Bauer, C. A. J., Amenitsch, H., and Holzapfel, G. A. (2002a). SAXS investigation of layer-specific collagen structures in human aortas during tensile testing. In *European Materials Research Society Spring Meeting, E-MRS 2002 SPRING MEETING - Synchrotron Radiation and Materials Science (Symposium I)*.

Schulze-Bauer, C. A. J., Regitnig, P., and Holzapfel, G. A. (2002b). Mechanics of the human femoral adventitia including high-pressure response. *Am. J. Physiol. Heart Circ. Physiol.* 282:H2427–H2440.

Shekhonin, B. V., Domogatsky, S. P., Muzykantov, V. R., Idelson, G. L., and Rukosuev, V. S. (1985). Distribution of type I, III, IV and V collagen in normal and atherosclerotic human arterial wall: Immunomorphological characteristics. *Coll. Relat. Res.* 5:355–368.

Sommer, G., Gasser, T. C., Regitnig, P., Auer, M., and Holzapfel, G. A. (2008). Dissection of the human aortic media: an experimental study. *J. Biomech. Eng.* in press.

Sonka, M., Hlavac, V., and Boyle, R. (1999). *Image Processing, Analysis and Machine Vision.* Pacific Grove, CA: Brook/Cole Publishing Company, 2nd edition.

Stary, H. C., Blankenhorn, D. H., Chandler, A. B., Glagov, S., Jr. Insull, W., Richardson, M., Rosenfeld, M. E., Schaffer, S. A., Schwartz, C. J., Wagner, W. D., and Wissler, R. W. (1992). A definition of the intima of human arteries and of its atherosclerosis-prone regions. A report from the committee on vascular lesions of the council on arteriosclerosis, American Heart Association. *Circulation* 85:391–405.

Stary, H. C. (2003). *Atlas of Atherosclerosis: Progression and Regression.* Boca Raton, London, New York, Washington, D.C.: The Parthenon Publishing Group, 2nd edition.

Sun, Y. L., Luo, Z. P., Fertala, A., and An, K. A. (2002). Direct quantification of the flexibility of type I collagen monomer. *Biochem. Biophys. Res. Commun.* 295:382–386.

Taber, L. A. (1995). Biomechanics of growth, remodelling, and morphognesis. *Appl. Mech. Rev.* 48:487–543.

Taber, L. A., and Humphrey, J. D. (2001). Stress-modulated growth, residual stress, and vascular heterogeneity. *J. Biomech. Eng.* 123:528–535.

Tam, A. S. M., Sapp, M. C., and Roach, M. R. (1998). The effect of tear depth on the propagation of aortic dissections in isolated porcine thoracic aorta. *J. Biomech.* 31:673–676.

Tower, T. T., Neidert, M. R., and Tranquillo, R. T. (2002). Fiber alignment imaging during mechanical testing of soft tissues. *Ann. Biomed. Eng.* 30:1221–1233.

Tronc, F., Wassef, M., Esposito, B., Henrion, D., Glagov, S., and Tedgui, A. (1996). Role of NO in flow-induced remodeling of the rabbit common carotid artery. *Arterioscl. Thromb. and Vasc. Biol.* 16:1256–1262.

van Baardwijk, C., and Roach, M. R. (1987). Factors in the propagation of aortic dissection in canine thoracic aortas. *J. Biomech.* 20:67–73.

Vande Geest, J. P., Sacks, M. S., and Vorp, D. A. (2006). The effects of aneurysm on the biaxial mechanical behavior of human abdominal aorta. *J. Biomech.* 39:1324–1334.

von der Mark, K. (1981). Localization of collagen types in tissues. *Int. Rev. Connect. Tissue. Res.* 9:265–324.

Whittaker, P., Schwab, M. E., and Canham, P. B. (1988). The molecular organization of collagen in saccular aneurysms assessed by polarized light microscopy. *Connect. Tissue. Res.* 17:43–54.

Wright, N. T., and Humphrey, J. D. (2002). Denaturation of collagen via heating: an irreversible rate process. *Annu. Rev. Biomed. Eng.* 4:109–128.

Zollikofer, C. L., Salomonowitz, E., Sibley, R., Chain, J., Bruehlmann, W. F., Castaneda-Zuniga, W. R., and Amplatz, K. (1984). Transluminal angioplasty evaluated by electron microscopy. *Radiology* 153:369–374.

Chapter 12
The Extracellular Matrix of Skeletal and Cardiac Muscle

P.P. Purslow

Abstract Well-organized and distinct extracellular matrix networks exist within both striated and cardiac muscles. Individual muscle cells are separated by a fine collagen fiber network embedded in a proteoglycan matrix (the endomysium). Larger groups or bundles of muscle cells are separated by a thicker connective tissue structure, the perimysium. The endomysium separating adjacent muscle cells joins together at the nodes between cells to form a continuous structure within the muscle fiber bundle, and perimysium similarly forms a continuous network throughout the entire organ.

In striated muscle, the networks of wavy (non-straight) collagen fibers in both endomysium and perimysium can easily reorientate and offer little tensile resistance to changing muscle length. However, forces can be transmitted efficiently by shear through the thickness of the endomysium. The endomysium fulfills the role of mechanically linking adjacent muscle fibers so as to coordinate their length changes and keep their sarcomere lengths uniform. Especially in series-fibered muscles, shear through the thickness of the endomysium is a key mechanism for transmission of forces generated by contraction of muscle fibers.

There is evidence that the perimysium of striated muscle can also play a role in muscle force transmission (myofascial force transmission), but there is also a clear role for the perimysial boundaries between adjacent muscle fiber bundles to accommodate shear deformations generated when muscles contract. Differences in shear strains generated in anatomically different muscles appear to be the main explanation of why the division of muscles into fiber bundles (fascicles) by perimysium varies so much from muscle to muscle.

The endomysium and perimysium in cardiac muscle shows strong similarities in structure and function to the same extracellular matrix structures in striated muscle.

The extracellular matrix in muscle is dynamically remodeled according to the loads imposed on it during muscle growth, exercise, and as a response to damage. This is especially relevant to the properties of cardiac muscle after ischemia. Matrix metalloproteinases responsible for remodeling extracellular matrix within muscle are secreted both by fibroblasts and by the muscle cells.

P. Fratzl (ed.), *Collagen: Structure and Mechanics*,
© Springer Science+Business Media, LLC 2008

The atrioventricular valves in the heart are special connective tissue structures well adapted to their function. Collagen fiber orientation in the cusps of these valves is closely modulated to reinforce the valves against predominant haemodynamic stresses.

The connective tissue structures within striated and cardiac muscles are an integrated part of the muscle as a whole tissue or organ and play key roles in their in vivo mechanical functions and properties.

12.1 Introduction

This chapter gives an overview of the main features of the structure, composition, mechanical properties and in vivo functions of the extracellular matrix (ECM) associated with skeletal and cardiac muscles, and heart valves. This review attempts to provide a synthesis of basic knowledge and current thinking so as to provide an informative overview and up-to-date understanding of intramuscular ECM. Intramuscular ECM is also often referred to as intramuscular connective tissue (IMCT) and both of these names will be used here interchangeably.

The structure–function relationship in intramuscular ECM is a dynamic one, as in other connective tissue structures, and the turnover and remodeling of intramuscular ECM involve transduction of mechanical signals and alterations in cell expression to modulate the ECM structures accordingly. One subtle difference in the case of intramuscular ECMs is that the turnover and synthesis of some ECM components are due to expression from muscle cells, in addition to expression from fibroblasts.

Bowman (1840) was the first to describe connective tissue structures separating and surrounding each muscle fiber in striated or voluntary muscle. In addition to anatomical studies, there is a century-long history of interest in the role that IMCT plays in determining the toughness of muscle when it is eaten as meat. A number of substantial reviews exist on the subject of IMCT, and it is not the intention here to exhaustively repeat these. Some are concerned with the structure and properties of IMCT as a component of living muscle, and so deal with in vivo properties, and some relate IMCT structure to the properties of postmortem muscle, in relation to the texture of meat as a food. The reader is referred to the reviews listed in Table 12.1 as substantial sources of detailed information. In particular, the basic mechanisms of mechanotransduction and cell signaling that result in the dynamic remodeling of intramuscular ECM and the response of these pathways to mechanical loading and exercise will not be covered in any depth here; the reader is referred to the reviews of Bishop and Lindahl (1999), Purslow (2002), Kjær (2004) and Chapter 9 of this volume for this topic.

In this chapter, the following abbreviations are being used:

ECM – ptextracellular matrix PG – proteoglycan
SEM   scanning electron microscopy
IMCT   intramuscular connective tissue
TEM   transmission electron microscopy

Table 12.1 Previous reviews of IMCT structure, properties, turnover and function

References	Main focus
Skeletal muscle	
Mayne and Sanderson (1985)	Skeletal muscle IMCT structure
Purslow and Duance (1990)	Morphology, composition and properties of IMCT
Light (1987)	Structure and biochemistry of IMCT in relation to meat texture
McCormick (1994)	Biochemistry and cross-linking of IMCT
Trotter et al. (1995)	Endomysial structure and tensile properties; force transmission in series-fibered skeletal muscles
Purslow (2002)	Variations in the structure and development of IMCT; mechanical roles in vivo; cell–matrix signaling
Sanes (2003)	Functions of basement membrane in skeletal muscle
Kjær (2004)	Turnover and adaptation of IMCT on loading
Purslow (2005)	Role of IMCT in meat quality; turnover of IMCT
Kjær et al. (2006)	Adaptation of IMCT to exercise
Cardiac muscle	
Bishop and Lindahl (1999)	Cardiovascular ECM synthesis and turnover in relation to loading
de Souza (2002)	Changes in myocardial ECM with age
Anderson et al. (2005)	Anatomy of the myocardium and its ECM
Lunkenheimer et al. (2006)	Anatomy of the myocardium and its ECM
Miner and Miller (2006)	Changes in cardiac ECM in heart failure

12.2 General Structure of IMCT

There are three general types of muscle tissues within the vertebrate body: smooth muscle, striated muscle and cardiac muscle. Cardiac muscle is the highly specialized contractile tissue of the heart. Smooth muscle is found in locations such as the bladder, arteries and in the gut wall. Although there are many features and signaling pathways in common between matrix signaling and turnover in smooth versus striated and cardiac muscles, the role of ECM in smooth muscle function will not be covered here, and the reader is referred to a number of texts and articles reviewing the role of ECM in smooth muscle growth and development, and signaling events in angiogenesis (Schwartz, 1995; Stegemann et al., 2005; Rhodes and Simons, 2007).

The commonest and biggest mass of muscle tissue in the body is the striated muscle, which attach to the bone via tendon (also known as skeletal muscles) and are responsible for the posture and movement of the body. Striated muscle is the only one of the three types under voluntary control (and so also known as voluntary muscle).

12.2.1 Striated Muscle: Gross Morphology of Intramuscular Connective Tissue

The following description of the general structure of intramuscular connective tissue (IMCT) associated with striated muscles summarizes the consensus of a number of sources (Bourne, 1973; Borg and Caulfield, 1980; Rowe, 1981; Schmalbruch, 1985; Purslow and Duance, 1990). The general architecture of IMCT is schematically shown in Fig. 12.1.

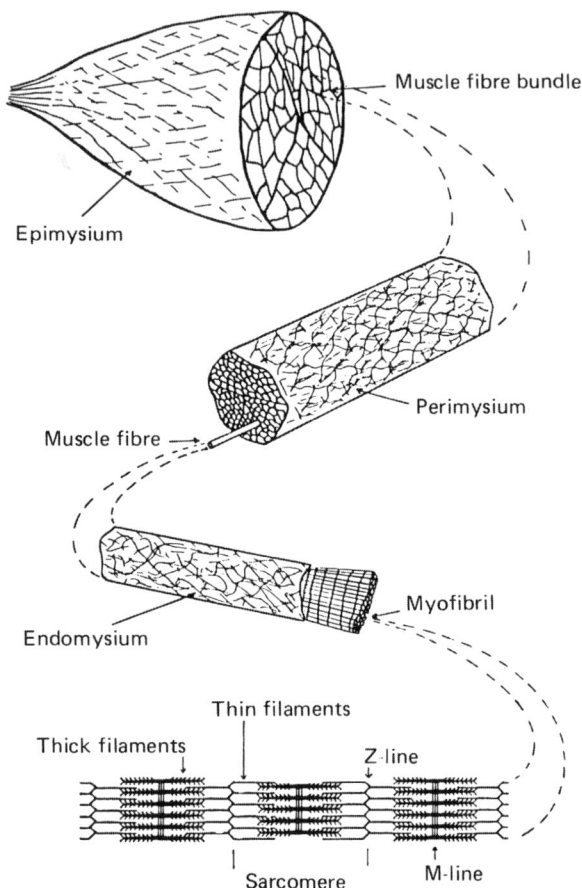

Fig. 12.1 Schematic diagram to show the arrangement of IMCT within muscle. Three distinct structures are shown. The whole muscle is surrounded by epimysium. Fascicles or fiber bundles are separated by perimysium. Individual muscle fibers are separated by endomysium. Adapted from Jolley and Purslow (1988)

Surrounding each individual muscle is a connective tissue layer, the *epimysium*. This is continuous with the tendons attaching the muscle to the bones. In some long strap-like muscles (e.g., bovine sternomandibularis muscle), the epimysium is comprised of a crossed-ply arrangement of two parallel sets of wavy collagen fibers at an angle of approximately ±55° to the long axis of the muscle fibers, embedded in a proteoglycan matrix (see Fig. 12.2). In other cases (e.g., bovine semitendinosus muscle), the arrangement of collagen fibers in the epimysium is parallel to the long axis of the muscle and forms a dense sheath that acts as an epitendon. This arrangement is also common in some pinnate muscles.

Attached to the epimysium at the surface of the muscle, the *perimysium* is a continuous network that divides the muscle up into fascicles or muscle fiber bundles. The perimysial layer separating two fascicles is comprised of two crossed-plies of

Fig. 12.2 Light micrograph showing the arrangement of collagen fibers in the epimysium of bovine sternomandibularis muscle. *Horizontal lines* show the edges of individual muscle fiber bundles in the surface of the muscle. The collagen fibers of the overlying epimysium lie in two plies of parallel fibers, at approximately ±55° to the muscle fiber direction. From Purslow (1999) with permission

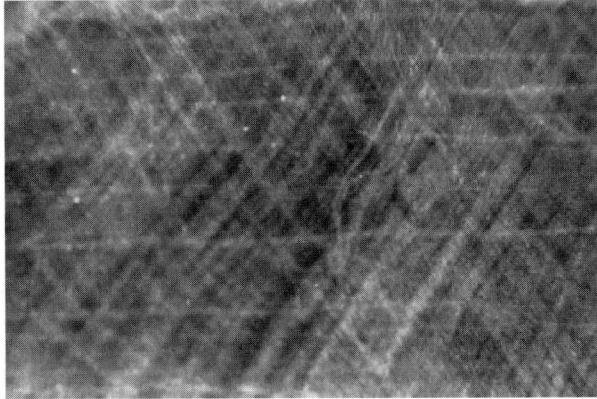

wavy collagen fibers in a proteoglycan matrix, forming a planar network. The long axis of each set of collagen fibers lies at ±55° to the longitudinal axis of the fascicle at the resting length of the muscle. (The significance of the 55° angle is explained in a later section.)

Attached at the margins of the fascicle to the perimysium by weak and intermittent junction plates (Passerieux et al., 2006), the *endomysium* is a continuous network of connective tissue that separates individual muscle fibers. As with the perimysium, the endomysial layer separating two adjacent muscle fibers is a planar network of collagen embedded in a proteoglycan matrix that runs the length of the fibers it separates. Bowman (1840) first described the endomysium (calling it the sarcolemma) as a "tubular membranous sheath of the most exquisite delicacy". We now refer to the basement membrane of the muscle fibers as the sarcolemma and use the term endomysium to denote the reticular layer of fibrous ECM that occupies the space between the sarcolemma of one fiber and the sarcolemma of its adjacent fiber.

The concept of endomysial connective tissue sheaths surrounding each fiber (or fascicle, in terms of the perimysium) is a common feature of descriptions of IMCT architecture. This concept has long been reinforced by the observation that when physiologically viable single muscle cells are dissected out (e.g., from the easily dissected frog leg muscles; Natori, 1954), some endomysial sheath is apparent around each fiber. However, it should be stressed that the endomysial layer separating the two muscle cells is a single layer, joined at the nodes between adjacent cells into a continuous network. Retention of some ECM material on dissected single fibers does not indicate that endomysium forms independent sheaths around each muscle fiber, but it speaks to the fact that attachment of the basement membrane to the interior cytoskeleton of the fiber via transmembrane proteins such as dystrophin and integrins, and costamere structures on the internal surface of the sarcolemma, are not easily broken. Similarly, the perimysial layer separating two fascicles is a single layer, joined at the nodes between fascicles to form a continuous network throughout the muscle cross-section.

This continuous network structure of IMCT is clearly revealed in Fig. 12.3. This is one of the many scanning electron micrographs in the literature for muscle treated with sodium hydroxide to dissolve away the muscle fiber proteins, cell membranes and proteoglycans, leaving only the collagen networks of the IMCT structures. In the low-magnification view of the muscle cross-section (Fig 12.3, upper), thick perimysial walls separating fascicles are clearly seen, as is the thinner endomysial network separating individual muscle fibers. A higher magnification view, at an oblique angle to the surface (Fig. 12.3, lower), clearly shows the endomysial network as a honeycombed continuous network of fine wavy collagen fibrils. The large lacunae defined by this network are the spaces occupied by adjacent muscle fibers. The wavy collagen fibrils form a quasi-random planar network parallel to the surface of the muscle fibers. No collagen fibers run through the thickness of the endomysium.

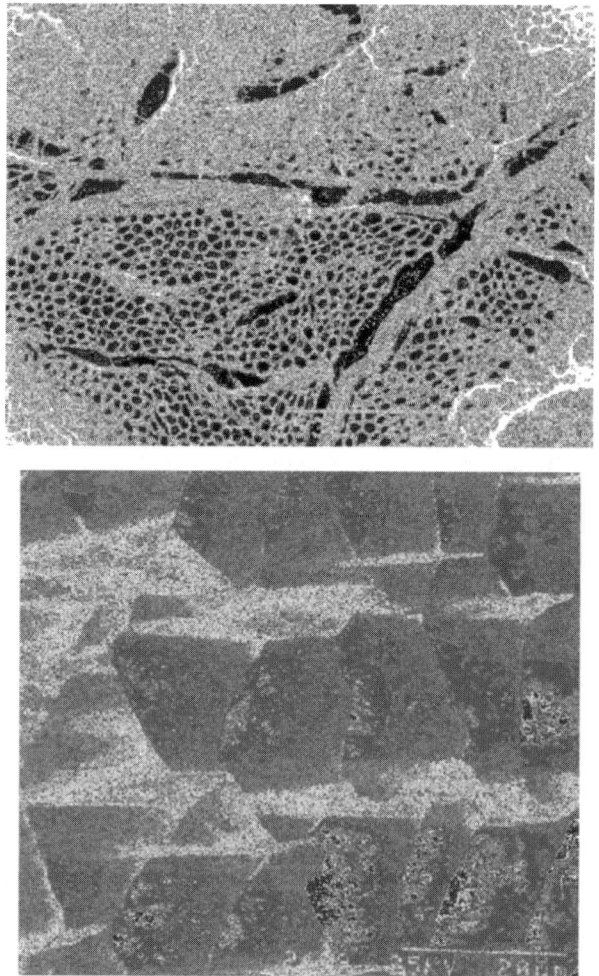

Fig. 12.3 SEM micrographs of IMCT following NaOH digestion of bovine sternomandibularis muscle. *Upper panel*: low-magnification view, showing thicker perimysial sheets surrounding fascicles. *Lower panel*: high-magnification oblique view, showing endomysial networks. From Purslow and Trotter (1994) with permission

Depending on the pH, temperature, strength of NaOH and time, sodium hydroxide digests the myofibrillar proteins, PGs and also some collagenous structures to a lesser or greater extent (Passerieux et al., 2007). Although it is a great technique for revealing the general structure of IMCT, it should not be used in an attempt to quantify the density or the stability of IMCT structures without exhaustive controls.

The general morphology of perimysium is common to all striated muscles, although its thickness, total amount and spatial distribution vary considerably from muscle to muscle, as discussed further below. The endomysium again has a common appearance, being similar in structure for both continuous-fibered muscles (muscle fibers running from tendon to tendon) or series-muscled fibers (muscle fibers are much shorter than the fascicles running from tendon to tendon).

12.3 Composition of the Perimysium and Endomysium

ICMT is generally composed of collagen fibers in an amorphous matrix, princi- pally containing proteoglycans (PGs). The amount of collagen typically accounts for 1–10% of the dry weight of striated muscle (Bendall, 1967). Fibers of elastin are also to be found in some IMCT (principally the perimysium), although the proportion of elastin to collagen is small in most muscles, and elastin is typically less than 1% of muscle dry weight (Bendall, 1967). Seven molecular types of collagen have been identified in muscle (types I, III, IV, V, VI, XII and XIV; Listrat et al., 1999, 2000) of which the fiber-forming types I and III are the most prevalent in mammalian striated muscle. In the major ECM structures within fish muscle, the myocommata, type V collagen is far more common than in mammalian striated muscle (Sato et al., 1998). The proportions of type I to type III collagen vary considerably from muscle to mus- cle within an animal (Light et al., 1985). Type IV collagen comprises the basement membrane layer of the muscle fibers. This non-fibrous network layer with its own PG components such as laminin and fibronectin forms the boundary between the cell and the collagen fiber networks of the overlaying endomysium.

Proteoglycans are associated with the collagen networks of both the perimysium and the endomysium. Heparin sulphate-containing PGs are associated with the base- ment membrane layer (Carrino and Caplan, 1982; Anderson et al., 1984; Nishimura et al., 1996). Chondroitin sulphate-containing and dermatan sulphate-containing PGs are found in greater amounts in the perimysium (Nishimura et al., 1996). Decorin has been identified in IMCT (Eggen et al., 1994; Nakano et al., 1997). Although it is recognized that the amorphous matrix of hydrated PGs plays a cru- cial role in mechanically linking together the collagen fibers of IMCT (Scott, 1990) and that there are variations in PG content during muscle development (Velleman et al., 1999), the differences from muscle to muscle in PG composition of IMCT are not well understood.

During embryonic muscle development, growth and differentiation of myoblasts into myotubes and the organization of these into orientated muscle fiber bundles are patterned, guided and stimulated by interactions between the muscle cells and

IMCT. The composition of IMCT changes in terms of collagens as well as PGs. Listrat et al. (1999) showed that the total collagen content and amount of types I and III collagen in two bovine muscles peak between halfway and two thirds through gestation, at a point of muscle development coincident with expression of adult forms of myosin within differentiated myotubes. After this point, the collagen concentration declines as the functionally contractile muscle fibers begin to hypertrophy. However, the type I content is always higher than the type III content throughout gestation and post-natally. Type XII (one of the FACIT collagens, which appears to regulate the fibril diameters of the fiber-forming collagen types) is expressed more in early stages of fetal development and reduces toward birth, whereas another FACIT collagen, type XIV, is more prevalent after birth (Listrat et al., 2000). Lawson and Purslow (2001) show variations in collagen type I content between the pectoralis and the quadriceps muscle of the chick from embryonic day 10 through hatching (day 21) to 8 weeks post-hatch. In this avian species, no dip in collagen type I is apparent in the second half of pre-hatch development and the amounts of type I collagen increases steadily, although there is always less in quadriceps than in pectoralis. The laminin content of the two muscle also increases over the same period, but the laminin content of chick quadriceps is lower than chick pectoralis until embryonic day 13, and then a little higher thereafter.

Throughout gestation and post-natally, there are substantial changes in the types and amounts of covalent cross-links that mechanically stabilize the collagen molecules within the fibrous networks of the endomysium and perimysium. The nature of these cross-links, how they change with maturation and physiological ageing and the mechanical consequences of this are covered in detail in Chapter 4 and will not be repeated here.

12.4 The Amount, Composition and Architecture of Endomysium and Perimysium Vary Between Different Striated Muscles

As shown in the cross-sections of various muscles in Fig. 12.4, the perimysium

(a) comprises a continuous network throughout the muscle cross-section,
(b) varies in thickness between muscles, and
(c) surrounds or separates fascicles of quite different sizes and shapes in different muscles.

The total collagen content varies between different muscles in an animal's body (Bendall, 1967). The relative amounts of endomysium and perimysium also vary from muscle to muscle (Light et al., 1985). Variations in the perimysial content appear much greater than variations in endomysial content. In a comparison of 14 bovine muscles, the endomysium collagen accounted for between 0.47 and 1.20% of dry weight, whereas the perimysial collagen content varied from 0.43 to 4.6%

Fig. 12.4 Cross-sections of three bovine muscles (*top*: pectoralis profundus, *mid*: sternocephalicus, *bottom*: rhomboideus cervicus) showing differences in fascicular size, shape and perimysial thickness. From Purslow (1999) with permission

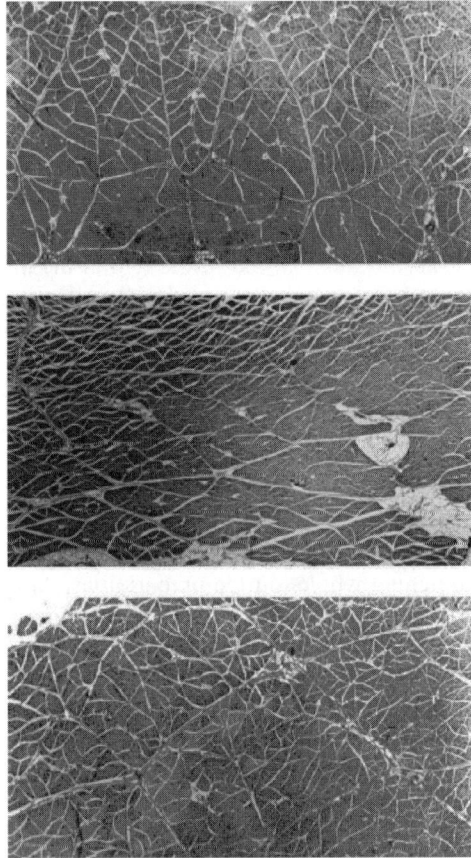

(Purslow, 1999). Brooks and Savell (2004) report that perimysium from bovine semitendinousus can be as much as 2.4 times the thickness of perimysium from psoas major muscles. These variations in the amount of endomysium and perimysium and in the spatial organization of perimysium in dividing up the whole muscle into fascicles have long been thought to be a consequence of the different mechanical roles and characteristics of different anatomical muscles. Some insight into the contributions of IMCT to muscle functioning and possible explanations of the reasons why perimysial content and spatial distribution varies between muscles is offered below.

Like most other ECM structures, IMCT is not a static structure defined in terms of amount and composition at birth by variations programmed in during embryogenesis and subsequently modified by ageing processes. Rather IMCT structure and composition are dynamic balances between deposition, growth, remodeling and degradation affected by the interplay between functional demands on the tissue and the mechanical environment. Mechanical adaptation and remodeling of ECM are covered in depth in Chapter 9, and some specific considerations for IMCT of both skeletal and cardiac muscles are reviewed below.

12.5 The Orientation of Collagen Fibers in Perimysium and Endomysium Changes with Muscle Length

(a) *Perimysium*. At the rest length of the muscle, the collagen fibers in the crossed-ply arrangement of the perimysium lie at $\pm 55°$ to the long axis of muscle fibers and are wavy, with a uniform orientation and waviness between collagen fibers in each ply. As muscle shortens or lengthens, the overall orientation of the collagen fibers and their degree of waviness both change (Rowe, 1981). Purslow (1989) documents how the overall orientation of the collagen fibers with respect to the muscle fiber direction (θ) varies with muscle sarcomere length in bovine sternomandibularis muscle. Over the range of muscle lengths experimentally possible with this muscle, θ varied from about 80° at an extremely short sarcomere length of 1.1 μm to approximately 20° at 3.9 μm, the longest sarcomere length that was possible to achieve by stretching whole muscle preparations. At rest length (around 2 μm for this muscle on excision), θ was 55°. A straightening of the waviness in the fibers was also documented at both short and long muscle lengths, with the maximum waviness at muscle rest length. Based on these experimental observations, Purslow (1989) built a geometrical model to explain these changes in the collagen network. Independently, Lepetit (1991) later published a model that is identical except that the end-points, where the waviness of the fibers becomes zero, are different. However Lepetit's model was not compared to any experimental data, unlike the end-points in Purslow (1989).

The geometrical basis of the model is shown in Fig. 12.5. Muscle fibers and fascicles are known to change length in vivo without changing volume. If they do this and retain the same cross-sectional shape (i.e., a roughly cylindrical fascicle remains cylindrical), then the diameter of the fascicle must go up as the length goes down. If the collagen fibers are assumed to be practically inextensible but free to change waviness and reorientate to follow length changes in a cylinder with fixed volume V but variable length L, then the following relationship is predicted:

$$\tan \theta = \left[\frac{4\pi V}{L^3} \right]^{0.5}.$$

The mean end-to-end length, x, of each collagen fiber is given by

$$x = \left[L^2 \left(1 + \frac{4\pi V}{L^3} \right) \right]^{0.5}.$$

Muscle fibre direction

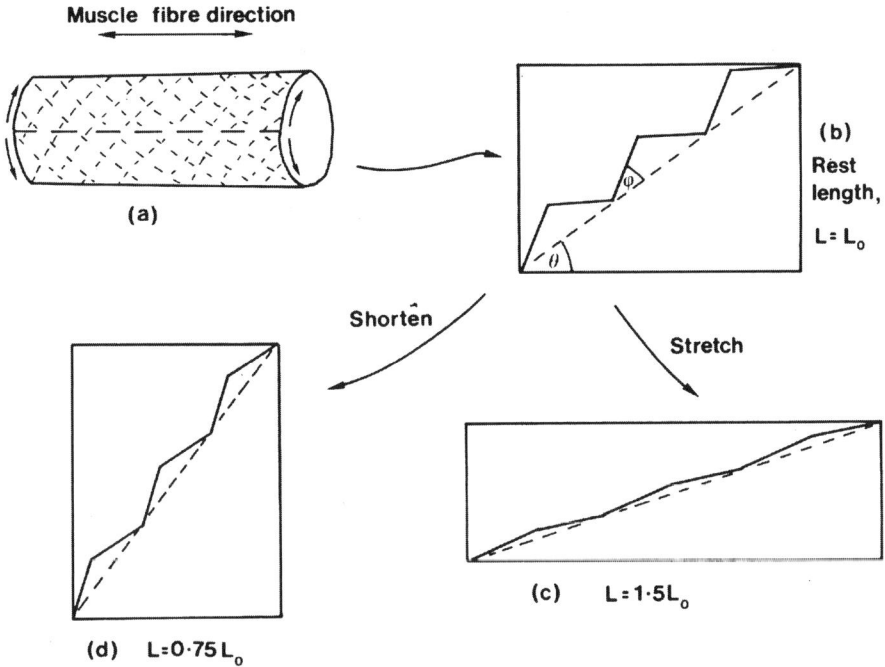

Fig. 12.5 Schematic diagram of the geometric model of perimysium. (**a**) The perimysial net surrounding a cylindrical muscle fiber bundle, cut along the muscle fiber axis (*dashed line*) and flattened out (*arrows*) to give (**b**) a flat sheet of perimysium at rest length. The orientation of one collagen fiber in one of the two plies is defined by the angle θ to the muscle fiber direction. ϕ is the crimp angle. The *dotted line* represents the mean end-to-end length of the fiber, x. If the muscle fiber bundle is stretched (**c**), or shortened (**d**), θ decreases or increases respectively. ϕ decreases in both cases. From Purslow (1989) with permission

If x has a minimum (x_0) at the muscle rest length (L_0), so that the fibers have maximum waviness at rest, it is possible to show that this model predicts $\theta - 54°44'$ at muscle rest length (L_0). This value of $\theta = 54°44'$ is a well-known result in fiber reinforcement of cylindrical vessels, as composite theory predicts that stiff fibers will have the maximum efficiency of reinforcement of a pressurized cylindrical vessel wall at this angle. Purslow (1989) shows that this model fits that change in network orientation experimentally measured in perimysium from muscle samples fixed at different lengths.

(b) *Endomysium.* The endomysium is usually described as a random feltwork or weave of collagen fibrils because the orientation of collagen fibrils in the endomysial network is far more disordered than in the perimysium and it is difficult to visualize any preferred directions. However, by making detailed image analysis measurements of the orientation of each fibril within the network, it is possible to calculate a numerically weighted average orientation angle of θ for the whole distribution and to show that this changes with muscle length.

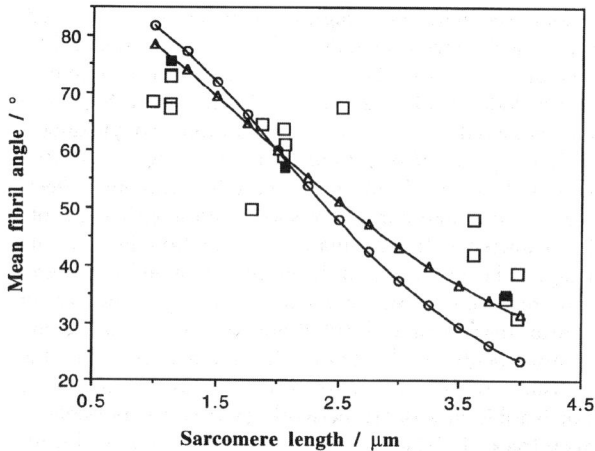

Fig. 12.6 Numerically weighted mean fibril orientation of endomysium versus muscle sarcomere length. The two lines shown are the predicted mean orientation versus sarcomere length from (*circles*) the isoareal model and (*triangles*) the constant shape model. Data points (*squares*) from measurements of micrographs show a rate of change with sarcomere length most consistent with the constant shape model. From Purslow and Trotter (1994) with permission

Figure 12.6, from Purslow and Trotter (1994), shows this numerically – averaged θ as a function of muscle sarcomere length for bovine sternomandibularis muscle. Here, the data for mean θ values are compared to two theoretical models: the "constant shape" model, which makes the same assumptions as the geometrical model for the perimysial network (i.e., that the network surrounds a constant volume cylinder of changing length and whose cross-sectional area changes to compensate, but remains circular in shape) and the "isoareal" model, which assumes that the network surrounds a constant volume cylinder of changing length but whose surface area also remains constant. The data appear to fit better with the "constant shape" geometrical model, in agreement with the previous perimysial model. The "isoareal" model is theoretically attractive because it predicts that the thickness of the endomysium would not change with sarcomere length (Trotter and Purslow, 1992). However, TEM images of muscle at different sarcomere lengths do show some evidence of the endomysium thinning at longer sarcomere lengths (Trotter and Purslow, 1992, Fig 12.4.).

Trotter and Purslow (1992) and Purslow and Trotter (1994) explain that because the value of strain along the overall direction of each collagen fibril for a given increase in muscle length is a function of its original orientation angle (θ) and the fibrils are distributed over a wide range of angles, many of the fibrils will actually become more wavy as the muscle length increases. In fact, at all practical muscle lengths, there will be a high proportion of the collagen fibrils in the network that are wavy. Accepting that the high tensile stiffness of the collagen fibers is only seen when the fibrils are straightened out, these observations suggest that these networks are very compliant and can easily be deformed to follow the changing working lengths of the muscle in vivo.

12.6 Mechanical Properties of the Perimysium: Models and Measurements

There are few direct measurements of the mechanical properties of isolated perimysium. Lewis and Purslow (1989) performed tensile tests on perimysial strips isolated from bovine semitendinosus muscle. In this study and subsequent work (Lewis and Purslow, 1991; Lewis et al., 1991), tensile stress–strain curves are shown for perimysium under a variety of conditions (pH, heating, age postmortem of muscle) relevant to muscle as meat, but the properties of perimysium from raw, unaged muscle are clearly most relevant to its properties in vivo. In tension, the perimysium is easily deformed near rest length, becoming progressively stiffer at high extensions. Figure 12.7 (from Purslow, 1999) shows the load-displacement curve of perimysium excised from bovine semitendinosus muscle. This experiment was conducted using the microscope-mounted tensiometer described by Lewis and Purslow (1989), which allows the orientation of collagen fibers in the perimysial strip to be monitored optically during the tensile test. The collagen fibers become aligned with the stretching direction at high extensions, accompanied by an increase in incremental tensile modulus. The rate of change of orientation angle with changing length in this uniaxial test on strips of perimysium agrees well with the model of Purslow (1989), described in Fig. 12.5. The perimysium is also a viscoelastic material, exhibiting creep and stress relaxation. Purslow et al. (1998) used simultaneous mechanical testing and X-ray diffraction to describe the non-linear relaxation behavior of perimysial strips from bovine semitendinousus muscle, and also show that the time-course of creep and stress–relaxation is unrelated to reorientation of the collagen fibers. The in-plane tensile properties of the perimysium were modeled by Purslow (1989) based on a simple fiber orientation model of Krenchel (1964). The modulus of the network parallel to the muscle fiber direction (E_L) is given by

$$E_L = (\cos^4 \theta) E_f V_f + V_m E_m,$$

where V_f is the volume fraction of the collagen fibers, V_m is the volume fraction occupied by the amorphous matrix, E_m the stiffness of the amorphous matrix (assumed to be small) and E_f is the incremental modulus of the collagen fibers. Because the waviness is a changing function of the muscle length, E_f is also a function of muscle length, and an approximate fit for this was calculated from a previous bent-elastica model. The predicted stress–strain curve from this model showed very low stiffness until strains of approximately 30% were reached, when the stiffness of the network increased dramatically.

Fig. 12.7 (**a**) Load–
displacement curve for
excised perimysium from
bovine semitendinousus
muscle. (**b**) Measurements
(*triangles*) of the angle θ
between the collagen fiber
and the muscle fiber axis
agree with the model of
Purslow (1989) (*fitted line*)
for reorientation of
perimysial fibers surrounding
a fascicle changing length at
constant volume. From
Purslow (1999) with
permission

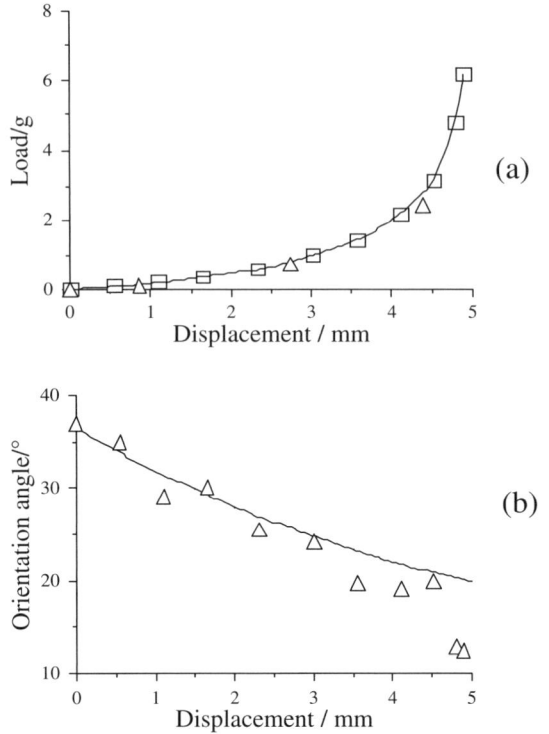

12.7 Mechanical Properties of the Endomysium: Models and Measurements

Using a similar approach based on Krenchel-type fiber orientation considerations, Purslow and Trotter (1994) modeled the in-plane tensile properties of the
endomysium. The model given for the modulus parallel to the muscle fiber direction
(E_L) is given as

$$E_L = \eta_{or}\eta_L E_f V_f + E_m V_m,$$

where E_m, V_m and V_f are defined as above. In this model, the varying stiffness along
the wavy collagen fibril is encompassed in the term ($\eta_L E_f$), and η_{or} describes the
orientation of the fibrils in the quasi-random network:

$$\eta_{or} = \int f(\theta)\cos^4\theta\, d\theta,$$

where $f(\theta)$ is the distribution of fibril orientations measured from their SEM images
of endomysium in NaOH-digested muscle fixed at a range of sarcomere lengths. As
with the perimysial model, the predicted modulus E_L is low at muscle rest length

and rises slowly at extended lengths as the network becomes gradually more aligned with the stretching direction.

The predicted in-plane tensile properties of the endomysial network are in accord with experimental force–length measurements for endomysium achieved by comparing the tensile properties of relaxed single muscle fibers with and without endomysium (Natori, 1954; Podolsky, 1964; Magid and Law, 1985). By looking at the difference that the presence of the endomysium makes to these passive elasticity measures on single fibers, it has long been recognized that the endomysium is extremely compliant in tension along the muscle fiber direction over normal working muscle lengths in vivo.

12.8 Mechanical Roles In Vivo for Perimysium and Endomysium in Striated Muscles

Why does the amount, composition and distribution of IMCT (and especially perimysium) vary between functionally different muscles? Although there is a common perception that these variations in some way reflect the different biomechanical functions of different muscles, direct links between IMCT structure and function and muscle function are tenuous. What precisely are the roles of endomysium and perimysium in functioning muscle, and why do these roles call for different amounts and distributions of some IMCT structures?

12.8.1 Endomysial Role in Force Transmission

The passive elasticity measurements of Natori (1954), Podolsky (1964) and Magid and Law (1985) on single muscle fibers are among many that have historically created the view that endomysial IMCT may have a role in preventing overstretching of non-active muscle by external forces or the action of antagonistic muscles, but that endomysium does not have any role in transmission of the forces produced in the muscle fiber upon active contraction of the tendons and hence the skeleton. At normal working lengths of actively contracting muscle, the in-plane tensile stiffness of the collagen network is very small and so tensile loads engendered in the endomysium by small contractions along the length of the muscle fiber would be negligible. Indeed, the tensile properties of the endomysium seem designed to allow the network to easily follow the changes in muscle fiber length and diameter generated during contraction and lengthening. If single muscle fibers are pinched so that the myofibrillar apparatus ruptures and retracts (Ramsay and Street, 1940), the empty endomysial tube that remains can carry the contractile forces generated by the undamaged segments of the muscle fiber, but the endomysial tube is extremely elongated and thin, leading to extreme reorientation of the collagen fibers along the stretching axis. Thus, although endomysium is capable of carrying high loads in tension in such experiments, it only does so under extremely non-physiological conditions.

There are a very considerable number of muscles in animals from many phyla where the individual muscle fibers do not run the entire length of fascicles, from tendon to tendon. These muscles with intrafascicularly terminating fibers, or series-fibered muscles, are in fact quite common (except perhaps in humans), as reviewed by Gans and Gaunt (1991), Trotter (1993) and Trotter et al. (1995). For example, the major flight muscle in birds (the pectoralis) is series-fibered. Muscle fibers in series-fibered muscles are relatively short compared to the length of the fascicle and generally taper gently at the ends (although other types of termination have been reported; Young et al., 2000). The fibers are staggered by about one quarter of their length with respect to their lateral neighbors, so that the tapering end of one fiber overlaps its neighbors. They have no direct connections to the tendons, and yet large contractile forces are generated in the muscle and transmitted to the tendons. The endomysium is the structure that links adjacent muscle cells. Although compliant in tension, the endomysium is thought to transmit forces by shear through its thickness with efficiency.

In their study of endomysium in (series-fibered) feline biceps femoris, Trotter and Purslow (1992) show that the tensile properties of the endomysium are unsuitable for force transmission, and point out that the passive elasticity measurements of Natori (1954), Podolsky (1964) and Magid and Law (1985) on single muscle fibers by their very design could only measure the in-plane tensile properties of the endomysium. They cite work of Street (1983) where small discontinuous "splints" of myofibrils attached to the surface of the endomysium of single muscle fibers after dissection were seen to passively follow the length changes in the neighboring intact cell. This implies that the endomysium provided a shear linkage of deformations from the intact fiber to the "splints". From a qualitative discussion of fibrous composites theory, Trotter and Purslow (1992) infer that although the in-plane tensile properties of the endomysium will vary greatly with muscle length due to collagen fibril reorientation, the through-thickness shear properties of the network were more likely to be nearly constant at all physiologically relevant muscle lengths. The notion of translaminar shear transmission of force was reiterated in their study of the endomysium in the (series-fibered) bovine sternomandibularis muscle (Purslow and Trotter, 1994). Purslow (2002) derives the relationships for the two through-thickness shear moduli (along and across the muscle fiber direction) and how these change as a function of θ due to reorientation in the endomysial collagen network. From this detailed composite theory treatment, it is evident that the two translaminar shear moduli are, for practical purposes, independent of the orientation of the collagen fibrils in the plane of the endomysium.

A requirement of any linkage that efficiently transmits force from muscle fibers to tendinous attachments is that it is non-compliant. Especially in isometric muscle contractions, any significant stretching in the length of the fascicle due to stretchy connections would result in a very poor transmission of contractile force. Trotter et al. (1995) point out that even if the shear modulus of the endomysium is low (so that local shear strains generated could be large), the effective longitudinal compliance in an endomysium 2 μm thick extending along the length of a muscle fiber for several millimeters to centimeters is insignificant. Purslow (2002) showed that the displacements along the long axis of the muscle due to translaminar shearing of the endomysium could be represented as a series compliance with stiffness E_{app} given by

$$E_{app} = G(L/H)^2,$$

where G is the shear modulus of the endomysium, H is its thickness and L the muscle fiber length. For fibers even as short as 1 cm, L/H is in the order of 2000, so that E_{app} is going to be in the order of 4×10^6 greater than the true translaminar shear modulus of the endomysium. In a "composite" of two parallel muscle cells with the endomysium sandwiched between them, the longitudinal stiffness of the composite (E_c) is given by

$$\frac{1}{E_c} = \frac{1}{E_f} + \frac{1}{E_{app}},$$

where E_f is the longitudinal stiffness of the muscle fibers. Due to the very high value of E_{app}, the longitudinal stiffness of the muscle fibers is going to dictate the longitudinal stillness of the entire assembly (E_c). The functional significance of this is that the endomysium provides a shear linkage of force from one muscle cell to its neighbors, which is efficient.

As Trotter et al. (1995) point out, the structure (and therefore presumably the mechanics) of endomysium separating and joining adjacent muscle fibers in series-fibered muscles shows no obvious differences from continuous fibered muscles. The conclusion is that load sharing between adjacent muscle fibers is a common function in all muscles. Lateral load sharing through the endomysium is actually an important concept, which explains why it is possible for muscles to grow and to repair damaged sarcomeres. Lateral load sharing means that the longitudinal connectivity along a fiber can be interrupted for the addition of new sarcomeres necessary for muscle lengthening during growth, without loss of function of an entire contractile column. Similarly, if the muscle fiber is damaged at any point along its length, the contractile capacity of the undamaged parts is not lost, and the weakness of a sarcomere with which damaged myofibrils are being broken down and remodeled does not lead to stretching or tearing of the fiber at this point, as lateral connections between the fibers serve to keep the strains uniform throughout the tissue. Most voluntary muscle contraction is sub-maximal, i.e., not all the motor units in the muscle are recruited. This means that many non-contracting fibers are usually adjacent to contracting fibers. However, lateral connections through the endomysium between adjacent mean that sarcomere lengths in non-contacting fibers keep register with contracting ones (as shown by the observations of Street, 1983), keeping sarcomere length uniform in the muscle.

Willems and Purslow (1997) performed extensions on either single fibers or small bundles of muscle fibers from porcine longissimus muscle in the postmortem rigor state. In agreement with other studies (Mutungi et al., 1995), single rigor fibers developed inhomogeneities as they extended, with small zones of sarcomeres becoming hyper-extended and eventually rupturing. In contrast, fibers within small bundles were far more uniform in sarcomere length and could be drawn out to far greater overall extensions before eventually rupturing. Again, this study shows that coordination of deformations between adjacent muscle fibers is an important function of endomysium. Linkages from the contractile apparatus (myofibrils) laterally via intermediate filaments and costameres attached to the internal surface of the sarcolemma connect through transmembrane molecules (such as dystrophin and integrins) to molecules such as laminin in the ECM (Bloch et al., 2002; Grounds et al., 2005)). As well as acting as sites of force transmission, these cell–matrix interactions are also of tremendous importance in mechanotransduction and cell signaling, processes that modulate muscle cell expression to adapt the tissue to the mechanical loads and demands placed upon it. The extensive topic of muscle cell–matrix interactions and mechanotransduction is a separate area beyond the scope of the present review, and the reader is referred to reviews of Kjær (2004), Bishop and Lindahl (1999) and Purslow (2002).

12.8.2 Myofascial Force Transmission

There is now a substantial and consistent body of research on the ability of IMCT to transmit the forces produced within muscle. In addition to considerations about load sharing between muscle fibers within a fascicle via endomysial shear, there is considerable experimental evidence to support the notion that contractile force can be transmitted between different fascicles and entire sections of a muscle by IMCT structures, principally the perimysium. The greatest body of evidence comes from the detailed studies of Huijing and coworkers. They show that tension developed in one "head" of multiple-headed muscles can be transmitted via shear to the adjacent "head" (Huijing et al., 1998), and that cutting of the aponeurosis in a pennate muscle does not preclude tension generation further along toward the tendon (Jaspers et al., 1999). Huijing and Baan (2001) show evidence for transmission of forces from one muscle to another within three muscles that normally act as a synergistic group in vivo, which they argue must be by fascia structures linking them (extramuscular myofascial force transmission).

Huijing and Jaspers (2005) provide a comprehensive review of studies in this area to which the reader is referred. These authors point out the considerable evidence for force transmission between regions within a muscle, between various natural divisions of muscle with multiple heads (i.e., separate tendinous attachment points) and between synergistic or agonistic muscles. The long-range transmission of forces between separate parts of a muscle or even between adjacent muscles must not only involve load sharing between fibers within a fascicle by shear through the endomysium but also involve transmission from fascicle to fascicle via the perimysium and from muscle to muscle via the epimysium and fascia. For myofascial force transmission to work between fascicles, the same mechanisms of shear through the thickness of the perimysium must occur as for the endomysium. Just like the endomysium, the wavy cross-ply of collagen fibers in the perimysium can easily reorientate and allow shape changes in a fascicle as it changes length. The negligible in-plane tensile modulus of this network at physiological muscle lengths shows that there will be little tension generated as a result of these accommodations. However, as can clearly be seen from the left panel of Fig. 12.3, the thickness of the perimysium is thicker by at least one order of magnitude than the endomysium. This means that the apparent longitudinal modulus (E_{app}) for perimysium, analogous to the analysis shown above for endomysium, will be at least 2 orders of magnitude less, if the value of the local through-thickness shear modulus (G) is comparable. Transmission of force through the thickness of the perimysium would therefore be expected to be less efficient.

12.8.3 Perimysium: Coordination of Shape Change on Muscle Contraction

In his comprehensive text on muscle structure and function, Schmalbruch (1985) cites a model by Feneis (1935), which proposes that perimysia provide "neutral" (i.e., non-extended) connections between muscle fascicles, which allows the muscle fascicles to slide past each other, for example as the geometry of a pennate muscle changes upon contraction. Purslow (1999) pointed out that all fan-shaped, fusiform and pennate muscles change shape when contracting, and this will necessitate the slippage of some elements past each other (i.e., shear deformations) to accommodate this. In fact, only a uniform cylindrical or thin strap-like muscle could be expected to show no slippage of longitudinally arranged elements past each other as it shortens. Purslow (1999) went on to propose that the division of muscles into fascicles by the perimysial network is related to the need to accommodate shear displacement (as Feneis (1935) originally suggested) and that the very obvious differences evident from Fig. 12.4 in fascicular size, shape and perimysial thickness between functional different muscles are related to the different amounts of shear displacements that each muscle has to accommodate.

   This proposed function for the perimysial tract as a slip plane specifically supposes that, whereas muscle fibers within a fascicle are bound together quite tightly in shear by the endomysium, the boundary between fascicles, defined by the perimysium, is altogether more compliant in shear, so that the majority of shear displacements in the contracting muscle would be concentrated here. Purslow (1999) demonstrates that shape changes imposed on a piece of muscle in the rigor state result in large shear deformations between fascicles, but relatively little shear displacements within a fascicle. The argument is developed more formally by Purslow (2002), specifically by considering the shear in pinnate muscles upon contraction. For an increase in pennation angle upon contraction from its initial value (θ_i) to its maximum value (θ_m), the shear strain γ is given by

$$\gamma = \frac{1}{\tan \theta_i} - \frac{1}{\tan \theta_m}.$$

   As Fig. 12.8 shows, this is a non-obvious sort of function: the amount of shear strain for a given change of angle depends very much on the initial starting position.

Fig. 12.8 Shear strain for a 5° increase in pennation angle shown as a function of final pennation angle. Note the highly non-linear distribution. A change of pennation from 10° (relaxed) to 15° (contracted) yields a shear strain of 1.939, whereas the same 5° change in angle, if occurring between 20° and 25° yields a shear strain of 0.985 (almost half as much)

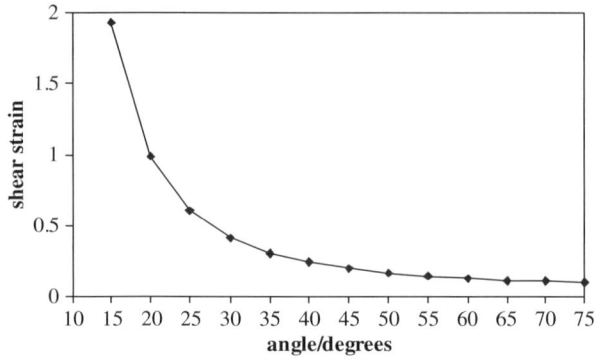

There is an extensive range of data in the literature on the change of pennation angle during in vivo contraction, mainly involving human subjects. Purslow (2002) lists seven such studies where initial and maximal pennation angles have been measured, mainly by non-invasive techniques such as MRI imaging. The change in pennation angle and the calculated shear strains resulting from this are highest in the gastrocnemius muscle, and slightly less in the vastus, quadriceps and tibilais anterior muscles. Predicted shear strains range form 0.93 to 2.62. The results from this theoretical model predict the shear displacements between fascicles in functionally different pinnate muscles to be (a) substantial and (b) variable form muscle to muscle.

This theory has the attraction that it seems to offer an explanation of why the amount and distribution of perimysium changes so very markedly from muscle to muscle, from thin perimysia surrounding small fascicles in long cylindrical muscles like psoas major and strap-like muscles, such as sternomandibularis, to much thicker perimysia separating bigger fascicles in fan-shaped muscles such as pectoralis. However, no detailed quantitative assessment of any possible relationship has yet been carried out, and so this theory remains just an interesting possibility.

12.9 Connective Tissue Networks Within Cardiac Muscle

12.9.1 The Structure of the Cardiac Wall

The heart is an organ with its own special architecture. It is surrounded by a connective tissue layer (the pericardium). On the external surface of the heart, there is an external connective tissue layer (the epicardium) and also an internal lining (the endocardium). The ventricular wall between the epi- and endocardium is termed the myocardium, which is a composite of cardiac muscle cells, each surrounded and separated by endomysium, and arranged in muscle fiber tracts by perimysium. However, in contrast to most skeletal muscles, the architecture of the myocardium is difficult to analyze, and subject to controversy (Anderson et al., 2006). A complicating issue is the branching nature of cardiac muscle cells. These relatively short cells are longitudinally joined together by intercalating discs, a mechanical interlocking junction, and also branch along their length, forming a 3-D network. Despite this, there is a local preferred muscle fiber direction in the myocardium, so that a "grain direction" can be discerned. There is a general pattern to this cardiac muscle fiber orientation; in the outer wall of the heart, the overall direction is roughly $+35°$ to the circumferential direction of the heart or $55°$ to the longitudinal direction (base–apex direction). In the inner wall of the heart, the overall direction is $-35°$ (or $-55°$ to the longitudinal direction). However, with respect to the transverse plane through the thickness of the heart wall, there is also a tendency for overall cardiac muscle fiber direction to show a tangential "spiraling" through the wall thickness from outer to inner surfaces. These overall trajectories are schematically represented in Fig. 12.9 (Bovendeerd et al., 1994). Another analysis by Harrington et al. (2005) broadly confirms this simple scheme, but shows the twisting of the fiber direction through the thickness to be a little more complicated (Harrington et al., 2005; Fig. 12.5). However, there is a great deal of variation in the local fiber orientation (in part due to the complications of cardiac muscle cell branching), and in reality, the picture is far more confused than the schematic diagram in Fig. 12.7 suggests (Dorri et al., 2007). Recent papers (Anderson et al., 2006; Lunkenheimer et al., 2006) make the case for a myocardium made up of a continuous network of cardiac myofibers, embedded within a continuous connective tissue matrix, and take issue with a previous interpretation (Torrent-Guasp et al., 2001, 2005) that there is a spiral construction of the heart, the so-called "myocardial band". Although Torrent-Guasp and colleagues offer the opinion that the heart can be peeled apart in a spiral unwinding to yield a linear strip of myocardium running from the pulmonary artery to the aorta, there seems to be no anatomical structures, and no ECM boundaries, that clearly define this "spiral construction", and no obvious consistency in cardiac muscle fiber orientation in the segments of myocardium separated in this spiral unwinding (Lunkenheimer et al., 2006).

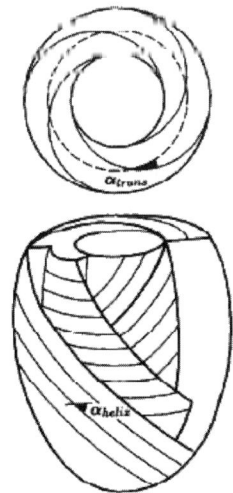

Fig. 12.9 Schematic of muscle fiber orientation, the myocardium; the *bottom panel* shows the helix angle α_{helix}, which is negative at the indicated epicardial site; the *upper panel* shows the transverse angle α_{tran}, in a top view of the basal plane. From Bovendeerd et al. (1994), with permission

12.9.2 Structure and Arrangement of ECM Within the Myocardium

Independent of the controversy in the question of cardiac myocyte orientation that exists, some details of the structure of both the endomysial and perimysial IMCT components of the myocardium are also open to some discussion, although there is a good consensus on the major features.

12.9.2.1 (Myocardial) Perimysium

Robinson et al. (1988) describe perimysium in the papillary muscle of rat heart as a branched, continuous network of collagen fibers that divide groups of cardiomyocytes into bundles or layers of myocytes typically about four cells thick. They describe the perimysial layers as weaves of collagen fibers of 1–10 μm in diameter, connected to each other by larger collagen fibers. The collagen fibers are not straight, but show a periodic waviness. Robinson et al. (1988) specifically describe the perimysial fibers as highly coiled structures (i.e., forming long helices), as opposed to forming only 2-D waveforms (i.e., a planar crimped or sine-wave structure). Hanley et al. (1999) used 3-D reconstructions from confocal microscopy studies on the ventricle of rat heart to re-examine this specific question and, conversely, showed that perimysial collagen fibers form 2-D planar wavy structures.

12.9.2.2 (Myocardial) Endomysium

Borg and Caulfield (1981) viewed the connections between cardiac myocytes in the hamster and rat heart by SEM. Their micrographs show networks of collagen fibrils that surround, separate and interconnect parallel muscle cells, which they term "weave networks". The quasi-random continuous network of fine collagen fibrils in their micrographs shows strong parallels with the endomysial structures shown in skeletal muscle by SEM (Rowe, 1981; Trotter and Purslow, 1992; Purslow and Trotter, 1994). In their description of the micrographs, however, they use the term "collagen struts" to describe bundles of collagen fibers connecting adjacent myocytes, and in a schematic diagram (Borg and Caulfield, 1981; Fig. 12.7), they depict these "collagen struts" as discrete thick twisted rope-like structures running perpendicular to the surface of the myocytes at intervals along their length and running straight across the space that separates the myocytes. This schematic interpretation is at odds with the SEM micrographs that show, as in striated muscle, a continuous network of fine fibrils that lie in the plane of the network, parallel to the surface of the muscle cells. Borg and Caulfield's description of connecting "collagen struts" has stuck in the literature. Intrigila et al. (2007) show two schematic diagrams of the connections between adjacent cardiac muscle cells. The first has a small number of thick, twisted collagen bundles running perpendicular to the cell surfaces across the space between them (the "strut" model) and the second shows a continuous network of random collagen fibrils lying parallel to the surface of each muscle cell. They state "to the best of our knowledge it is still a matter of debate which of the two models of the endomysial collagen microarrangement is to be preferred".

Fig. 12.10 SEM of
NaOH-digested rabbit
ventricle.
ml = myocyte lacuna,
cl = capillary lacuna. These
lacunae are left when the
structures are digested away
by the NaOH. From
Macchiarelli and
Ohtani (2001), with
permission

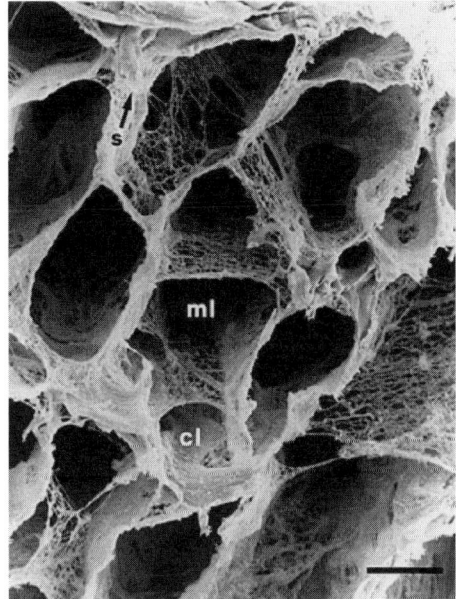

Figure 12.10 is an SEM image of the ECM of the left ventricle of a rabbit heart after NaOH digestion (Macchiarelli and Ohtani, 2001). It is similar to the SEM images of Icardo and Colvee (1998), which is a NaOH-digested papillary muscle. The space labeled "ml" marks the position of a myocyte before its removal by the NaOH digestion, and similarly "cl" denotes the lacuna left behind by the digestion of a capillary. The authors describe a continuous network of collagen fibrils that overlie and enwrap each myocytes to form connections between them. In a later publication showing similar SEM micrographs following NaOH digestion of ventricular wall, Macchiarelli et al. (2002) describe again this laminar network of collagen fibrils, which runs continuously along the length of myocytes, and say that this view of the connections between myocytes is "emphasised, rather than a type of 'strut connection' anchored to defined loci, as usually described".

Not withstanding the heritage of the "strut" interpretation that seems to have colored the description of endomysial linkages between cardiac muscle cells, a reasonable comparison of Figs. 12.3 and 12.10 shows the remarkable similarity in endomysial architecture between striated and cardiac muscles. Fine, wavy collagen fibrils form a planar network in which the long axis of the fibrils is parallel to the cell surface, at a variety of orientations to the long axis of the muscle cells. These appear to be no bundles of collagen fibrils twisted into skeins or struts. There are no collagen fiber bundles running out of the plane of the network, at right angles to the cell surface. So although Intrigila et al. (2007) describe a debate within the cardiac field about the existence or otherwise of "struts", examination of the SEM data shows an arrangement in cardiac endomysium that is perfectly consistent with the continuous laminar networks seen and described in studies on striated muscle endomysium.

The perimysium and the endomysium of cardiac muscle contain both type I and type III fibrous forms of collagen. de Souza (2002) reviews evidence suggesting that type I is approximately four times more prevalent than type III in the heart. The amounts of myocardial collagen, and especially type I collagen, increase with age (Debessa et al., 2001; de Souza, 2002).

12.10 Mechanical Roles for ECM in the Myocardium

The mechanical properties and patterns of deformations in the wall of the ventricle during the cardiac cycle are both important and complex. A systematic review of studies of ventricular mechanics is beyond the scope of this present chapter, and we shall focus instead on a select few publications that serve to illuminate the role and properties of cardiac ICMT.

Costa et al. (1999) measured the 3-D patterns of strain in the canine heart between the end of diastole (low pressure) and the end of systole (high pressure) in the cardiac cycle. This detailed study reveals the following major features of changes in the ventricular wall during contraction of the heart:

(a) Myocytes are organized into branching laminar sheets, approximately four cells thick. The endomysial coupling between them appears tight, whereas the perimysial coupling between the sheets appears "looser".

(b) When the heart contracts to produce maximal (systolic) blood pressure, the circumference of the heart decreases and it shortens along its length (base to apex axis). At the base of the heart, the negative strains (contractions) in these two directions are about equal, and at the apex of the ventricle circumferential shortening exceeds longitudinal. Both circumferential and longitudinal strains get progressively greater moving from the outside (epicardial) surface of the wall through its thickness to the endocardial suface (the radial direction). Although longitudinal and circumferential strains are negative, the thickness of the wall greatly increases (radial strain is large and positive). Radial strain also increases with depth in the wall from epicardial to endocardial surfaces. This stain gradient is greater at the apex than the base of the ventricle.

(c) These changes in length, circumference and thickness of the ventricular wall also involve patterns of shear strain. Torsional (longidudinal–cicumferental plane) shear is uniform from base to apex and also through the wall thickness. Circumferential–radial shear strains change sign (i.e., direction) from base to apex but is uniform across the wall depth, and the longitudinal–radial shear strains change sign through the wall thickness.

(d) The sheets of myocytes are aligned (see schematic diagram of directions in Fig. 12.9) such that contraction along their length produces the patterns of the strains seen. Comparing the overall patterns of extensional and shear strains described in (b) and (c) above to local muscle fiber orientation, Costa et al. (1999) found that there is a uniform shortening of 5–8% along the muscle fiber axis. There is a substantial (10–20%) extension in the myocyte sheets in the direction transverse to the fiber direction but parallel to the ventricular wall, and this positive strain increases with depth through the wall from epicardial to endocardial surfaces. There is also thinning of the myocyte sheets in the radial direction. Within the myocyte sheets, this means that there are large shear strains in the plane cross-sectional to the fiber direction, and these change in sign (direction) from base to apex, but increase with wall depth.

(e) Putting all of (a)–(d) together, it is obvious that in order to have a big increase in the thickness of the ventricular wall while the myocyte sheets locally get thinner in the radial direction, there has to be a reorientation of the myocyte sheets and substantial sliding of sheets past each other.

These changes in the local myocyte sheets on contraction are shown schematically in Fig. 12.11 (this is a much simplified and highly exaggerated version of the schematic depiction by Costa et al. (1999) in their Fig. 12.10).

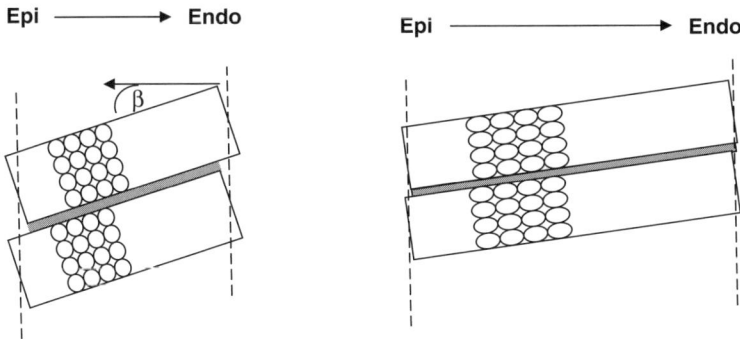

Epi ⟶ Endo   Epi ⟶ Endo

Fig. 12.11 Schematic diagram of a small part of the apical ventricular wall at approximately 60% of the depth from the epicardial surface showing two sheets of myocytes, approximately four cells thick. Arrangement at diastole (**A**) and at systole (**B**), upon contraction of the fibers. Individual myocytes are shown in cross-section (*circles*). Upon contraction, they shorten in the direction perpendicular to the plane of the diagram. The local orientation of the sheet has a negative angle β with respect to the through-wall axis (*arrow*). Distance along the endo- to epicardial direction (wall thickness) increases due to rotation of the sheets; the sheets widen transverse to the fiber direction and get thinner, to lie at a less negative value of β. Examination of the shaded area (occupied by perimysium) shows that there is considerable shear strains engendered in these "slip planes" between the myocyte sheets. The contracted myocytes in **B** have a greater individual cross-sectional area. For sheets to thin on contraction, one possibility is that myocytes become elliptical (shown here). Another possibility is that the myocytes remain circular in cross-section but rearrange so that the cell count through the thickness of the sheet decreases (not shown here). Redrawn in exaggerated and simplified form from Costa et al. (1999)

The pattern of strains during the cardiac cycle observed by Costa et al. (1999) in canine hearts was also found by Cheng et al. (2005) in sheep hearts. Arts et al. (2001)

refined techniques used to measure β and showed that there are two broad populations of orientations for myocyte sheets lying approximately 90° apart for most of
the wall thickness, with one population at approximately $\beta = 45°$ and another at
135°. Arts et al. (2001) again make the interpretation that "Layers of perimysial
connective tissue, separating the myofiber sheets, could be effective is permitting
large shear strains without large shear stresses". The reference to lack of large
shear stresses supposes that the perimysial connections are quite compliant. Dokos
et al. (2002) measure shear stress – deformation curves for porcine ventricular wall.
In all directions tested, shear properties are highly non-linear, with high stiffness
at larger displacements, but low stiffness at lower deformations near rest length.
The initial shear deformation therefore engenders only small loads. Shear properties
were anisotropic, with greatest resistance to shear deformations causing extension
of the myocytes, and least resistance to shear displacements involving slippage
between myocyte sheets. These findings were confirmed by a further study by
Schmid et al. (2006) using an analysis based on Costa's work to model shear force
deformation curves for the ventricular wall from the data of Dokos et al. (2002).
The findings of Dokos et al. (2002) confirm that cardiac perimysium provides slip
planes in the working ventricular wall.

The parallels in interpretation of function of endomysial and perimysial IMCT in
cardiac muscle and the roles described above for these IMCT structures in striated
muscle are very apparent from this. In this model, endomysium provides tight shear
linkage between adjacent cardiac myocytes, keeping them in register within the
myocyte sheets, while the perimysium defines "slip planes", regions of high shear
compliance that allow the necessary rearrangements as the whole tissue contracts.
Relating the values of the angle β to the analysis for shear strain caused by a change
in "pennation angle" given in Fig. 12.8 and the accompanying text, it can be seen
that the actual shear strain for a small reorientation of the myocyte sheet can be
large, depending on initial orientation. For an initial value of $\beta = 45°$, a shear
strain of about 19% would be predictedfor a 5° reduction of β due to myocyte sheet
reorientation on ventricular contraction. Estimates of the change in β during the
cardiac cycle would permit a prediction of the amount of shear displacements seen
in the cardiac perimysium, but this is technically difficult to measure accurately in
vivo. However, the principle is clear that, like striated muscle, cardiac muscle must
change shape and so accommodate shear displacements within it during contraction.
Perimysial boundary layers seem to provide this function. We can conclude that
comparisons between skeletal muscle and cardiac muscle in terms of structure and
function may illuminate commonalities of the form and function of IMCT in both
cases.

The ECM of the heart, like the ECM of skeletal muscle, is not a static structure,
but is constantly remodeled as an adaptation to mechanical loads that the muscle
experiences. The review of Bishop and Lindahl (1999) listed in Table 12.1 provides a good starting point for understanding the principles of mechanotransduction
in cardiac muscle due to changes in blood pressure (hypertension) and resulting
changes in the expression of collagen. Brower et al. (2006) also review this. One
special consideration for remodeling in the heart is the response to heart failure, a

condition of high medical relevance. It is well known that the kind of damage done to cardiac muscle by a significant and chronic hypertension (ventricular hypertrophy) and myocardial infarction causes a large deposition of collagen as a wound repair response (fibrosis). Deposition of collagen essentially as scar tissue reduces myocardial compliance and alters ventricular mechanics. There has been considerable interest in manipulating the degradation of cardiac fibrosis using MMPs. Matrix metalloproteinases (MMPs) are the major group of proteases normally expressed by fibroblasts in ECM and by muscle cells to degrade collagen and other ECM components during tissue remodeling, turnover and growth. MMP-1, -2, -9, -14 and -16, together with tissue inhibitors of matrix metalloprotease (TIMP)-1, -2 and -3, are known to be expressed by both fibroblasts and myoblasts in skeletal muscle (Balcerzak et al., 2001). The possibilities and problems in using MMPs as an intervention strategy for heart failure are the subjects of a huge amount of work that is well beyond the scope of this review. There is also clinical interest in using MMP inhibitors as a therapeutic treatment for pathologies where there is abnormal degradation of collagen in the cardiovascular system. The reader is referred to the following reviews in addition to Bishop and Lindahl (1999) as good starting points for this area: Kieseier et al. (2001), Heeneman et al. (2003), Deschamps and Spinale (2005), Holmes et al. (2005), Miner and Miller (2006), Brower et al. (2006) and Peterson (2006).

12.11 Heart Valves are Special ECM Structures

There are four valves in the mammalian heart. Two are atrioventricular, closing when the ventricles contract to prevent backflow into the atria. These are anchored into the ventricles by chordae tendinae to prevent inversion. The papillary muscle is attached to the chordae tendinae to help tension the anchoring chordae. The atrioventricular valve between left ventricle and left atrium is the mitral valve, so called because it has two leaflets or cusps, like a Bishop's Miter (hat). The atrioventricular valve between right ventricle and atrium has three cusps and is called the tricuspid valve. The other two valves of the heart are the semilunar valves. They lie in the two major arteries exiting each ventricle (the pulmonary artery exiting the right ventricle and the aorta exiting the left ventricle), to prevent backflow of blood from arteries into the ventricles as they re-fill. Semilunar valves are similar to the venous valves (comprised of two simple flaps) and do not have chordae tendinae to anchor them.

Failure of the leaflets of the mitral valve in humans is a serious condition, which may be addressed by replacement of the valve. In order to construct artificial or tissue-engineered heart valves and valve leaflets, knowledge of the structure and mechanical properties of the normal valves is an important area of study. A great deal of work in this area has been performed by Michael Sacks and colleagues.

Human mitral valves contain types I and III collagen, PGs and elastin. Collagen accounts for 57% of the dry weight in mitral cusps, and 29% of this is type III collagen. Elastin comprises 10% of the dry weight of the cusps. Chordae contain slightly more collagen and slightly less elastin than the cusps, and the type III content is slightly lower (Lis et al., 1987). Floppy or ballooning cusps removed in surgery were found to have increased collagen content, due to secondary surface fibrosis, and an elevated PG content in the cusps. In rheumatic valves, the water content of the cusps was greatly reduced (Lis et al., 1987). Bigi et al. (1982) reported that collagen fibers are aligned along the long axis of the chordae tendinae but form a more 3-D network in the cusps. The cusps have three layers: the spongiosa, fibrosa and ventricularis layers, of which the fibrosa layer is the thickest, accounting for just over two thirds of the total cusp thickness. The base of the cusps forms an annular structure; the direction parallel to the base of this annulus is defined as the circumferential direction and, within the plane of the leaflets, the direction moving out to the tips of the cusps is defined as the radial direction. Sacks et al. (1998) studied the orientation of collagen fibers in the cusps at a range of transvalvular pressures. They report a circumferential orientation of collagen in the thick fibrosa layer and a more random orientation in the ventricularis at low transvalvular pressures, but a reorientation at higher pressures so that both layers were more circumferentially orientated. Reorientation of collagen fibers in the cusp as a function of transvalvular pressure is expected to change the mechanical properties of the valve between diastolic and systolic pressures.

He et al. (2005) studied the complex pattern of stains in vivo in the posterior leaflet of porcine mitral valves. As the valve closes and transvalvular pressure builds, there is a rapid increase in strain in the plane of the leaflet. The strains are highly anisotropic, with much higher strains in the radial direction than circumferentially. This is consistent with observed collagen fiber orientations; the major orientation of collagen is in the circumferential direction, and so strains are less in this direction than at 90° to the fibers, in the radial direction. Strains in both directions ceased to change at the end of valve closing. The change in area of the valve leaflet (areal strain) shows a highly non-linear relation to transvalvular pressure, with a compliant "toe region" and a much higher stiffness at higher pressures.

Liao et al. (2007) used simultaneous in-plane biaxial loading and small-angle X-ray diffraction to study how changes in collagen orientation in mitral values relate to mechanical properties. Liao et al. (2007) studied the orientation of collagen and strain within collagen fibers under a variety of biaxial loading conditions, including stress–relaxation and creep. Very little creep or stress–relaxation behavior was in fact observed in the tissue, which is unusual for a soft connective tissue. Strain within the collagen fibers themselves is small compared to the macroscopic strains in the tissue.

The anisotropic and non-linear properties of the normal heart valve leaflets make the development of a viable tissue-engineered bioprosthetic heart valve challenging, but increasing knowledge of structure–function relationships in native leaflets is presenting realistic targets to emulate (Sun et al. 2003; Merryman et al. 2006).

12.12 Conclusions

Intramuscular ECM is an integral part of the functioning of both skeletal and cardiac muscle tissues. Load transmission between tightly linked adjacent muscle fibers within bundles allows for coordination of forces, protection of damaged areas of fibers against over-extension and, in series-fibered muscle at the very least, is a major pathway for the transmission of contractile force. Perimysial tracts defining the boundaries between muscle fascicles of myocyte sheets seem to have a role in allowing the whole tissue accommodate large shear displacements. Although this may seem paradoxical, substantial evidence supports the notion that perimysium and epimysium can also act as pathways for myofascial force transmission. Exciting discoveries lay ahead in defining the exact parameters that make the composition and architecture of IMCT in skeletal muscles so hugely variable. Although not reviewed here, these IMCT structures are fine-tuned in a dynamic balance between synthesis and remodeling so as to be continually adapted for the changing mechanical environment both generated by muscle as a force-producing tissue and by mechanical loads imposed on them from external sources. Understanding the control of growth and turnover in IMCT will in future ultimately provide benefits in control of cardiac myopathies, muscle injury and repair, and combating age-related degradations in function.

References

Anderson MJ, Klier FG, Tanguay KE (1984) Acetylcholine receptor aggregation parallels the deposition of a basal lamina proteoglycan during development of the neuromuscular junction. J. Cell Biol. 99: 1769–1784.

Anderson RH, Ho SY, Redmann K, Sanchez-Quintana D, Lunkenheimer PP (2005) The anatomical arrangement of the myocardial cells making up the ventricular mass. Eur. J. Cario-thoracic Surg. 28: 517–525.

Anderson RH, Ho SY, Sanchez-Quintana D, Redmann K, Lunkenheimer PP (2006) Heuristic problems in defining three-dimensional arrangement of the ventricular myocytes. Anat. Rec. 288A: 579–586.

Arts T, Costa KD, Covell JW, McCulloch AD (2001) Relating myocardial laminar architecture to shear strain and muscle fiber orientation. Am. J. Physiol. Heart Circ. Physiol. 280: H2222–H2229.

Balcerzak D, Querengesser L, Dixon WT, Baracos VE (2001) Coordinate expresión of matrix-degrading proteinases and their activators and inhibitors in bovine skeletal muscle. J. Anim. Sci. 79: 94–107.

Bendall JR (1967) The elastin content of various muscles of beef animals. J.Sci. Food Agric. 18: 553–558.

Bigi A, RipamontiI A, Roveri N, Compostella L, Roncon L, Schivazappa L (1982) Structure and orientation of collagen-fibers in human mitral-valve. Int. J. Biol. Macromol. 4: 387–392.

Bishop JE, Lindahl G (1999) Regulation of cardiovascular collagen synthesis by mechanical load. Cardiovasc. Res. 42: 27–44.

Bloch RJ, Capetanaki Y, O'Neill A, Reed P, Williams MW, Resneck WG, Porter NC, Ursitti JA (2002) Costameres: repeating structures at the sarcolemma of skeletal muscle. Clin. Orthoped. Rel. Res. 403S: S203–S210.

Borg TK, Caulfield JB (1980) Morphology of connective tissue in skeletal muscle. Tissue Cell 12:197–207.

Borg TK, Caulfield JB (1981) The collagen matrix of the heart. Federation Proc. 40: 2037–2041.

Bourne GH (1973) The structure and function of muscle Vol. II Structure part 2. Academic Press, NY.

Bovendeerd PHM, Huyghe JM, Arts T, Van Campen DH, Reneman RS (1994) Influence of endocardial–epicardial crossover of muscle fibers on left ventricular wall mechanics. J. Biomech. 27:941–951.

Bowman W (1840) On the minute structure and movements of voluntary muscle. Phil. Trans. Roy. Soc. Lond. 130: 457–501.

Brooks JC, Savell JW (2004) Perimysium thickness as an indicator of beef tenderness. Meat Sci. 67: 329–334.

Brower GL, Gardner, JD, Forman MF, Murray DB, Voloshenyuk T, Levick SP, Janicki JS (2006) The relationship between myocardial extracellular matrix remodelling and ventricular function. Eur. J. Cardio-thoracic Surg. 30: 604–610.

Carrino DA, Caplan AI (1982) Isolation and preliminary characterization of proteoglycans synthesized by skeletal muscle. J. Biol. Chem. 257: 14145–14154.

Cheng A, Langer F, Rodriguez F, Criscione JC, Daughters GT, Miller DC, Ingels NB. (2005) Transmural sheet strains in the lateral wall of the ovine left ventricle. Am. J. Physiol. Heart Circ. Physiol. 289: H1234–H1241.

Costa KD, Takayama Y, McCulloch AD, Covell JW (1999). Laminar fiber architecture and three-dimensional systolic mechanics in canine ventricular myocardium. Am. J. Physiol. Heart Circ. Physiol. 276:H595–H607.

Debessa CRG, Maifrino LBM, de Souza RR (2001) Age related changes in the collagen network of the human heart. Mech. Ageing Dev. 122: 1049–1058.

Deschamps AM, Spinale FG (2005) Matrix modulation and heart failure: new concepts question old beliefs. Curr. Opinion Cardiol. 20: 211–216.

de Souza RR (2002) Ageing of myocardial collagen. Biogerontology 3: 325–335.

Dokos S, Smaill BH, Young AA, LaGrice IJ (2002) Shear properties of passive ventricular myocardium. Am. J. Physiol. Heart Cierc. Physiol. 283: H2650–H2659.

Dorri F, Niederer PF, Redmann K, Lunkenheimer PP, Cryer CW, Anderson RH (2007) An analysis of the spatial arrangement of the myocardial aggregates making up the wall of the left ventricle. Eur. J. Cardio-thoracic Surg. 31: 430–437.

Eggen KH, Malmstrom A, Kolse, SO (1994) Decorin and a large dermatan sulfate proteoglycan in bovine striated muscle. Biochim. Biophys. Acta 1204: 287–297.

Feneis H (1935) Uber die Anordnung und die Bedentung des indegewebes fur die Mechanik der Skelettmuskulatur. Morph. Jb. 76: 161–202.

Gans C, Gaunt AS (1991) Muscle architecture in relation to function. J. Biomech. 24: 53–65.

Grounds MD, Sorokin L, White J (2005) Strength at the extracellular matrix – muscle interface. Scand. J. med. Sci. Sports 15: 381–391.

Hanley PJ, Young AA, leGrice IJ, Edgar SG, Loiselle DS (1999). 3-Dimensional configuration of perimysial collagen fibres in rat cardiac muscle at resting and extended sarcomere lengths. J. Physiol. 517: 831–837.

Harrington KB, Rodriguez F, Cheng A, Langer F, Ashikaga H, daughters GT, Criscione JC, Ingels, NB, Miller DC (2005) Direct measurement of transmural laminar architecture in the antero-lateral wall of the ovine left ventricle: new implications for wall thickening mechanics. Am. J. Physiol. Heart Circ. Physiol. 258: H1324–H1330.

He ZM, Ritchie J, Grashow JS, Sacks MS, Yoganathan AP (2005) In vitro dynamic strain behavior of the mitral valve posterior leaflet. J. Biomech. Eng. (Trans ASME) 127: 504–511.

Heeneman S, Cleutjens JP, Faber BC, Creemers EE, van Suylen R-J, Lutgens E, Cleutjens KB, Daemen MJ (2003) The dynamic extracellular matrix: intervention strategies during heart failure and atherosclerosis. J. Pathol. 200: 516–525.

Holmes JW, Borg TK, Covell JW (2005) Strucutre and mechanics of heraling myocardial infarcts. Ann. Rev. Biomed. Eng. 7: 223–253.

Huijing PA, Baan GC, Rebel G (1998) Non-myotendinousforce transmission in rat extensor digitorum longus muscle. J. Exp. Biol. 201, 682–691.

Huijing PA, Baan, GC (2001) Extramuscular myofascial force transmission within the rat anterior tibial compartment; proximo-distal differences in muscle force. Acta Physiol. Scand. 173: 297–311.

Huijing, PA, Jaspers, RT (2005) Adaptation of muscle size and myofascial force transmission: a review and some new experimental results. Scand. J. Med. Sci. Sports 15: 349–380.

Icardo JM, Colvee E (1998) Collagenous skeleton of the human mitral papillary muscle. Anat. Rec. 252: 509–518.

Intrigila B, Melatti I, Tofani A, Macchiarelli G (2007) Computational models of myocardial endomysial collagen arrangement. Computer Methods and Programs in Biomedicine 86: 232–244.

Jaspers RT, Brunner R, Pel JMM, Huijing PA (1999) Acute effects of intramuscular aponeurotomy on rat gastrocnemius medialis: force transmission, muscle force and sarcomere length. J. Biomech. 32, 71–79.

Jolley PD, Purslow PP (1988) Reformed meat products – fundamental concepts and new developments. In: Mitchell J, Blanshard JMV (Eds.) Food Structure – Its Creationand Evaluation. Butterworths, London 231–264.

Kieseier BC, Schneider C, Clements JM, Gearing AJH, Gold R, Tokya KV, Hartung H-P (2001) Expression of specific matrix metalloproteinases in inflammatory myopathies. Brain 124: 341–351.

Kjær, M (2004) Role of extracellular matrix in adaptation of tendon and skeletal muscle to mechanical loading. Physiol. Rev. 84: 649–698.

Kjær M, Magnusson P, Krogsgaard M, Moller JB, Olesen J, Heinemeier K, Hansen M, Haraldsson B, Koskinen S, Esmarck B, Langberg H (2006) Extracellular matrix adaptation of tendon and skeletal muscle to exercise. J. Anat. 208: 445–450.

Krenchel. H. (1964) Fibre Reinforcement. Akademisk Forlag, Copenhagen.

Lawson MA, Purslow PP (2001) Development of components of the extracellular matrix, basal lamina and sarcomere in chick quadriceps and pectoralis muscles. Br. Poult. Sci. 42: 315–320.

Lepetit J (1991) Theoretical strain ranges in raw meat. Meat Sci. 29: 271–283.

Lewis GJ, Purslow PP (1989) The strength and stiffness of perimysial connective-tissue isolated from cooked beef muscle. Meat Sci. 26: 255–269.

Lewis GJ, Purslow PP (1991) The effect of marination and cooking on the mechanical properties of intramuscular connective tissue. J. Muscle Foods 2:177–195.

Lewis GJ, Purslow PP, Rice AE (1991) The effect of conditioning on the strength of perimysial connective-tissue dissected from cooked meat. Meat Sci. 30: 1–12.

Liao J, Yang L, Grashow J, Sacks MS (2007) The relation between collagen fibril kinematics and mechanical properties in the mitral valve anterior leaflet J. Biomech. Eng. (Trans ASME) 129. 78–87.

Light ND (1987) The role of collagen in determining the texture of meat. In Pearson AM, Dutson TR, Bailet AJ (Eds.) Advances in Meat Research Vol. 4: Collagen as a Food. Van Nostrand Reinhold, NY 87–107.

Light N, Champion AE, Voyle C, Bailey AJ (1985) The role of epimysial, perimysial and endomysial collagen in determining texture in six bovine muscles. Meat Sci. 13: 137–149.

Lis Y, Burleigh MC, Parker DJ, Child AH, Hogg J, Davies MJ (1987) Biochemical-characterization of individual normal, floppy and rheumatic human mitral-valves. Biochem. J. 244: 597–603.

Listrat A, Picard B, Geay Y (1999) Age-related changes and location of type I, III, IV, V and VI collagens during development of four foetal skeletal muscles of double muscles and normal bovine muscles. Tissue Cell 31: 17–27.

Listrat A, Lethias C, Hocquette JF, Renand G, Menissier F, Geay Y, Picard B (2000). Age related changes and location of types I, III, XII and XIV collagen during development of skeletal muscles from genetically different animals. Histochem. J. 32: 349–356.

Lunkenheimer PP, Redmann K, Westermann P, Rothaus K, Cryer CW, Niederer P, Anderson RH (2006) The myocardium and its fibrous matrix working in concert as a spatially netted mesh: a critical review of the purported tertiary structure of the ventricular mass. Eur. J. Cardio-thoracic Surg. 295: 541–549.

Macchiarelli G, Ohtani O (2001). Endomysium in left ventricle. Heart 86: 416–416.

Macchiarelli G, Ohtani O, Nottola SA, Stallone T, Camboni A, Prado IM, Motta PM (2002) A micro-anatomical model of the distribution of myocardial endomysial collagen. Histol. Histopath. 17: 699–706.

Magid A, Law DJ (1985) Myofibrils bear most of the resting tension in frog skeletal muscle. Science 230: 1280–1282.

Mayne R, Sanderson RD (1985) The extracellular matrix of skeletal muscle. Collagen Relat. Res. 5: 449–468.

McCormick RJ (1994) The flexibility of the collagen compartment of muscle. Meat Sci. 36: 79–91.

Merryman WD, Engelmayr GC, Liao J, Sacks MS (2006) Defining biomechanical endpoints for tissue engineered heart valve leaflets from native leaflet properties. Progress Pediatric Cardiol. 21: 153–160.

Miner EC, Miller WL (2006) A look between the cardiomyocytes: the extracellular matrix in heart failure. Mayo Clin. Proc. 81: 71–76.

Mutungi G, Purslow P, Warkup C (1995) Structural and mechanical changes in raw and cooked single porcine muscle-fibers extended to fracture. Meat Sci. 40: 217–234.

Nakano T, Li X, Sunwoo HH, Sim JS (1997) Immunohistochemical localization of proteoglycans in bovine skeletal muscle and adipose connective tissues. Can. J. Anim. Sci. 77: 169–172.

Natori R. (1954) The role of myofibrils, sarcoplasma and sarcolemma in muscle contraction. Jikeikai Med. J. 1, 18–28.

Nishimura T, Hattori A, Takahashi K (1996) Arrangement and identification of proteoglycans in basement membrane and intramuscular connective tissue of bovine semitendinousus muscle. Acta Anat. 155: 257–265.

Passerieux E, Rossignol R, Chopard A, Carnino A, Marini JF, Letellier T, Delage JP (2006) Structural organization of the perimysium in bovine skeletal muscle: Junctional plates and associated intracellular subsomains. J. Struct. Biol. 154: 206–216.

Passerieux E, Rossignol R, Letellier T, Delage JP (2007) Physical continuity of the perimysium from myofibers to tendons: Involvement in lateral force transmission in skeletal muscle. J. Struct. Biol. 159: 19–28.

Peterson JT (2006) The importance of estimating the therapeutic index in the development of matrix metalloproteinase inhibitors. Cardiovasc. Res. 69: 677–687.

Podolsky RJ (1964) The maximum sarcomere length for contraction of isolated myofibrils. J. Physiol. 170, 110–123.

Purslow PP (1989) Strain-induced reorientation of an intramuscularconnective tissue network: Implications for passive muscle elasticity. J. Biomech. 22: 21–31.

Purslow PP (1999) The intramuscular connective tissue matrix and cell-matrix interactions in relation to meat toughness. Proceedings of the 45th International Congress Meat Science And Technology. Yokohama, Japan 210–219.

Purslow PP (2002) The structure and functional significance of variations in the connective tissue within muscle. Comp. Biochem. Physiol. Part A 133: 947–966.

Purslow PP (2005) Intramuscular connective tissue and its role in meat quality. Meat Sci. 70: 435–447.

Purslow PP, Duance VC (1990) The structure and function of intramuscular connective tissue. In Hukins DWL (Ed.) Connective Tissue Matrix Vol 2. MacMillan, London 127–166.

Purslow PP, Trotter JA, (1994) The morphology and mechanical properties of endomysium in series-fibred muscles; variations with muscle length. J. Muscle Res. Cell Motil. 15: 299–304.

Purslow PP, Wess TJ, Hukins DWL (1998). Collagen orientation and molecular spacing during creep and stress–relaxation in soft connective tissues. J. Exp. Biol. 201: 135–142.

Ramsay RW, Street SF (1940) The isometric length-tension diagram of isolated skeletal fibers of the frog. J. Cell Comp. Physiol. 15: 11–34.

Rhodes JM, Simons M (2007) The extracellular matrix and blood vessel formation: not just a scaffold. J. Cell. Molec. Med. 11: 176–205.

Robinson TF, Geraci MA, Sonnenblick EH, Factor SM (1988). Coiled perimysioal fibres of papillary muscle in rat heart: morphology, distribution and changes in configuration. Circ. Res. 63: 577–592.

Rowe RWD (1981) Morphology of perimysial and endomysial connective tissue in skeletal muscle. Tissue Cell 13: 681–690.

Sacks MS, Smith DB, Hiester ED (1998) The aortic valve microstructure: Effects of transvalvular pressure. J. Biomed. Mats. Res. 41: 131–141.

Sanes JR (2003) The basement membrane/basal lamina of skeletal muscle. J. Biol. Chem. 278: 12601–12604.

Sato K, Ohashi C, Muraki M, Itsuda H, Yokoyama Y, Kanamori M, Ohtsuki K, Kawabata M (1998) Isolation of intact type V collagen from fish intramuscular connective tissue. J. Food Biochem. 22: 213–225.

Schmalbruch H. (1985) Skeletal Muscle. Springer. Berlin.

Schmid H, Nash MP, Young AA, Hunter PJ (2006) Myocardial material parameter estimation – a comparative study for simple shear. J. Biomech. Eng. (Trans. ASME) 128: 742–750.

Schwartz SM (Ed) (1995). The Vascular Smooth Muscle Cell: Molecular and Biological Responses to the Extracellular Matrix. Academic Press, NY. ISBN-10: 0126323100.

Scott JE (1990) Proteoglycan: collagen interactions and subfibrillar structure in collagen fibrils. Implications in the development and ageing of connective tissues. J. Anat. 169: 23–35.

Street SF (1983) Lateral transmission of tension in frog myofibers: A myofibrillar network and transverse cytoskeletal connections are possible transmitters. J. Cell. Physiol. 114:346–364.

Stegemann JP, Hong H, Nerem RM (2005) Mechanical, biochemical, and extracellular matrix effects on vascular smooth muscle cell phenotype. J. Appl. Physiol. 98: 2321–2327.

Sun W, Sacks MS, Sellaro TL, Slaughter WS, Scott MJ (2003) Biaxial mechanical response of bioprosthetic heart valve biomaterials to high in-plane shear. J. Biomech. Eng. (Trans ASME) 125: 372–380.

Torrent-Guasp F, Ballester M, Buckberg GD, Carreras F, Flotats A, Carrio I, Ferreira A, Samuels LE, Narula J (2001) Spatial orientation of the ventricular muscle band: Physiologic contribution and surgical implications. J. Thorac. Cardiovasc. Surg 122: 389–392.

Torrent-Guasp F, Kocica MJ, Corno AF, Komeda M, Carreras-Costa F, Flotats A, Cosin-Aguillar J, Wen H (2005) Towards new understanding of the heart structure and function. Eur. J. Cardio-thoracic Surg. 27: 191–201.

Trotter JA (1993) Functional morphology of force transmission n skeletal muscle. Acta Anat. 146: 205–222.

Trotter JA, Purslow PP, (1992) Functional morphology of the endomysium in series fibered muscles. J. Morphol. 212:109–122.

Trotter JA, Richmond FJR, Purslow PP (1995). Functional morphology and motor control of series fibred muscles. In: Holloszy, JO (Ed.), Exercise and Sports Sciences Reviews Vol 23. Williams and Watkins, Baltimore, 167–213. ISBN 0-683-00037-3.

Velleman SG, Liu XS, Eggen KH, Nestor KE (1999) Developmental downregulation of proteoglycan synthesis and decorin expression during turkey embryonic skeletal muscle formation. Poult. Sci. 78, 1619–1626.

Willems MET, Purslow PP (1997) Mechanical and structural characteristics of single muscle fibres and fibre groups from raw and cooked pork Longissimus muscle. Meat. Sci. 46: 285–301.

Young M, Paul A, Rodda J, Duxson M, Sheard P (2000) Examination of Intrafascicular Muscle fiber terminations: Implications for tension delivery in series-fibered muscles. J. Morphol. 245:130–145.

Chapter 13
The Cornea and Sclera

K.M. Meek

Abstract The cornea and sclera make up the outer tunic of the eye. Each is a connective tissue containing collagen fibrils embedded in a proteoglycan-rich extrafibrillar matrix, but whereas the cornea is uniquely transparent, the sclera is totally opaque. Both tissues require strength to maintain the excess pressure within the eye and to resist external knocks and the forces applied by the extraocular muscles during eye movement. This mechanical strength is provided by the deposition of collagen in a lamellar structure, where the lamellae run parallel to the surface of the tissue rather than through its thickness. The cornea is the main refractive element in the eye's optical system, and it transmits over 90% of the incident light at visible wavelengths. Transparency is achieved because, at the nanoscopic level, the corneal collagen fibrils within the lamellae have a small, uniform diameter and are positioned with respect to each other with a high degree of lateral order. This exquisite arrangement causes destructive interference of scattered light and constructive interference of directly transmitted light throughout the visible wavelengths. As a lens, the cornea also has to be precisely curved, almost spherical near the visual axis but flattening in the periphery. Although the basis of this contour is not fully understood, corneal shape is likely achieved by the arrangement of the collagen at the microscopic level, and it is therefore not surprising that the lamellae have different preferential orientations centrally and peripherally.

The sclera (the white part of the eye) constitutes the rest of the globe. It is a tough connective tissue and is continuous with the cornea. Scleral collagen is, in composition and arrangement, more similar to that seen in skin, with wider fibrils and a much more interwoven structure than cornea. It has no optical role other than to provide a support for the retina on the back of the eye but has important physiological functions (it contains fluid outflow channels to prevent excessive pressure within the eye) and mechanical functions (it maintains eye shape during ocular movement).

This chapter describes the structure of the corneal stroma from the macroscopic level to the nanoscopic level and focuses on the role of collagen in determining the mechanical and optical properties of this fascinating connective tissue. The chapter ends with a section describing the sclera and what is currently known about the changes in collagen that accompany the development of shortsightedness (myopia).

P. Fratzl (ed.), *Collagen: Structure and Mechanics*,
© Springer Science+Business Media, LLC 2008

13.1 Introduction

The cornea is the clear anterior section of the eyeball and forms about 15% of the outer tunic of the eye. The opaque sclera comprises the remaining 85%, and the two regions meet at a region called the limbus. Together, cornea and sclera provide a tough shell that contains the other contents of the eye, protecting these from infection and injury. The cornea is essentially a connective tissue containing collagen in the form of fibrils. In this sense, it is similar to other tissues such as cartilage, skin and bone, but whereas cartilage has evolved to withstand compression, skin to provide an extensible covering and bone to withstand loading, the cornea has to fulfil several roles simultaneously. First, it has to withstand any external insult as well as contain the ocular pressure from within the eye. This is achieved by collagen fibrils embedded in a gel-like matrix, which form a strong yet resilient scaffold. Secondly, it has to be precisely curved. The basis of corneal curvature is not understood but, as we will see, there is evidence that, as in other connective tissues, form is provided by the specific arrangement of the collagen in different parts of the tissue. Thirdly, the cornea must have a smooth surface. The cornea is coated by a 4–7 μm thick tear film – a layer that transports metabolic products to and from the cornea, prevents the cornea from drying, lubricates the eyelids and has bactericidal properties. Optically, the tear film provides a smooth surface over the cornea; the precise curvature of the cornea and the optical smoothness of the tear film together produce a very efficient converging lens. In fact, over two thirds of the eye's focusing occurs at the air–tear film interface, making the cornea and its tear film the main refracting component in the eye. Finally, the cornea must be transparent to visible light; the cornea is the window of the eye, and if light cannot pass through undeviated, blindness ensues. As we will see, the unique arrangement of the collagen fibrils in the cornea allows the tissue to transmit nearly all the visible light that passes through.

13.1.1 Macroscopic Structure

All dimensions in the eye vary greatly between individuals. The human cornea is about 0.5 mm thick at the center, increasing to about 0.7 mm at the periphery. When viewed from the front, it is smaller in the vertical direction (11 mm diameter) than in the horizontal direction (12 mm diameter). It is more curved than the rest of the eyeball, with a radius of curvature of about 7.7 mm (measured at the front surface) compared with a radius of curvature of about 12 mm in the rest of the eye (Fig. 13.1).

However, because of the peripheral thickening, the radius of curvature is smaller if measured on the back of the cornea, with a value closer to 6.8 mm. Furthermore, in most people, the corneal curvature on the front surface is different in the vertical (superior–inferior) and horizontal (nasal–temporal) directions. This toricity leads to corneal astigmatism. In young corneas, the horizontal meridian is more curved, but this usually reverses as corneas age.

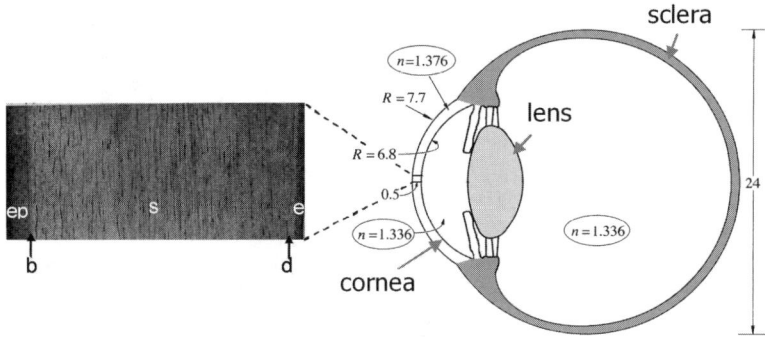

Fig. 13.1 Cross-section through a human eye (Hogan et al., 1971). The clear cornea is continuous with the white sclera. The radius of curvature of the cornea, R (in mm), is greater on the front surface than on the back surface. The diameter of the eyeball itself is about 24 mm. The cornea has a greater refractive index, n, than the air in front and the aqueous humor behind, and this leads to refraction of incoming light. From front to back, the cornea consists of five layers: the epithelium (ep), Bowman's membrane (b), the stroma (s), Descemet's membrane (d) and the endothelium (e)

13.1.2 Microscopic Structure

The cornea is generally regarded as being composed of five parallel layers (Fig. 13.1). The outermost layer, on which sits the superficial tear film, is composed of about five stacks of cells known as epithelial cells. Below this is a thin (8–12 μm) layer known as Bowman's membrane. This is not a basement membrane, but a modified acellular region of the corneal stroma beneath it. Bowman's membrane consists of randomly arranged collagen fibrils, mainly types I, III, V and VII. It is not present in some animals, such as rabbit, and it has been suggested that, where present, it stabilizes corneal shape by providing an anchoring point for the most anterior layers of the collagenous stroma. The stroma makes up about 90% of the cornea and is a dense layer of connective tissue. Beneath this is a thin limiting lamina called Descemet's membrane, a modified basement membrane of the underlying corneal endothelium. The endothelium itself is a single layer of non-dividing cells that play a crucial role in maintaining the hydration of the tissue, which is vital for the maintenance of corneal transparency.

Most of the collagen in the cornea is found in the stroma. Unlike skin, where collagen fibrils are interwoven, in the corneal stroma, they are found within structures called lamellae (Fig. 13.2) that occur parallel to the tissue surface at different angles, producing a plywood-like structure. Lamellae vary in size, but they are typically about 2 μm thick and up to 0.2 mm broad (Polack, 1961; Komai and Ushiki, 1991). These lamellae, except those near the front of the cornea (i.e., the outermost layers), seem to run in belts across the cornea from limbus to limbus (Maurice, 1969). Adjacent lamellae make angles with each other so that, overall, they are found at all angles within the plane of the cornea. Although no fibril ends have been recorded, calculations suggest a mean fibril length within a lamella of about 940 μm (Holmes and Kadler, 2005). There is a distinct difference between the lamellae in the outer,

Fig. 13.2 Scanning electron micrograph of stromal lamellae (Courtesy R. Radner). Adjacent lamellae make large angles with each other. Lamellae sometimes split and interweave (*arrowhead*), particularly toward the front of the stroma. Bar = 10 μm. (Meek and Fullwood, 2001)

middle and inner cornea. Outer and middle lamellae bifurcate and interweave and many outer lamellae course frontward, ending up anchoring in Bowman's layer (Fig. 13.3). Inner lamellae tend to lie in stacked layers.

Fig. 13.3 Three-dimensional reconstructions from forward-scattered second-harmonic-generated signals taken using a confocal microscope. **A** Cross-sectional slice through 3-D data set showing that the outer 1/3 of cornea contains transverse-oriented collagen bundles (*TL, arrows*), while the middle 1/3 contains interwoven lamellae (*IL*) and the inner 1/3 contains orthogonally oriented (*OL*) collagen lamellae. **B** 3-D reconstruction cross-section of the cornea showing multiple transverse collagen (*arrow*). Insert shows a cross-sectional slice through a 3-D data set depicting transverse collagen lamellae, Bowman's membrane shown by backscattered signal (*asterisk*) underlying corneal epithelium. Bar = 50 μm, inset 20 μm. (Morishige et al., 2006)

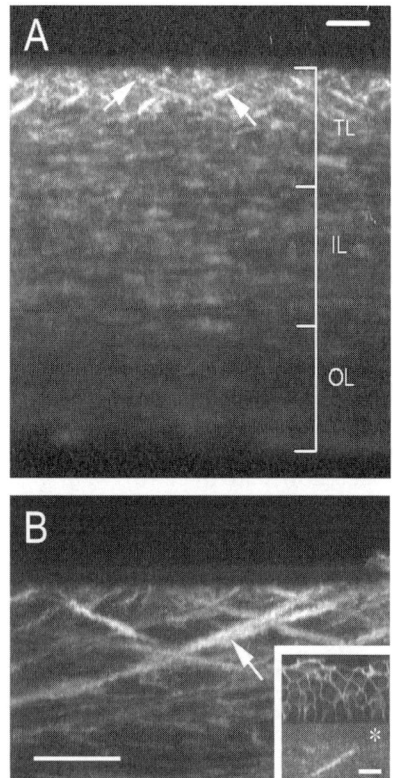

The stroma contains cells called keratocytes. These are flattened fibroblasts that normally lie quiescently between the lamellae but are activated in response to injury. They are responsible for the slow turnover of the collagen and other components (the half-life of type I collagen in the human cornea is not known, but in other tissues, it varies between 45–244 days) (Rucklidge et al., 1992).

13.1.3 Nanoscopic Structure

In the electron microscope, the lamellae are seen to be composed of narrow, uniform diameter collagen fibrils. The fibrils are spaced with a degree of lateral order and run parallel to the direction of the lamellae (Fig. 13.4), an arrangement that confers radial strength to the tissue. No fibrils run through the thickness of the cornea (except in certain cartilaginous fish corneas). Other constituents are occasionally seen such as long-spacing collagen (Fig. 13.4), which may represent abnormal aggregation of type VI collagen.

Collagen fibril diameters are highly regulated in the cornea. In the human cornea, x-ray scattering reveals that diameters are in the region of 31 nm, increasing to nearer 34 nm with age (Meek and Leonard, 1993; Daxer et al., 1998). The fibril diameter remains constant across most of the cornea and then rises abruptly at about 4 mm from the center, increasing to nearly 50 nm at the limbus (Boote et al., 2003). Fibril diameters do not vary with depth in the cornea (Freund et al., 1995).

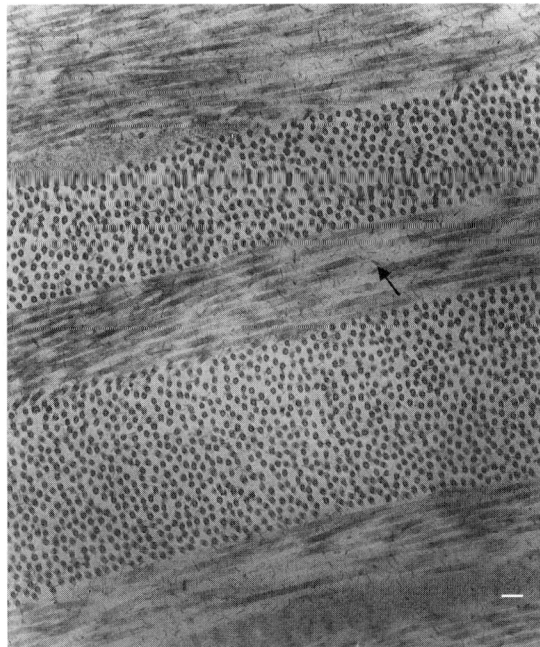

Fig. 13.4 Transmission electron micrograph showing a cross-section through five lamellae in the human cornea. Fibrils in adjacent lamellae make large angles with one another, so some fibrils are seen in cross-section and others in transverse section. The interfibrillar space is filled with proteoglycan molecules (*arrow*) that appear as fine filaments following cuprolinic blue staining. Long-spacing collagen (near *bottom* of image) is sometimes seen in older human corneas. Scale bar = 100 nm

Small-angle x-ray scattering has provided data on the average center-to-center spacing of the fibrils. This also increases away from the center of the cornea, but in a two-stage fashion. It is fairly constant (~57 nm) in the central 4 mm diameter cap of the cornea, then increases to 62 nm at the edge of the limbus (4.5–5 mm away from the center). In the limbus itself, spacings increase even more rapidly.

The fibrils in the stroma comprise mostly type I collagen, with a small amount of type V collagen (see Chapter 3) and have an axial stagger similar to other type I collagen-containing tissues (Meek and Holmes, 1983). As in other collagenous tissues, the molecules are stabilized by covalent cross-links (see Chapter 4), which increase as a function of age. The major reducible cross-links are dehydro-hydroxylysinonorleucine (deH-HLNL) and dehydro-histidinohydroxymerodesmosine (deH-HHMD). In bovine corneas, the former rapidly diminishes after birth; however, the latter persists in mature animals. A non-reducible cross-link, histidinohydroxylysinonorleucine (HHL) is another major cross-link in mature cornea. The presence of cross-links in the cornea similar to those found in skin suggests a similar mode of molecular packing (Fig. 13.5), and it has been suggested that HHL and deH-HHMD stabilize aggregates composed of three collagen molecules within the corneal collagen fibrils (Yamauchi et al., 1996). This arrangement would facilitate fibril swelling and would allow the refractive index of the fibrils to decrease, thus reducing light scattering. It is known that fibrils in corneal stroma are more hydrated than those in sclera or tendon, with a lateral spacing of molecules of about 1.8 nm. Molecules then group together to form microfibrils with a diameter of about 4 nm (Holmes et al., 2001), which are tilted

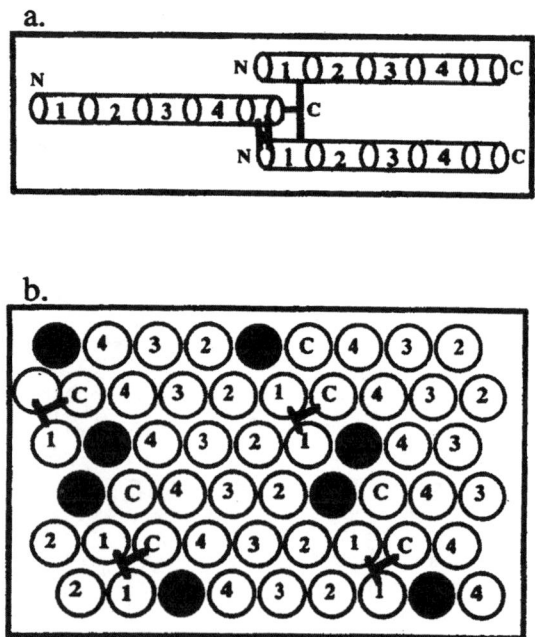

Fig. 13.5 Cross-linking and molecular packing in corneal collagen. The type I collagen molecule is represented by a cylinder having four sequential segments (1–4) and one 38 nm segment at the COOH terminal end. HHL cross-linking is indicated by ⊥ and deH-HHMD cross-links by H. a Scheme showing molecular loci of HHL and deH-HHMD, thought to be identical in skin and cornea. b Cross-section of fibril through COOH terminus of molecules. The possible location of type V collagen molecules in cornea is indicated by solid circles (Yamauchi et al., 1996)

by about $15°$ to the fibril axis (Baldock et al. 2002). This tilting leads to a reduced axial D-periodicity of 65 nm compared to 67 nm in tendon, where the molecules are essentially parallel to the fibril axis.

13.1.4 Composition of the Corneal Stroma

Table 13.1 shows the approximate composition of the stroma on the assumption that 11.4% of the water is contained in the keratocytes (Huang, 1995). The collagen in the stroma is mostly type I (58%), with smaller amounts of types V (15%), VI (24%) and XII and XIV (Drubaix et al., 1996). Type V co-aggregates with type I (Fig. 13.5), buried within fibrils with only its N-propeptide protruding on the surface. The protruding NH_2-terminal domains (Fig. 13.6) may cause steric hindrance to prevent the accretion of more molecules onto the fibril surface, thus limiting the diameter of the fibrils in the cornea. In vitro studies have confirmed that fibril diameters are limited when type V is included during collagen fibrillogenesis, thus supporting this idea.

Small amounts of type III have occasionally been reported in cornea (Newsome et al., 1981; Marshall et al., 1991) but others have failed to find this collagen in mature tissue. Type VI collagen forms 100 nm periodic filaments throughout the extracellular matrix, and its association with type I collagen fibrils suggests it acts as a bridging structure, possibly contributing to the tissue's overall biomechanical properties. In older corneas, type VI sometimes appears to aggregate into long spacing structures (Fig. 13.4). Types XII and XIV are FACIT collagens. Type XII is aligned on the surface of type I fibrils in a head-to-tail pattern, and this arrangement is thought to confer stability (Wessel et al., 1997). Type XIV has been located throughout the avian stroma associated with the surfaces of types I/V collagen (Ricard-Blum and Ruggiero, 2005), where it is thought to play a role in the compaction of the cornea, critical to its transparency (Gordon et al., 1996). Its presence in the mammalian cornea has not yet been demonstrated.

The interfibrillar spaces within the stroma are rich in molecules called proteoglycans (Fig. 13.4). These molecules consist of a single glycosaminoglycan chain attached to a small protein core. Since glycosaminoglycans are long chains of highly sulphated repeating disaccharides, they are very negatively charged and therefore hydrophilic. By attracting and binding water, they produce a gel that resists compression of the tissue. The protein part of the molecule is thought to interact at specific sites along the collagen fibrils (Scott and Haigh, 1985),

Constituent	Wet weight (%)
Collagen	14.6
Other proteins including cellular components	8.2
Proteoglycans	1.0
Cellular water	11.4
Matrix water	64.8

Table 13.1 Composition of the corneal stroma

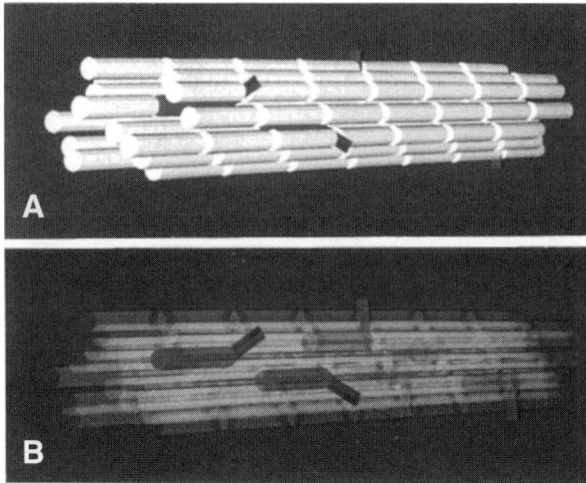

Fig. 13.6 A proposed model for the arrangement of type V collagen molecules within a heterotypic type I + V collagen fibril. **A** and **B** show the fibril in longitudinal view; the type I molecules are light colored; the type V molecules are dark. The type V molecules have the retained NH$_2$-terminal domain extending perpendicular to the long axis. The type V molecules are arranged such that their triple-helical domain lies within the interior of the fibril, and their NH$_2$-terminal domain extends through a hole zone to the fibril surface. In **B**, the type I molecules have been rendered transparent, thus, allowing better visualization of the interior type V molecules (Linsenmayer et al., 1993)

and there is ample evidence that proteoglycans are involved in diameter regulation (probably by preventing lateral fusion of neighboring fibrils) and fibril spacing. There are different glycosaminoglycans in the cornea, keratan sulphate (about 65%), chondroitin-4-sulphate and dermatan sulphate (Soriano et al., 2000), the latter two occurring together in the same chains. Keratan sulphate is seen attached to three distinct protein cores, producing three different proteoglycans, lumican (Blochberger et al., 1992), keratocan (Corpuz et al., 1996) and mimecan (Funderburgh et al., 1997). Chondroitin sulphate and dermatan sulphate are found together in the proteoglycans decorin (Li et al., 1992) and biglycan (which becomes significant when the cornea is wounded).

13.2 The Basis of Corneal Shape – Collagen Lamella Organization

A flexible membrane of high tensile strength is a very efficient way of producing a sphere with high resistance to impact. In such a spherical surface, the radial stress σ within the surface plane resulting from the internal pressure is equal to

$$\sigma = pr/2t$$

where r is the radius of curvature, t is the thickness of the membrane and p the excess internal pressure over the outside pressure (in the case of the eye, this is called the intraocular pressure or IOP). Intraocular pressure is constant, but r is smaller in the cornea than in the rest of the eyeball, so the tension in the cornea is less than that in the sclera (Fig. 13.7). The intraocular pressure in normal eyes is about 1.4×10^3–2.8×10^3 Pa above atmospheric pressure. Thus the stress in the cornea is about 1.5×10^4 Nm^{-1}. Clearly, as in an inflated football, this will tend to stretch the cornea in all directions in its plane and, to resist this, collagen in the body of the cornea has to be disposed in all radial directions.

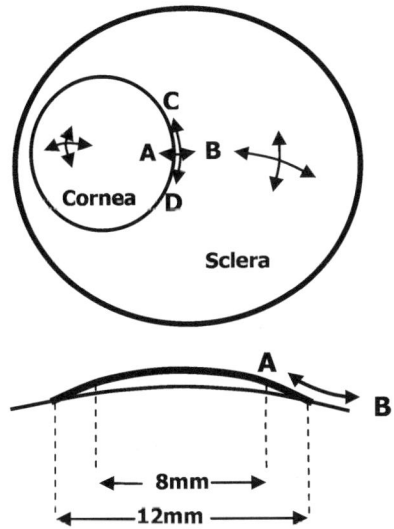

Fig. 13.7 The *top* diagram shows that the curvature of the human cornea is greater than that of the sclera. To maintain this change in, curvature requires twice the tension along CD than along AB. The *bottom* diagram shows that the "optical" zone of the cornea occupies the central two thirds of the diameter

Figure 13.7 also demonstrates that the curvature changes from cornea to sclera, and this is important for the cornea to be able to focus light onto the retina. It may be likened to a blister on a bubble, and to sustain such a change in curvature (along AB in Fig. 13.7) requires withstanding a tension along CD that is twice the tension along AB (Maurice, 1969). Collagen at the limbus therefore has to be arranged so as to maintain this change of curvature.

Although the cornea has been likened to a blister on a larger bubble (the eye), it is, in fact, rather more subtly shaped, possessing as it does both toricity and asphericity. Asphericity is a term used to describe surfaces that deviate from being perfect spheres. The "optical" zone is the maximum region through which light passes to contribute to a foveal image on the retina and occupies the central 8 mm or so of the tissue (Fig. 13.7). The asphericity, Q, of the optical zone may be described in terms of an ellipsoid as follows:

$$Q = (b^2/a^2) - 1$$

where a and b are ellipse axes' semi-lengths. Values for Q are usually negative, indicating that the cornea flattens away from the vertex. For the human cornea, values between -0.01 and -0.29 have been recorded for Q (Atchison and Smith, 2000). This range probably reflects differences in measurement technique as well as variation between individuals. The reasons for this asphericity are not known. Although flattening would reduce spherical aberrations, the effect would be very small, and it seems more likely that the flattening is to enable a smoother transition from the cornea to the less-curved sclera.

So how is the shape described above maintained? The role of collagen is to provide strength and form to connective tissues, and cornea is no exception. But, while microscopical methods have been invaluable for understanding the lamellar organization of the corneal collagen, to date, x-ray scattering has provided the only quantitative information on collagen organization in the corneal stroma.

13.2.1 X-ray Scattering Used to Determine Lamellar Organization in the Cornea

A beam of x-rays that pass through an ordered structure such as the cornea are scattered, and this leads to a so-called fiber diffraction pattern or x-ray scatter pattern. Wide-angle scattering occurs from the molecules within the fibrils (see Chapter 3), and small-angle scatter arises from interference of scattered radiation from the fibrils themselves. The angle of scatter of the different reflections can be used to determine structural parameters in the cornea, such as the intermolecular and interfibrillar spacing, and the fibril diameter (see Section 13.1.3). However, scattering can also provide information about the orientation of the molecules, fibrils and therefore lamellac.

Figure 13.8 shows how x-rays will scatter from an array of parallel cylinders that are not in contact. If the cylinders are tilted by an angle φ to a reference direction, equatorial diffraction maxima will occur at an angle ($\varphi \pm 90°$) because these arise

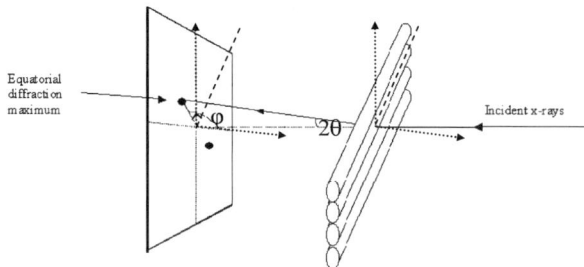

Fig. 13.8 The relationship between collagen orientation and the corresponding equatorial x-ray diffraction maxima. The positions of the scattering maxima are characterized by the scattering angle 2θ, which depends on the lateral center-to-center spacing of the collagen molecules or fibrils, and the rotation angle, φ, which is in the plane of the detector and depends on the orientation of the collagen axes (Aghamohammadzadeh et al., 2004)

from interference of x-rays scattered at right angles to the long axis of the cylinder. The Bragg angle θ will depend on the spacing of the cylinders. In the case of collagen, the cylinders could represent the fibrils themselves ($\theta \approx 0.1°$) or the molecules within the fibrils ($\theta \approx 3°$). It follows that, if there is a spread of angles ($\varphi \pm \Delta\varphi$) in which the molecules/fibrils are disposed, the diffraction maxima will spread out into arcs of angular spread $2\Delta\varphi$. Thus, by measuring the angular spread subtended by the arcs in the diffraction patterns, it is possible to determine the angular spread of the fibrils, a quantity often referred to as the fibril orientation distribution function. This has been done for collagen using both wide-angle scatter (Kirby et al., 1988) and small-angle scatter (Daxer and Fratzl, 1997).

A beam of x-rays passed through the center of the human cornea along the optical axis produces an equatorial reflection that is essentially a ring, but which is dominated by four distinct arcs or lobes (Meek et al., 1987). The ring indicates that there are collagen molecules, and hence fibrils, running in all radial directions at the center of the cornea. The lobes, however, suggest that there are two orthogonal directions in which an excess of preferentially aligned fibrils exist. In Fig. 13.9, the top and bottom lobes arise from an excess of collagen fibrils running in the horizontal direction (the so-called nasal–temporal direction in the eye) and the lobes on the left and right come from an excess of collagen in the vertical direction (inferior–superior). This arrangement has so far not been seen in any other species except the marmoset (Boote et al., 2004) and the chick (Quantock et al., 2003). Daxer and Fratzl (1997) measured the intensity of scatter as a function of rotation angle φ to quantify the proportion of collagen in each direction within the corneal plane. They found that about two thirds of the fibrils are orientated in a $45°$ sector around the vertical and horizontal meridians, whereas only one third is found in the oblique sectors between.

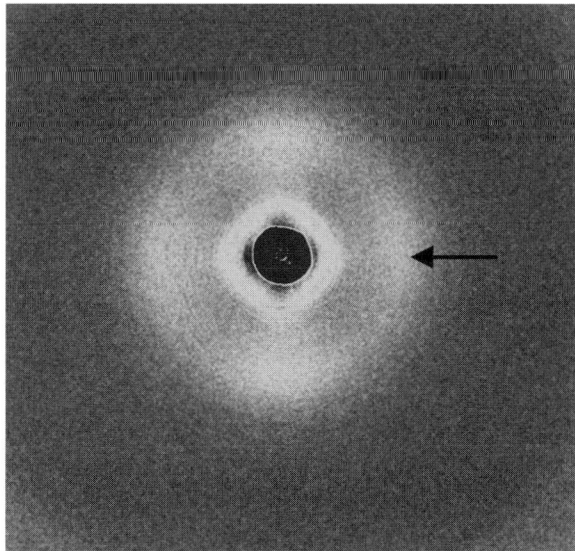

Fig. 13.9 Wide-angle x-ray scatter pattern from the center of the human cornea. The equatorial reflection corresponds to an intermolecular Bragg spacing of 1.6 nm (*arrow*) and has four maxima that arise from an excess of collagen axes in the vertical and horizontal directions

These authors, and Boote et al. (2005) found that on average, equal numbers of lamellae occur vertically and horizontally.

The intense x-ray beams available from synchrotron radiation sources have allowed x-ray scatter patterns to be obtained in minutes rather than hours and have facilitated mapping collagen orientation across the whole cornea and adjacent sclera. X-ray patterns are recorded at regular intervals across an imaginary grid covering the tissue, and each is analyzed to produce information about the orientation of the collagen averaged throughout the tissue at the sampling point, and about the relative amount of collagen that is preferentially aligned. The general method is outlined for the two-lobed x-ray pattern shown in Fig. 13.10. The wide-angle scatter pattern is shown in Fig. 13.10A. The intensity is integrated around the reflection in the direction of the arrow, leading to a plot of intensity versus rotation angle (Fig. 13.10B). This plot can be divided into a uniform background scatter coming from a population of collagen equally disposed in all radial directions in the cornea

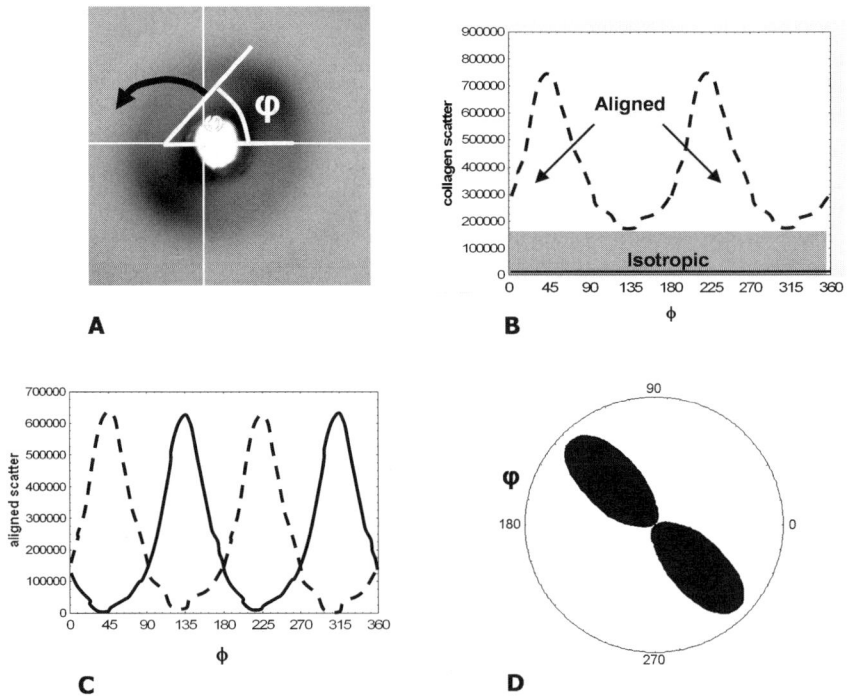

Fig. 13.10 **A** shows a two-lobed diffraction pattern indicating a single preferred orientation of the collagen. When this reflection is integrated in the direction of the *arrow*, a plot of integrated intensity versus rotation axis is obtained (**B**). This scatter can be divided into two contributions: one from isotropically arranged fibrils and the other from preferentially aligned fibrils. On removal of the isotropic component, the broken line in **C** is obtained, which when phase shifted by 90° (*full line*) becomes a good approximation to the orientation distribution of the collagen axes. This orientation can then be represented in polar coordinates (**D**). In **D**, the asymmetry gives a visual indication of the preferential orientation of the collagen, that gives rise to the pattern in **A**, and the size of the lobes is proportional to the relative amount of preferentially aligned collagen

(isotropic scatter) and a second population of collagen that adopts a preferential orientation (aligned scatter) (Daxer and Fratzl, 1997). On removal of the isotropic scatter and a 90° phase shift to account for the fact that the equatorial scatter is at right angles to the fibrillar axes (Fig. 13.10C), we obtain a good approximation to the fibril orientation distribution (full line in Fig. 13.10C). Finally, to obtain a visual representation of this distribution, the plot is converted to polar coordinates (Fig. 13.10D). In this polar plot, the asymmetry indicates the preferred orientation of the collagen that gave rise to the pattern in Fig. 13.10A, and the radial size of the plot in a given direction is proportional to the intensity of scattered x-rays.

The preferred collagen orientation throughout a human cornea is displayed in Fig. 13.11, superimposed on a picture of an eye, and several features become immediately apparent. First, it is clear from the cross-shaped polar plots that the orthogonal arrangement of collagen persists throughout the central 8 mm "optical" zone of the cornea. The plots become larger, when moving from the center, because more collagen occurs in these and other directions as the cornea thickens away from the center. By dissecting the cornea into layers, this orthogonal arrangement was found to be confined to the inner lamellae. Inspection of the plots in the sclera in Fig. 13.11 suggests that these vertical and horizontal lamellae may continue through the limbus (where they are obscured by other collagen), into the sclera. Kokott (1938) and Daxer and Fratzl (1997) have suggested that these two orthogonal populations of lamellae may help to take up the mechanical forces exerted by the extraocular rectus muscles with which they line up further into the sclera.

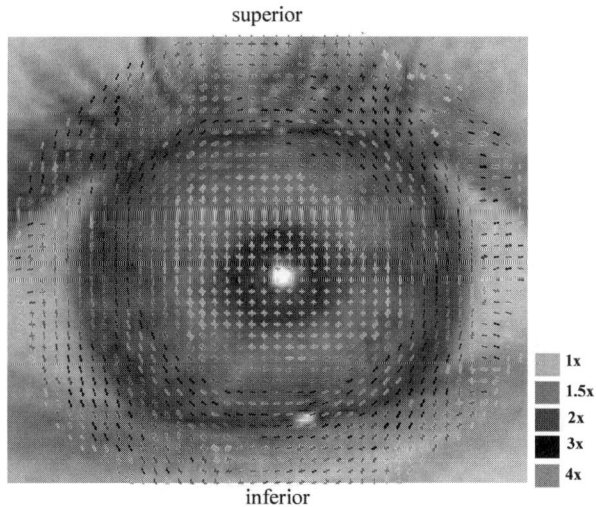

Fig. 13.11 Polar plots from x-ray scatter patterns taken at different positions across the cornea and sclera. These plots represent a 2-D projection of the preferentially aligned collagen distributions at different points in the tissue, averaged throughout the depth of the tissue. The plots are superimposed on an image of the eye and are scaled down by the factors indicated in the key so that they can all fit onto the same grid

Figure 13.11 also shows that a different collagen arrangement exists at the limbus. Here the plots are many times larger (taking into account the scaling shown in the color key) and are directed tangentially. As pointed out earlier, the change in curvature from cornea to sclera requires twice the tension circumferentially to radially (Fig. 13.7) and it is, therefore, not surprising that extra collagen in this region has evolved to maintain the curvature change, or perhaps just to support the curvature change if the pressure within the eye increased.

The integrated intensity of the 1.6 nm equatorial Bragg reflection may be regarded as being proportional to the amount of collagen molecules in the path of the x-ray beam, since the molecular environment does not change with position in the cornea. Thus, the area under the graph in Fig. 13.10B may be taken as a measure of the amount of collagen, and this quantity can therefore be mapped throughout the cornea (Aghamohammadzadeh et al., 2004). Similarly, the isotropic and aligned collagen can be mapped separately. Figure 13.12A and B shows the

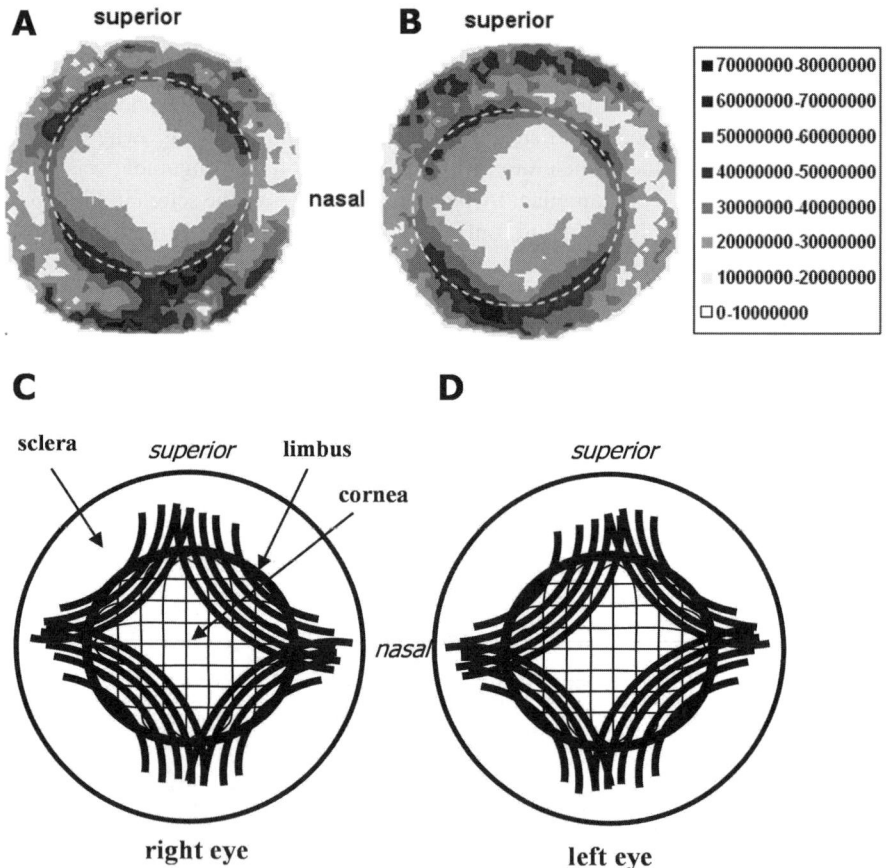

Fig. 13.12 **A** and **B** show the distribution of scatter intensity from the preferentially aligned collagen in left and right corneas from a human donor. The scale represents the x-ray scatter intensity in arbitrary units. The broken white line indicates the limbus. The aligned collagen that constitutes a limbal annulus is evident, and it can be seen that this is not a uniform structure, but shows the highest alignment in the infero-nasal sector. Within the cornea, there appears to be additional preferentially aligned collagen in the inter-cardinal segments giving rise to a rhombic shape. **C** and **D** show a possible arrangement of collagen (as a 2-D projection) that could give rise to this. Thus, in addition to the vertical and horizontal lamellae and the limbal annulus, a third population of lamellae enter and leave the cornea near the cardinal points (Boote et al., 2006)

aligned scatter from a pair of eyes and several observations can be made. First, it is clear that in the central region of each cornea, the distribution of preferentially aligned collagen is not radially symmetrical. Rather the distribution is rhomboid, though interestingly, it is uniform in the central cornea. Microfocus x-ray scattering studies (Meek and Boote, 2004) have shown that moving away from the center of the cornea vertically or horizontally, there are more collagen fibrils in a radial direction in the outer zone. In other words, the tissue is reinforced radially. Inspection of Figs. 13.11 and 13.12A and B suggest that the rhombus shape is produced by a third population of preferentially aligned reinforcing lamellae that enter and leave the cornea at the cardinal points without crossing the central zone (Fig. 13.12C and D). Such an arrangement has not been proved to exist, but if it does, its purpose may be to help to flatten the peripheral cornea and thus aid the transition to the less-curved sclera. In Section 13.2.3, we will see that support for the existence of peripheral oriented lamellae that are somehow connected with corneal shape comes from studying the corneal disease keratoconus, a disease that is accompanied with major changes in corneal shape.

13.2.2 Corneal Biomechanics

The factors that govern corneal topography, and in particular, the maintenance of asphericity are not known. The normal shape assumed by the cornea may be a passive consequence of the distension of the corneal tissue by the IOP such that the adopted shape is favorable with respect to the elastic energy content (McMonnies and Schief, 2006), or it may be a direct consequence of the precise distribution and organization of the collagen. Collagen fibrils in the cornea are narrower than those in most connective tissues, which may impart greater resistance to plastic deformation because a larger surface area of fibrils is in contact with the interfibrillar matrix. But the cornea nevertheless shows some degree of flexibility, though it quickly returns to its normal shape when the distending forces are removed. In fact, this elasticity probably serves as a buffering mechanism for microvolumetric changes in the eye, thus protecting the eye from IOP changes in vivo (Johnson et al., 2007). Although subtle permanent shape changes occur as a function of age, particularly in the horizontal and vertical radii of curvature, in certain situations, shape changes can be extreme and pathological. These will at least cause irregular astigmatism and, in severe cases, can lead to impaired vision requiring a corneal transplant. To understand these shape changes, one needs to relate the biomechanical properties of the cornea to the structure of the tissue. In this respect, it is worth noting that it is known that the only layer of mechanical importance for the in vivo cornea is the collagenous stroma, and that the whole of the intraocular pressure gradient is dropped across the stroma (Sjontoft and Edmund, 1987).

The cornea has very low strength from front to back (i.e., through its thickness). The tissue is resisting compression in that direction so there is no need for collagen fibers to run this way. The lamellae themselves are easily teased apart, particularly the inner ones. These are thought to be held together by proteoglycan bonds. The outer lamellae are more interwoven, and this adds to their cohesive strength. In fact, it has been suggested that these lamellae, many of which insert into Bowman's layer, play an important role in determining corneal surface shape (Muller et al., 2001). Tissue water content influences the pressure-induced radial straining of the tissue and hence also plays a part in determining corneal shape. It is known that the cornea distributes tensile stress unequally through its thickness as a function of the level of hydration. The extensibility of corneas with increased hydration is higher than that of normally hydrated corneas, and this has been attributed to relaxation of innermost collagen fibrils in hydrated corneas (Hjortdal, 1995). Corneal hydration is thus a key factor in the biomechanical behavior of this tissue.

Interlamellar strength varies not only with corneal depth but also from the center to periphery, in a non-uniform manner. Nasal/temporal lamellae require a separating force of 2.9×10^{-1} N mm^{-1}, whereas those in the superior cornea require 2.34×10^{-1} N mm^{-1}. The inferior cornea is the weakest part, with a lamella cohesive strength of 1.96×10^{-1} N mm^{-1} (Smolek, 1993). These results support the idea that collagen microstructure is anisotropic across the cornea, so that the cornea's physical properties will depend on position and direction.

Within the plane of the cornea, the tissue is under stress as a result of the IOP, stress that is resisted by the collagen fibrils. Collagen has high tensile strength and low extensibility. The cornea, therefore, has high tensile strength along the directions of the collagen fibril axes – an isolated cornea can withstand a fluid pressure of 30 atm (Fatt and Weisseman, 1992). There is no biomechanical requirement for collagen fibrils to run from limbus to limbus to provide this strength, so long as their length is greater than the critical length l_c:

$$l_c = \frac{d\sigma_f}{2\tau},$$

where d is the fibril diameter, σ_f is the tensile strength of the fibril and τ is the sheer stress exerted on the fibril by the interfibrillar matrix. This is because when the tissue is stressed, the fibril–matrix interface is sheared causing the collagen fibrils to become strained, thus supporting the applied load. Typical values for σ_f and τ give values for l_c of the order of micrometers, considerably less than the length of collagen fibrils in the cornea.

Stress–strain measurements on the cornea show a characteristic J-shaped curve with a toe, heel and steep region. The basis of this behavior is not fully understood. At small strains, there may be a straightening of the collagen fibrils (Jue and Maurice, 1986) or of the constituent molecules. However, it is likely that this low stiffness phase is matrix regulated. At higher strains, the tissue is much stiffer. In this phase, as in other collagenous systems, there may be molecular gliding within the fibrils and ultimately disruption of the fibril structure (Fratzl et al., 1998). Many attempts have been made to calculate Young's modulus for the human cornea in order to represent the viscoelastic response of the intact tissue due to IOP increases.

Much work has been carried out by cutting and stretching strips of cornea, but there are a number of problems with this such as

- in which direction was the strip cut? The x-ray scattering results above show preferential directions of the collagen lamellae;
- was tissue hydration controlled during the measurements?
- was tissue straightening considered? The corneal strip was part of a sphere, so will have non-uniform stress distribution across its width, and different tensile strains across its thickness, as it is extended and flattened.

Inflation testing avoids cutting and extending corneal strips. Different methods have been employed; stress is provided by increasing the IOP (ex vivo, of course), whereas strain has been calculated by dividing the extension of the cornea (determined by the displacement of two small drops of mercury on the corneal surface) by its original length (Hjordtal, 1995), or by using a laser to determine the displacement of the center of the cornea as IOP is increased and then using mathematical analysis based on shell theory to relate this to IOP and Young's modulus (ElSheikh and Anderson, 2005). Because of the disparity of the techniques used, it is therefore not surprising that there are a wide variety of quoted values for Young's modulus, which range from 2.45×10^4 to 5.7×10^7 Pa. It should also be noted that any non-linear stress–strain system does not allow a definition of a unique modulus value, but instead, requires a definition as a function of load or as a mean value over a specified loading interval.

As with interlamellar cohesive strength, it has been found that there are also regional differences in elasticity as one moves across the surface away from the center of the cornea (Hjordtal, 1996). Radially as viewed from the front, the cornea is stiffest in the central region out to about 4 mm, with maximum strength in the inferior/superior and nasal/temporal directions. The reasons are clear from Fig. 13.11 – it is seen that there is an excess of collagen vertically and horizontally in this zone. Circumferentially, the cornea is stiffest at the limbus, with a tensile modulus of about 13 MPa, up to 13 times greater than in the horizontal direction (Ruberti et al., 2007). Again, this is not surprising if one examines the circumferential (or possibly tangential) arrangement of fibrils near the limbus.

Finally, it should be noted that the cornea is a viscoelastic structure. This means that the strain following a given stress depends on the time for which the stress acts, and there is a certain amount of creep if the stress is prolonged. Creep rate varies between species, but human corneas have a long-term creep component when put under prolonged stress in vitro. One must therefore be careful to distinguish between static and dynamic stress–strain; creep may be very slow (up to months), and this may influence the tissue's biomechanical behavior in certain pathological conditions such as ectasia.

13.2.3 Corneal Ectasia

The term ectasia means stretching of an organ beyond its normal limits. In the cornea, this means progressive non-inflammatory thinning leading to irregular topographic steepening and resultant irregular astigmatism, and this can occur either pathologically or post-operatively. The commonest form of pathological corneal ectasia occurs in the condition keratoconus (Fig. 13.13). Keratoconus is associated with thinning of the central cornea, and protrusion leading to severe, irregular astigmatism. The causes of the condition are not known but are likely to be a combination of genetics and environment, since it tends to run in families but has also been associated with eye rubbing or mechanical injury. There is a loss of collagen from the thinned ("cone") region, which has been ascribed to enzymatic degradation or to a reduction in the number of lamellae as they slide away from the central area.

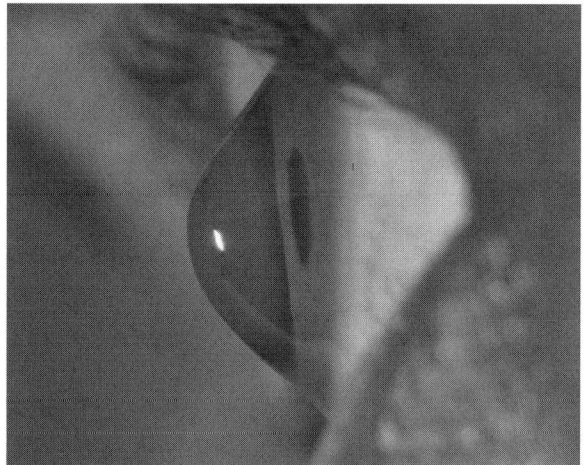

Fig. 13.13 Keratoconus ("conical cornea"). Courtesy S. Tuft

Using x-ray scattering, Daxer and Fratzl (1997) demonstrated that the orthogonal arrangement of lamellae seen in the central cornea is severely altered in keratoconus. Building on this, Meek et al. (2005) mapped the lamella arrangement in keratoconus specimens, and the results strongly supported the idea that interlamella (and possibly intralamella) shearing occurs due to a breakdown in the cohesive strength of the collagen–proteoglycan linking. Furthermore, in severe cases, the collagen distribution as determined by x-ray scattering (Fig. 13.14) shows clear changes in the rhombic pattern believed to be caused by reinforcing peripheral lamellae (Fig. 13.12). This suggests that during the course of development of the disease (usually first detected in the late teens), creep effects dominate the biomechanical behavior of the cornea, leading to lamellar sliding and reorganization of the collagen, changes in stress patterns and ultimately to changes in shape (ectasia)

Ectasia can also occur if the cornea is weakened by surgery. Laser refractive surgery is now a common method to restore vision in mild to moderate shortsightedness. In particular, laser in situ keratomileusis (LASIK) is a currently popular

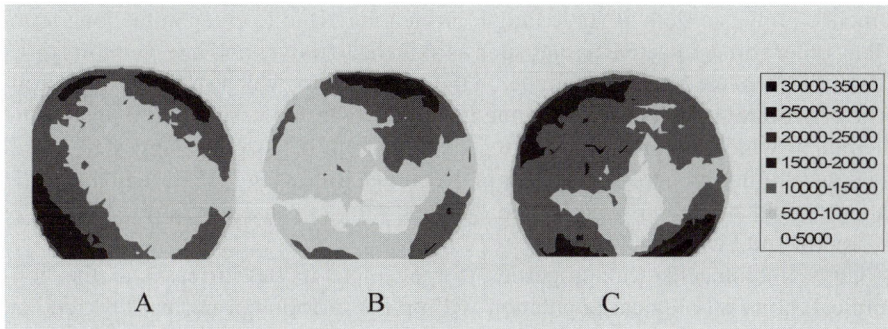

Fig. 13.14 A shows the relative distribution of scatter intensity (in arbitrary units indicated by the scale on the right) from the preferentially aligned collagen in the central 7.5 mm of a normal human cornea. The **B** and **C** show the same region from two keratoconus corneas. Notice in **B** and **C** the distortion of the rhombus shape that is visible in **A**. This suggests that a redistribution of the peripheral lamellae occurs in keratoconus (Meek et al., 2005)

treatment with over 30 million operations carried out worldwide at the time of writing. LASIK involves creating a corneal flap and then removing (ablating) collagen from the mid-stroma with an excimer laser (Fig. 13.15). The flap is then repositioned. Clearly, the technique involves severing collagen fibrils in the outermost stromal layers to create the flap. The question is, does this weaken the cornea and predispose to post-surgical ectasia? This is currently the subject of much research, but there are an increasing number of reports of post-LASIK ectasia (estimated to occur in 0.2–0.66% of LASIK treatments). Many of these have been attributed by supporters of the technique to the fact that poor precautions were taken to ensure the patient did not have early keratoconus, and hence had a weaker or thinner cornea. However, the very fact that tensioned collagen fibers in the outer layers are severed in the procedure suggests that the stress produced by the action of the IOP has to be supported solely by the remaining inner lamellae. The increased stress on these

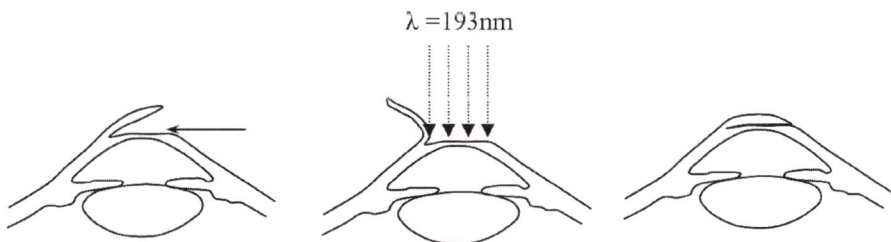

Fig. 13.15 Laser in situ keratomileusis (LASIK) is a refractive surgical procedure to correct myopia (shortsightedness). A hinged flap is made in the cornea (*left picture, arrow*) and an excimer laser is used to flatten the exposed stromal bed by ablating away some of the connective tissue (*middle picture*). This flattening increases the focal length of the cornea. The laser is controlled by a computer pre-programmed with a suitable algorithm. The flap is finally repositioned, without stitches (*right picture*)

lamellae implies a state of static fatigue, which may lead to creep in the long term. Many other soft connective tissues such as skin, heart valves and intervertebral discs can suffer from the effects of fatigue. In the pericardium, which shares several properties in common with cornea, fatigue is known to hasten tissue failure, increasing the rate at which enzymatic processes are triggered. It has been suggested, therefore, that similar processes can occur in the cornea following LASIK, and that static fatigue could be exacerbated by the dynamic fatigue caused by any eye rubbing (Comaish and Lawless, 2002).

Corneal biomechanical properties (for example tissue stiffness) in the non-surgical, non-pathological population will probably follow a normal distribution. This will be influenced by age (corneal collagen becomes more cross-linked and therefore stiffer as a function of age). It is likely, therefore, that corneas with a lower-than-normal stiffness preoperatively will be more prone to ectasia, and it follows that increasing corneal stiffness artificially should help to protect against the occurrence of ectasia. It is interesting, therefore, that intrafibrillar cross-linking of corneas with UVA/riboflavin treatment has been shown to be beneficial, both in the case of keratoconus and post-LASIK ectasia (Wollensak et al., 2003).

13.3 The Basis of Corneal Transparency – Collagen Fibril Organization

13.3.1 Transparency in the Normal Cornea

The corneal stroma is a unique connective tissue by virtue of its remarkable transparency. The basis of this transparency is still not fully understood, but below, I will give a brief summary of the main models put forward to explain this vital property of the tissue. The tissue is avascular and contains no pigments or molecules that would absorb in the visible part of the spectrum. However, *absorption* is not the only cause of transparency loss, transparency will be compromised if the light is *scattered* away from its intended direction of travel.

The cells between the collagen lamellae are flattened so as to minimize the path of the light through them and hence reduce scatter. They are also believed to contain molecules called corneal crystallins, which match their cytoplasm to the refractive index of the extracellular material (Jester et al., 1999). The collagen fibrils, however, do scatter light and, even though this is minimal on a per fibril basis (the diameter of the fibrils is very much less than the wavelength of visible light), scattering would be considerable simply because of the vast number of fibrils in the path of the light.

Classically, the scattering of light as it passes through a medium may be expressed by the fraction of energy that remains after the light has traveled a distance t:

$$(F/F_0) = \exp(-\alpha_{\text{sca}}t)$$

where α_{sca} is the scattering coefficient, analogous to an absorption coefficient. For a parallel arrangement of identical collagen fibrils, such as in the cornea, α_{sca} can be written as the product $\rho\sigma$, where ρ is the number of fibrils per unit area in a cross section, and σ is the scattering cross section. If the tissue had a uniform refractive index, σ would equal zero and the tissue would be 100% transparent. However, a number of studies have shown that the refractive index of the collagen fibrils is greater than the surrounding proteoglycan-rich matrix (Maurice, 1969; Leonard and Meek, 1997). σ is therefore non-zero and theoretical calculations show that this would result in about 94% of the incident light being scattered at 500 nm (Maurice, 1957).

The fact that about 95% of the light is transmitted has been attributed to the effects of destructive interference. By approximating the collagen fibrils to perfect, infinitely long cylinders arranged on a crystalline lattice, Maurice (1957) theorized that scattered light would interfere destructively in all directions except forward. He likened the tissue to a 3-D diffraction grating, whose spacing is less than the wavelength of light. This was an attractive hypothesis as it explained not only why the cornea is transparent but also its birefringence properties and the reasons why, when the lattice is distorted, transparency is lost. Feuk (1970) showed that random displacements of fibrils with a root mean displacement of about 15 nm around their lattice sites would also lead to a transparent cornea. Ameen et al. (1998) also considered long-range order but used photonic band structure methods to explain light transmission through corneal lattices.

However, to date, analysis of electron micrographs from the cornea does not reveal such underlying long-range order, and x-ray scattering suggested that order extends only to about 120 nm from any given fibril (Sayers et al., 1982). Hart and Farrell (1969) showed that such short-range order is sufficient to obtain the required interference effects. They used a classical approach considering each collagen fibril as a dielectric cylinder embedded in a medium of different refractive index. Applying a dielectric needle approximation to this system (valid because the diameter of a fibril is small compared to the wavelength of light), they expressed the total scattering cross-section of an isolated fibril in terms of its radius, the refractive indices of the fibril and of the interfibrillar matrix, and the light wavelength. They then used techniques analogous to those developed to analyze x-ray scattering from liquids to predict light scatter from the arrays of fibrils seen in electron micrographs. The fibril arrangement in such micrographs may be characterized by the radial distribution function, $g(r)$, a quantity that measures the likelihood of finding a fibril center within a circular ring at a distance r from a reference fibril and is normalized by the average number of fibrils in a ring of the same area. The radial distribution function corresponding to the micrograph in Fig. 13.4 is shown in Fig. 13.16. The function is zero at small r because fibrils cannot approach closer than touching. The first peak indicates the "nearest neighbor" separation. At large r, the function approaches unity because the correlation between widely separated fibrils is lost. The smallest distance at which $g(r)$ levels off is called the correlation distance, r_c, usually about 200–250 nm.

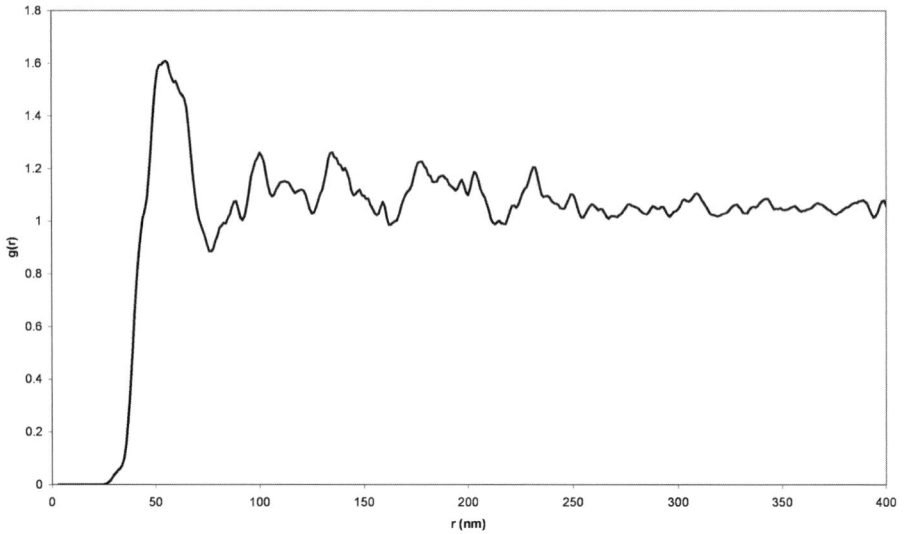

Fig. 13.16 Radial distribution function from the electron micrograph shown in Fig. 13.4. The large peak corresponds to a collagen fibril nearest-neighbor distance of about 55 nm, slightly less than that typically observed using x-ray scattering, possibly because of the effects of dehydration during processing for electron microscopy (Courtesy J. Doutch)

The scattering cross section for all the fibrils, taking interference effects into account is given by Hart and Farrell (1969):

$$\sigma = \left\{ (1 - f_1) + \frac{2(1 - f_2)}{(m^2 + 1)^2} \right\} \left\{ \frac{\sigma_0}{1 + 2/(m^2 + 1)^2} \right\},$$

where

$$f_1 = 2\pi\rho \int_0^{r_c} r \, dr [1 - g(r)] J_0^2(kr)$$

and

$$f_2 = 2\pi\rho \int_0^{r_c} r \, dr [1 - g(r)] [J_0^2(kr) + J_2^2(kr)].$$

In this equation, J_i is the Bessel function of the first kind of order i, $k = 2\pi/\lambda$, λ is the wavelength, m is the ratio of the refractive indices of the fibrils to that of the interfibrillar matrix, σ_0 is the scattering cross section per unit length of an isolated fibril. The terms $(1 - f_1)$ and $(1 - f_2)$ take account of the effects of interference. This equation therefore shows that scattering depends on the density of fibril packing, the wavelength of incident light, the ratio of the refractive indices and on the manner of fibril packing. For a random distribution, $g(r) = 1$, so $f_1 = f_2 = 0$, and the cross section equals that of an isolated fibril, σ_0, which of course depends on the diameter of the fibril.

Twersky (1975) developed an alternative model in which fibrils have a composite structure consisting of an inner core surrounded by an outer coating of a material that matches the refractive index of the interfibrillar matrix. This idea was supported by the x-ray scattering data of Fratzl and Daxer (1993). These authors showed that corneal drying is a two-stage process that can be explained if the fibrils are surrounded by a proteoglycan-rich coating (Fig. 13.17). In stage 1 of the drying, the coating releases water, but in stage 2, the fibrils release their water because the coating has become completely dehydrated. The x-ray data also indicated that the coating is formed by a fractal network with dimension 2.7 ± 0.1.

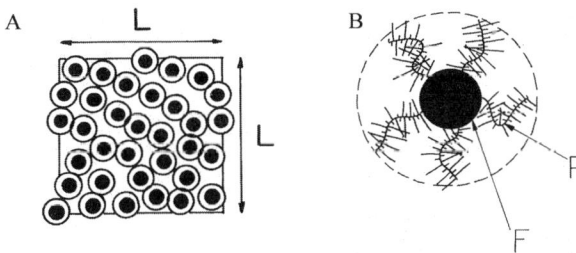

Fig. 13.17 **A** The black disks indicate the cross section of fibrils with a diameter of about 26 nm, inside a box with side length $L = 0.25\,\mu$m. The *open circles* around each disk have a diameter of 36.5 nm. **B** Model of the structure of each coated fibril (F) in the stroma. The coating consists of a network of mainly proteoglycans (P) attached with one end to the fibril forming a porous network with fractal dimension. The *broken line* indicates the outer limit of the coating which can vary from 36.5 to 80 nm depending on the degree of hydration. (Fratzl and Daxer, 1993)

The presence of such a coating may be a crucial factor for transparency since, in order to achieve the required destructive interference of scattered light, it is important that collagen fibrils in the cornea do not touch, the high negative charge of such a coating would set up repulsive forces that may prevent fibrils from aggregating.

13.3.2 Light Scattering in Swollen Corneas

A number of pathological or post-surgical conditions can lead to increased light scatter and a cloudy cornea. In many cases, this is due to abnormalities that cause cells to scatter light, or to the presence of deposits within the cornea that either scatter light themselves or disrupt the collagen organization and thus increase light scatter. A detailed account of these is beyond the scope of the current chapter, but one example will be given to illustrate what can happen if the collagen organization is severely disrupted.

Pathologically, the cornea can swell if the endothelial cells coating its back surface, which normally pump excess fluid into the aqueous chamber, are damaged.

This swelling or edema can occur following surgery, or as a consequence of endothelial pathologies such as Fuch's dystrophy. The swelling also takes place in vitro – an isolated cornea placed in a bathing medium, such as water or salt solution, will swell to many times its original weight within a few hours. The only effect of the swelling is to increase tissue thickness – the other dimensions of the cornea are unaltered. As fluid enters the cornea, attracted by the hydrophilic glycosaminoglycans surrounding the collagen, the collagen fibrils move apart and redistribute. Figure 13.18 shows that at hydrations up to $H = 1$ (where hydration is defined as the weight of water/dry weight of the cornea), water goes uniformly into the tissue; in other words, the fibrils increase in diameter at the same rate that they move apart. Beyond this

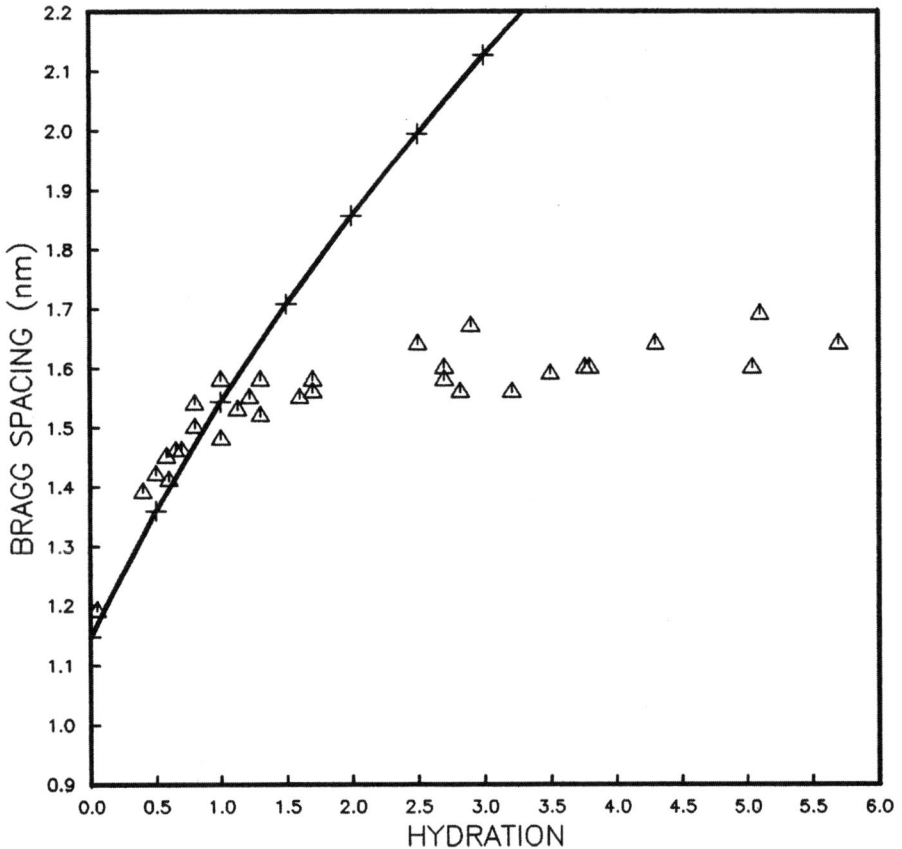

Fig. 13.18 Plot showing the change in the intermolecular Bragg spacing as a function of tissue hydration in the bovine cornea (*triangles*). After $H = 1$, very little fluid enters the fibrils and the intermolecular spacing (and hence the fibril diameter) soon becomes constant. This stage in the swelling is analogous to stage 2 in the drying process described in Fig. 13.17. The change in intefibrillar Bragg spacing with hydration is plotted on the same graph (*solid line*), scaled down so that it is normalized at $H = 0$ to the intermolecular data. This allows us to see that the interfibrillar spacing increases at the same rate as the intermolecular spacing to $H = 1$. Above $H = 1$, the fibrils continue to move apart but their diameters remain essentially constant (Meek et al., 1991)

hydration, the fluid gradually stops entering the fibrils, which swell very little above physiological hydration ($H = 3.2$). Instead, the additional water is taken up by the interfibrillar space, and this can continue in vitro until the cornea has absorbed many times its own weight in water.

When water enters the area between the collagen fibrils, they separate equally in all directions within their cross-sectional plane (this involves some redistribution of fibrils because we know that the cornea only swells in the front-to-back direction).

There should thus be a linear relationship between interfibrillar spacing *squared* and tissue hydration. By plotting the measured interfibrillar spacing squared (determined by x-ray scattering) as a function of tissue hydration, one finds that the gradient of the graph falls below that expected on the assumption that the additional water is distributed evenly between all the collagen fibrils (Fig. 13.19). This implies

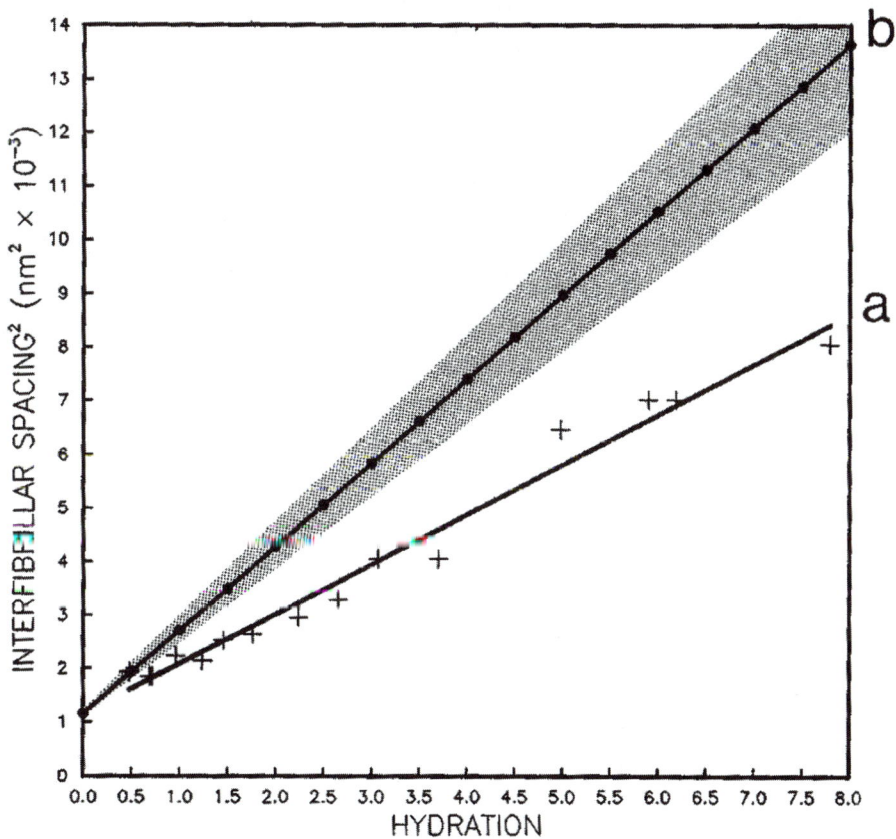

Fig. 13.19 Corneal interfibrillar volume per unit fibril length plotted as a function of tissue hydration. Line a is the least squares best fit to the experimental x-ray scatter data. Line b is the theoretical interfibrillar volume calculated by assuming the water to be absorbed uniformly throughout the tissue; calculated points are shown as *black dots*. The two sets of data are normalized at hydration $H = 0$. The confidence limits of the calculated data are indicated by the shaded area (Meek et al., 1991)

that the additional water moves into spaces devoid of collagen fibrils, spaces usually referred to as "lakes". These lakes can be seen in electron micrographs of corneas swollen in vivo (Fig. 13.20). Such lakes will interfere with the destructive interference necessary to eliminate light scatter, particularly if they approach half the wavelength of light (Benedek, 1971), and this is the cause of corneal blindness that can accompany severe cases of corneal edema. It should also be noted that corneal edema often accompanies ectatic disorders such as keratoconus. The increased tissue hydration, which is often accompanied by a loss of proteoglycans, will reduce the extensibility and shear modulus of the tissue and may thus play a role in the mechanical failure of the cornea.

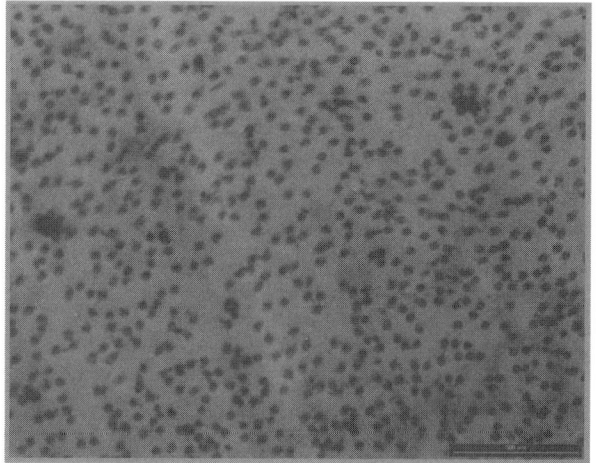

Fig. 13.20 Collagen fibril packing is severely disrupted when the stroma swells in Fuch's dystrophy, a condition that leads to severe visual impairment (Meek et al., 2003)

13.4 Artificial Corneal Constructs

Corneal disorders are currently second only to cataract as the most common cause of blindness in the world population. The factors responsible for these disorders can be hereditary, age related, traumatic or infectious, the latter being especially important in underprivileged areas. There is a particularly poor outcome when patients are affected by specific conditions such as severe chemical burns, connective tissue destroying diseases, such as Stevens Johnson's syndrome, and vascularized traumatic injuries. In many of these cases, corneal grafting fails because the injury destroys the stem cells located in the limbus, or because the infection reoccurs in the donor tissue. An alternative for curing these patients is prosthokeratoplasty, a procedure in which the damaged cornea is replaced with an artificial cornea, known also as a keratoprosthesis (KPro). The idea of replacing the cornea was first suggested in 1796 (for a historical review see

Hicks et al., 1997) and has been revisited throughout the nineteenth and twentieth centuries. In the 1950s, poly (methylacrylate) implants were first used, but with only limited success. One problem was the poor integration of the KPro with the surrounding tissue; to overcome this, the KPro was surrounded with autologous tissue such as tooth, bone or cartilage, but even these met with little success. One problem is that rigid components cause mechanical stresses in the cornea and can lead to ulceration or opacification of the optical interface. Glued or stitched components lead to sites where epithelial downgrowth, leakage of intraocular fluid and infection can occur. The ideal KPro is an epithelialized, flexible, transparent button that could be permanently joined to the surrounding tissue just like a donor cornea.

In many cases where a corneal graft is required, the ideal scenario would be to use collagen as the scaffold. Collagen has biocompatibility, low toxicity and well-known physical, chemical and immunological properties, and it can be easily purified and processed in bulk. Furthermore, when cells are cultured on collagen films, parallel orientations of cell populations occur, and the tension exerted by the cells can orient both the cells and their extracellular matrix. These properties have stimulated research into the possibility of using culture techniques to create a biological artificial cornea.

The key to constructing such an artificial cornea is to recreate the different layers known to be present in the normal cornea. Several groups have been working toward producing such a device, and a full account can be found in Ruberti et al. (2007). Many attempts are based on populating collagen gels with keratocytes and relying on the cells to remodel the gels into a cornea-like stroma, which can then be populated with superficial layers of epithelial and endothelial cells. The first problem with such collagen-based scaffolds relates to producing gels with the required stability. In the in vivo cornea, we have seen that the collagen is stabilized by a complex interfibrillar matrix consisting of proteoglycans and type VI collagen filaments. The fibrils themselves are stabilized by intermolecular cross-links, so to recreate this stability using reformed collagen substrates, covalent cross-links must be introduced. Glutaraldehyde has been used but it is cytotoxic and may not be suitable for in vivo applications. Others have used dehydrothermal treatment, ultraviolet radiation, diisocyanates and, more recently, 1-ethyl-3 (3-dimethyl aminopropyl) carbodiimide (EDC), which has less toxicity and links adjacent collagen molecules directly, without forming long bridges.

Using such collagen substrates, Griffith et al. (1999) produced what they termed "human corneal equivalents" comprising the three main layers of the cornea (epithelium, stroma and endothelium). The cell layers were fabricated from immortalized human corneal cells that were screened on the basis of morphological, biochemical and electrophysiological similarity to their natural counterparts. The scaffold for the cell lines was a collagen–chondroitin sulphate substrate cross-linked with 0.4% glutaraldehyde, then treated with glycine to remove the unbound glutaraldehyde. Stromal, epithelial and endothelial layers were created by mixing cells into and layering cells above and below this substrate and then culturing the tissue. The resulting construct had some cornea-like properties with respect morphology, transparency, ion and fluid transport and gene expression, but the resulting stromal architecture was not investigated, nor were the mechanical properties.

Ruberti et al. (2007) point out that despite the success in populating collagen-based gels with epithelium, fibroblasts/keratocytes and endothelium, no corneal construct has yet been produced that is suitable for clinical use. The resulting constructs are either too weak to bear the load, too opaque to transmit the light or too irregular to form a suitable refractive surface. The real challenge remains, therefore, to produce an artificial corneal stroma. Unfortunately, the attempts to recreate many of the subtleties of the corneal stroma have failed to produce the highly organized 3-D collagen architecture which, as we have seen, is essential for the biomechanical and optical properties of the native cornea. Many attempts to achieve this have relied on populating collagen scaffolds with cultured corneal keratocytes. However, in the cornea, keratocytes are quiescent, and when activated to remodel tissue, as happens in corneal scars, their phenotype changes to myofibroblast, and they scatter light and render the tissue opaque. Remodeling of scar tissue can take many years and is rarely complete, so it seems that this will pose a barrier to relying on keratocytes introduced into collagen gels to resorb the scaffold and remodel the tissue to create an artificial stroma.

The ability of untransformed human fibroblasts to produce organized arrays of collagen when stimulated by the addition of ascorbic acid is another approach being pursued in the creation of an artificial stroma. This has the advantage that no scaffold is involved, so no tissue remodeling is necessary. Guo et al. (2007) expanded human corneal stromal keratocytes in culture in the presence of fetal bovine serum and ascorbic acid, and the cells were allowed to synthesize a matrix for up to 5 weeks. The stratified constructs comprised parallel arrays of small, polydisperse fibrils alternating in direction in a manner similar to the developing corneal stroma. The study showed that under the correct conditions, the keratocytes themselves are capable of assembling a stroma-like array of narrow (38.1 ± 7.4 nm), parallel collagen fibrils with alternating lamellae, and that, therefore, seeding onto a collagen scaffold is not necessary.

Recently, Torbet et al. (2007) have produced a stroma-like scaffold by reconstituting type I collagen in a 7T horizontal magnetic field and combining this with a series of gelation–rotation–gelation cycles (Fig. 13.21). In this way, each layer of collagen gel is specifically oriented at an angle to the next layer. Interestingly, when keratocytes are introduced, they align by contact guidance along the direction of the collagen fibrils and respect the orthogonal design of the matrix as they penetrate. The addition of selected proteoglycans greatly improved transparency, probably by regulating fibril diameters and spacing. No attempts have yet been made to include other collagen types normally found in the cornea, into these artificial matrices. However, the early results show that it is possible to produce a layered collagenous structure in which the angle between successive layers can be manipulated as required, and so is an important step forward in the search for a biomechanically and optically viable corneal scaffold that mimics the complex lamellar orientation found in vivo.

Fig. 13.21 A–E Optical micrographs using a polarizing microscope descending vertically through a magnetically aligned reconstructed gel consisting of three (each ~0.5 mm thick) lamellae supported on a glass-bottomed culture dish. Images show successive switches in orientation; the middle layer is perpendicular to the other two (Torbet et al., 2007)

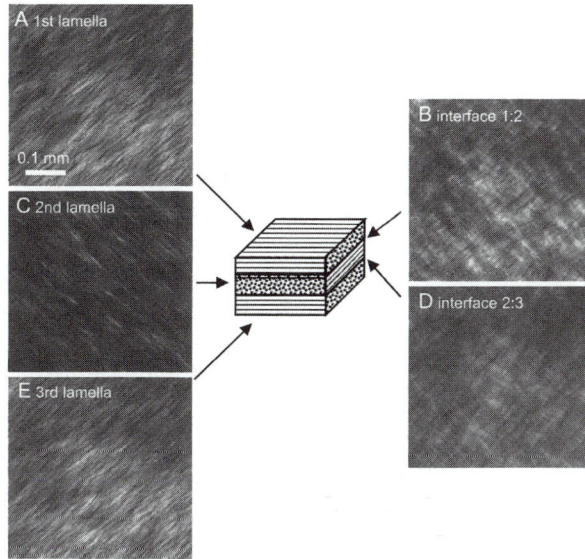

13.5 The Sclera

13.5.1 Scleral Structure

The sclera is the connective tissue that is contiguous with the cornea at the limbus, and it constitutes the major part of the eye globe. It is enclosed by the episclera, a loose connective tissue connecting it with the overlying conjunctiva anteriorly and is generally continuous with the tissue of Tenon's capsule elsewhere. It is an opaque structure and this prevents internal light scattering, and it is highly resilient compared to cornea, thus providing the strength to transfer the action of the extraocular muscles into eye movement without distorting the shape of the eye. The sclera is approximately spherical with an average vertical diameter of 24 mm (Fig. 13.1). Its thickness is not uniform varying from 0.53 mm at the limbus decreasing to 0.39 mm near the equator,[1] but increasing to 0.9 mm near the optic nerve (Olsen et al. 1998).

The stromal lamellae from the cornea, upon crossing the limbus, begin to interweave; within each lamella, the size and arrangement of the constituent collagen fibrils become less uniform, with diameters between 25 and 230 nm (Komai and Ushiki, 1991). Fibrils are themselves grouped into bundles about 0.5–6.0 μm thick. Elastic fibers and microfibrils are occasionally seen between or within the collagen bundles, together with fibroblasts. The water content also changes, from 76% in the cornea to 68% in the sclera. The orientation of the collagen fibrils in sclera is highly dependent on the region and is governed by the intraocular tension and the pull

---

[1] In the eye, the equator is the circumference separating the anterior or front half of the eye from the posterior half.

of the extraocular muscles (Kokott, 1938). There are also variations in the collagen with tissue depth; in the outer sclera, fibrils are thicker, and the bundles are narrower (about 1–5 μm) and thinner (0.5–2 μm) than those in deeper regions. These bundles are lamellar in nature (Fig. 13.22) and form a reticular structure, whorls, loops or arches depending on the position in the sclera. This means that they can take up tensile forces by being aligned in the direction corresponding to intraocular and extraocular tension. Below this, collagen fibrils are thinner and are aligned either meridionally or circularly and are densely interwoven. In the inner sclera, the bundles show a wide range of widths (1–50 μm) and thicknesses (0.5–6 μm), running in various directions and intertwining in a complex fashion. In these deep layers, the collagen fibrils are smaller in diameter and are not parallel within the lamellar bundles, but rather, are wavy and intermingled. In the posterior sclera, which is the region where eye growth occurs, this variation in diameter with depth gives rise to a trans-scleral diameter gradient that has been extensively studied in the tree shrew, an animal model frequently used to monitor corneal development and axial elongation in myopia. The tree shrew has collagen properties very similar to the human except that fibril dimensions are scaled down. Using this model, McBrien et al. (2001) showed that the posterior trans-scleral gradient is minimal at birth and becomes more accentuated with age, with the appearance of more large diameter fibrils in the outer, relative to the inner, sclera.

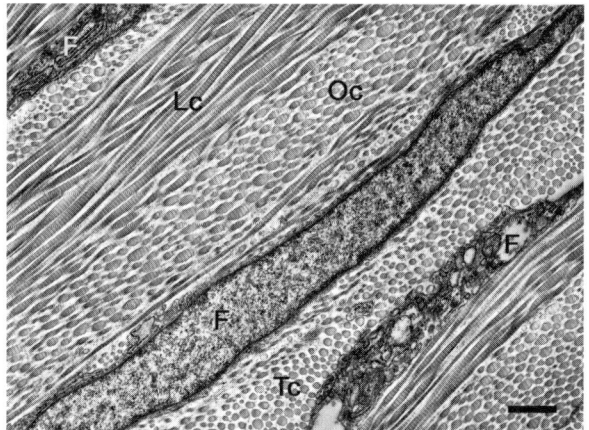

Fig. 13.22 Outer layers of normal, supero-temporal sclera showing a lamellar structure. Collagen fibrils are present in longitudinal (*Lc*), transverse (*Tc*) and oblique (*Oc*) sections and exhibit a wide variation in diameter. Fibroblasts (*F*) are visible. Bar represents 1.5 μm (Courtesy Dr. R.D. Young)

13.5.2 Scleral Composition

As in other connective tissues, the sclera is composed mainly of collagen, interfibrillar proteoglycans, glycoproteins, and the tissue is populated by fibroblasts (Table 13.2). It also contains about 2% elastin (Watson and Young, 2004). Collagen occurs in its type I, type III, type V and type VI and type XII forms, but 99% is type I. The turnover rate of scleral collagen is not known for human sclera, but in the tree

Table 13.2 Approximate
composition of the sclera

Constituent	Wet weight (%)
Collagen	28.8
Elastin	0.64
Other proteins including cellular components	<3
Proteoglycans	<1
Water	68

shrew, it is in excess of 75 days (McBrien and Gentle, 2003). The proteoglycans in sclera are mainly decorin and biglycan, small proteoglycans that are believed to regulate collagen fibril assembly and interactions. Human sclera also contains smaller amounts of the large proteoglycan aggrecan, which probably plays an important role in the regulation of scleral hydration. All three proteoglycans are present throughout the tissue thickness, though aggrecan predominates in the posterior sclera (Rada et al., 1997, 2000). Sclera also contains hyaluronan, a non-sulphated glycosaminoglycan that does not associate with a core protein. Thus, the proteoglycan composition of sclera is distinctly dissimilar to that in cornea. This leads to differences in the fixed negative charge of the interfibrillar matrix that accounts for the lower water content in sclera, as well as to different collagen fibril surface interactions with the matrix, which play a role in producing fibril diameter variations between the tissues.

13.5.3 Scleral Biomechanics and the Development of Myopia

The biomechanical properties of the sclera depend on several properties of the tissue. Obviously, structural parameters such as tissue thickness and organization of collagen will play a role. But, as in cornea, hydration is also important, although in the case of the sclera, this must be governed solely by the hydrophilic proteoglycans as there are no epithelial or endothelial barriers. The sclera is a viscoelastic structure that shows a typical two-phase response when deformed. First there is a rapid but brief lengthening followed by a semi-fluid phase that results in slow stretching. The modulus of elasticity is reported as 2.9×10^6 N m^{-2} for the anterior sclera and 1.8×10^6 Nm^{-2} for the posterior sclera at stress levels from 2×10^5 to 2.6×10^6 N. Sclera is less extensible in the anterior and equatorial regions and most extensible in the posterior regions, which is consistent with observations that it is more mature anteriorly and that collagen production is more aggressive posteriorly.

Stress–strain graphs for sclera display a degree of hysteresis such that when increasing pressure is applied, the sclera gradually deforms. When the pressure is removed, the tissue tends to recover but the final conformation will not be the same as that before the pressure was applied. Even if the applied pressure is constant, over time there is an increasing deformation. In such studies, when loads are applied that approximate normal or elevated IOP, the sclera undergoes elastic expansion, after which it displays a slower time-dependent extension (creep). Scleral creep has been the subject of several studies as it is implicated in axial length changes that accompany the development of myopia (see later). A study by Siegwart and Norton (1999) showed that scleral creep rate in tree shrews increased and decreased in concert with increasing and decreasing axial elongation rate. The authors suggested that these changes in creep rate were due to scleral tissue remodeling. Changes in collagen organization were not noted, but they proposed an alteration in the levels of pro-teoglycans between the lamellae or at their edges, which would affect the ability of collagen bundles to slide relative to one another, in much the same way as has been proposed for the development of keratoconus in the cornea.

It is thought that the collagen fibril diameter gradient occurs in response to the circumferential stresses imparted on sclera by the IOP, which will cause differential stresses across the scleral thickness. There is a known relationship between colla-gen fibril diameter and load resistance, and it has been suggested that tissues with a greater proportion of smaller diameter fibrils resist creep better, whereas those with more larger diameter fibrils withstand stress better (Parry and Craig, 1988). It is likely, therefore, that the trans-scleral gradient evolves to resist the expansive force of the IOP once the sclera has developed (McBrien and Gentle, 2003). Creep extensibility of normal sclera reduces slightly with age, which can be attributed to scleral thickening during development as well as developmental changes in the scleral tensile properties. Collagen fibril diameters also increase with age. These changes are accompanied by an increase in the modulus of elasticity, showing that sclera becomes less extensible with age.

Myopia is a common refractive error in which light is focused in front of the retina, and in most cases, this is due to abnormal ocular elongation. Severe myopia is of major concern because the incidence of myopia-related pathology, often in the form of retinal degeneration or detachment, is significantly increased. The causes of myopia, and specifically of the collagen structural changes in the sclera, are not fully understood. Although the condition may have a genetic basis, it is possible to simulate myopia in animal models by light occlusion, so environment may also play a role. Far from being an inert tissue as has previously been supposed, the sclera undergoes constant remodeling during eye growth and to some extent throughout life, and this activity has been associated with the development of myopia in many individuals. However, it is known that with axial elongation, there is a change in the distribution and content of collagen and glycosaminoglycans as well as an increased number of smaller diameter collagen fibrils (Curtin et al., 1979). Furthermore, there is evidence that the posterior sclera thins, and this has led some to suppose that the elongation is due to stretching rather than to remodeling. Though this debate con-tinues, animal studies have shown that in the early stages of myopia development, there is a net tissue loss from whole sclera of up to 7%, indicating that tissue is lost rather than simply redistributed (McBrien and Gentle, 2003). This tissue loss is accompanied some time later by changes in the trans-scleral collagen fibril diam-eter gradient (Fig. 13.23), which is almost absent in eyes with long-standing high myopia, reinforcing the hypothesis that the gradient has a functional role in eye growth. However, the early tissue loss, and the accelerated eye growth, is probably not related to collagen changes, but to changes in glycosaminoglycan charge density and/or tissue hydration, as the collagen changes occur when the initial spurt in eye growth has slowed down. The presence of smaller fibril diameters throughout the

Fig. 13.23 Reduction in trans-scleral collagen fibril diameter gradient in eyes with progressive myopia. **A** Electron micrographs show transverse sections through collagen fibrils in the inner, middle and outer posterior sclera of highly myopic, fellow control and age-matched normal eyes of 9–9.5-month-old tree shrews. **B** Graphic representation of the trans-scleral collagen fibril diameter gradient in highly myopic ($n = 3$), fellow control ($n = 3$) and age-matched normal ($n = 8$) eyes of 9–9.5-month-old tree shrews (McBrien et al., 2001)

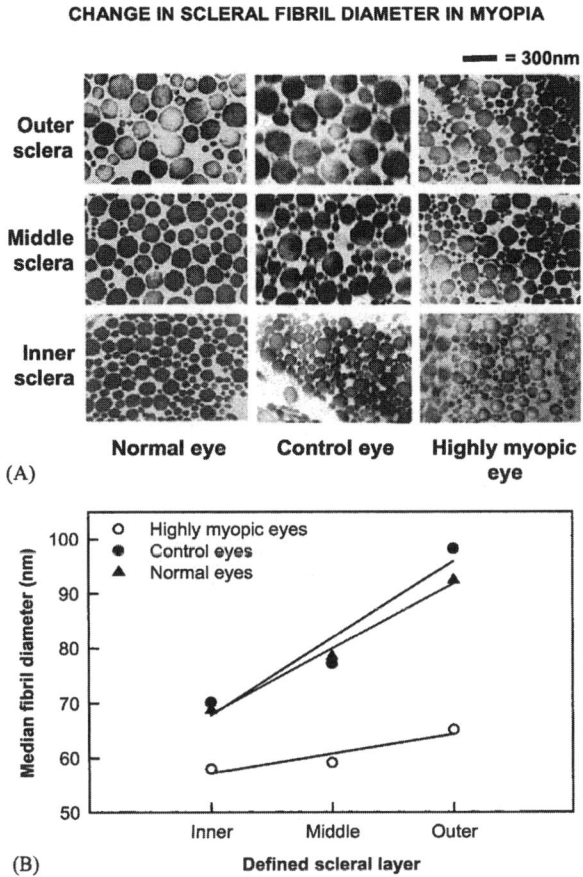

CHANGE IN SCLERAL FIBRIL DIAMETER IN MYOPIA

= 300nm

Outer sclera
Middle sclera
Inner sclera

Normal eye   Control eye   Highly myopic eye

(A)

Median fibril diameter (nm)

○ Highly myopic eyes
● Control eyes
▲ Normal eyes

Inner   Middle   Outer
Defined scleral layer

(B)

scleral thickness suggests that the myopic tissue would be more susceptible to creep, and measurements have confirmed an increased creep rate and correlated this with the degree of myopia.

The posterior sclera of highly myopic eyes can weaken and bulge, a condition termed staphyloma. As in the cornea, the collagen fibril diameters in sclera are likely to be controlled by the presence of type V collagen within fibrils, as well as by interfibrillar proteoglycans. In the case of type V collagen, it is of interest that the abundance of more narrow fibrils in myopic sclera is accompanied by a loss of type I collagen but no change in the levels of types III and V, thus there is relatively 20% more type V to type I collagen present, sufficient to account for the smaller fibril diameters. The markedly thinned posterior region of the myopic sclera coupled with reduced glycosaminoglycan content and small collagen fibrils imply that the tissue will have a reduced resistance to IOP and a greater predisposition to mechanical failure.

Normal eye formation depends on the normal development of scleral biome-chanical properties. As the eye develops, there is scleral thickening (especially posteriorly), accumulation of collagen and elastin accompanied by a trans-scleral collagen diameter gradient, and a lower glycosaminoglycan content in the equato-rial region, all of which are associated with an increase in both tensile strength and the modulus of elasticity. If any of these events do not take place, the individual is more at risk of developing myopia and even more severe conditions such as myopic staphyloma.

13.6 Conclusion

Collagen is responsible for determining the gross anatomy as well as the biome-chanical properties of both the cornea and sclera. Although both tissues contain predominantly type I collagen, their properties are not the same, and this results from differences in collagen structure at all hierarchical levels. At the nanoscopic level, corneal collagen fibrils are more hydrated than those of sclera, the fibrils are much narrower and are arranged in a more ordered array. At the microscopic level, the collagen fibrils in the cornea are packed into lamellae, most of which run parallel to the tissue surface, whereas a lamella-like arrangement is far less apparent throughout the sclera. Presently, very little is known about how nanoscopic features are related to microscopic ones. Some of these tissue differences relate to the composition of the interfibrillar matrix. In cornea, there is a higher content of glycosaminoglycans, which leads to an increased tissue hydration and hence greater spaces between collagen fibrils. However, many of these differences may be attributed to more subtle effects such as the association of type I collagen with other collagens within the tissue and to the surface interactions between collagen and the matrix components, such as proteoglycans and glycoproteins. In vitro stud-ies implicate type V collagen in the control of collagen fibril diameters, and details of the roles of specific collagen–matrix interactions are becoming clearer through the study of genetically modified murine models. For example, lumican-deficient mice show ample evidence of corneal collagen fibril fusion, suggesting that this proteoglycan plays a role in keeping corneal collagen fibrils spaced apart. However, there is still no consensus on the nature of the various interactions between the tissue components, and even less on how failure in one or more of the array of complex associations between collagens and other molecules can lead to gross anatomical changes in pathological situations, both in cornea and sclera. Finally, much effort is currently being devoted to understanding more about the relationship between collagen structure and biomechanical function – in the cornea, this has been led by the explosion of new surgical refractive correction techniques and in the sclera by the need to understand the cause of eye elongation that accompanies the develop-ment of myopia. In both cases, it is likely that many answers will be found once the connections between the various structural hierarchies of the collagen, from the nanoscopic to the macroscopic level, have been elucidated.

References

Aghamohammadzadeh H, Newton RH and Meek KM (2004) X-ray scattering used to map the preferred collagen orientation in the human cornea and limbus. Structure 12. 249–256

Ameen DB, Bishop MF and McMullen T (1998) A lattice model for computing the transmissivity of the cornea and sclera. Biophys. J. 75. 2520–2531.

Atchison DA and Smith G (2000) Optics of the Human Eye. Butterworth Heinemann, Oxford, UK.

Baldock C, Gilpin CJ, Koster AJ, Ziese U, Kadler KE, Kielty CM and Holmes DF (2002) Three-dimensional reconstructions of extracellular matrix polymers using automated electron tomography. J. Struct. Biol. 138. 130–136

Benedek GB (1971) Theory of transparency of the eye. Appl. Opt. 10. 459–473.

Blochberger TC, Cornuet PK and Hassell JR (1992) Isolation and partial characterization of lumican and decorin from adult chicken corneas. A keratan sulfate-containing isoform of decorin is developmentally regulated. J Biol Chem. 267. 20613–20619.

Boote C, Dennis S, Newton RH, Puri H and Meek KM (2003) Collagen fibrils appear more closely packed in the prepupillary cornea – optical and biomechanical implications. Invest. Ophthalmol. Vis. Sci. 44. 2941–2948.

Boote C, Dennis S and Meek KM (2004) Spatial mapping of collagen fibril organisation in primate cornea – an X-ray diffraction investigation. J. Struct. Biol. 146. 359–367

Boote, C, Dennis S, Quantock AJ and Meek KM (2005) Lamellar orientation in human cornea in relation to mechanical properties. J. Struct. Biol. 149. 1–6

Boote C, Hayes S, Abahussin M and Meek KM (2006) Mapping collagen organization in the human cornea: left and right eyes are structurally distinct. Invest. Ophthalmol. Vis. Sci. 47. 901–908.

Comaish IF and Lawless MA (2002) Progressive post-LASIK kerectasia. Biomechanical instability or chronic disease process? J. Cataract Refract. Surg. 28. 2206–2213

Corpuz LM, Funderburgh JL, Funderburgh ML, Bottomly GS, Prakash S and Conrad GW (1996) Molecular cloning and tissue distribution of keratocan. Bovine corneal keratan sulphate proteoglycan 37A. J. Biol. Chem. 271. 9759–9763.

Curtin BJ, Iwamoto T and Renaldo DP (1979). Normal and staphylomatous sclera of high myopia. Arch. Ophthalmol. 97. 912–915.

Daxer A and Fratzl P (1997) Collagen fibril orientation in the human corneal stroma and its implications in keratoconus. Invest. Ophthalmol. Vis. Sci. 38. 121–129.

Daxer A, Misof K, Grabner B, Ettl A and Fratzl P (1998) Collagen fibrils in the human corneal stroma: structure and ageing. Biophys. J. 39. 644–647.

Doubrair I, Legeais J M, Muluk Chobira N, Savoldelli M, Menasche M, Robert L, Renard G and Pouliquen Y (1996) Collagen synthesised in fluorocarbon polymer implant in the rabbit cornea. Exp. Eye Res. 62. 367–376.

ElSheikh A and Anderson K (2005) Comparative study of corneal strip extensometry and inflation tests. J. R. Soc. Interface. 2. 177–185.

Fatt I and Weisseman B (1992) Physiology of the Eye: An Introduction to the Vegitative Functions. 2nd ed. Butterworth-Heinemann, Boston.

Feuk T (1970) On the transparency of the stroma in the mammalian cornea. IEEE Trans. Biomed. Eng. BME17. 1866–1890

Fratzl P and Daxer A (1993) Structural transformation of collagen fibrils in corneal stroma during drying. An x-ray scattering study. Biophys. J. 64. 1210–1214.

Fratzl P, Misof K, Zizak I, Rapp G, Amenitsch H and Bernstorff S (1998) Fibrillar structure and mechanical properties of collagen. J. Struct. Biol. 1122. 119–22

Freund DE, McCally RL, Farrell RA, Cristol SM, L'Hernault NL and Edelhauser HF (1995) Ultra-structure in anterior and posterior stroma of perfused human and rabbit corneas – relation to transparency. Invest. Ophthalmol. Vis. Sci. 36. 1508–1523.

Funderburgh JL, Corpuz LM, Roth MR, Funderburgh ML, Tasheva ES and Conrad GW (1997) Mimecan, the 25 KDa corneal keratan sulphate proteoglycan, is a product of the gene producing osteoglycin. J. Biol. Chem. 272, 28089–28095.

Griffith M, Osborne R, Munger R, Xiaijuan X, Doillon CJ, Laycock NLC, Hakim M, Song Y and Watsky MA (1999) Functional human corneal equivalents constructed from cell lines. Science 286. 2169–2172.

Gordon MK, Foley, JW, Linsenmayer TF and Fitch JM (1996) Temporal expression of types XII and XIV collagen mRNA and protein during avian corneal development. Dev. Dyn. 206. 49–58.

Guo X, Hutcheon, AEK, Melotti SA, Zieske JD, Trinkhaus-Randall V and Ruberti JW (2007) Morphologic characterisation of organized extracellular matrix deposition by ascorbic acid-stimulated human corneal fibroblasts. Invest. Ophthalmol. Vis. Sci. 48. 4050–4060.

Hart RW and Farrell RA (1969) Light scattering in the cornea. J. Opt. Soc. Am. 59. 766–774.

Hicks CR, Fitton JH, Chirila TV, Crawford GJ and Constable IJ (1997) Keratoprostheses: Advancing toward a true artificial cornea. Surv. Ophthalmol. 42. 175–189

Hjordtal JO (1995) Biomechanical studies of the human cornea: Development and application of a method for experimental studies of the extensibility of the intact human cornea. Acta Ophthalmol Scand 73. 364–365.

Hjortdal JO (1996) Regional elastic performance of the human cornea. J. Biomech. 29. 931–942.

Hogan MJ, Alvarado JA and Weddell J (1971) Histology of the human eye. W.B. Saunders Company, Philadelphia.

Holmes DF, Gilpin CJ, Baldock C, Ziese,U, Koster AJ and Kadler KE (2001) Corneal collagen fibril structure in three dimensions: structural insights into fibril assembly, mechanical properties, and tissue organization. PNAS 98. 7307–7312.

Holmes DF and Kadler KE (2005) The precision of lateral size control in the assembly of corneal collagen fibrils. J. Mol. Biol. 345. 773–784.

Huang Y (1995) The effects of alkali burns and other pathological conditions on the ultrastructure of the cornea. PhD Thesis. The Open University, Milton Keynes, UK

Jester JV, Moller-Pedersen T, Huang J, Sax CM, Kays WT, Cavanagh HD, Petrol WM and Piatigorsky J (1999) The cellular basis of corneal transparency: evidence for "corneal crystallins". J. Cell Sci. 112. 613–622.

Johnson CS, Mian SI, Moroi S, Epstein D, Izatt J and Afshari NA (2007) Role of corneal elasticity in damping intraocular pressure. Invest. Ophthalmol. Vis. Sci. 48. 2540–2544.

Jue B and Maurice DM (1986) The mechanical properties of the rabbit and human cornea. J. Biomech. 19. 847–853

Kirby MC, Aspden RM and Hukins DWL (1988) Determination of the orientation distribution function for collagen fibrils in a connective tissue site from a high-angle x-ray diffractin pattern. J. Appl. Cryst. 21. 929–934.

Kokott W (1938) Ubermechanisch-funktionelle Strikturen des Auges. Albrecht von Graefes. Arch. Ophthalmol. 138. 424–485.

Komai Y and Ushiki T (1991) The three-dimensional organisation of collagen fibrils in the human cornea and sclera. Invest. Ophthalmol. Vis. Sci. 32. 2244–2258.

Leonard DW and Meek KM (1997). Estimation of the refractive indices of collagen fibrils and ground substance of the corneal stroma using data from X-ray diffraction. Biophys. J. 72. 1382–1387

Li Y, Vergnes JP, Cornuet PK and Hassell JR (1992) cDNA clone to chick corneal chondroitin/dermatan sulfate proteoglycan reveals identity of decorin. Arch. Biochem. Biophys. 296. 190–197.

Linsenmayer TF, Gibney E, Igoe F, Gordon MK, Fitch JM, Fessler LI and Birk DE (1993) Type V collagen: Molecular structure and fibrillar organization of the chicken $\alpha 1(V)$ NH_2-terminal domain, a putative regulator of corneal fibrillogenesis. J. Cell Biol. 121. 1181–1189

Marshall GE, Konstas AG and Lee WR (1991) Immunogold fine structural localization of extracellular matrix components in aged human cornea. I. Types I-IV collagen and laminin. Graefes Arch. Clin. Exp. Ophthalmol. 229. 157–163.

Maurice DM (1957) The structure and transparency of the cornea. J. Physiol. 136. 263–286.

Maurice DM (1969) The cornea and sclera. In Davson H ed. The Eye, Academic Press, New York. pp. 489–599.

McBrien NA, Cornell LM and Gentle A (2001) Structural and ultrastructural changes in the sclera in a mammalian model of high myopia. Invest. Ophthalmol. Vis. Sci. 42. 2179–2187.

McBrien NA and Gentle A (2003) Role of sclera in the development and pathological complications of myopia. Prog. Ret. Eye Res. 22. 307–338.

McMonnies CW and Schief WK (2006) Biomechanically coupled curvature transfer in normal and keratoconus corneal collagen. Eye Contact Lens. 32. 51–62.

Meek KM and Holmes DF (1983) Interpretation of the electron microscopical appearance of collagen fibrils from corneal stroma. Int. J. Biol. Macromol. 5. 17–25

Meek KM, Blamires T, Elliott GF, Gyi T and Nave C (1987). The organisation of collagen fibrils in the human corneal stroma: A synchrotron X-ray diffraction study. Current Eye Res. 6, 841–846.

Meek KM and Fullwood NJ (2001) Corneal and scleral collagens – a microscopist's perspective. Micron 32. 261–272.

Meek KM, Fullwood NJ, Cooke PH, Elliott GF, Maurice DM, Quantock AJ, Wall RS and Worthington CR (1991) Synchrotron X-ray diffraction studies of the cornea with implications for stromal hydration. Biophys. J. 60. 467–474

Meek KM and Leonard DW (1993) Ultrastructure of the corneal stroma – a comparative study. Biophys. J. 64. 273–280.

Meek KM, Leonard DW, Connon C, Dennis S and Khan S (2003) Transparency, swelling and scarring in the corneal stroma. Eye 17. 927–936.

Meek KM and Boote C (2004) The organization of collagen in the corneal stroma. Exp. Eye Res. 78. 503–512.

Meek KM, Tuft SJ, Huang Y, Gill P, Hayes S., Newton RH and Bron AJ. (2005) Changes in collagen orientation and distribution in keratoconus corneas. Invest. Ophthalmol. Vis. Sci. 46. 1948–1956.

Morishige N, Petroll WM, Nishida T, Kenney MC and Jester JV (2006) Non-invasive stromal collagen imaging using two-photon-generated second-harmonic signals. J Cataract Refract. Surg. 32. 1784–1791.

Muller LJ, Pels E and Vrensen GFJM (2001) The specific architecture of the anterior stroma accounts for maintenance of corneal curvature. Br. J. Ophthalmol. 85. 437–443.

Nakao H, Matsuda T, Nakayama Y, Hara Y and Saishin M (1993) Design concept and construction of a hybrid lamellar keratoprosthesis. ASAIO J. 39. M257–M260

Newsome DA, Foidart JM, Hassell JR, Krachmer JH, Rodrigues MM and Katz SI (1981) Detection of specific collagen types in normal and keratoconus corneas. Invest. Ophthalmol. Vis. Sci. 20. 738–750.

Olsen TW, Aaberg SY, Geroski DH and Edelhauser HF (1998) Human sclera: thickness and surface area. Am. J. Ophthalmol. 125. 237–241

Parry DAD and Craig AS (1988) Collagen fibrils during development and maturation and their contribution to the mechanical attributes of connective tissue. In Nimni ME, ed. Collagen: biochemistry, biomechanics, biotechnology. CRC Press, Boca Raton. Vol. II pp 1–40.

Polack FM (1961) Morphology of the cornea. Am. J. Ophthalmol. 51. 179–184.

Quantock AJ, Dennis S, Adachi W, Kinoshita S, Boote C, Meek KM, Matsushima Y, Tachibana M. (2003) Annulus of collagen fibrils in mouse cornea and structural matrix alterations in a murine-specific keratopathy. Invest. Ophthalmol. Vis. Sci. 44. 1906.

Rada JA, Achen VR, Perry CA and Fox PW (1997) Proteoglycans in the human sclera. Evidence for the presence of aggrecan. Invest. Ophthalmol. Vis. Sci. 38. 1740–1751

Rada JA, Achen VR, Penugonda S, Schmidt RW and Mount BA (2000) Proteoglycan composition in the human sclera during growth and aging. Invest. Ophthalmol. Vis. Sci. 41. 1639–1648

Ricard-Blum S and Ruggiero F (2005) The collagen superfamily: from the extracellular matrix to the cell membrane. Pathol. Biol. 53. 430–442.

Ruberti JW, Zieske JD and Trinkaus-Randall V (2007) Corneal-tissue replacement. In Lanza RP, Langer R, and Vacanti J eds. Principles of Tissue Engineering 3rd Ed. () Elsevier, Inc, New York.

Rucklidge GJ, Milne G, McGaw BA, Milne E and Robins SP (1992) Turnover rates of different collagen types measured by isotope ratio mass spectrometry. Biochim. Biophys. Acta 1156. 57–61.

Sayers Z, Whitburn SB, Koch MHJ, Meek KM and Elliott GF (1982). Synchrotron X-ray diffraction study of corneal stroma. J. Mol. Biol. 160. 593–607.

Scott JE and Haigh M (1985) 'Small'-proteoglycan:collagen interactions: keratan sulphate proteoglycan associates with rabbit corneal collagen fibrils at the 'a' and 'c' bands. Biosci. Rep. 5. 765–774

Siegwart JT and Norton TT (1999) Regulation of the mechanical properties of the tree shrew sclera by the visual environment. Vision Res. 39. 387–407

Sjontoft E and Edmund C (1987) In vivo determination of Young's modulus for the human cornea. Bull. Math. Biol. 49. 217–232.

Smolek MK (1993) Interlamellar cohesive strength in the vertical meridian of human eye bank corneas. Invest. Ophthalmol. Vis. Sci. 34. 2962–2969.

Soriano ES, Campos MS and Michelacci YM (2000) Effect of epithelial debridement on glycosaminoglycan synthesis by human corneal explants. Clinica. Chemica. Acta. 295. 41–62.

Torbet J, Malbouyres M, Builles N, Justin V, Roulet M, Damour O, Oldberg A, Ruggiero F and Hulmes DJS (2007) Orthogonal scaffold of magnetically aligned collagen lamellae for corneal stromal reconstruction. Biomaterials 28. 4268–4276

Twersky V (1975) Transparency of pair-correlated, random distributions of small scatterers, with applications to the cornea. J. Opt. Soc. Am. 65. 524–530.

Watson PG and Young RD (2004) Scleral structure, organisation and disease. A review. Exp. Eye Res. 78. 609–623.

Wessel H, Anderson S, Fite D, Halvas E, Hempel J and SundarRaj N (1997) Type XII collagen contributes to diversities in human corneal and limbal extracellular matrices. Invest. Ophthalmol. Vis. Sci. 38. 2408–2422.

Wollensak G, Sporl E and Seiler T (2003) Behandlung von Keratokonus durch kollagenvernetzung. Ophthalmologe 100. 44–49

Yamauchi M, Chandler GS, Tanzawa H and Katz EP (1996) Cross-linking and the molecular packing of corneal collagen. Biochem. Biophys. Res. Comm. 219. 311–315.

Chapter 14
Collagen and the Mechanical Properties of Bone and Calcified Cartilage

J. Currey

Abstract In bone type I collagen is mineralized by very small crystals of carbonated hydroxyapatite. There is usually some water present. These three materials together produce a composite whose mechanical properties are unlike that of any of the constituents. The mechanical behavior of bone is not strange and will eventually be explained in terms of standard composite theory. However, that time is not yet, particularly because there is still considerable dispute about some fundamental features of bone, for instance the size and shape of the mineral crystals and their topographical relationship to the collagen. Calcified cartilage, made by the calcification of type II collagen, is the stiff structural element in the skeleton of many chondrichthyean fish. It shows interesting similarities to and differences from bone.

14.1 Introduction

This book is concerned with the structure and mechanics of type I collagen tissues. Bone, and dentine (the subject of Chapter 15) have employed the deposition of mineral crystals within and between the collagen fibrils to produce tissues mechanically quite unlike those of other collagenous tissues. They are much stiffer and much less tough. This chapter will discuss collagen and mineral as two sides of the same coin, and the emphasis will not be on collagen so much as on its interaction with mineral. Also, the emphasis on collagen will be less than in other chapters. Furthermore, there will be discussion of higher level structures, where the fact that the bony tissue is made of collagen will be almost irrelevant. This is because there is no level at which it is reasonable to consider that one is dealing with the "unit" of bone. To be understood properly, all levels of bony structure must be understood.

14.2 Structure of Bone

Bone is one of a number of vertebrate mineralized tissues that use a version of calcium phosphate as their mineral. Most of the tissues have as their principal organic component type I collagen. These are bone itself, dentine (Chapter 15) and enameloid. Enameloid, covering the teeth of many fish, usually seems to be

P. Fratzl (ed.), *Collagen: Structure and Mechanics*,
© Springer Science+Business Media, LLC 2008

a very highly mineralized collagen-based tissue (Sasagawa et al., 2006), similar to the highly mineralized dentine – petrodentine – of the lungfish *Lepidosiren* (Currey and Abeysekera, 2003). Next there is calcified cartilage. This occurs as two main manifestations. One is a temporary calcification of type II collagen-based cartilage in growing bones, particularly clearly seen in the metaphysis of long bones, which is soon eroded and replaced by bone. The other, much more interesting from the mechanical point of view, is the permanent skeletal structures of well-mineralized type II collagen-based cartilage structures found almost entirely in the chondrichthyean fishes (sharks, rays, etc.). This type of calcified cartilage is discussed briefly at the end of this chapter.

The history of the development of the huge variety of collagen-calcium phosphate-based tissues seen in the early fishes is beyond the scope of this book. Nevertheless, their study is mildly instructive in that it shows what variations of the well-known bone–dentine model are possible (Nian-Zhong et al., 2005; Reif, 2002). Kawasaki et al. (2004) provide an introduction to the genetics involved in the early evolutionary development of bone and dentine. All these different types of mineralized collagens are an embarrassment to people who like neat classifications, but biology is like that (Donoghue et al., 2006). Lastly there is enamel. This is very highly mineralized, but its organic component, such as it is, is not collagen at all.

Bone's organic material is about 90% by mass collagen type I. The other organics are various non-collagenous proteins and glycoproteins. The function of these other organics is the subject of much research. Some of them have "biological" rather than purely structural functions; for instance bone sialoprotein and bone morphogenetic protein have roles in the initiation and control of mineralization. It has been suggested that a glycoprotein is necessary for the determination of apatite nucleation sites and so on. The extent to which all these non-collagenous proteins have any mechanical effect once the bone is formed is even more obscure, but will be explored a bit below. From now on in this chapter I shall usually write as if the only relevant organic in bone is type I collagen.

Bone mineral is an impure version of calcium phosphate called hydroxyapatite, whose unit cell contains $Ca_{10}(PO_4)_6(OH)_2$. There are various impurities. In particular there is about 4–6% of carbonate replacing the phosphate groups, making the mineral more truly a carbonate apatite (dahllite). The overall shape of the crystal is usually far from well known; partially this is because the shape is different in different bony types. In one direction the crystals are small, of the order of 5 nm. They are about 40 nm wide, but sometimes they are hardly wider than 5 nm. What is less clear is the size of the crystals in their long direction (which is the c axis of the crystal). They can be at least 50 nm long, and it is quite possible that they can join, or grow, till they are several hundreds of nanometers long (Ziv and Weiner, 1994). The small size in one direction may have, as we shall see, profound mechanical implications.

One particularly hazy area is how the mineral relates topographically to the collagen and is, surprisingly, still a matter of considerable dispute. Some mineral crystals lie within the fibrils, somewhat disrupting them as they grow, and some lie between the fibrils. The crystals lying within the fibrils are oriented with their

Table 14.1 A classification of bony types. It must be admitted that "parallel-fibered bone" is a moving target that no one seems quite sure about. The main purpose of this table is to emphasize the hierarchical nature of bone. The table should be read across at each level

Ultramicroscopic type of bone	Woven	Parallel-fibered	Lamellar
Fibrils (join to make fibers)	Woven: 0.1–3 μm wide in a felt	Parallel-fibered: intermediate	Lamellar: fibrils 2–3 μm in lamellae 2–6 μm thick
Bone cells (within the bone tissue)	These contain bone cells (osteocytes) Beware: most bony fishes (the majority of vertebrate species) have acellular bone		
Lacunae (which contain the bone cells) connected by:	In woven: roughly isodiametric ca 20 μm in diameter		In lamellar: oblate spheroids 5:1 greatest to least axes. Major axis about 20 μm long
Canaliculi	Osteocytes connected to each other and, indirectly, to blood channels, by cell processes in tubes (canaliculi) 0.2–0.3 μm wide. About 50–100 canaliculi per osteocyte		
Microscopic types of bone	**Woven** In large lumps in young animals and in fracture calluses	**Lamellar** In large lumps in reptiles and in circumferential lamellae in mammals and birds	**Fibrolamellar** Alternating sheets of lamellar and woven bone, with 2-dimensional sheets of blood vessels. About 200 μm between repeats	**Secondary osteones** (Haversian systems) Cylinders of lamellar bone. Solid save central tube for blood vessels. About 200 μm in diameter
Primary or secondary	*Primary*	*Primary and secondary*	*Primary*	*Secondary*
Macroscopic types of bone	**Compact bone** Solid, only porosity for canaliculi, osteocyte lacunae and erosion cavities		**Cancellous (trabecular) bone** Porosity visible to the naked eye. Rods and plates of lamellar bone, never forming closed cells

long axes along the same axis as that of the collagen fibril. Those lying between the fibrils are not constrained in this way. It seems that the mineral lying within the fibrils first nucleates in the "Hodge–Petruska" gaps, probably initially at the "e" band (details in Chapter 2). It then coalesces and extends along the long axis of the collagen fibrils, disrupting them to some extent (Landis et al., 1996). Some people, for instance Weiner et al. (1999), Jäger and Fratzl (2000) and Gao et al. (2003), think that most of the crystals are within the collagen fibrils. Others, for instance Pidaparti et al. (1996) and Fritsch and Hellmich (2007), think that they mostly lie outside the collagen or rather that the mineral and collagen form interpenetrating phases. Sasaki et al. (2002) suppose that about three quarters of the mineral lies outside the collagen, while Hellmich and Ulm (2002) suppose that almost all of the mineral lies outside the collagen. It is not possible to reconcile these ideas, and no doubt the matter will be resolved sooner or later. Certainly, the methods that involve direct visualization of bone would suggest that much of the mineral is within the fibrils. However, when it comes to modeling bone's stiffness and fracture, it is important that we know the truth!

Table 14.1 gives some idea of the kinds of bone one is likely to see during an ordinary working life. However, to see what really exists one need to consult something like the article by Francillon-Vieillot et al. (1990).

One of the features of many bones is that they remodel, that is some of the bone material is destroyed and then new bone is laid down in the hollow cylinder or gutter so formed. The original bone is called "primary". The bone that is laid down later in the cavities is called "secondary". The resulting structure is called an Haversian system or secondary osteone. It is important for understanding the mechanical properties of bone to distinguish primary osteones from secondary osteones, which superficially appear rather similar.

14.3 Mechanical Properties of Compact Bone

It is easiest to discuss the mechanical properties of bone by starting with the "type specimen" of bone mechanical studies – the tensile load–deformation curve of compact bone. Such a curve is shown in Fig. 14.1. It has two main components: an initial straight, steadily rising part, followed by a much flatter part that ends in fracture. In the straight part the bone behaves linearly elastically, that is, the load–deformation curve is nearly straight, and if it is unloaded it will return (almost) to its original dimensions. The extent to which it does not return is the result of viscoelastic effects, which are seen much more prominently in non-mineralized collagens (Chapters 5 and 6). At the end of the straight part the curve bends over, and the specimen is said to have yielded. What happens is that microdamage occurs; lots of little cracks and bits of damage are formed. This increases the compliance of the specimen and accounts for the flatness of the post-yield part of the curve. Eventually some cracks join up, and a big "fatal" crack travels through the specimen and it breaks.

Fig. 14.1 Load–deformation (stress–strain) curve for compact bone loaded in tension

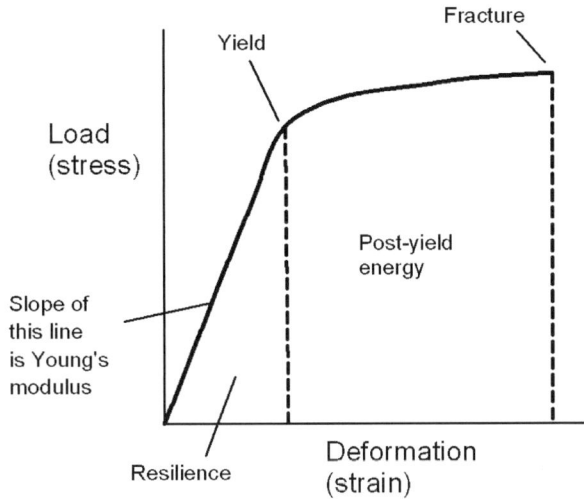

The amount of load borne and the deformation undergone will depend on the size and shape of the specimen (the larger the specimen the larger the load it can bear without breaking and the less its proportional deformation for a given load), and so the load–deformation curve has to be normalized into a stress–strain curve (Chapter 1). For a simple tensile specimen this is easily done by dividing the load by the cross-sectional area to get the stress, and dividing the extension by the original length to get the strain. (Such ease of transformation does not apply for other more complicated loading modes, like bending or torsion, or for specimens that do not have a completely uniform shape.)

The slope (stress/strain) of the straight rising part is the Young's modulus of elasticity and is a measure of the stiffness of the bony material. The stress and the strain just below yield are the levels that would be better for the bone to be confined to, because here the tissue is not, or is hardly at all, damaged. (Obviously in life the bone should not be routinely damaged.) Also, the area under this part of the stress/strain curve is the amount of energy that can be absorbed elastically per unit volume and is called the "resilience". It is found that during fast running or other extreme activities the bone is loaded to a stress of about one third of the yield stress. In this way bone is said to have a safety factor of about three. Of course this safety factor varies in different animals and in different bones. However, it is rarely less than two or greater than about five (Alexander, 1981; Blob and Bewiener, 1999).

The flatter part of the stress/strain curve after the yield region might seem to be unimportant, because the stress hardly increases at all and the stress at fracture is hardly greater than the stress at yield. However, it is very important, because it is a measure of how much work has to be done on the specimen before it breaks. A brittle material will hardly show any post-yield deformation. Very roughly, bone that shows a great deal of post-yield deformation is tougher than bone that hardly shows any. Toughness is the ability to make it difficult for cracks to travel through the material. A tough material may have quite long cracks but not be seriously inconvenienced by them, whereas a brittle material may be greatly weakened by quite short cracks. As we shall see, bone varies considerably in its toughness. Toughness is a very important feature of most materials, and a whole science of "fracture mechanics" has grown up to determine the importance of cracks in structures of different shapes and under different loading conditions. I shall say little about fracture mechanics properties here, because it has not been greatly studied in bone, and many studies have concerned themselves with cracks traveling in improbable directions. Nevertheless, it is an important matter.

When bone is loaded in compression the situation is more complicated, because whereas in tension the fractured parts pull away from each other, in compression the bone is forced in on itself as it were, and fails in shear, that is, in directions at considerable angles to the direction of the load. The situation is similarly complicated in bending, because in bending the strain varies throughout the specimen, and much of the bone may have gone well past its yield point, while the specimen still holds together. This matter is discussed at some length in Currey (2002, pp. 93–98) but is beyond the scope of this chapter. The important point is that the bending strength of bone will be its tensile strength (because bone is weaker in tension than compression) plus some addition caused by post-yield deformation. The more the post-yield deformation the greater the bending strength, though this does not increase indefinitely.

There are other features of bone's mechanical properties. One is viscoelastic effects. When bone is strained and then held at some constant strain, it is found that the stress necessary to hold it there decreases, in a negative exponential kind of way. This is called stress relaxation and results in bone loaded very slowly having a different apparent Young's modulus than bone loaded very quickly. Similarly, if a piece of bone is loaded to some stress and then held at that stress, the bone continues to undergo strain, that is, it "creeps". Although the rate of strain declines initially, and may possibly fall to zero, if the load is great enough the rate of straining will start to accelerate again, and the bone eventually ruptures. This is called "creep rupture". (Not only bone but all materials behave like this to a greater or lesser extent, particularly wet collagen – Chapters 5 and 6.) There are other effects on mechanical properties caused by viscoelastic effects, but I shall not consider them here.

Finally there is the matter of fatigue. (This is discussed fully in Chapter 5.) Bone loaded repeatedly to a stress (or strain) that it could bear successfully if loaded just once may break. It is unclear whether it is the repetition that is important or whether fatigue is simply a creep rupture event with varying stress levels (Caler and Carter, 1989; Zioupos et al., 2001) In tension it seems to be a matter of time only; in compression the matter is more complicated. Fatigue damage (often called "stress fractures" by clinicians) is of considerable importance in real life, because many otherwise healthy young people suffer debilitating pain or even fracture when undergoing extreme or unwonted exercise, as a result of fatigue damage to their bones.

Table 14.2 gives some idea of the mechanical properties of compact bone. To people who come fresh to mechanics, without any context, this may not be

Table 14.2 Some values of mechanical properties of bone. This is just to give an idea of the values and their variation. Original work should be consulted for definitive values. "Modal value". This is the value near which most "ordinary" bone would lie. "Upper limit": Approximately the largest value for the property that has been recorded. "Lower limit": Approximately the lowest value for the property that has been reported. All values (except fatigue) are for quasi-static tests, that is at low strain rates, using wet compact bone. The fatigue tests were carried out at about 10 Hz. The orientation was that giving the highest values. Some bone can be very anisotropic, both structurally and mechanically. The numbers in parentheses refer to the (more or less) original sources for these data

Property	Modal value	Upper limit	Lower limit	Comments
Young's modulus of elasticity (E)	20 GPa (1)	45 GPa (2): $M.$ $densirostris$ rostrum	6 GPa (1): deer antler	Probably nearly the same in tension, compression and bending
Tensile yield stress	120 MPa (3)			Higher in compression
Tensile strength	150 MPa (1)	300 MPa (1): some deer antler	20 MPa: ear bones (1)	Often much lower in "bad" orientations
Ultimate tensile strain	0.03 (1)	0.12: deer antler (1)	0.002: ear bones (1)	
Compressive strength	250 MPa (3, 4)			
Bending strength	250 MPa (5)		50 MPa: ear bone (5)	
Work of fracture	3500 J m^{-2} (6)		90 J m^{-2}: $M.$ $densirostris$ rostrum (6)	Fracture mechanics. Not many bones studied
Critical stress intensity (K_c)	5 MN m$^{-1.5}$ (6)		1.3 MN m$^{-1.5}$: $M.$ $densirostris$ rostrum (6)	Fracture mechanics. Not many bones studied
Fatigue life at 0–100 MPa tension	1300 (young human) (7)		150: antler (7)	Very variable according to method of testing
Fatigue life at 0–50 MPa tension	50,000 (8)			Bovine humerus. Tested to fracture

(1) Currey (2002, p. 130); (2) Zioupos et al. (1997); (3) Cezayiroglu et al. (1985); (4) Reilly and Burstein (1975); (5) Currey (1999); (6) Currey et al. (2001); (7) Zioupos et al. (1996); (8) Kim et al. (2007)

particularly helpful, but it does give some idea of the basic mechanical properties of bone and a little idea of how variable they may be.

14.4 Mechanical Properties of Cancellous (Trabecular) Bone

Cancellous bone is the bone whose porosity is easily visible to the naked eye. It consists of interconnecting rods and plates of bone. The amount of bone per unit volume of tissue can vary from about 20%, below which the tissue would probably be considered to be compact bone, down to a very low value, with just the occasional strut passing across the lumen of a long bone shaft. The bone, and the space between the bony elements, is always filled with marrow in mammals, though sometimes with air in some bones of some birds and some pterodactyls, and is always totally interconnecting. That is to say, it is not bone with pockets of marrow, or marrow with isolated lumps of bone. The general fine histology of the cancellous material is similar to that of compact bone. In fact, in "compact-coarse-cancellous bone" the interstices of the cancellous bone are filled in, producing compact bone during the increase in length of long bones.

There is still considerable disagreement as to whether the bone material in cancellous bone has the same mechanical properties as that of the neighboring compact bone. This dispute continues because it is very difficult to test the mechanical properties of cancellous material directly, as the individual trabeculae and plates of which it is composed are so small. Back-calculation from finite element models suggests that the Young's modulus of cancellous material is somewhat lower than that of the neighboring bone. However, it is certain that some of the low values found in the literature, for instance by Bini et al. (2002) who tested single trabeculae and suggested values for Young's moduli as low as 1.9 GPa, must be too low. In my laboratory we have found values for Young's modulus for whole blocks of cancellous bone, holes and all, of 6 GPa, implying that the material modulus must be considerably higher (Hodgskinson and Currey, 1992).

Cancellous bone is somewhere between being a material and being a structure. If one considers the mechanical properties of a block of cancellous bone, say 1 cm on a side, then it is sufficient to know that the mechanical properties of the material are similar to those of compact bone. This is because the mechanical behavior of the block will be determined overwhelmingly by three features: the proportion of the block that is occupied by bone and not marrow, the anisotropy of the trabecular struts and their relationship to the direction of loading. In a classic meta-analysis Rice et al. (1988) showed that the strength and Young's modulus of bone were both proportional to ρ^2 where ρ is the overall bone volume fraction; doubling the volume of bone material in a block of cancellous bone will multiply the strength and stiffness by four. Cancellous bone usually has a bone volume fraction of about 10–20% implying that its strength and stiffness are about 1–4% that of compact bone. Clearly, small (say 20%) differences in Young's modulus between cancellous and compact bone are completely trivial compared with the differences in overall density.

Cancellous bone almost never appears on its own, being covered with a layer, often very thin, of compact bone. It appears in three characteristic places: underneath synovial joints; in flat plate-like structures like the scapula and ilium; and running end to end in short bones like vertebral centra and the bones of the wrist. It has rather different functions in these three places, but explication of these functions would run well outside the subject matter of this book. Interested readers could read pages 158–173 and 212–220 of Currey (2002).

14.5 Collagen–Mineral Interactions and the Effect of Different Collagen/Mineral Ratios on the Mechanical Properties of Bone

There are many ways in which collagen and the hydroxyapatite-like mineral could interact. Walsh and Guzelsu wrote that the combined electrostatic interactions between cationic and anionic sites would be powerful, and that even direct bond formation between the two types of material could occur. Various other methods of bonding have been proposed (Walsh and Guzelsu, 1994). Wilson et al. (2006) argue, with some experimental evidence, that water plays an important role, both in stabilizing the imperfect lattice of the hydroxyapatite and in possibly coupling the mineral and the bone. Certainly tight bonding between the collagen and apatite is possible, and most people when making models to describe the behavior of bone assume perfect bonding between the two types of material.

For dry bone, collagen $\approx$ (total mass − mineral). Usually we are ignorant of the proportion of other organics present besides collagen. Collagen is by far the major component and organics are in general called "collagen". Figure 14.2 is a ternary diagram showing the relative amounts, by mass, of collagen, mineral and water in various vertebrate bony tissues. The points fall more or less in a line between least highly mineralized bones and most highly mineralized, water disappearing more rapidly than collagen. The variation is considerable. For instance deer antler is about 35% collagen, 45% mineral and 20% water whereas whales' ear bones (de Buffrénil et al., 2004) and, even more extreme, the enigmatic rostrum of the toothed whale *Mesoplodon densirostris* are almost entirely mineral, though they are still bone, having some residual collagen (Rogers and Zioupos, 1999; Zylberberg et al., 1998). There is no reason for thinking that other mineralized collagenous tissues would lie far away from the distribution shown here.

Figures 14.3 and 14.6 (Currey et al., 2004), referring to dry weights, show Young's modulus (stiffness) and total work under the curve (a fairly good measure of toughness) vs. mineral. The data set of which this graph is a representation includes a large variety of different bones. As collagen decreases, and mineral increases, stiffness increases markedly and work of fracture decreases even more sharply. Mineral/collagen ratio has a profound effect on pre- and post-yield mechanical properties. Young's modulus is, of course, affected by other things apart from the mineral/collagen ratio. Particularly important is the porosity of the specimen.

Composition of bones

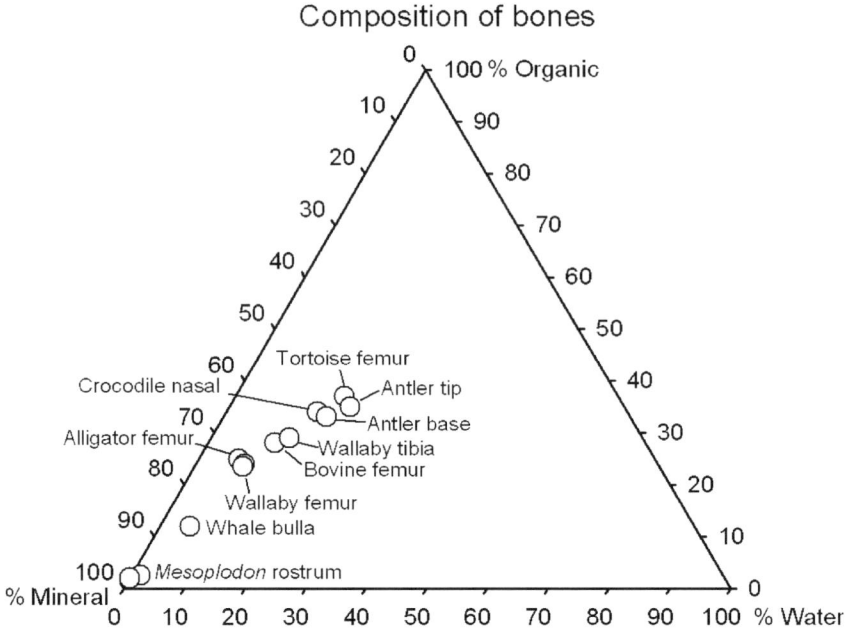

Fig. 14.2 Ternary diagram of the constituents of various bones, showing the relative proportion, by weight, of the three main constituents, organic (mainly collagen), mineral and water

E vs Mineral

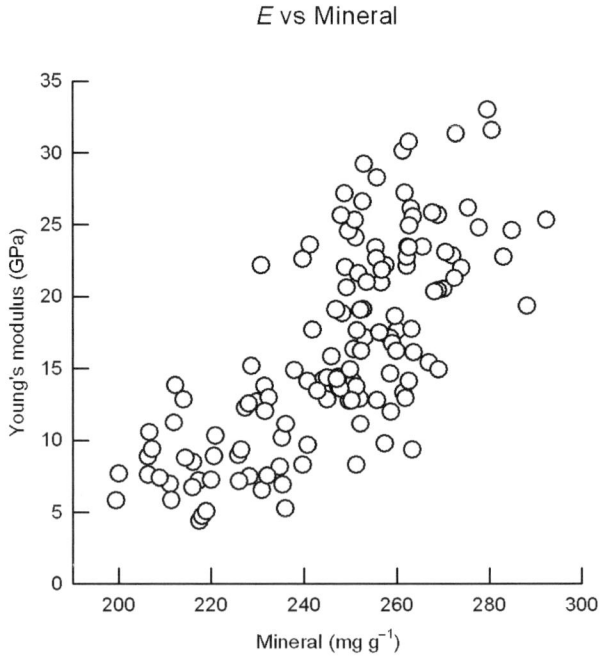

Fig. 14.3 The relationship, for a large variety of different bones of different species, between the mineral content, in mg calcium per gram of dry bone, and Young's modulus. The very highly mineralized ear bones of the whale are not included. The relationship is clearly positive but very messy. Linear $R^2 = 53\%$

Figure 14.4 shows how, for the same specimens shown in Fig. 14.3, Young's modulus declines as porosity increases. These two effects can be combined statistically, and Fig. 14.5 shows the effect of this. It is interesting that porosity has virtually no effect on work under the curve, apart from its effect in reducing the bony constituent of the specimen cross-section (not shown).

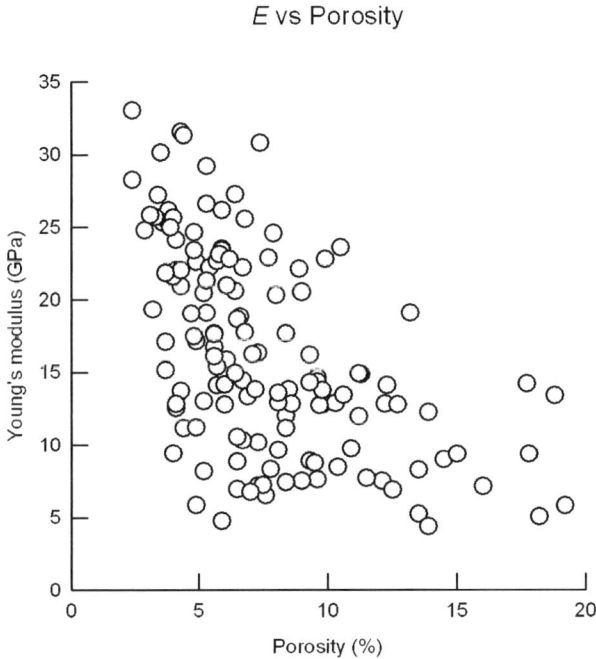

Fig. 14.4 Relationship between Young's modulus for the same specimens as in Fig. 14.3 and the porosity of the specimens. A few notionally "compact" specimens have more than 20% porosity. They are excluded from all this analysis. The relationship is clearly negative, but again is very messy. Linear $R' = 30\%$

Because mineral is so much stiffer than collagen and water, it is not surprising that the modulus of elasticity increases with mineralization. Unfortunately, despite many attempts, there seem to be no good analytical accounts of why the stiffness increases at the rate it does. The very early attempt of Katz (1971) showed that the stiffness of bone lay between that predicted by the so-called Reuss and Voigt models, but these bounds were so far apart that this was not very helpful. Somewhat more recent attempts, by, e.g., Sasaki et al. (1991) and Wagner and Weiner (1992), can predict the stiffness in one direction, but fail to predict properly the mechanical anisotropy of bone – the extent to which its stiffness is different when it is loaded in different directions. Probably the reasons for the lack of analytical solutions are that (1) the direction of the long axis of the mineral crystals varies markedly throughout the bone, over a few microns, and (2) we are really ignorant of the aspect ratio of the mineral crystals – the ratio of the greatest to the least dimension. These two

Fig. 14.5 Effect of taking
both mineral content and
porosity into account in the
determination of Young's
modulus. *Abscissa*: Predicted
Young's modulus, using an
equation having both mineral
content and porosity as
explanatory variables.
Ordinate: Observed Young's
modulus. Linear $R^2 = 63\%$

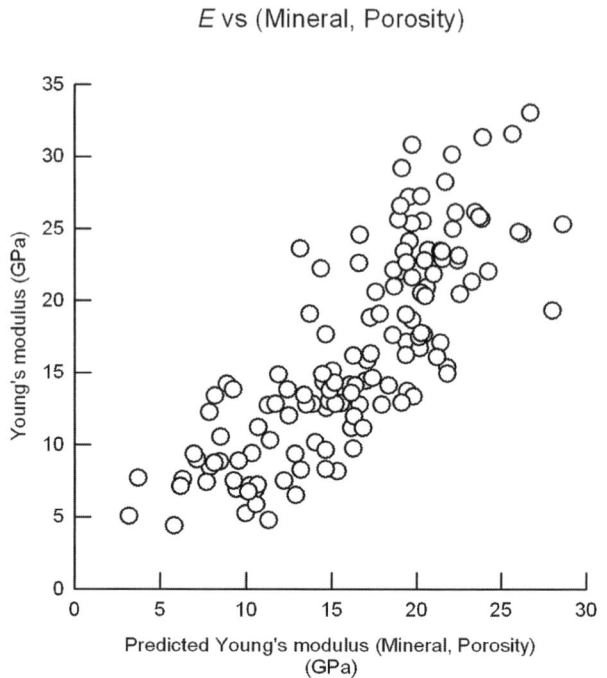

E vs (Mineral, Porosity)

variables are very important in any equation relating mineral volume fraction to Young's modulus of elasticity (Jäger and Fratzl, 2000). Fritsch and Hellmich (2007), on the other hand, claim that most mineral is outside the collagen and oriented every which way and can therefore be modeled as isotropic spherical objects in a mechanically anisotropic collagen matrix. They propose a micromechanical model of bone in which the anisotropy is predicted well and in which there are different anisotropies at different length scales. It is difficult to know what to make of such a (to me) counterintuitive treatment of mineral, but much of their modeling does seem to predict values close to those produced by ultrasonic testing.

What has been shown clearly is that when bone is strained, the strain in the mineral is less than the overall strain of the specimen. The ratio of whole-tissue strain to collagen fibril strain to mineral particle strain in wet bone has been studied in detail by Gupta and colleagues and is discussed fully in Chapter 7. For instance Gupta et al. (2006) found the ratios of these three strains to be about 12:5:2. Similar results, albeit on dry specimens, were found by Fujisaki and Tadano (2007). The considerable differences in the strains are to be expected, of course, but it is good to see the common-sense expectation to be borne out experimentally. On the other hand papers by Almer and Stock (2005, 2007) produce results that are less easy to understand. Using similar x-ray methods, but on larger specimens, these authors found quite large residual strains and calculated stresses when no overall stress was applied. For instance the mineral had a calculated compressive prestress of about −60 MPa. It is interesting that the idea of prestressing of bone (with very little real

Fig. 14.6 Total work under the tensile stress/strain curve as a function of mineral content. Note logged scales

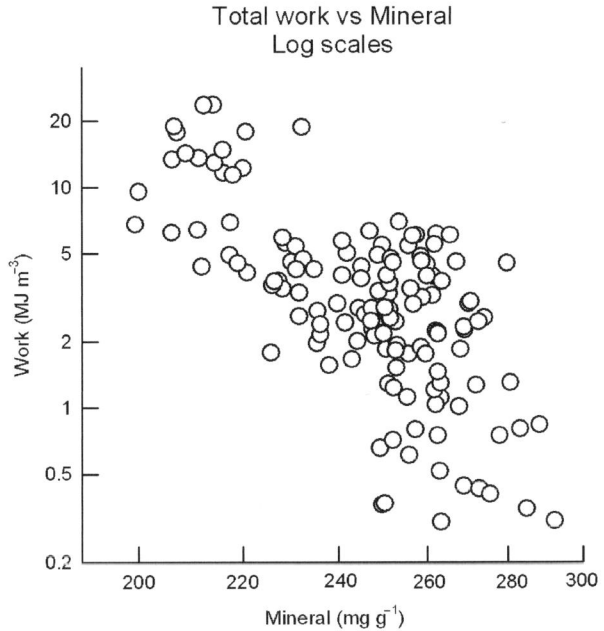

Total work vs Mineral
Log scales

evidence to back it up) has been in the literature for 50 years, since Knese (1958) proposed it. If the prestressing were real it would have important implications for the mechanical properties of bone. I have to admit that the mechanism by which such a large prestress could be induced is difficult to comprehend.

The effects of changes of mineral/collagen ratio on fracture are even more obscure than their effect on stiffness. What happens when bone yields (if it yields, rather than breaking completely brittlely) is that microfractures develop. These are a myriad of initially tiny fractures, or at least linear "disturbances" of the bone (Burr et al., 1985; Zioupos, 2001). They can be very short, say 5 μm long, though many workers concentrate on larger, more obvious cracks, say 100 μm long. These cracks initiate, and therefore increase the compliance of the specimen, but do not increase in length if the load is increased. The work done on extending the specimen is used up in multiplying the number of microfractures and thereby increasing the compliance of the specimen. Eventually some fuse, become really long and dangerous, and the specimen breaks into two. In bone of greater mineralization the process of initiation of microfractures is inhibited, and the bone becomes progressively more brittle. In hand-waving terms one can explain this by suggesting that as the mineralization increases there is a tendency for the mineral crystals to join up to form larger lumps, which are weaker than small lumps. Gao et al. (2003) point out that very small sizes of the mineral crystals do not allow them to develop fatal "Griffith" flaws, and therefore they will be very strong, and that any lack of cohesion will be either in the organic material or at the organic–mineral interface. Gao et al. also calculated that there is an optimum aspect ratio of the mineral crystals, and this relates to the organic content. Mother of pearl, in mollusc shell, has about 3% organic material and a rather low aspect ratio, about 10, while bone has a high organic content and a much higher aspect ratio, possibly as high as 30–40 in well-developed bone. These values for bone and for mother of pearl will tend to make the mineral and the organic material fracture at the same stress, the optimum situation. Ballarini et al. (2005), on the other hand, consider that the small size of the crystals is not important, that small crystals break and that bone depends on other features to prevent catastrophic failure.

A feature of the amount of collagen in mineralized tissues is that there is a gap in the collagen/mineral ratio in mineralized tissues between pure organic collagen-based materials (such as cartilage and tendon) and collagen-based mineralized tissues such as bone. This is shown in the ternary diagram, Fig. 14.2. This is an extract from many assessments that deliberately included both highly mineralized and poorly mineralized tissues. There are no mineralized tissues with an organic amount of more than 40% wet weight. The mechanically adaptive reason for this may be seen in Fig. 14.7 (Currey et al., 2004) showing the impact energy absorption and notch sensitivity of various mammalian mineralized bony tissues. As the mineral decreases the impact energy absorption increases, and the notch sensitivity (the reduction in mechanical properties induced by the presence of a notch) decreases. Indeed, antler cannot be shown on this diagram because, although antlers do break in the wild, it was virtually impossible to break wet laboratory specimens in impact. Since antler bone, the least mineralized of all known mammalian mineralized tissues, seems to be notch-insensitive in impact, little adaptive purpose would

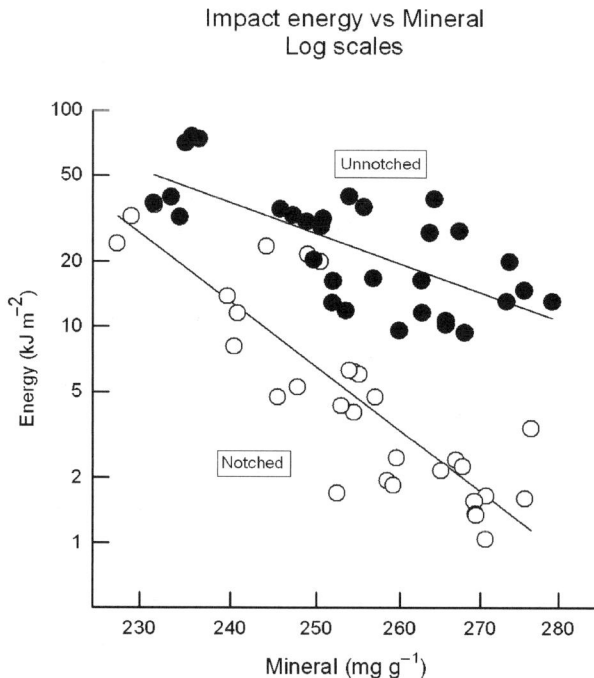

Fig. 14.7 Impact energy absorption as a function of mineral content (mg Ca g^{-1}). Notched and unnotched specimens. Note axes are logged. The antler specimens are not shown because they did not reliably break. The specimens are not the same as in Figs. 14.3, 14.4, 14.5 and 14.6

be served by having mineralized tissues of a lower mineralization than antler. It would hardly be any better in impact, yet would have a lower stiffness, which might cause problems in the pushing match that follows the initial clash of antlers. This may explain the cut-off in mineralization seen in mammals.

14.6 The Effect of Remodeling on Mechanical Properties

The bone of many vertebrates undergoes considerable modeling and remodeling. "Modeling" refers to the changing of the external form of the bone, by either the addition or removal of bone from external surfaces. The function of modeling is fairly clear: adapting the bone's shape to the loads falling on it. Roughly, bone is added where strains are higher, and removed where strains are lower. How this is brought about is extremely clinically important, still unclear and anyhow beyond the scope of this chapter. In contrast, internal "remodeling" refers to the production of new, secondary, bone within the old bone by the formation, within long cavities eaten out by osteoclasts, of cylindrical secondary osteones (often called Haversian systems). Internal remodeling has many functions attributed to it – taking out microcracks, improving the grain of the bone, preventing bone from becoming too highly mineralized and so on (Currey, 2002, pp. 368–277) – and there is no consensus yet as to the causes, although taking out microcracks is certainly one function.

For our purposes it is sufficient to note that newly laid down bone (always lamellar) is less highly mineralized than older bone. The process of mineralization in bone is rapid at first but then slows, approaching its final state asymptotically. This is because as the bone becomes mineralized the rate of diffusion of ions through the tissue of the bone inevitably declines. As a result secondary osteones, after they have been in being for some time, are more highly mineralized near their core, where the blood channel is, and are less highly mineralized at the periphery. There is then a sudden upward jump in mineralization as one passes out of the Haversian system into the "interstitial lamellae" which are pre-existing and therefore necessarily older than the secondary osteone. This variation in mineralization is reflected in the elastic properties as shown by nanoindentation (Rho et al., 1999). Whether these mechanical heterogeneities will make the bone stronger or weaker is unclear. Probably, by introducing sharp strain gradients, they will make the initiation of cracks easier. However, the perimeter of the secondary osteone, where there is a "cement sheath" of different properties from the rest of the bone, seems to be a line of weakness, which breaks up as the crack approaches and will make it difficult for cracks to spread into, and therefore through, the secondary osteone (Yeni and Norman, 2000). O'Brien et al. (2007) have loaded bone in fatigue and show that as bone starts to accumulate damage there is initially a rapid multiplication of microcracks, but these are prevented from spreading by running into the cement lines round secondary osteones. However, once the cracks have reached a length of about 300 μm the secondary osteones seemed to act as volumes of weakness, and actually helped the cracks to spread.

Like compact bone, cancellous bone undergoes modeling/remodeling. However, because the trabeculae and sheets are rather thin, it is not often that complete secondary osteones appear. Instead hemi-osteones are often formed, in which a gutter is scooped out of the bone by osteoclasts and then filled in with new bone. This all means that the difference between modeling and remodeling is not as clear-cut as it is in compact bone. Cancellous bone is probably more intensively remodeled than compact bone, and its half-life will be lower. This is probably the main reason why it has a lower Young's modulus than nearby compact bone.

14.7 A Natural Experiment

To round off this account of the effect of mineral on the properties of bone I show the results of an old experiment (Currey, 1979) that nevertheless makes a point very well. The mechanical properties, the density and the mineral content of three bony tissues were compared. These tissues are the antler bone of the red deer *Cervus elaphus*, the femur of a cow and the tympanic bulla of the fin whale *Balaenoptera physalus*. These bones vary considerably in their mineral content (and therefore in their collagen content). However, the difference in the mechanical properties is greater than the difference in mineral (Table 14.3).

Table 14.3 Physical properties of three bony tissues, derived from Currey (1979)

Property	Antler	Femur	Bulla
Work of fracture ($J m^{-2}$)	6190	2800	20
Bending strength (MPa)	179	247	33
Young's modulus (GPa)	7.4	13.5	31.1
Mineral content (% ash)	45	65	86
Density ($10^3 kg m^{-3}$)	1.9	2.1	2.5

Table 14.3 shows that the mineral content of the bulla is almost twice as great as that of the antler, but the density of the bulla is only just over 30% greater. Young's modulus of the bulla is over four times as great, whereas the bending strength of the bulla is only one fifth as great as that of the antler, and its work of fracture (a measure of the work necessary to drive a crack through the material) is only 0.3% of the antler value. Large, but not huge, differences in density are associated with very large differences in mechanical properties. Particularly notable is that the greater mineral and density are associated with a large increase in stiffness, but an even greater decline in work of fracture.

These differences are appropriate for the functions of the various bones. The high collagen content of the antler makes it tough without reducing its stiffness completely, and this is what is needed for a structure that is to be used in fights in which the critical moment is when the antlers collide in impact, but where some stiffness is required for the pushing match that follows. Indeed, as mentioned above, it is almost impossible to break antler specimens in impact. The bulla with little collagen needs, for acoustical reasons (Currey, 2002, pp. 127–129), to be very stiff and is hardly loaded at all. Therefore its strength and toughness properties are unimportant. The cow's femur, being a relatively unspecialized long bone, is intermediate between the antler and the bulla, though having nothing like the pathetic strength and toughness characteristics of the bulla.

14.8 The Effect of Differences of Collagen on Bone's Mechanical Properties

14.8.1 Aging

The changes that take place in the collagen with aging and with the development of osteoarthritis are described by Bailey and Avery (Chapter 4) The non-enzymic glycation in which sugar molecules are involved (Knott and Bailey, 1998) is, in some soft tissues, particularly in diabetes, very deleterious (Paul and Bailey, 1996) but its effect in bone is less clear. In bone bonds produced by glycation will be continually removed by remodeling, in which older bone is replaced by young bone. During ordinary, non-osteoporotic, non-osteoarthritic aging of cancellous bone, although the amount of collagen decreases relative to the mineral, the various biochemical variables seem fairly stable (Bailey et al., 1999). Recent developments in spectroscopic methods are making it possible to investigate the distribution and development of these collagen cross-links over small distances within bony tissue (Ou-Yang et al., 2002), which holds great promise for greater resolution of mechanical/collagen interactions.

A problem for any mechanical analysis of aging effects is that the collagen may change in deleterious ways and the bone's mechanical properties may also decline as a function of age, but this association may not be causal. It may be possible to distinguish the effects, but one usually needs large sample sizes or tricky experimentation. For instance, Wang et al. (2002) examined 30 human femora with a large age range. They showed that various mechanical properties declined with age. They also showed that the mechanical properties of the demineralized bone declined, that is, that the "collagen network" became less stiff, strong and tough. They were able to factor out the effect of bone's increasing porosity that occurs with aging, and when this had been done reduction in the strength of the collagen network appeared to be particularly related to reduction in the post-yield "toughening" part of the curve of whole bone. At a common-sense level this is convincing. It remains possible (though unlikely), nevertheless, that both the collagen network behavior and the whole bone's behavior were covarying with some other undiscovered feature that was also changing with age. These authors also suggested that the glycation of the collagen that occurs with aging may be degrading the mechanical properties of the network, but possibly again this glycation could simply be a function of time and have no important effect on the mechanical properties.

Vashishth et al. (2001) applied a ribose treatment to mature bovine compact bone specimens, which increased mature-type cross-linking. In apparent distinction to the results of Wang et al. (2002) demineralized glycated specimens were much stiffer and stronger than controls. The mineralized specimens showed much less difference. Yield stress and strain were slightly higher in the glycated group, but post-yield strain and work and damage accumulated before fracture were lower in the glycated group, though not always significantly so. These results are from artificially glycated specimens. In a more recent paper from this group (Tang et al., 2007) similar and rather more consistent mechanical results were found for the effects of ribosylation on human cancellous bone. However, glycation of tendon collagen in vitro produces very similar morphological changes, such as an increase in mean collagen fibril diameter, to those seen in diabetic tendon (Odetti et al., 2000); so, although collagen fiber diameter presumably remains unchanged in bone, in vitro glycation is probably a valid model for investigating glycation-induced changes in bone and, importantly, there is no aging effect to contend with.

Sometimes it is possible to factor out the effect of time itself. For instance Zioupos et al. (1999) showed that the shrinkage temperature of the collagen in demineralized bone decreased with age, as did the force at which it contracted during shrinkage, suggesting that the collagen fibers were less stable. Both these changes correlated with the toughness. However, once the effect of age had been "partialled out" only the force of the contraction, but not the shrinkage temperature, remained significantly correlated with toughness.

These results together suggest that there is something about aging collagen that affects the post-yield properties of whole bone, though whether this is glycation, which presumably increases collagen stability, or a reduction in collagen stability, as suggested by the results of Zioupos et al., is unclear.

14.8.2 Osteoporosis and Osteoarthritis

In osteoporosis the collagen has a reduced level of the immature cross-links and the collagen fibrils tend to be narrower and apparently more disorganized (Tzaphlidou, 2005). Nevertheless, the evidence that this has much effect on the mechanical properties is scanty. It would be very interesting to know whether the anomalous collagen structure becomes more marked as osteoporotic osteopenia develops.

In osteoarthritis the main focus of interest is on changes in the cartilage collagen. However, there are changes in the subchondral bone, a homotrimer is present in much more than usual quantities, and probably as a result the collagen fibers are narrower, more disorganized, and less stable (Bailey et al., 2002).

14.8.3 Osteogenesis Imperfecta

This disease is a case study in how difficult it is to separate out the effects of collagen changes from other changes in their effect on mechanical properties. Osteogenesis imperfecta (brittle bone disease) is a heritable type I collagen disorder which results in weak bones, though whether they are more brittle is less clear. The mouse osteo-genesis imperfecta model (oim) which seems to have similar phenotype to severe OI in humans is frequently studied. In these mice, post-yield deformation is reduced, and the fracture surface is smoother, as is characteristic of brittle fracture (Jepsen et al., 1996). However, care is needed, because the effect may be partially indi-rect. Although the collagen is obviously seriously deranged, this itself leads to a derangement of mineral size and shape; the mineral crystals are more variable in size and orientation (Grabner et al., 2001). In composite materials technology it is known that the size and distribution of the stiffer component (in bone the mineral) is usually much more important for mechanical virtue than the properties of the more compliant material (in bone the organics) (Lucchinetti, 2001). Furthermore, the ratio of woven to lamellar bone increases (Jepsen et al., 1996), and the mineral/organic ratio increases leading to a greater microhardness (Grabner et al., 2001). An increase in the mineral/organic ratio in a set of normal bones reduces the post-yield strain, and therefore this itself will make the bone more brittle.

Nevertheless, Misof et al. (1997) showed that the type I collagen in the unmin-eralized tail tendon of oim mice has the same stiffness as that of normals, but a lower yield point, and a sharply reduced strength and toughness. If this effect is also found in collagen of the bones, it could explain many of the changes found in OI bone material. A difficulty with this explanation is that the differences in the rat tail tendon collagen appear only after a strain of about 0.04, by which time bone would have already fractured and, even earlier, yielded. However, this difficulty might possibly be resolved if we knew the local strain occurring when microcracks develop in a bone specimen. It could well be that this is considerably greater than the overall strain (0.004–0.01) at which yield starts in ordinary bone. This might be the case, as is shown by the digital imaging correlation studies of Nicolella et al. (2006).

These studies show that the weakness of OI bones is probably caused by the derangement of the collagen molecule. However, analyzing the separate mechanical effects of weakness of the collagen network, changes in the size and shape of the mineral crystals, in the histology and in the organic/mineral ratio have barely started.

14.9 Calcified Cartilage

There are two types of bone ossification, intramembranous and endochondral. Intramembranous ossification occurs in membranes and need not concern us here. Endochondral ossification takes place in a cartilage anlage (or model). The cartilage mineralizes with calcium phosphate. When completely mineralized, the cartilage cells undergo apoptosis, blood vessels invade the calcified cartilage, then osteo-clasts invade and destroy the calcified cartilage model, and bone is laid down by osteoblasts. The typical place where this process occurs is in the metaphyses of long bones. Here, during long bone growth in length, calcified cartilage forms by proliferation of chondrocytes and mineralization of the rather watery matrix pro-duced by these chondrocytes. The calcified cartilage extends in long columns which are replaced very quickly by "proper" bone. Although the cartilage is nearly all replaced by bone, isolated spicules of unresorbed cartilage remain, which are found by histochemical means to have type II collagen (Horton et al., 1983). The mechanical properties of this calcified cartilage are hardly of interest, since it occurs so ephemerally (though fracture through the epiphyseal plate, where the columns originate, is a reasonably common one in growing children). It is also possible that the isolated spicules remaining may somewhat degrade the mechanical properties of the metaphysis, through stress-concentrating effects. Nothing is known about this possibility, however.

Since the cartilage that undergoes mineralization has type II collagen, it is therefore not the main concern of this book. However, it is appropriate to mention it because in one group of animals, the chondrichthyean fish, mineralized cartilage forms proper skeletal elements, very like bones in some ways. Although the fine structure of these tissues is at present unknown, it seems that the relationship between the (type II) collagen and the mineral is less intimate than the collagen–mineral relationship in bone (Dean and Summers, 2006). One says "type II", though in fact the evidence that it is type II collagen that mineralizes is fairly scanty. Summers says that his preliminary SDS-Page studies strongly suggest that both the unmineralized and mineralized parts of jaws from various elasmobranchs have type II and not type I collagen, but there seem to be no studies that show this unequivocally (Adam Summers, personal communication). Nevertheless, it would be strange if the type II collagen became replaced by type I collagen just before calcification.

The cartilage of chondrichthyean fish may become partially calcified, particularly with a thin surface layer of mineralized blocks called "tesserae". In some fish the tesserae become quite thick and may also increase in number, so that there are several layers of thick tesserae. These and similar arrangements are described by Dean and Summers (2006). Of particular interest are the quite complex structures that can be made from this rather unpromising material. This is perhaps best shown in the stingrays (Summers, 2000). Many species of this group of fish are "durophagous", crushing hard-shelled prey such as snails between cylindrical tooth plates and necessarily imposing large loads on their own skeleton. This becomes heavily mineralized with many layers of tesserae, sometimes up to six. Even more interesting is the fact that mineralized trabeculae are formed, taking loads from underneath the tooth plates to the thickened parts of the jaw. These trabeculae, which perform the same kind of function as the trabeculae in bone, are often hollow. This is an advance on bony trabeculae, because they produce a greater stiffness per unit mass than do solid struts. Producing hollow trabeculae seems not to be part of the adaptive repertoire of bone.

The trabeculae of these durophagous fish transmit large compressive loads. However, there is another potential use of trabeculae, seen in sandwich structures in bones such as the skull vault and the iliac crest. This is to keep the two rinds of the sandwich apart, and so maintain a large second moment of area. The loads taken by the trabeculae are much less in this case. Trabeculae performing such a function are also found in Chondrichthyes, for instance in *Narcine brasiliensis*. This fish does not crush its prey but protrudes its jaw into the soft substrate to snap up worms. The jaw structures are such that buckling could be a problem. The design of the trabeculae is quite different from that found in the durophagous fish and is appropriate to prevent this buckling (Dean et al., 2006).

Unfortunately, it seems that at the moment we have no good idea of the mechanical properties of calcified cartilage tissue. This is partially because rather few people have worked on it and they have more urgent things to find out, and partially because one finds in calcified cartilage the mineral and the remnant cartilage in close proximity, so it is difficult to know what is being tested. This is shown in the interesting paper by Porter et al. (2006) who examined the mechanical properties of the calcified vertebrae of a variety of chondrichthyean species. They found very low values, about 20 MPa, for compressive strength (as opposed to say 250 MPa for compact bone) and 0.5 MPa for stiffness (a remarkably low value, as opposed to say 20 GPa for compact bone). However, as the authors state, they were testing whole vertebral centra, whose cross-section is a mixture of calcified and non-calcified cartilage. It is not surprising, therefore, that the values they obtained were like those obtained from cancellous bone material. Probably nanoindentation could be used to sort out this business of modulus.

There remains the question, which I shall pose as a question only, as to why type II collagen seems unable to produce proper bone-like material. Unfortunately, as Dean and Summers (2006) state, referring to the endoskeletons of Chondrichthyes, "There is no nanostructural data on the orientation and size of these [hydroxyapatite] crystals". Nevertheless it is clear from the studies of these and other workers that the crystals come in rather large lumps, considerably larger than those of bone. The small size of the crystals in bone is the key to bone's excellent mechanical properties, and one supposes that, were it possible, the hydroxyapatite associated with type II collagen would be adaptive were it similarly small. Somehow the mineral cannot have the intimate relationship with the collagen fibrils that is so characteristic of bone.

References

Alexander RMcN (1981) Factors of safety in the structure of mammals. Sci Prog 67:109–130.

Almer JD, Stock SR (2005) Internal strains and stresses measured in cortical bone via high-energy X-ray diffraction. J Struct Biol 152:14–27.

Almer JD, Stock SR (2007) Micromechanical responses of mineral and collagen phases in bone. J Struct Biol 157:365–370.

Bailey AJ, Sims TJ, Ebbesen EN, Mansell JP, Thomsen JS, Mosekilde Li (1999) Age-related changes in the biochemical properties of human cancellous bone collagen: relationship to bone strength. Calcif Tissue Int 65:203–210.

Bailey AJ, Sims TJ, Knott L (2002) Phenotypic expression of osteoblast collagen in osteoarthritic bone: production of type I homotrimer. Int J Biochem Cell Biol 34:176–182.

Ballarini R, Kayacan R, Ulm F-J, Belytschko T, Heuer AH (2005) Biological structures mitigate catastrophic fracture through various strategies. Int J Fract 135:187–197.

Bini F, Marinozzi A, Marinozzi F, Patanè F (2002) Microtensile measurements of single trabeculae stiffness in human femur. J Biomech 35:1515–1519.

Blob RW, Bewiener AA (1999) In vivo locomotor strain in the hindlimb bones of *Alligator mississipiensis* and *Iguana iguana*: implications for the evolution of limb bone safety factor and non-sprawling limb posture. J Exp Biol 202:1023–1246.

Burr DB, Martin RB, Schaffler MB, Radin EL (1985) Bone remodeling in response to in vivo fatigue microdamage. J Biomech 18:189–200.

Caler WE, Carter DR (1989) Bone creep-fatigue damage accumulation. J Biomech 22:635–635.

Cezayirlioglu H, Bahniuk E, Davy DT, Heiple KG (1985) Anisotropic yield behavior of bone under combined axial force and torque. J Biomech 18:61–69.

Currey JD (1979) Mechanical properties of bone tissues with greatly differing functions. J Biomech 12:313–319.

Currey JD (1999) What determines the bending strength of compact bone? J Exp Biol 202: 2495–2503.

Currey JD (2002) Bones: Structure and Mechanics. Princeton, Princeton University Press.

Currey J (2004) Incompatible mechanical properties in compact bone. J Theor Biol 231:569–580.

Currey JD, Abeysekera RM (2003) The microhardness and fracture surface of the petrodentine of Lepidosiren (Dipnoi), and of other mineralised tissues. Arch Oral Biol 48:439–447.

Currey JD, Zioupos P, Davies P, Casinos A. (2001) Mechanical properties of nacre and highly mineralized bone. Proc R Soc Lond B 268:107–111.

Currey JD, Brear K, Zioupos P (2004) Notch sensitivity of mammalian mineralized tissues in impact. Proc R Soc Lond B Biol Sci 271:517–522.

Dean MN Summers AP (2006) Mineralized cartilage in the skeleton of chondrichthyean fishes. Zoology 109:164–168.

Dean MN, Huber DR, Nance HA (2006) Functional morphology of jaw trabeculation in the lesser electric ray *Narcine brasiliensis*, with comments on the evolution of structural support in the Batoidea. J Morph 267:1137–1146.

de Buffrénil V, Dabin W, Zylberberg L (2004) Histology and growth of the cetacean petro-tympanic bone complex. J Zool 262:371–381.

Donoghue PCJ, Sansom IJ, Downs JP (2006) Early evolution of vertebrate skeletal tissues and cellular interactions, and the canalization of skeletal development. J Exp Zool B Mol Dev Evol 306B:278–294.

Francillon-Vieillot H, de Buffrénil V, Castanet J, Géraudie J, Meunier FJ, Sire JY, Zylberberg L, de Riqlès A (1990) Microstructure and mineralization of vertebrate skeletal tissues. In Carter (1990), Vol. I, pp. 471–530.

Fritsch A, Hellmich C (2007) 'Universal' microstructural patterns in cortical and trabecular, extracellular and extravascular bone materials: micromechanics-based prediction of anisotropic elasticity. J Theor Biol 244:597–620.

Fujisaki K, Tadano S (2007) Relationship between bone tissue strain and lattice strain of HAp crystals in bovine cortical bone under tensile loading. J Biomech 40:1832–1838.

Gao H, Baohua JI, Jäger IL, Arzt E, Fratzl P (2003) Materials become insensitive to flaws at nanoscale: Lessons from nature. PNAS 100:5597–5600.

Grabner B, Landis WJ, Roschger P, Rinnerthaler S, Peterlik H, Klaushofer K, Fratzl P (2001) Age- and genotype-dependence of bone material properties in the osteogenesis imperfecta murine model (oim). Bone 29:453–457.

Gupta HS, Seto J, Wagermaier W, Zaslansky P, Boeseke P, Fratzl P (2006) Cooperative deformation of mineral and collagen in bone at the nanoscale. PNAS 103:17741–17746.

Hellmich CH, Ulm FJ (2002) Are mineralized tissues open crystal foams reinforced by crosslinked collagen? some energy arguments. J Biomech 35:1199–1212.

Hodgskinson R, Currey JD (1992) Young's modulus, density and material properties in cancellous bone over a large density range. J Mater Sci Mater Med 3:377–381.

Horton WA, Dwyer C, Goering R, Dean DC (1983) Immunohistochemistry of type I and type II collagen in undecalcified skeletal tissues. J Histochem Cytochem 31:417–425.

Jäger I, Fratzl P (2000) Mineralized collagen fibrils: a mechanical model with a staggered arrangement of mineral particles. Biophys J 79:1737–1746.

Jepsen KJ, Goldstein SA, Kuhn JL, Schaffler MB, Bonadio J (1996) Type-I collagen mutation compromises the post-yield behavior of Mov13 long bone. J Orthop Res 14:493–499.

Katz JL (1971) Hard tissue as a composite material. I. Bounds on the elastic behavior. J Biomech 4:455–473.

Kawasaki K, Suzuki T, Weiss KM (2004) Genetic basis for the evolution of vertebrate mineralized tissue. PNAS 101:11356–11361.

Kim JH, Niinomi M, Akahori T, Toda H 2007 Fatigue properties of bovine compact bones that have different microstructures. Int J Fatigue 29:1039–1050

Knese K-H (1958) Knochensruktur als Verbundbau. In: Zwanaglose Abhandlungen aus dem Gebiet der normalen und pathologischen Anatomie. Editors W Bargmann, DW Stuttgart. Georg Thieme, pp. 1–56.

Knott L, Bailey AJ (1998) Collagen cross-links in mineralizing tissues: a review of their chemistry, function and clinical relevance. Bone 22:181–187.

Landis WJ, Hodgens KJ, Arena J, Song MJ, McEwen BF (1996) Structural relations between collagen and mineral in bone as determined by high voltage electron microscopic tomography. Microsc Res Tech 33:192–202.

Luchinetti E (2001) Composite models of bone properties. In: Bone Mechanics Handbook. Editor S.C. Cowin. CRC Press, Boca Raton, pp. 12–19.

Misof K, Landis WJ, Klaushofer K, Fratzl P (1997) Collagen from the osteogenesis imperfecta mouse model (oim) shows reduced resistance against tensile stress. J Clin Invest 100: 40–45.

Nian-Zhong W, Donoghue PCJ, Smith MM, Sansom IJ (2005) Histology of the galeaspid dermoskeleton and endoskeleton, and the origin and early evolution of the vertebrate cranial endoskeleton. J Vert Paleont 25:745–756.

Nicolella DP, Moravits DE, Gale AM, Bonewald LF, Lankford J (2006) Osteocyte lacunae strain in cortical bone. J Biomech 39:1735–1743.

O'Brien FJ, Taylor D, Lee TC (2007) Bone as a composite material: the role of osteons as barriers to crack growth in compact bone. Int J Fatigue 29:1051–1056.

Odetti P, Aragno I, Rolandi R, Garibaldi S, Valentini S, Cosso L, Traverso N, Cottalasso D, Pronzato MA, Marinari UM (2000) Scanning force microscopy reveals structural alterations in diabetic rat collagen fibrils: role of protein glycation. Diabetes Metab Res Rev 16:74–81.

Ou-Yang H, Paschalis EP, Boskey AL, Mendelsohn R (2002) Chemical structure-based three-dimensional reconstruction of human cortical bone from two-dimensional infrared images. Appl Spectrosc 56:419–422.

Paul RG, Bailey AJ (1996) Glycation of collagen: the basis of its central role in the late complications of ageing and diabetes. Int J Biochem Cell Biol 28:1297–1310

Pidaparti RMV, Chandran A, Takano Y, Turner CH (1996) Bone mineral lies mainly outside collagen fibrils: predictions of a composite model for osteonal bone. J Biomech 29:909–916.

Porter ME, Beltrán JL, Koob TJ, Summers AP (2006) Material properties and biochemical composition of mineralized vertebral cartilage in seven elasmobranch species (Chondrichthyes). J Exp Biol 209:2920–2928.

Reif WE (2002) Evolution of the dermal skeleton of vertebrates: concepts and methods. Neues Jahrb Geol Palaontol Abh 223:53–78.

Reilly DT, Burstein AH (1975) The elastic and ultimate properties of compact bone tissue. J Biomech 8:393–405.

Rho JY, Zioupos P, Currey JD, Pharr GM (1999) Variations in the individual thick lamellar properties within osteons by nanoindentation. Bone 25:295–300.

Rice JC, Cowin SC, Bowman JA (1988) On the dependence of elasticity and strength of cancellous bone on apparent density. J Biomech 21:155–168.

Rogers KD, Zioupos P (1999) The bone tissue of the rostrum of a Mesoplodon densirostris whale: a mammalian biomineral demonstrating extreme texture. J Mater Sci Lett 18:51–654.

Sasagawa I, Ishiyama M, Akai J (2006) Cellular influence in the formation of enameloid during odontogenesis in bony fishes. Mat Sci Engng C Biomimet Supramolec Syst 26:630–634.

Sasaki N, Ikawa T, Fukuda A (1991) Orientation of mineral in bovine bone and the anisotropic mechanical properties of plexiform bone. J Biomech 24:57–61.

Sasaki N, Tagami A, Goto T, Taniguchi M, Nakata M, Hikichi K (2002) Atomic force microscopic studies on the structure of bovine femoral cortical bone at the collagen fibril-mineral level. J Mater Sci Mater Med 13:333–337.

Summers AP (2000) Stiffening the stingray skeleton an investigation of durophagy in myliobatid stingrays (Chondrichthyes, Batoidea, Mylioatidae). J Morphol 243:113–126.

Tang SY, Zeenath U, Vashisth D (2007) Effects of non-enzymatic glycation on cancellous bone fragility. Bone 40:1144–1151.

Tzaphlidou M. (2005) The role of collagen in bone structure: an image processing approach. Micron 36:593–601.

Vashishth D, Gibson GJ, Khoury JI, Schaffler MB, Kimura J, Fyrhie DP (2001) Influence of nonenzymatic glycation on biomechanical properties of cortical bone. Bone 28:195–201.

Wagner HD, Weiner S (1992) On the relationship between the microstructure of bone and its mechanical stiffness. J Biomech 25:1311–1320.

Walsh WR, Guzelsu N (1994) Compressive properties of cortical bone – mineral organic interfacial bonding. Biomat 15:137–145.

Wang X, Shen X, Li X, Agrawal CM (2002) Age-related changes in the collagen network and toughness of bone. Bone 31:1–7.

Weiner S, Traub W, Wagner HD (1999) Lamellar bone: structure-function relations. J Struct Biol 126:241–255.

Wilson EE, Awonusi A, Morris MD, Kohn DH, Tecklenburg MMJ, Beck LW (2006) Three structural roles for water in bone observed by solid-state NMR. Biophys J 90:3722–3731.

Yeni YN, Norman TL (2000) Calculation of porosity and osteonal cement line effects on the effective fracture toughness of cortical bone in longitudinal crack growth. J Biomed Mater Res 51:504–509.

Zioupos P (2001) Accumulation of in-vivo fatigue microdamage and its relation to biomechanical properties in ageing human cortical bone. J Microsc-Oxford 201:270–278.

Zioupos P, Wang X-T, Currey JD (1996) Experimental and theoretical quantification of the development of damage in fatigue tests of bone and antler. J. Biomech 29:989–1002

Zioupos P, Currey JD, Casinos A, de Buffrénil V (1997) Mechanical properties of the rostrum of the whale Mesoplodon densirostris, a remarkably dense bony tissue. J Zool 241:725–737

Zioupos P, Currey JD, Hamer AJ (1999) The role of collagen in the declining mechanical properties of aging human cortical bone. J Biomed Mater Res 45:108–116.

Zioupos P, Currey JD, Casinos A (2001) Tensile fatigue in bone: are cycles-, or time to failure, or both, important? J Theor Biol 210:389–399.

Ziv V, Weiner S (1994) Bone crystal sizes: a comparison of transmission electron-microscopic and X-ray-diffraction line-width broadening techniques. Connect Tissue Res 30:165–175.

Zylberberg L, Traub W, de Buffrénil V, Allizard F, Arad T, Weiner S (1998) Rostrum of a toothed whale: Ultrastructural study of a very dense bone. Bone 23:241–247.

Chapter 15
Dentin

P. Zaslansky

Abstract Dentin, composed mainly of dahllite and fibrillar collagen, has a complex irregular microstructure. It has several distinct features such as micron-sized tubules that traverse the entire thickness, a large hollow pulp cavity and an external stiff cap. Dentin forms the foundation of teeth and provides a mechanical backing for the enamel cap. Being vital and highly innervated, it is a sensitive tissue, capable of responding to both mechanical and chemical stimulations from the environment. High permeability to fluid-flow through the tubules as well as a directional design suggest that dentin has sensory functions related to pressure that may be exerted on the outer surfaces.

Average mechanical properties have been reported for dentin: E = modulus of about 20 GPa, compressive and shear strengths amounting to 250 MPa whereas tensile strength is only about 50 MPa. Fracture toughness measurements found a range of 1.5 to 3.5 MPa$\sqrt{}$m, and work of fracture estimates span from 250 to 550 J/m^2. Some anisotropy exists, yet values of all properties seen to vary with both location and orientation.

Much of the mineral in crown dentin is found in peritubular dentin-encircling tubules, where no collagen is found. Between tubules, intertubular dentin is seen as a mesh of mineralized collagen fibrils. Fibril arrangements are more regular in the root, where they are set in incremental layers, orthogonal to the tubules. In the crown however, many fibrils are arranged at angles to and also along the orientation of the tubules.

Various types including of dentin are known mantle dentin – confined to a narrow region beneath the junction with enamel and slow-forming secondary dentin – appearing over the years in the pulp; Non-normal dentin types such as reactionary/reparative dentin, interglobular dentin and caries display significant variations in their structure, resulting in altered mechanical properties.

15.1 Introduction

Dentin is a collagen and calcium phosphate-based biocomposite that forms the bulk of mammalian teeth and is found in almost all vertebrates (birds being a major exception). It is partially covered by a thin, stiff and brittle layer of enamel in areas

P. Fratzl (ed.), *Collagen: Structure and Mechanics*,
© Springer Science+Business Media, LLC 2008

of the tooth crown, protruding into the oral cavity (Fig. 15.1). Dentin also forms the bulk of one or several roots and houses the pulp tissue. Blood vessels and nerve bundles enter the pulp chamber through canals within the roots. This allows fully grown teeth to react or respond to external stimuli by sensation or tissue formation. Root dentin is physiologically lined on the surface by a material termed cementum (50–200 μm thick), which has a central role in the attachment of teeth to the jaw and gums. Cementum is structurally similar to the bone, and forms an anchoring layer for a thin 150–400 μm periodontal ligament of connective tissue holding the roots in sockets in the surrounding alveolar bone. An excellent reference for the histology, development and structure of teeth is Ten Cate's oral histology (Nanci 2003).

Dentin in human teeth is the most extensively researched type of dentin. Much is now known about its development and characteristics, following almost 200 years of investigation. Nevertheless, many open questions remain, awaiting resolution in the context of both mechanical and biological fields of research. The purpose of this chapter is to introduce the structure and function of human dentin, which is considered as a remarkable variant of the collagen-based mineralized tissues. This is however by no means a comprehensive summary of the enormous body of work and the multitude of publications in the field. For example, only few aspects of the complex biology of dentin are discussed. The interested reader is referred to extensive reviews such as Pashley (1996) and those found in Linde (1984) for further information. It is hoped that this chapter can be used as a reasonable starting point for anyone wanting to walk the complex path of understanding human tooth dentin.

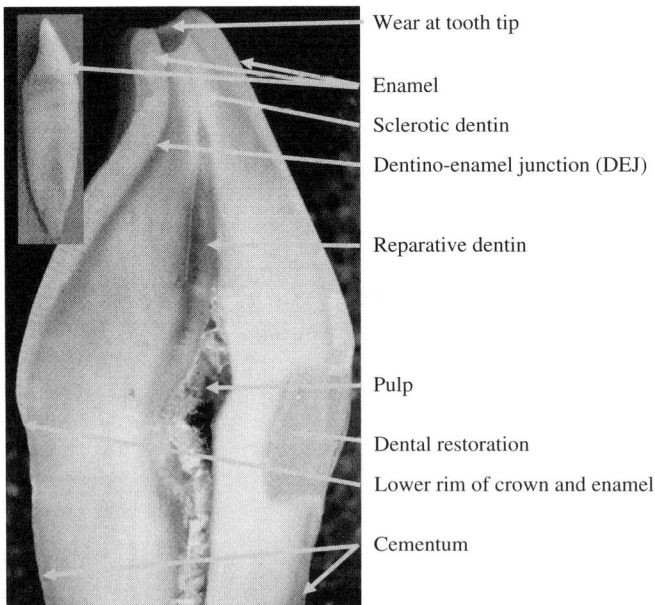

Wear at tooth tip

Enamel

Sclerotic dentin

Dentino-enamel junction (DEJ)

Reparative dentin

Pulp

Dental restoration

Lower rim of crown and enamel

Cementum

Fig. 15.1 Tissues of the tooth (sliced and exposed slice; inset whole human tooth sized 25 mm)

15.2 Composition and Main Features

Dentin is a member of the bone family of materials, which include deer antler, mineralized turkey tendon and ivory (Weiner and Wagner 1998). Its basic constituents are apatite mineral, organic macromolecules and water, with corresponding volume fraction ratios of about 45:33:22, respectively (corresponding weight fraction ratios of 65:22:13). The mineral is a calcium phosphate salt identified as dahllite, composed of carbonated hydroxyl-apatite nano-crystals, sized $\sim 36 \times 25 \times 10$ nm (Schroeder 85). The organic component is mostly type I collagen (about 90%) in the form of a highly cross-linked fibrous mesh. Small amounts of other types of collagen (IV, V and even VI) have also been reported (see review by Goldberg et al. 1995). Other organic components are mostly non-collagenous proteins classified as being either phosphorylated or non-phosphorylated. Additionally, small amounts of proteoglycans, neutral and acidic mucopolysaccharides and lipids are found.

Dentin in mammals is formed by connective-tissue derived cells termed, odontoblasts. These cells lay out a complex three-dimensional arrangement of thin, unmineralized predentin in a process that is followed by the production of enamel and cementum tissues (reviewed in Dean 2006) on the outer part of the tooth (long before eruption into the mouth. On average these tissues develop at a rate of about 4 µm/day (Nanci 2003). As dentin is mineralized, the odontoblasts continuously retract into the pulp region, where they integrate into an internal soft connective-tissue chamber surrounded by mature circumpulpal dentin. After tooth formation has ended, these very same cells continue to produce dentin at a slow but steady pace. Odontoblasts are distinct in mineralized tissues in that they have very long cellular protrusions (odontoblast processes). These processes (at least 500 µm long; Garberoglio and Brannstrom 1976, Goracci et al. 1999) extend far out into dentin surrounding the pulp, radiating outward through long and narrow tubules that they created. The tubules, many of which are lined by a dense mineral collar, have internal diameters of about 1 µm for most of their length, which may extend up to several millimeters. Near the pulp, the tubules converge and their internal diameters double. Away from the pulp, the tubules traverse the entire dentin thickness, thus ending adjacent to cementum at the dentino-cemental junction in the root, or near the dentino-enamel junction (DEJ) in the crown (Fig. 15.1). Near these junctions, each tubule branches and terminates, although on occasion, they may penetrate into the adjacent tissue (Jones and Boyde 1984).

Dentin is thus characterized by being a thick, a-cellular, highly organized tubular (porous) bulk, tightly attached to and riddled with living cell extensions. Consequently, dentin in vital teeth is part of an active biological system, often referred to as the dentin–pulp complex (Pashley 1996).

15.3 Functions of Dentin

The major volume of dentin is found in the roots, and perhaps, the most basic of its functions is to *form the foundation* of teeth. Variations are well known for the size, shape and number of roots and root canals. For example, front teeth situated close to the midline such as incisors and canines have only one root, whereas upper molars usually have three different sized roots (see, for example, Ash and Nelson 2003). Identical dentin material is thus used in different geometries, presumably well suited for redistribution of very different load requirements during function (front teeth serve mainly for piercing and cutting, whereas posterior teeth, including premolars and molars, serve more for crushing and grinding). This is particularly intriguing because once fully formed, the shapes of the teeth are fixed and there is no regrowing (or remodeling) of fractured tooth structure. Thus, dentin obviously excels in providing long-lasting structural foundations which – unlike most other biological tissues in humans – do not appear to undergo biological replacement. Some increase in mineral content does occur over time, and also small amounts of dentin might fill in the pulp cavities over many years (Murray et al. 2002), but no cellular activity deliberately aimed at replacing tissue is known.

Structural durability of teeth is not due only to properties of dentin but rather – it seems – is a consequence of how the complete structures are built. In humans, as in many other species, the long-term mechanical function of teeth, namely transmitting mastication forces from the jaw into foods so as to break them up, relies heavily on the existence of an external enamel layer. One of the main functions of dentin is to *provide mechanical support* for this layer. Enamel appears to be constructed so as to be stiff and abrasion resistant, yet it deforms appreciably during function (Hood 1991 and others, see review Zaslansky and Weiner 2007). The bunching of unusually long apatite crystals in enamel into woven prisms entails fracture resistance to this hard tissue (substantial prism decussations stop advancing cracks at crystal boundaries and in the many water-containing pores – see review Rensberger 2000). Cushioned by dentin that is known to arrest cracks if they were to develop (Imbeni et al. 2005), the enamel does not usually break. In humans, it does not usually completely wear off either (it does in other animals, however). Thus teeth are built from long-lasting components and, there seems to be little need for routine tissue turnover. Indeed, dentin is not biologically replaced unlike bone. Long-lasting mechanical durability must surely be a consequence of exquisite underlying engineering, whereby the dentin infrastructure elastically and repeatedly supports enamel without fatigue-fracture under regular use. An excellent interface design overcomes problems associated with the large differences of properties between dentin and enamel. It is difficult to ensure good binding between such very different materials. However, in teeth, this has been resolved by an elaborate dentin–enamel junction (Weidenreich 1925) which functions as a mechanical interphase (Zaslansky et al. 2006a). A thick transition zone of 200–300 μm, in which fiber orientations and mineral arrangement vary, seems to be a key factor contributing to human whole-tooth longevity.

Mammals seem to have evolved different sized and shaped teeth, presumably better suited to their diets and living conditions (Lucas 2004). One such adaptation is the maintenance of shape and *self-sharpening by active wear*. Thus, in some animals, dentin seems to be good for more than simple passive support beneath enamel. Dentin in the front teeth of rodents is designed to be worn down through use. Situated on the tongue aspect of these teeth (providing support for enamel on the front), dentin continuously grows as the whole tooth continuously erupts. The hardness of this dentin varies in a well-controlled manner, having values somewhat lower in a soft zone adjacent to the enamel as well as a gradual decrease in hardness at greater distances toward the tongue. Osborn (1969) showed that in beavers, differences in dentin hardness in upper and lower incisors are important to maintaining functional longevity of the front teeth. Continual growth combined with gradual variations in wear-rates brings about a balance that maintains the overall shape and hence mechanical functionality of the teeth. A tough fibrous diet ensures a self-sharpening mechanism: when enamel chips away and fragments break off, differential wear of dentin allows the teeth to resume chisel-like sharpness. The softer dentin restricts the extent of damage by providing lateral stabilization for the stiffer but more brittle enamel, while possibly protecting the inner soft gum tissue from inadvertent damage due to chewing of sharp bristles.

The concept of active wear and dentin regrowth through use, while supporting a hard stiff and brittle outer layer, is also found in lower vertebrates. The teeth of sharks and many fish are covered by enameloid – which differs from enamel both structurally (contains significant amounts of collagen) and developmentally (enameloid is made by odontoblasts, see Sasagawa 2002). Enameloid supposedly functions similarly to enamel, and in many fish, the strategy of dealing with fracture and wear is by replacement with new grown teeth. In a detailed relevant example, Carr et al. (2006) studied mastication in parrotfish, analyzing the continuous wear of opposing upper and lower teeth. As the teeth wear down, they move and new teeth emerge in the front of the upper jaw as well as in the back of the lower jaw, slowly migrating as if situated on conveyor belts. The enameloid is worn down soon after the teeth begin to function, and much of the grinding surface of the teeth is quickly overtaken by exposed dentin. Using simple jaw movements, the fish are able to crush coral and then finely grind down the broken particles by shearing, so as to free the algae that they ingest. The teeth appear to be designed to wear down and be replaced, pre-engineered to expose dentin. One may wonder if the exposed vital and presumably sensitive dentin might not be used for another purpose in this case. Is it possible that dentin may be used for *sensing the state of food*?

A deliberate strategy of wear of the thin external enamel layer is found in many herbivores, large as rhinoceroses, horses or deer or small as mice and guinea pigs. In many teeth, the enamel coverage is such that it is designed to expose dentin almost as soon as teeth start to function. Within weeks after eruption, teeth in young rats display significant wear (Lovschall et al. 2002). Figure 15.2 shows an image of a slice through a newly erupted molar of a 1.5-year-old fallow deer (*Dama dama*). The thin almost-unused enamel is missing just above the dentin over the pulp horns, appearing to expose dentin inside small cones within enamel. The wear is induced by the animal feeding on seeds, grasses and woods that contain abrasives such as silica particles. As a result of the wear, rough chewing tooth surfaces develop, well suited to grind down the plant food before passing it on to the gut. But the proximity to the vital pulp – seen also in rats (Crooks et al. 1983) – raises, at least in principle, the possibility that the dentin–pulp complex might be providing a sensory service:

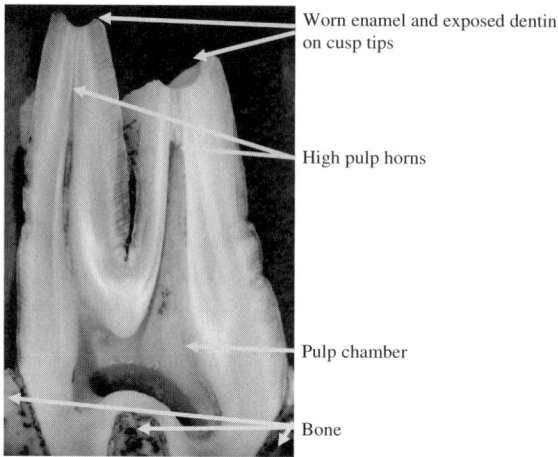

Worn enamel and exposed dentin
on cusp tips

High pulp horns

Pulp chamber

Bone

Fig. 15.2 A slice through a newly erupted lower posterior molar of fallow deer (*Dama dama*)

being a living tissue, sensitivity of exposed dentin to pressure at multiple points on the occlusion surface might provide tactile sensation of the fragmentation-state of the food. It is not impossible that – analogous to sensation in the skin or, conversely, the tactile activation of mouse track-pads in computer notebooks – differences in contact forces due to soft, hard, fine or coarse food particles might be sensed. An active, tooth-borne sensing mechanism could definitely provide important mastica-tory "real-time" feedback.

It is not simple to directly test for the existence of such a sensory mecha-nism experimentally, and demonstrating this in human teeth is probably technically impossible. Still, an hypothesis of tactile functional sensitivity ought to be consid-ered, in light of the striking circumstantial evidence of its existence. A mechanism providing direct sensory feedback from the working surfaces of teeth would allow for example, to modify bite forces "on the fly". If also goes well with the notion that, at least in mammals, knowledge of the state of fragmentation of food is important for the decision of when to swallow (Lucas 2004). Naturally, a crucial component of such sensory device that would provide the backbone of 'feeling the food' in teeth would be extensive and widespread innervation. Could this be one explanation for the extensive innervation in the pulp (see review in Avery et al. 1984), which is thought to be primarily associated with pain?

Sensory functions of dentin have been advocated by paleontologists for a long time, as part of theories concerned with the development and evolution of the earli-est dentins, dating back to fish in the late Cambrian period some 500 million years ago (see review Smith and Sansom 2000). Primitive and diverse forms of dentin are thought to have evolved first in the skin of jawless (agnathan) fish as part of protec-tive dental armor, and they are assumed to have assisted in sensing the environment.

Through evolution, dentin became covered with hard wear-resistant layers (be it enameloid, enamel or petrodentin-as is found in extant and fossil lungfish) significantly enhancing the durability of teeth, but not necessarily preventing sensory tactile roles: as mentioned above, wear routinely exposes dentin in many animals, but even when it does not, enamel deforms substantially under physiological load (see for example Zaslansky et al. 2006b) and probably compresses dentin locally. Dentin might therefore facilitate sensing the state of food or mastication trajectories, possibly helping to know how to apply or alter subsequent mastication strokes. Tactile performance of this type has partially been demonstrated in human teeth, where the participants of experiments lost their ability to discriminate between hard or soft food, if the enamel of their teeth was capped by hard plastic (Paphangkorakit and Osborn 1998). Structures within dentin that may be involved in this function are discussed further in Section 15.5.4.

Tactile sensation delivered directly from the working surfaces of teeth would presumably be useful to augment other sensory input from the oral soft tissue and tooth-supporting periodontal ligaments. Although what is known about tooth sensitivity in humans is primarily associated with pain, it is important to consider that such an extreme sensation is a very rare response to regular daily use of teeth. (Pain is defined as "an unpleasant sensory and emotional experience associated with actual or potential tissue damage, or described in terms of such damage". One may wonder how likely it is that pure pain-reporting is the main intended evolutionary purpose of tooth innervation!) Pashley (1989) reported that dentin in the crown, particularly above the pulp horns (the extremities in the pulp that point up into enamel), exhibits high permeability to liquid. Fluid movement through selected tubules beneath dentin and enamel due to compression during mastication can, in principle, provide a simple, precise and reliable mechanism of pressure sensation. It has been shown that the odontoblasts have cilia capable of deflecting in response to fluid flow (Magloire et al. 2004). Extensive fluid movement is known to be associated with dentinal pain, and may be elicited with thermal as well as osmotic fluctuations. The fact remains that in healthy, normal functioning teeth, pain is almost never sensed (but might be, in patients who tightly clench their teeth, often due to psychological stress). On the other hand, the presences of sand in the salad or broken food particles are routinely encountered and sensed. Could our teeth be complex sensors, as well as cutting tools? The mechanism for pressure sensation seems to exist. Is this an additional function of dentin? These questions remain unanswered at this time.

Dentin seems to have an even more unusual, almost bizarre function as well. Brear et al. (1993) studied design and properties of dentin in the long and slender tusks of narwhals (whales). Based on the findings of a high-toughness design, they suggested an adaptation well suited for withstanding lateral blows during *aggressive sexual display*. The tusks in these marine animals are designed not for stabbing or longitudinal loading, as is the case for horns, but rather for rubbing or "tusking". Tactile sensory functions were not considered as part of these narwhal experiments, but other researchers have found the tusks to be highly sensitive structures, indicating that tactile sensation is probably an important functional mode in this dentin.

The diversity of uses found for dentin all center on mechanical function. The mechanical properties of dentin are the subject of the next sections.

15.4 Dentin as a Material

The properties of dentin have been the subject of many studies, and much of the information available has been extensively reviewed by Braden (1976) and Waters (1980) with significant extensions in Kinney et al. (2003). The following is therefore not an additional review but rather a discussion about selected concepts of what might be considered to be the main mechanical properties of dentin. The reader is referred to the above-mentioned, important reviews, as well as Currey (2002) for additional in-depth perspectives.

When considering the basic constituents of dentin, the mineral crystals and protein filaments form mineralized collagen fibrils, similar to those found in bone. At this very basic level, dentin is a biological composite material, and Katz (1971) as well as other workers explored the possibilities and limitations of modeling dentin (and similar hard tissues) as a simple composite of two phases: "collagen infused with hydroxyapatite". With estimates of volume fractions of each, they examined the relevance of deriving bulk, shear and Young's modulus as well as Poisson's ratios based on simple rules of mixtures. Particularly with respect to Young's modulus of elasticity, the calculations lead to estimates of bounds within very broad ranges of what might be considered to be "reasonable stiffness estimates": almost anywhere between the modulus of collagen ($\sim$1.2 GPa) and that of hydroxyapatite (114 GPa). Katz attributed the problem of applying Voigt or Reuss models as upper and lower bounds of dentin elasticity to the large discrepancy between the modulus of mineral and the organics, concluding that the heterogeneity, anisotropy and structural organization are extremely important factors. These factors greatly affect the true mechanical properties that must be known for any meaningful modeling. He also noted that the effects of microstructure as well as origin and orientation of the samples were critical, and he emphasized that the experimental strain rate due to viscoelasticity had to be considered. Thus, only limited insight into dentin properties arises from a simple two-component composite model.

The unidirectional tubular arrangements of dentin make it appear as an anisotropic material, but the degree to which the mechanical properties follow the ordered tubular microstructure or other prominent anatomical features is – surprisingly – still uncertain. Furthermore, how different are different zones in the teeth? Stanford et al. (1960) noted that the stiffness of root differed substantially from crown dentin (with moduli of 7.8 versus 12.5 GPa, respectively). Many researches in the middle of the twentieth century appear to have attributed the differences in modulus reports to errors in the measurement techniques or problems associated with storage conditions as well as differences due to loading mode and strain rates. This might be a consequence of the fact that a large number of studies were aimed at quantifying dentin stiffness so as to elucidate the demands on dental restorative materials, essentially requiring a single correct value to be "true". However, the broad range of values reported for the modulus of elasticity might serve to illustrate the unexpected complexity of the task faced by researchers of dentin properties: already in 1957, Craig and Peyton expressed their concerns about the large differences in reported modulus values for dentin, spanning from 5.5 to 28.3 GPa. Viewed from an engineering perspective, this large range was suspected to represent gross inaccuracies of the measurement experiments or methods rather than structural variations, a perspective that seems to have prevailed in dentin research.

In their extensive review of dentin mechanical properties' literature, Kinney et al. (2003) address, among other issues, the broad range of modulus values published in the literature, pointing to the key experimental parameters as possible causes of discrepancy between measurements. They show how issues that might, at first glance, seem minor such as sample storage conditions or strain rate variation may have profound influence on the reported mechanical properties. Age also seems to matter. It has been suggested by various researchers such as Beust (1931) and others that dentin becomes harder with time, and Tonami and Takahashi (1997) reported that fracture surfaces of young bovine dentin differ from those of old dentin. These results and many later studies suggest that important age-related changes occur in the mechanical properties of dentin, adding to the experimental parameters that need to be interpreted. Undoubtedly, the mechanical properties of dentin are complicated to interpret.

15.4.1 Average Materials Properties of Dentin

Braden's review (1976) appears to be the first systematic treatment of the properties of bulk dentin. He noted his concern about the deviation of dentin from being a homogeneous, elastic and isotropic material and that strain rates need to be considered for the correct interpretation of dentin properties. He also noted the surprising finding that apparently the elastic constants of dentin do not depend on tubule orientation. Braden's review describes many technical aspects of performing measurements of dentin, systematically listing the methods and results for elastic (compression, tension, shear), fracture and hardness, as well as thermal- and electrical-property testing. He cites many of the observations that still largely dominate the current understanding of the mechanical properties of dentin. Braden's observations are included in the larger but less technical review by Waters (1980), from which the following general conclusions can be drawn:

- Tooth availability, small sample sizes and storage conditions pose serious experimental difficulties, suggesting in a way, that some of the results reported in the literature might not be valid.
- Samples in more than 50 experiments, by different researchers, were mostly tested at room temperature at low strain rates, exhibiting at least partially linear

stress–strain curves. Some time-dependent or non-linear responses were however noted.

- The reported elastic moduli for compression, tension, bending and indentation ranged from about 8 to 21 GPa (mean 13.5, standard deviation of the mean 4.1).
- Shear modulus based on Gilmor et al. 1969 and Renson and Braden 1975), was reported to be in the range of 5–8 GPa.
- Small Poisson's ratios below 0.2, such as 0.014 reported by Haines (1968) or 0.12 reported by Renson and Braden (1971), were thought to be miscalculations or statistically insignificant. Waters considered the correct range to be between 0.23 and 0.3, assuming that higher values such as 0.32 reported by Lees and Rollins (1972) were overestimates produced by the high strain rates of the acoustic method of measurement.
- Flexural and compressive strengths of 230–350 MPa were noted. Tensile strength values were much lower, ranging from 30 to about 60 MPa and amounting to 10–15% of the compression strength. Shear strength tests, mostly punch tests, ranged from 60 to about 150 MPa, with higher values found away from the pulp and near the DEJ.
- Work of fracture estimates was based on tests performed at low fracture rates by Rasmussen et al. (1976), where, similar to bone, values of $270 \, J/m^2$ and $550 \, J/m^2$ were reported across (orthogonal to) and along (parallel to) dentin tubules, respectively. Fractography studies of high fracture rate experiments revealed brittle fractures with no specific influences of the orientation of dentin microstructure (for example, tubules) on the path of fracture.
- Time-dependent measures of viscoelastic stress relaxation were based mainly on the work of Duncanson and Korostoff (1975), where an estimated relaxation modulus $E(\infty) = 12 \, GPa$ was found. This is well within but on the lower side of quasi-static stiffness estimates of other researchers, indicating an important influence of strain rate.

Both Braden and Waters discussed hardness measurements, reporting values in Knoop hardness numbers and noting that these tests were not standardized. The hardness values they report are equivalent to the range of 500–750 MPa. Little significance was attributed to the fact that large variations known to exist in dentin (Craig and Peyton 1958, Osborn 1969) showed systematic variation at different distances from the enamel margin at the DEJ. Teeth are now known to be graded structures, with their properties varying significantly (Tesch et al. 2001), and thus single "correct" properties – do not exist.

15.4.2 Variation of Properties as a Design Concept – Elastic Properties

Micro- and naonindentation-based measurements of stiffness have added greatly to the understanding of the structure–property relations in many materials, following the methodological developments of Doerner and Nix (1986) and later Oliver and

Pharr (1992). With these newer indentation methods, it has become possible to probe variations in dentin properties associated with details of the structure. Nanoindentation has disclosed large differences in modulus across distances of several micrometers, with reported values apparently spanning the entire range that Craig and Peyton discussed back in 1957: dentin surrounding the tubules (peritubular dentin, up to several microns thick) has a stiffness of almost 30 GPa, which is at least 50% higher than dentin not directly adjacent to tubules (intertubular dentin) (Kinney et al. 1996). Hardness at the submicron level presents an even greater variation, whereby a 20-fold increase in hardness is seen in some regions. Thus, somewhat differently to bone, hardness and elastic modulus in dentin do not scale linearly (Currey and Brear 1990). The hardness of intertubular dentin, first reported by vanMeerbeek et al. (1993), was found to be about 500 MPa. Kinney et al. (1996) found a range of values spanning 130–510 MPa for intertubular dentin (near the pulp as compared to the bulk) values of almost 2.5 GPa were reported for peritubular dentin. Microstructural variations thus clearly result in mechanical property variations. It is possible that the small decrease in the modulus of intertubular dentin reported by these authors (from 21.1 GPa near enamel down to 17.7 GPa near the pulp) also represents some sort of microstructural variation.

The bulk of crown dentin is characterized by high-modulus co-aligned tubules within a softer matrix of intertubular dentin. This microstructure, discussed further in Section 15.5, seems inconsistent with the experimental findings that dentin deforms as an isotropic material. The tubules and associated peritubular dentin certainly make the material itself structurally anisotropic, however, only infrequently has the mechanical anisotropy actually been demonstrated. Currey et al. (1994) reported a significant dependence of modulus on collagen fibril orientation in the dentin of the narwhal tusk. Due to the large more regular microstructure of this tissue, Currey et al. were able to test and measure both modulus and strength as a function of fibril orientation. They then used their experimental results to model bone–lamella properties using fiber-reinforced composite models. Although the Young's modulus values they report (12 GPa along and about 6 GPa across the fibril orientation) are on the lower range of those reported for human dentin (Kinney et al. 2003), two important conclusions can be drawn: (a) the properties of dentin vary with fibril orientation and (b) fibril orientations vary substantially over the thickness of dentin. Kinney et al. (2003) paid special attention to the subject of dentin anisotropy and proposed that human dentin is slightly (about 10%) anisotropic. They emphasize the importance of a report by Palamara et al. (2000), who found that both elastic and failure properties of crown dentin were lower by about 10% (determined from measurements of deformation of grids sputtered on the surfaces of compressed samples) along the direction of the tubules as compared with the orthogonal orientation. The moduli values that Palamara et al. reported however are very low (similar to those found for the narwhal tusks): 10.7 GPa along the tubules and 11.9 GPa across the tubules. Kinney et al. (2004) reported a similar 10% difference but with much higher values of modulus based on large (2 mm long) samples that were measured by resonant ultrasound spectroscopy: they obtained Young's modulus estimates of 23.2 GPa along the tubules and 25.0 GPa across the tubules when wet. Although this small difference vanished when the dentin was dried, the anisotropic findings of fracture strength in shear/tension hint to the existence of an underlying orientation dependence. Kinney et al. (2003) reevaluated some of the literature and concluded that quasi-static measurements, in general, and in particular measurements of samples stored for long periods of time in water were bound to be too low. They proposed that the modulus of human dentin is 25.5 GPa if stress relaxation of indentation measurements was accounted for. Katz et al. (2005) considered the implications of the small anisotropy, determined by the resonant modes for dentin, and suggested that scalar estimates of compressive and shear anisotropy may help to understand and predict variations in experimentally measured elastic constants in many mineralized tissues. They showed that small differences in the degree of anisotropy (0% for dry dentin and 1.3% for wet dentin) completely explained the findings of Kinney et al. (2004) and others. They concluded that the degree of anisotropy varies in different regions of dentin, similar to what seems to occur in bone, and they use this to interpret conflicting results of different experimental measurements that are difficult to reconcile.

Admittedly however, sonic determinations of elastic constants rely on precise knowledge of the tissue density, which to a large extent means definite knowledge about the mineral density distribution. Mineral densities as well as fiber orientations in the microstructure vary substantially across dentin (Tesch et al. 2001). Consequently, dentin moduli determined from mid-crown bulk sonic measurements must differ from those obtained from other regions in the tooth. As pointed out by Currey et al. (1994), the large samples of narwhal dentin contain extended regions of regular collagen fibril arrays. Their samples clearly displayed anisotropy of both modulus and strength values, both being about twice as high along the fiber orientation. With a clear knowledge of structure and properties, they could successfully model off-angle elasticity as well as a variety of other parameters, such as the crystal form. As human dentin does not exhibit such anisotropy, it may be concluded that at least in crown dentin, much of the fibril microstructure is not arranged in uniform parallel arrays. The large number of human crown dentin samples of narrow age range, studied by Arola and Reprogel (2006), provide direct evidence to the small degree of anisotropy of this tissue. They report flexural moduli of 15.5 GPa along the tubules and 18.7 GPa across the tubules with almost no dependence on distance from the pulp. The degree of anisotropy for human dentin is thus on the order of 20%, suggesting that the organization collagen fibril is by no means simple.

15.4.3 Anisotropic Failure Properties

Unlike elastic properties, anisotropy and dependence on tubule orientation have been easier to demonstrate for the failure properties of human dentin. Surprisingly however, this has not been the case for indentation measurements although such measurements are tightly associated with failure and estimates of strength. Thus it is interesting to note the scantiness of evidence of anisotropy and orientation dependence by indentation methods. Renson and Braden (1971) showed that by use of conical indenters, yield strength of dentin could be derived – well matching bulk measurements of yield in compression (260 MPa). A dependence on orientation was presumably not investigated, despite the fact were that gradients easily demonstrated by indentation (Osborn 1969 at the time and later, Meredith et al. 1996, Wang and Weiner 1998b) for which significant variations in strength (at least in compression) are to be expected.

A dependence of shear strength on dentin site and tubule orientation within the crown was clearly demonstrated by Watanabe et al. (1996). Anisotropy of tensile strength was reported by Carvalho et al. (2001) and other workers, using micro-tensile tests. All studies revealed higher resistance to fracture for cracks running along the path of the tubules. Wang (2005) distinguished between the marked anisotropy of bovine root dentin and crown dentin. His fracture toughness estimates for crown dentin ($2.1 \, \text{MPa}\sqrt{\text{m}}$) were similar to other reports ($1.8 \, \text{MPa}\sqrt{\text{m}}$ was reported by Imbeni et al. 2002, however values exceeding $3 \, \text{MPa}\sqrt{\text{m}}$ were reported by El-Mowafy and Watts 1986, Lin and Douglas 1994, Xu et al. 1998 and others). In the root, significant anisotropy was found both by three-point bending and by a non-conventional impression test, where a sphere is pressed into dentin to produce cracks. Wang demonstrated a difference in the crack patterns within root versus patterns in crown dentin, and he attributed the difference in anisotropy to the ordered incremental fibril arrangement in the root, compared to the tubular reinforced structure of the crown.

Arola and Reprogel (2006) demonstrated significantly reduced strains at fracture along tubule orientations in crown dentin (about 0.005% compared with 0.015% across tubule orientation). They reported the flexural strength across the tubules ($\sim$160 MPa) to be more than 30% higher than the strength when flexing along the tubules – a ratio that matches other reports of strength anisotropy. They noted a somewhat exponential decay of strength with increasing tubule density closer to the pulp. Thus, unlike elastic properties, the tubules seem to have a marked influence on failure properties of dentin, as was suggested by Imbeni et al. (2002).

The large range of values reported for strength measurements in the literature (shear, tension) led Kinney et al. (2003) to highlight within an extensive review the issue of *flaw distribution* in the samples. By use of measures of the Weibull distribution function and modulus, they provide ideas about the nature of flaws in the structure of dentin that govern its strength. They suggest that large variations in the Weibull modulus for different sample batches reflect differences in flaw distributions and defects that are produced during sample preparation. They contend that such distribution measurements may be useful for quantification of changes that occur with age (such as a decrease in fracture resistance as was shown by Tonami and Takahashi (1997) and other researchers). Kinney et al. (2003) draw attention to the tight connection between the fracture toughness K_c and the critical flaw size in diverse materials and discuss how variations in reports of fracture toughness may reflect differences in flaw distribution in the dentin samples.

Kinney et al. (2003) also review dentin *fatigue*, which is a topic covered in great detail in more recent research. Although measures of cyclic loading and fatigue are better suited for flaw-free materials, they have use in dentin when predicting the number of cycles that are needed to induce failure at different repeated loads. Under regular conditions in the mouth, small normal physiological loads (20 MPa) at a frequency of 1 Hz, repeated a little more than 2500 times per day (10^6 per year), were examined. Experimental and numerical estimates of expected lifetime of dentin (fatigue limit) undergoing repeated stress cycles suggest that there exists a threshold stress–intensity factor, ΔK, below which cracks in dentin will not grow (equivalent to 1.1 MPa$\sqrt{m}$). When exposed to 10^6 cycles of stresses of 20 MPa, only very large cracks (0.9 mm sized flaw!) are expected to extend, but this critical size drops (to 0.3 mm flaw size) if higher average cyclic loads – 30 MPa – are imposed. The reports by Arola and Reprogel (2006) of fatigue endurance strengths of 44 MPa across the tubules versus 24 MPa along their orientation (for 10^7 cycles) provide additional evidence for an orientation dependence of these mechanical property estimates.

The analysis of Kinney et al. (2003) extensively reviews the properties of dentin, expanding in many ways on what was described by Braden (1976) and Waters (1980). Their review, however, did not consider corrections that may be needed for predictions accounting for the redistribution of loads in dentin due to the function of the enamel cap (Wang and Weiner 1998b, Zaslansky et al. 2006b), or for possible protective responses and reduced mastication forces resulting from pain due to damage accumulation in dentin. Both contributions would result in prolonging the life expectancy of teeth, which appear to often last for tens of years (possibly 10^8 chewing cycles).

Presumably by now, there is little difficulty to convince the reader that the mechanical properties of dentin are significantly affected by the underlying microstructure. Much can be inferred about these properties from the textural details, and the next section deals with this subject further.

15.5 Dentin Microstructure as the Basis for Mechanical Properties

Dentin has been explored extensively as a biological tissue and also as a material with mechanical function, for more than a century. Different and occasionally conflicting interpretations of the microstructure have been published in hundreds of studies by biologists, clinicians, material scientists and others. The premise is simple: teeth are hard structures that need to be able to repeatedly withstand the application of sufficient load to perform their task (food procurement and/or mastication). This is coupled with the fact that (with the exception of rodents) the entire structure is basically formed once, and undergoes only minor, if any, biological changes over its lifetime. Such a scenario is very different from tendons or bones, two tissues that also function under load but undergo remodeling and restructuring in response to external mechanical stimulation. The structures in teeth are thus minimally modulated by active biological modifications in situ in the oral cavity environment.

Consequently, tooth design, which to a large extent means dentin microstructure, represents an excellent (or at least "good-enough") solution to the overwhelming task of withstanding repeated tearing and compression load cycles during feeding. It is hard to imagine that a simple architecture can result in the exceptional durability which, when little or no wear occur, lasts for many years. It is highly probable however that only the combined and optimized balance between the tissues in the entire tooth ensemble (i.e., dentin bulk, enamel cap and cementum lining, Fig. 15.1), make it well matched to its function. This balance is so good that man-made alternatives (to date) offer only short-lived and partial compensation for tooth material loss: teeth outperform even the best dental treatment, be it restorations (fillings) crowns or implants. The large discrepancy between the mechanical properties of the different components (for example, enamel is about four times as stiff as dentin) might well be a major design principle of teeth. The gradual and hierarchical transitions in properties along different orientations is probably another principle. Some elaborate findings and theories about function of whole teeth and relations to diet, structure and form can be found in Lucas (2004).

In the dental and materials literature, dentin found in human teeth is often referred to as one single material. The marked variations in mechanical properties suggest that dentin is composed of more than one type of microstructure. A survey of the different tissues that are classified under this name shows that "dentin" loosely refers to a family of dahllite-based composites, which are *not* enamel, bone or cementum. As unhelpful as this definition might be, it does reflect the fact that knowledge about the characteristics of dentin accumulated over many decades, when details of the structure were poorly understood. Along the root and throughout the crown, histological sections soon reveal several distinct and quite different dentin types. Although much is now known, there still seem to be many open questions about dentin microstructure.

15.5.1 Root Dentin

In the root, which is formed only after most of the crown is complete, primary dentin dominates. The root ultra-structure (Fig. 15.3) is characterized by aligned patches of cross-linked collagen fibers, and mineral is found both within fiber bundles and in spaces between fibers (Wang and Weiner 1998a). The tubules run outward laterally, extending from the living cells in the pulp out to the cement layer (Nanci 2003). The mineralized collagen fibers lie in planes that are orthogonal to the tubules, with a preferred orientation that is along incremental growth lines at a small angle to the long (root-crown) axis of the tooth. Within neighboring collagen bundles, many fibrils appear to be mutually rotated along the fiber axis, so that the crystals in different fibers are not aligned. For each mineralized fibril however, the crystalline mineral c-axis corresponds with the fibrilar long axis (Schmidt and Keil 1971), whereas between fibers, randomly oriented crystals are often observed. Similar to crown dentin, across root primary dentin, large variations in hardness and mineral density have been reported (Wang and Weiner 1998a, Kishen et al. 2000)

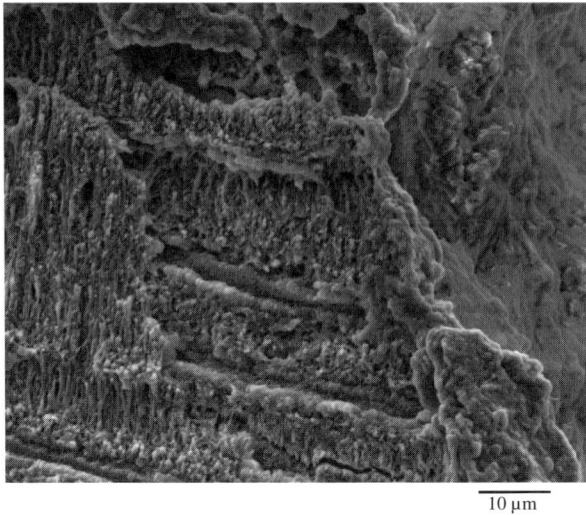

10 µm

Fig. 15.3 Root dentin: parallel arrays of collagen fibrils are arranged orthogonal to the almost horizontal tubule paths. Fibril preferred orientation (and mineral c-axis) is parallel to incremental lines approximately parallel to the root long axis (which is vertical in this specific image). Pulp chamber edge seen on right-hand side.

such that a strategy of gradual mineral content variations occur. Variations in the organic component might accompany this, but have not been reported to date.

Over long periods of time (years), secondary dentin develops adjacent to the pulp tissue of vital teeth (Karjalainen 1984). In a normal physiological process, pulp chambers and root canals in adults undergo a slow but gradual narrowing (Murray et al. 2002, Oi et al. 2004) that might lead to complete pulp obliteration. The slow-forming dentin appears less organized and is softer than primary dentin. It is known to reduce the overall permeability of the tissue, thickening and stiffening the root to some extent, and occurs in addition to increased mineralization of intertubular dentin and intra-tubular crystal precipitation (see overview in Porter et al. 2005 and Kinney et al. 2005).

15.5.2 Crown Dentin

At the micron lengthscale, crown dentin is much like root dentin, resembing a fiber-reinforced composite made of thin and highly mineralized hollow tubules. However unlike root dentin, the tubules in the crown are surrounded by a 2–6 µm dense cuff of *peritubular dentin* (Fig. 15.4). The tubules are embedded in a matrix of poorly oriented mineralized collagen fibers. While densely packed near the pulp, the tubules radiate out and become more sparsely packed near enamel, where their pertitubular sheaths become thinner.

Peritubular dentin has no collagen (Kramer 1951, Johnson and Poole 1967, Jones and Boyde 1984, Weiner et al. 1999, Gotliv et al. 2006) and has well-developed

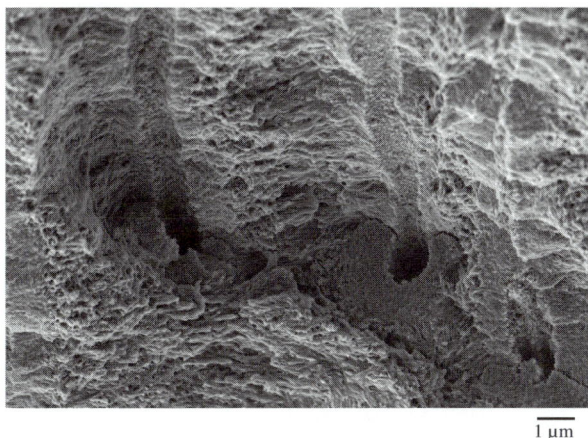

1 µm

Fig. 15.4 Peritubular dentin: the peritubular dentin-lined tubules are seen to be embedded in a surrounding matrix of highly mineralized collagen fibrils

hydroxyapatite crystals (Schroeder and Frank 1985). Often confused with secondary occlusion of tubules (sclerosis) due to exposure to oral irritants (Weber 1974), it is now well established that the highly mineralized tubules form during tooth development as part of the structure (Takuma and Eda 1966). The numerical density increases from about 20,000/mm^2 near the enamel to about 50,000 near the pulp (Pashley 1989). Interestingly, the tubules have symmetric and uniform thickness around their circumference only on trajectories heading up to the chewing surface of the crown, in the opposite direction to the root. Tubules on the outer sides of the crown (the lateral sides: closer to neighboring teeth, the tongue or the cheek) show an uneven C-shaped thickness, systematically arranged such that the thickest part lies on the side closer to the root (Weber 1968). This curious observation has no reported explanation, although it is clearly a design feature of human teeth.

Intertubular dentin forms the matrix of the bulk of crown dentin, filling the gaps between the dentinal tubules. Intertubular dentin is characterized by a fine mesh of 50–100 nm thick mineralized collagen fibrils similar to primary dentin in the root. The microstructure appears denser than root dentin, heavily loaded with mineral and more disordered with fibers being aligned only occasionally. Fierce debates between prominent researches dating back to the beginning of the twentieth century following vonEbner (1909) have led to a widespread view that fiber arrangement in intertubular dentin is primarily orthogonal to the tubules. In some areas of the crown, the fibers indeed present such a preferred orientation, however observations by Weidenreich (1925) and Orban in the 1920s (reviewed in Orban 1929) and by Kramer (1951) and Frank (1965) show this often not to be the case: mineralized fibers are seen to run orthogonal, oblique and even along the dentinal tubules (see Fig. 15.5). The dense network of very thin mineralized fibers appears to be quite random. If is not known if fibers have an *average* orientation that is orthogonal to the tubules. Conflicting early reports of fibril orientations may have been confused with reports of odontoblast cell branching projecting laterally from the tubules. Many reports clearly show that

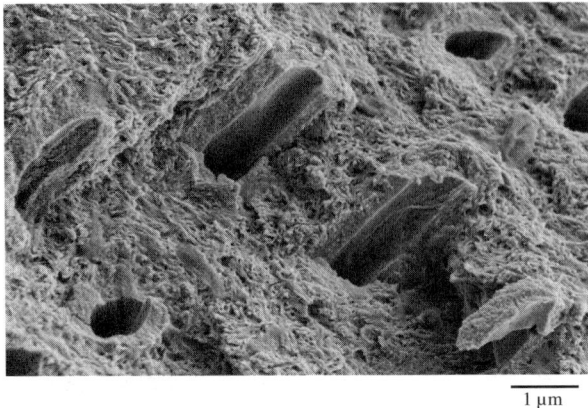

1 µm

Fig. 15.5 Various fibril arrangements in fracture surfaces of intertubular dentin: in crown intertubular dentin, fibrils appear to be arranged across, at angles to and along tubule orientations

collagen fibers also lie parallel and at an angle to the tubules adjacent to or within the highly mineralized peri tubular sheaths where they are often concealed, being embedded in the mineral of peritubular dentin. The elastic near-isotropy of crown dentin (Kinney et al. 1999) further attests to a design with almost isotropic, random fiber orientation. This is different from the rather well-aligned microstructure of root dentin (Wang and Weiner 1998a) or the narwhal tusk (Currey et al. 1994). Curiously, already in 1900, Gebhardt raised the fascinating possibility that the arrangement of fibers might differ between teeth of different animals due to their different functional demands. To date however, little has been done to clarify the matter further.

15.5.3 Variations of "Normal" Dentin Structure

The outer 100–300 µm of crown dentin abutting with enamel at the DEJ is known to be an irregular fibrous softer type of dentin. The mineral density observed by X-ray or backscattered electron microscopy is lower than bulk dentin (Tronstad 1972, Wang and Weiner 1998b) and an open reticulate microstructure is seen (described in detail in detail Bevelander 1941). The outermost ~50 µm layer near the DEJ corresponds to the first-produced dentin in the tooth and is termed *mantle dentin* (Moss 1974, Jones and Boyde 1984). Mantle dentin appears differently in different species, and it is continuous with enameloid in various lower vertebrates where the latter material is found. In mantle dentin, thick collagen fibrils (80–120 nm in diameter), sometimes termed von-Korff fibers, are seen to run orthogonal to the DEJ, extending out of the dentin and becoming embedded in the enamel. The varying degree to which collagen fibrils are bunched (making them more apparent in histological tissue preparations) is one of the big differences of the microstructure when compared with bulk crown dentin (Kramer 1951, Johnson and Poole 1967, Jones and Boyde 1984 see Nanci 2003). Deeper beneath the DEJ, the large fiber bundles splay out and merge with horizontally orientated 30–50 nm fibers of collagen (Lin et al. 1993). Promoted by a scalloped irregular DEJ topography, an intimate attachment is formed between the highly mineralized and crystalline enamel on the one side and collagen-based dentin on the other.

In what appears to be an *interphase* encompassing mantle dentin (White et al. 2000, Zaslansky et al. 2006a), dentin in a 200–300 μm zone beneath enamel appears to be well designed to withstand and transmit mechanical load. This allows dentin and enamel to function together, and indeed, they exhibit superb bonding that is difficult to fracture under physiological conditions (Rasmussen 1984, Urabe et al. 2000, Imbeni et al. 2005). A similar layer of splaying fibrils layer is occasionally seen beneath cementum (Jones and Boyde 1984) contributing to a transitional zone between dentin and cementum in the root (Ho et al. 2004).

15.5.4 *"Abnormal" Forms of Dentin*

Several types of dentin are thought to be associated with pathology. These include caries, sclerotic dentin, reparative dentin and dentinogenesis imperfecta, and they are only mentioned here in the context of their altered mechanical properties. Interglobular dentin is also thought to be pathologic, although it seems not to be.

Caries is probably the most abundant form of all collagen-based hard tissue diseases. It is characterized by the demineralization of the tooth hard tissues by acids produced by oral bacteria (summarized in vanHoute 1994) and the breakdown of organic matrix by host matrix metalloproteinases (enzymes found in the normal saliva and pulp, Tjaderhane et al. 1998). Several altered layers of different microstructures are typical for most carious lesions (reviewed in Marshall et al. 1997) differing by the amount of mineral dissolution and organic degradation, and corresponding to the extent of tissue damage. As the infected dentin becomes soft (this is what is sensed and excavated by dental surgeons during routine restorative dental treatment), deeper dentin may become hypo- or hypermineralized, occasionally becoming transparent (see below). Marshall et al. (2001) reported values of 36 GPa for modulus of peritubular dentin beneath carious lesions with values dropping precipitously within the lesions. While normal peritubular dentin exhibited hardness values of about 2.8 GPa, peritubular dentin in transparent dentin had a hardness of 1.9 GPa. Thus, degradation in one region of a carious lesion might result in increased mineral content in other, deeper regions.

Sclerotic dentin is a transparent type of dentin (a similar form to that seen beneath caries) in which the dentinal tubule lumens mineralize, becoming occluded (Nanci 2003). It is thought to be a phenomenon related to aging of dentin, not often studied in depth. Marshall et al. (2001) reported almost normal values of nanoindentation hardness and Youngs modulus for sub-carious sclerotic dentin. Kinney et al. (2005) reported higher mineral content relative to normal dentin in particular, near the pulp within secondary dentin. Average unremarkable shear and Young's moduli were reported, but all properties were consistent with a highly mineralized type of dentin, as compared to normal dentin tissue. Thus, lower yield strength, brittle patterns of fracture, reduced resistance to fatigue at moderate loads and a lower fracture toughness ($1.5\,\mathrm{MPa}\sqrt{m}$ as compared with $1.8\,\mathrm{MPa}\sqrt{m}$) were demonstrated, in addition to a decrease in the critical flaw size sustainable by this tissue. Various findings of age-related structural and compositional variations are discussed further in Porter et al. (2005).

Reparative dentin is thought to be a lower quality dentin, greatly disordered and often hastily produced from with in the pulp as a response to pathology. It is less organized, softer (at least in young teeth) and is known to reduce dentin permeability, forming at a rate that is related to the extent of external irritation from caries or attrition. Additional information on this type of dentin, often termed tertiary dentin, is given by Karjalainen (1984) and Pashley (1996).

Dentin in *dentinogenesis imperfecta* type I is associated with collagen-structural abnormalities and is dealt with in another section of this book. A second form, which is a genetic heritable syndrome of type II, is associated with a disorder of the non-collagenous proteins (MacDougall et al. 2006). Dentinogenesis imperfecta type II dentin has fewer tubules, lower mineral content and an abnormal appearance. Kinney et al. (2001) noted that when allowed to dehydrate, experimental samples buckled, unlike normal dentin. They reported larger crystal sizes and notable malformation of the tissue. Mineral density was found to be only about 67% that of normal dentin, although in some regions near the pulp, it was found to be higher than normal dentin. The same authors subsequently showed that although dry samples were only moderately softer with a moderate reduction of modulus, wet samples of the altered dentin displayed a hardness of only 200 MPa and Young's modulus of about 5 GPa. Thus, the wet tissue exhibited hardness and Young's modulus amounting to only 25% of normal dentin. These findings might explain the poor longevity of such teeth, suggesting an important role for the intrafibrillar mineral in normal dentin.

Interglobular dentin is perhaps the least understood of what is considered to be pathological dentin. It frequently forms in the crown in limited regions, during the early stages of crown formation (Jones and Boyde 1984). It is hypomineralized and appears to be a zone of poorly fused globular dentin spheres. Such spheres are typical of developmental dentin mineralization (they are termed calcospherites) whereby globules in predentin nucleate, grow and fuse along mineralizing fronts to form the bulk of dentin. It has been proposed that fast growth rates (perhaps too many nucleation sites) early in dentin maturation might prevent complete calcospherite fusion, leading to the formation of this dentin type. (Non-fused patches in human dentin are prevalent in certain metabolic deficiencies such as rickets or overexposure to fluoride, which is why this dentin is thought to be abnormal). It is noteworthy however that no peritubular dentin is seen in interglobular dentin (Nanci 2003) unlike the metabolic diseased forms. Furthermore, in humans, interglobular dentin forms only thin layers that are usually found beneath mantle dentin. Although it is not present in all human teeth it is frequently seen in such as the narwhal tusk and thus interglobular dentin is probably not pathological. Currey (2002) emphasized the high toughness of narwhal dentin, being resistant to impact and manifesting notch insensitivity. Thus, at least for narwhal dentin, impact energy of $36\,\mathrm{kJ/m}^2$ was measured for un-notched samples, whereas almost $31\,\mathrm{kJ/m}^2$ was measured for notched samples. Curry demonstrated deformation concentration in the interglobular dentin, which is perhaps to be expected due to the lower mineral content. In bovine bone for comparison, which has a much higher Young's modulus relative to its mineral content, no such structures exist, and much lower values of impact resistance were measured, amounting to only 12.6 and 1.5 kJ/m^2 for un-notched and notched samples, respectively. Although Currey (2002) emphasized the importance of interglobular dentin for its higher fracture and inelastic properties, interglobular regions probably deform appreciably during elastic loading, possibly serving as defined strain concentrators. It is possible that this less-dense tissue functions to increase tooth dentin toughness (essentially protecting dentin bulk from loads in enamel) as part of the soft zone discussed by Wang and Weiner (1998b) and others. Indeed, the small dimensions of these poorly mineralized regions suggest that they are much below the critical flaw sizes analyzed by Kinney (2003) and co workers. As shown by Zaslansky et al. (2006a), dentin in a zone beneath enamel is softer, has a lower modulus (in compression) and undergoes markedly higher strain relative to the bulk when loaded. Although the presence of interglobular dentin was not tested in these measurements, it was probably there since it is often found in premolars (Sato et al. 2000). Tsuchiya et al. (2001) studied the compositional difference between interglobular dentin and other regions of poor mineralization in dog tooth dentin. They demonstrated reduced mineral content and abundant proteoglycans, possibly inversely correlated. They suggested a mechanical role of this tissue - functioning to support load in teeth - similar to proteoglycan function in joint cartilage. It is fascinating to consider an even more exotic potential function for interglobular dentin: The fact that interglobular regions contain little peritubular dentin may facilitate the flow of fluid down the tubules when the tissue is compressed, and the lower mineral content ensures a lower stiffness. If so, then when teeth are loaded, fluid forced down the tubules may be sensed in the pulp. Such a function for interglobular dentin (serving as a source of fluid which can be pressed down the tubules) makes the concept of dentin being a part of a tactile sensing system even more feasible (see Section 15.3). Like many of the un-answered questions of dentin, this too, awaits further investigation.

15.6 Conclusion

Compositionally, dentin is similar to other members of the bone family of materials. However, at closer look, it has peculiarities not found in other mineralized collagen-based materials. It is a living tissue with no living cells within; it is designed to withstand many years of cyclic loading yet has no replacement mechanism for biologic turnover; it is attached to a very different protective material that is brittle and stiff (enamel) and thus forms a bi-material interface that rarely fails – very unlike man-made mechanical tools where such an interface would be prone to fracture. Dentin contains significant amounts of mineral that is not associated with collagen yet does not participate in the physiological calcium biochemistry of the body. It combines high-density regions with tubular voids, and mechanically, it is not nearly as anisotropic as the microstructure would suggest. Both elastic and plastic material properties vary significantly from region to region and along different orientations within very small distances. Indeed, it is not surprising that structure–function relations in dentin are not fully understood. It is hoped that the basic principles of design and properties described in this chapter allow comparison to other collagen-based materials with possible insights to the effects of pathology.

Dentin appears to be more an array of architectural designs than a simple material. Still, it is the mineralized tissue most frequently treated for disease, and adhesive methods of treatment include bonding to exposed surfaces within teeth. Modern restorative dental methods are based on acid-etching of the mineral phase to reveal the collagen fibril network, which is subsequently embedded with resins that may easily be cross-linked and polymerized (Eliades et al. 2005). Extensive dental-materials research has yielded fast-acting bonding systems that allow clinicians to reliably bond particle-filled dental composites to sound and partially modified wet tooth structure. Although much has been achieved, it seems prudent to assume that with a growing understanding of the natural dentin microstructure, its properties and their distribution, new chemistries and techniques will emerge that will allow dentists to better match and restore lost tooth tissue. Furthermore, perhaps what we learn can be used to design new materials and devices that may provide superior performance for mechanically demanding applications.

References

Arola, D. D., & Reprogel, R. K. (2006). Tubule orientation and the fatigue strength of human dentin. Biomaterials 27(9): 2131–2140.

Ash, M. M., & Nelson, S. J. (2003). Wheeler's Dental Anatomy, Physiology and Occlusion (8 ed.). St. Louis: Saunders.

Avery, J. K., Cox, C. F., & Chiego, D. J. (1984). Structural and physiologic aspects of dentin innervation. In A. Linde (Ed.), Dentin and Dentinogenesis (Vol. 1, pp. 19–46). Boca Raton: CRC Press.

Beust, T. B. (1931). Physiologic changes in the dentin. J. Dent. Res. 11: 267–275.

Bevelander, G. (1941). The development and structure of the fiber system of dentin. Anat. Rec. 81(1): 79–97.

Braden, M. (1976). Biophysics of the tooth. In Y. Kawamura (Ed.), Physiology of Oral Tissues (Vol. 2, pp. 1–37). Basel, New York: S. Karger.

Brear, K., Currey, J. D., Kingsley, M. C. S., & Ramsay, M. (1993). The mechanical design of the tusk of the narwhal (Monodon-Monoceros, Cetacea). J. Zool. 230: 411–423.

Carr, A., Tibbetts, I. R., Kemp, A., Truss, R., & Drennan, J. (2006). Inferring parrotfish (Teleostei: Scaridae) pharyngeal mill function from dental morphology, wear, and microstructure. J. Morph. 267(10): 1147–1156.

Carvalho, R. M., Fernandes, C. A. O., Villanueva, R., Wang, L., & Pashley, D. H. (2001). Tensile strength of human dentin as a function of tubule orientation and density. J. Adhes. Dent. 3: 309–314.

Craig, R., & Peyton, F. (1958). Elastic and mechanical properties of human dentin. J. Dent. Res. 37: 710–718.

Crooks, P. V., Oreilly, C. B., & Owens, P. D. A. (1983). Microscopy of the dentin of enamel-free areas of rat molar teeth. Arch. Oral Biol. 28(2): 167–175.

Currey, J. D. (2002). Bones: Structure and Mechanics. Princeton: Princeton University Press.

Currey, J. D., & Brear, K. (1990). Hardness, Young Modulus and Yield Stress in Mammalian Mineralized Tissues. J. Mater. Sci. Mater. Med. 1(1): 14–20.

Currey, J. D., Brear, K., & Zioupos, P. (1994). Dependence of mechanical properties on fiber angle in Narwhal tusk, a highly oriented biological composite. J. Biomech. 24: 57–61.

Dean, M. C. (2006). Tooth microstructure tracks the pace of human life-history evolution. Proc. Royal Soc. B. 273(1603): 2799–2808.

Doerner, M. F., & Nix, W. D. (1986). A method for interpreting the data from depth-sensing indentation instruments. J. Mater. Res. 1: 601–609.

Duncanson, M. G., & Korostoff, E. (1975). Compressive viscoelastic properties of human dentin.1. Stress-relaxation behavior. J. Dent. Res. 54(6): 1207–1212.

El-Mowafy, O. M., & Watts, D. C. (1986). Fracture toughness of human dentin. J. Dent. Res. 65(5): 677–681.

Eliades, G., Watts, D. C., & Eliades, T. (2005). Dental Hard Tissues and Bonding: Interfacial Phenomena and Related Properties. Berlin: Springer.

Frank, R. M. (1965). Ultrastructure of human dentine. In H. Fleisch, H. J. J. Blackwood, M. Owen, M. P. Fleisch-Ronchetti (Eds.), Calcified Tissues: Proceedings of the Third European Symposium on Calcified Tissues Held at Davos (Switzerland), April 11th-16, 1965 1966 (PP. 259–272). Springer-Verlag Berlin, Heidelberg, New York.

Garberoglio, R., & Brannstrom, M. (1976). Scanning electron-microscopic investigation of human dentinal tubules. Arch. Oral Biol. 21(6): 355–362.

Gebhardt, W. (1900). On the functional construction of some teeth. Arch. f. Entwcklngsgcd. Organ. 10(1): 135–243.

Gilmore, R. S., Pollack, R. P., & Katz, J. L. (1969). Elastic properties of bovine dentine and enamel. Arch. Oral Biol. 15(8): 787–796.

Goldberg, M., Septier, D., Lecolle, S., Chardin, H., Quintana, M. A., Acevedo, A. C., et al. (1995). Dental mineralization. Int. J. Dev. Biol. 39(1): 93–110.

Goracci, G., Mori, G., & Baldi, M. (1999). Terminal end of human odontoblast process: a study using SEM and confocal microscopy. Clin. Oral. Invest. 3: 126–132.

Gotliv, B. A., Robach, J. S., & Veis, A. (2006). The composition and structure of bovine peritubular dentin: Mapping by time of flight secondary ion mass spectroscopy. J. Struct. Biol. 156(2): 320–333.

Haines, D. J. (1968). Physical properties of human tooth enamel and enamel sheath material under load. J. Biomech. 1(2): 117–125.

Ho, S. P., Balooch, M., Marshall, S. J., & Marshall, G. W. (2004). Local properties of a functionally graded interphase between cementum and dentin. J. Biomed. Mater. Res. A 70A(3): 480–489.

Hood, J. A. A. (1991). Biomechanics of the intact, prepared and restored tooth: some clinical implications. Int. Dent. J. 41: 25–32.

Imbeni, V., Kruzic, J. J., Marshall, G. W., Marshall, S. J., & Ritchie, R. O. (2005). The dentin–enamel junction and the fracture of human teeth. Nat. Mater. 4(3): 229–232.

Imbeni, V., Nalla, R. K., Bosi, C., Kinney, J. H., & Ritchie, R. O. (2002). In vitro fracture toughness of human dentin. J. Biomed. Mater. Res. 66(A): 1–9.

Johnson, N. W., & Poole, D. F. G. (1967). Orientation of collagen fibres in rat dentine. Nature 213(5077): 695–696.

Jones, S. J., & Boyde, A. (1984). Ultrastructure of dentin and dentinogenesis. In A. Linde (Ed.), Dentin and Dentinogenesis (Vol. 1, pp. 81–134). Boca Raton: CRC Press.

Karjalainen, S. (1984). Secondary and reparative dentin formation. In A. Linde (Ed.), Dentin and Dentinogenesis (Vol. 2, pp. 107–120). Boca Raton: CRC Press.

Katz, J. L. (1971). Hard tissue as a composite material – bounds on the elastic behavior. J. Biomech. 4: 455–473.

Katz, J. L., Kinney, J. H., Spencer, P., Wang, Y., Fricke, B., Walker, M. P., et al. (2005). Elastic anisotropy of bone and dentitional tissues. J. Mater. Sci. Mater. Med. 16(9): 803–806.

Kinney, J. H., Balooch, M., Marshall, G. W., & Marshall, S. J. (1999). A micromechanics model of the elastic properties of human dentin. Arch. Oral. Biol. 44: 813–822.

Kinney, J. H., Balooch, M., Marshall, S., Marshall, G., & Weihs, T. (1996). Young's modulus of human peritubular and intertubular dentine. Arch. Oral Biol. 41(1): 9–13.

Kinney, J. H., Gladden, J. R., Marshall, G. W., Marshall, S. J., So, J. H., & Maynard, J. D. (2004). Resonant ultrasound spectroscopy measurements of the elastic constants of human dentin. J. Biomech. 37(4):437–441.

Kinney, J. H., Marshall, S., & Marshall, G. (2003). The mechanical properties of human dentin: a critical review and re-evaluation of the dental literature. Crit. Rev. Oral Biol. Med. 14(1): 13–29.

Kinney, J. H., Nalla, R. K., Pople, J. A., Breunig, T. M., & Ritchie, R. O. (2005). Age-related transparent root dentin: mineral concentration, crystallite size, and mechanical properties. Biomaterials 26(16): 3363–3376.

Kinney, J. H., Pople, J. A., Driessen, C. H., Breuning, T. M., Marshall, G. W., & Marshall, S. J. (2001). Intrafibrillar mineral may be absent in detinogenesis imperfecta type II (DI-II), J. Dent. Res. 80(6):1555–1559.

Kishen, A., Ramamurty, U., & Asundi, A. (2000). Experimental studies on the nature of property gradients in the human dentine. J. Biomed.Mater. Res. 51: 650–659.

Kramer, I. R. H. (1951). The distribution of collagen fibrils in the dentine matrix. Brit. Dent. J. 91: 1–7.

Lees, S., & Rollins, F. R. (1972). Anisotropy in hard dental tissues. J. Biomech. 15(6): 557–566.

Lin, C. P., & Douglas, W. H. (1994). Structure–property relations and crack resistance at the bovine dentin–enamel junction. J. Dent. Res. 73(5): 1072–1078.

Lin, C. P., Douglas, W. H., & Erlandsen, S. L. (1993). Scanning electron microscopy of type I collagen at the dentino-enamel junction of human teeth. J. Histochem. Cytochem. 41: 381–388.

Linde, A. (1984). Dentin and Dentinogenesis. Boca Raton: CRC Press.

Lovschall, H., Fejerskov, O., & Josephsen, K. (2002). Age-related and site-specific changes in the pulpodentinal morphology of rat molars. Arch. Oral Biol. 47(5): 361–367.

Lucas, P. W. (2004). Dental Functional Morphology: How Teeth Work. Cambridge: Cambridge University. press.

MacDougall, M., Dong, J., & Acevedo, A. C. (2006). Molecular basis of human dentin diseases. Am. J. Med. Genet. 140A(23): 2536–2546.

Magloire, H., Couble, M.-L., Romeas, A., & Bleicher, F. (2004). Odontoblast primary cilia: facts and hypotheses. Cell Biol. Int. 28: 93–99.

Marshall, G. W., Habelitz, S., Gallagher, R., Balooch, M., Balooch, G., & Marshall, S. J. (2001). Nanomechanical properties of hydrated carious human dentin. J. Dent. Res. 80(8): 1768–1771.

Marshall, G. W., Marshall, S. J., Kinney, J. H., & Balooch, M. (1997). The dentin substrate: structure and properties related to bonding. J. Dent. 25(6): 441–458.

Meredith, N., Sheriff, M., Setchell, D., & Sivanson, S. (1996). Measurements of the microhardness and Young's modulus of human enamel and dentin using an indentation technique. Arch. Oral. Biol. 41(6): 539–545.

Moss, M. L. (1974). Studies on dentin I: mantle dentin. Acta. Anat. 87: 481–507.

Murray, P. E., Stanley, H. R., Matthews, J. B., Sloan, A. J., & Smith, A. J. (2002). Age-related odontometric changes of human teeth. Oral Surg. Oral Med. Oral Pathol Oral Radiol Endod. 93(4): 474–482.

Nanci, A. (2003). Ten Cate's Oral Histology: Development, Structure and Function (6th ed.). St. Louis: Mosby.

Oi, T., Saka, H., & Ide, Y. (2004). Three-dimensional observation of pulp cavities in the maxillary first premolar tooth using micro-CT. Int. Endod. J. 37(1): 46–51.

Oliver, W. C., & Pharr, G. M. (1992). An improved technique for determining hardness and elastic-modulus using load and displacement sensing indentation experiments. J. Mater. Res. 7(6): 1564–1583.

Orban, B. (1929). The development of the dentin. J. Am. Dent. Assoc. 16(9): 1547–1586.

Osborn, J. W. (1969). Dentin hardness and incisor wear in the beaver (Castor fiber). Acta. Anat. 72: 123–132.

Palamara, J. E. A., Wilson, P. R., Thomas, C. D. L., & Messer, H. H. (2000). A new imaging technique for measuring the surface strains applied to dentine. J. Dent. 28(2): 141–146.

Paphangkorakit, J., & Osborn, J. W. (1998). Discrimination of hardness by human teeth apparently not involving periodontal receptors. Arch. Oral Biol. 43(1): 1–7.

Pashley, D. H. (1989). Dentin: a dynamic substrate: a review. Scanning Microsc. 3(1): 161–174.

Pashley, D. H. (1996). Dynamics of the pulpo-dentin complex. Crit. Rev. Oral Biol. Med. 7(2): 104–133.

Porter, A. E., Nalla, R. K., Minor, A., Jinschek, J. R., Kisielowski, C., Radmilovic, V., et al. (2005). A transmission electron microscopy study of mineralization in age-induced transparent dentin. Biomaterials 26(36): 7650–7660.

Rasmussen, S. T. (1984). Fracture properties of human teeth in proximity to the dentinoenamel junction. J. Dent. Res. 63: 1279–1283.

Rasmussen, S. T., Patchin, R. E., Scott, D. B., & Heuer, A. H. (1976). Fracture properties of human enamel and dentin. J. Dent. Res. 55: 154–164.

Rensberger, J. M. (2000). Pathways to functional differentiation in mammalian teeth. In M. F. Teaford, M. M. Smith & M. W. J. Ferguson (Eds.), Development, Function and Evolution of Teeth (pp. 252–268). Cambridge: Cambridge University Press.

Renson, C. E., & Braden, M. (1971). The experimental deformation of human dentine by indenters. Arch. Oral Biol. 16: 563–572.

Renson, C. E., & Braden, M. (1975). Experimental determination of rigidity modulus, poissons ratio and elastic limit in shear of human dentin. Arch. Oral Biol. 20(1): 43.

Sasagawa, I. (2002). Mineralization patterns in elasmobranch fish. Microsc. Res. Tech.. 59(5): 396–407.

Sato, H., Kagayama, M., Sasano, Y., & Mayanagi, H. (2000). Distribution of interglobular dentine in human tooth roots. Cells Tissues Organs 166(1): 40–47.

Schmidt, W. J., & Keil, A. (1971). Polarizing Microscopy of Dental Tissues. Oxford: Pergamon Press.

Schroeder, L., & Frank, R. (1985). High resolution transmission electron microscopy of adult human peritubular dentine. Cell Tissue Res. 242: 449–451.

Smith, M. M., & Sansom, I. J. (2000). Evolutionary origins of dentine in the fossil record of early vertebrates: diversity, development and function. In M. F. Teaford, M. M. Smith & M. W. J. Ferguson (Eds.), Development, Function and Evolution of Teeth (pp. 65–81). Cambridge: Cambridge University Press.

Stanford, J. W., Weigel, K. V., Pafenberger, G. C., & Sweeney, W. T. (1960). Compressive properties of hard tooth tissues and some restorative materials. J. Am. Dent. Assoc. 60(746–756).

Takuma, S., & Eda, S. (1966). Structure and development of peritubular matrix in dentin. J. Dent. Res. 45(3P1S): 683–692.

Tesch, W., Eidelman, N., Roschger, P., Goldenburg, F., Klaushofer, K., & Fratzl, P. (2001). Graded microstructure and mechanical properties of human crown dentin. Calcif. Tissue Int. 69. 147–157.

Tjaderhane, L., Larjava, H., Sorsa, T., Uitto, V. J., Larmas, M., & Salo, T. (1998). The activation and function of host matrix metalloproteinases in dentin matrix breakdown in caries lesions. J. Dent. Res. 77(8): 1622–1629.

Tonami, K., & Takahashi, H. (1997). Effects of aging on tensile fatigue strength of bovine dentin. Dent. Mater. J. 16(2):156–169.

Tronstad, L. (1972). Optical and microradiographic appearance of intact and worn human coronal dentin. Arch. Oral Biol. 17(5): 847–858.

Tsuchiya, M., Sasano, Y., Kagayama, M., & Watanabe, M. (2001). Characterization of interglobular dentin and Tomes' granular layer in dog dentin using electron probe microanalysis in comparison with predentin. Calc. Tissue Int. 68(3): 172–178.

Urabe, I., Nakajima, M., Sano, H., & Tagami, J. (2000). Physical properties if the dentino-enamel junction region. Am. J. Dent. 13: 129–135.

vanHoute, J. (1994). Role of microorganisms in caries etiology. J. Dent. Res. 73(3): 672–681.

vanMeerbeek, B., Willems, G., Celis, J. P., Roos, J. R., Braem, M., Lambrechts, P., et al. (1993). Assessment by nano-indentation of the hardness and elasticity of the resin-dentin bonding area. J. Dent. Res. 72(10): 1434–1442.

vonEbner, V. (1909). Ueber scheinbare und wirkliche Radiarfassern des Zahnbeines. Anat. Anz. 34: 289–309.

Wang, R. (2005). Anisotropic fracture in bovine root and coronla dentin. Dent. Mater. 21: 429–436.

Wang, R., & Weiner, S. (1998a). Human root dentin: structural anisotropy and Vickers microhardness isotropy. Conn. Tissue Res. 39: 269–279.

Wang, R., & Weiner, S. (1998b). Strain-structure relations in human teeth using Moire fringes. J. Biomech. 31(2): 135–141.

Watanabe, L. G., Marshall, G. W., & Marshall, S. J. (1996). Dentin shear strength: Effects of tubule orientation and intratooth location. Dent. Mater. 12(2): 109–115.

Waters, N. (1980). Some mechanical and physical properties of teeth. In Symposia of the Society for Experimental Biology. Mechanical Properties of Biological Materials. (Vol. 34, pp. 99–135). London: Cambridge University Press.

Weber, D. F. (1968). Distribution of peritubular matrix in human coronal dentin. J. Morph. 126(4): 435–446.

Weber, D. F. (1974). Human dentin sclerosis – microradiographic survey. Arch. Oral Biol. 19(2): 163–169.

Weidenreich, F. (1925). Ueber den Bau und die Entwicklung des Zahnbeines in der Reihe der Wirbeltiere. Ztschr. f. Anat. u. Entwcklngsges. 76: 218.

Weiner, S., Veis, A., Beniash, E., Arad, T., Dillon, J. W., Sabsay, B., et al. (1999). Peritubular dentin formation: crystal organization and the macromolecular constituents in human teeth. J. Struct. Biol. 126(1): 27–41.

Weiner, S., & Wagner, H. D. (1998). The material bone: structure–mechanical function relations. Ann. Rev. Mat. Sci. 28: 271–298.

White, S. N., Paine, M. L., Lou, W., Sarikaya, M., Fong, H., Yu, Z., et al. (2000). The Dentin–enamel junction is a broad transitional zone uniting dissimilar bioceramic composites. J. Am. Chem. Soc. 83(1): 238–240.

Xu, H. H. H., Smith, D. T., Jahanmir, S., Romberg, E., Kelly, J. R., Thompson, V. P., et al. (1998). Indentation damage and mechanical properties of human enamel and dentin. J. Dent. Res. 77(3): 472–480.

Zaslansky, P., Friesem, A. A., & Weiner, S. (2006a). Structure and mechanical properties of the soft zone separating bulk dentin and enamel in crowns of human teeth: Insight into tooth function. J. Struct. Biol. 153(2): 188–199.

Zaslansky, P., Shahar, R., Friesem, A. A., & Weiner, S. (2006b). Relations between shape, materials proprtics, and function in biological materials using laser speckle interferometry: In situ tooth deformation. Adv. Funct. Mat. 16(15): 1925–1936.

Zaslansky, P., & Weiner, S. (2007). Design strategies of human teeth: biomechanical adaptations. In M. Epple & E. Bauerlein (Eds.), Handbook of Biomineralization: Medical and Clinical Aspects (pp. 183–202). Weinheim: WILEY-VCH.

Chapter 16
Genetic Collagen Diseases: Influence of Collagen Mutations on Structure and Mechanical Behavior*

R.D. Blank and A.L. Boskey

Abstract Mutations in type I collagen as well as those in the enzymes involved in the processing of type I collagen and in proteins that associate with type I collagen have significant effects on the structure and mechanical properties of the collagen-containing tissues. Osteogenesis imperfecta, also known as "brittle bone disease" provides an illustration of how site-specific mutations and alterations can affect tissue integrity. Examples of other type I collagen diseases are also discussed.

16.1 Introduction

Diseases associated with collagen abnormalities have severe consequences because the collagens are major structural proteins within most human body tissues. Mutations resulting in loss of collagen expression or in abnormal folding of the collagen molecules have been reported for each of the fibrillar collagen types (Kuivaniemi et al. 1997). The majority of the mutations are single-base and either change the codon of a critical amino acid or lead to abnormal RNA splicing. Most of the amino acid substitutions are those of a bulkier amino acid for the obligatory glycine of the repeating Gly X Y sequence of the collagen triple helix. The mutations result in a variety of diseases of bone, tendon and ligament, and blood vessels, including osteogenesis imperfecta, and, some forms of osteoporosis.

The first described mutations in collagen were those associated with "brittle bone disease" or osteogenesis imperfecta (OI) (Peltonen et al. 1980). Characterization of the underlying basis for the formation of fragile bones defined the importance of type I collagen for bone growth and development, and provided the first example of genetic abnormalities in a structural protein that had consequences as life-determining as those reported earlier for the globular protein, hemoglobin (sickle cell disease). In this chapter, we focus on OI as a model for demonstrating the impact of changes in type I collagen and type I collagen-related proteins on tissue structure and mechanical integrity.

---

* This chapter is Madison, Wisconsin GRECC manuscript 2008–05.

P. Fratzl (ed.), *Collagen: Structure and Mechanics*,
© Springer Science+Business Media, LLC 2008

16.2 Osteogenesis Imperfecta

Osteogenesis imperfecta (OI) is a condition associated with bone fragility and decreased bone mass. In humans, severity varies, ranging from a lethal form with multiple fractures in utero, to milder forms with few fractures. While presentation may be variable, each of the type I collagen-containing tissues (bone, skin, eyes, dentin, tendon, ligament, blood vessels, and ears) are afflicted to some extent. Thus the "OI phenotype" includes brittle bones that fracture easily, translucent skin, blue sclera, dentin abnormalities (dentinogenesis imperfecta), hyperlaxity of ligaments and skin, vascular complications, and hearing abnormalities (Marini 2006).

16.2.1 General Description – OI Types

Although it is of limited utility in gauging the pathogenetic mechanism by which OI arises in a particular patient, the Sillence classification (Sillence et al. 1979) forms the basis for clinicians' categorization of OI. The four original categories have been supplemented by additional types defined by mutations in genes other than *COL1A1* and *COL1A2*, as discussed more fully below in Section 16.2.3.

A modified Sillence classification reflecting mainstream clinical practice is summarized in Table 16.1. While this scheme is widely used, in our opinion it is of limited utility because it provides little insight into the pathogenesis or genotype/phenotype correlation. It is also idiosyncratic insofar as type I is the mildest and type II the most severe, with types III and IV intermediate in severity. Inclusion of the phenocopies (types V–VIII) in the modified classification further limits the utility of the overall numbering scheme, even though it reflects common clinical practice.

16.2.2 Genotype/Phenotype

Limited improvement in OI classification is possible by considering the molecular pathophysiology of specific mutations. This approach accounts for broad trends of genotype/phenotype correlation, but exceptions to the general principles exist. Moreover, a pathophysiological classification scheme still falls short of providing a satisfactory explanation for phenotypic variability among individuals harboring the same type I collagen mutation. Most of the inferences that can be made regarding genotype/phenotype correlations are based on human disease, as the number of mutations in humans is so much larger than the number of mutations in model organisms. Some of the principles underlying the general genotype/phenotype trends are also reinforced by OI phenocopies, as discussed in Section 16.2.3. The general relationships between genotype and phenotype are as follows.

Table 16.1 Modified sillence classification of human OI

Type	Genes	Inheritance*	Cardinal clinical features
I	*COL1A1,* *COL1A2*	AD	Mildly to moderately increased skeletal fragility, normal stature, variable presence of blue sclerae, hearing loss, dental fragility, dental discoloration, easy bruisability
II	*COL1A1,* *COL1A2*	AD and AR	Severe, lethal skeletal fragility in perinatal period, with impaired calvarial mineralization, flattened vertebrae, beaded ribs, and multiple long bone fractures resulting in shortening and deformity
III	*COL1A1,* *COL1A2*	AD and AR	Short stature with normal head size, progressive limb shortening and deformity due to fractures, long bone metaphyseal flaring, often wheelchair-confined by maturity, triangular facies, blue sclerae, hearing loss, dental fragility, dental discoloration, easy bruisability
IV	*COL1A1,* *COL1A2*	AD and AR	Disease severity intermediate between types I and III. Some authors consider this to be a "wastebasket" classification for cases that don't fit neatly into any of the other categories
V	Unknown	AD	Skeletal features similar to type IV, distinguished by non-collagenous origin, formation of hyperplastic callus following fracture, calcification of the interosseous membrane of the forearm, and resulting restriction of forearm pronation and supination, and presence of a distinctive radiodense band adjacent to the growth plates during childhood and adolescence
VI	Unknown	AD	Skeletal features similar to severe type IV, distinguished by non-collagenous origin, elevated alkaline phosphatase, absence of lamellar bone architecture and its replacement by a "fish scale" appearance, and impaired mineralization without rickets
VII	*CRTAP*	AR	Skeletal features similar to types II and III, distinguished by non-collagenous origin, rhizomelia, coxa vara, impaired long bone modeling, first encountered in an inbred Native American population. Some authors classify some cases as class IIB
VIII	*LEPRE1*	AR	Skeletal features similar to types II and III, but distinguished by round facies and barrel-shaped chest

*AD = autosomal dominant, AR = autosomal recessive. Adapted from the following sources: Sillence et al. (1979); Marini (2006); Glorieux et al. (2000, 2002); Ward et al. (2002); Barnes et al. (2006); Morello et al. (2006); Cabral et al. (2007).

16.2.2.1 Mutations that Cause a Quantitative Deficiency of Protein Synthesis Result in Milder Disease than Those that Cause Synthesis of an Abnormal Protein

Even before the mechanism was understood, it was recognized that quantitative deficiency of type I collagen synthesis is a common feature of Sillence class I OI (Sykes et al. 1977; Francis et al. 1981; Rowe et al. 1985; Morike et al. 1992). Most nonsense and frameshift mutations of *COL1A1* and *COL1A2* generate unstable transcripts that are degraded prior to export from the nucleus

(Willing et al. 1992, 1994, 1996; Slayton et al. 2000). Indeed, a deletion of the entire *COL1A1* gene results in Sillence class I OI (Pollitt et al. 2006). This finding suggests that abnormal proteins exert a dominant-negative effect on the bone matrix, hence leading to more severe disease than mutations that are effectively null at the protein level. Dominant-negative actions may be exerted intracellularly both by interference with protein processing and secretion (termed "procollagen suicide" by Prockop (1984)) and by interference with higher-order assembly of collagen following secretion. *COL1A1* mutations listed in the collagen database (http://www.le.ac.uk/ge/collagen/) (Dalgleish 1997, 1998) illustrate the pattern. There are 15 reported nonsense mutations in the gene, 14 of which result in Sillence class I OI and the last results in class IV OI. Conversely, 14 *COL1A1* missense mutations result in class I disease, 25 result in class IV disease, 29 result in class III, 56 result in class II disease, and 12 result in overlap syndromes or otherwise unclassifiable disease. Frameshift and splicing mutations that alter the reading frame are also heavily skewed toward class I disease, with class I OI resulting from 58 such mutations and all other phenotypes from 18 mutations. Comparing the proportion of class I disease in missense mutations to that in nonsense and frameshift mutations (pooled), there is a highly significant difference ($\chi^2 = 109$, 1 df, $P < 10^{-5}$).

16.2.2.2 Mutations of COL1A1 Have More Severe Phenotypic Consequences than Comparable Mutations of COL1A2

This trend follows from the probability of assembling normal and abnormal heterotrimers. The collagen triple helix contains 2 $\alpha 1$ and 1 $\alpha 2$ chains. If an individual is heterozygous for an $\alpha 1$ mutation, assuming that both the mutant and wild-type alleles are expressed equally and equally likely to be incorporated into the nascent triple helix, then three-fourths of the molecules will contain a mutant $\alpha 1$ chain. In the case of an $\alpha 2$ mutation, only half of the molecules will contain a mutant chain. Therefore, heterozygosity for $\alpha 1$ missense mutations is expected to affect more molecules than heterozygosity for $\alpha 2$ missense mutations, resulting in a more severe phenotype when the $\alpha 1$ chain harbors the mutation.

The theoretical expectation is reflected by the phenotypic distributions of glycine-substitution mutations in *COL1A1* and *COL1A2*. Of 125 such *COL1A1* mutations listed in the collagen database (http://www.le.ac.uk/ge/collagen/) (Dalgleish 1997, 1998), 20 (16%) result in Sillence class I OI, 21 (17%) in type IV OI, 32 (26%) in type III OI, and 52 (41%) in type II OI. Thus, the distribution of phenotypes is clearly skewed toward severe disease. In contrast, the phenotype distribution for 99 glycine-substitution mutations of *COL1A2* is 12 (12%) class I or "osteoporosis", 31 class IV (31%), 31 class III (31%), and 25 (25%) class II. The difference of phenotype proportions differs significantly ($\chi^2 = 10.53$, 3 df, $P = 0.015$), with *COL1A1* displaying a substantially greater frequency of Sillence class II disease.

It is also tempting to speculate that the much lower frequency of known nonsense mutations in *COL1A2* than in *COL1A1* (Table 16.2) reflects a phenotype so mild that OI is not diagnosed.

Table 16.2 Summary of human *col1a1* and *col1a2* mutations

	Missense*	Nonsense	Frameshift †	Insertion/Deletion	Splice
COL1A1	152 (24)	15	76	19	64
COL1A2	107 (6)	1 + 1 anti-stop	1	17	38

* Total missense mutations (missense mutations in which glycine is NOT the normally present residue).

† Splice mutations that result in frameshifts are listed under frameshift.

Adapted from the Database of Human Type I and Type III Collagen Mutations (http://www.le. ac.uk/ge/collagen/).

16.2.2.3 Substitution of COL1A1 Glycines by Serine, Cysteine, and Alanine Is Better Tolerated than Substitution by Valine, Arginine, or Aspartate

Collagen's triple helical structure cannot accommodate a side chain where the three polypeptide chains cross. Thus, the Gly-X-Y structure incorporates glycine, the smallest amino acid, at this position. Large, charged amino acids presumably disrupt the triple helical structure more profoundly than do smaller amino acids such as serine. The distribution of each possible amino acid for glycine along the helical region of *COL1A1* is nicely displayed on the map available at the collagen database website (Dalgleish 1997, 1998).

The pattern for *COL1A2* is less informative, in part because fewer mutations cause lethal disease. Most of these result from substitution of aspartate for glycine (Dalgleish 1997, 1998).

16.2.2.4 There Are Critical Residues at Which Mutation Is Very Poorly Tolerated

Sites at which lethal missense mutations occur in *COL1A1* and *COL1A2* are clustered, and the pattern of clustering differs between the two chains (Marini et al. 2007b). The most important of these are located near the C-terminal end of *COL1A1* and includes interaction sites for other extracellular proteins, including integrins, fibronectin, and cartilage oligomeric matrix protein (COMP) (Dalgleish 1997, 1998). Moreover, the single proline residue that is the target of cartilage-associated protein (CRTAP)-mediated hydroxylation lies within this critical region (see phenocopies section below). The concept of critical domains for collagen assembly is considered in greater detail in the discussion.

16.2.2.5 Mutations that Cause Skipping of Exon 6 or Interfere with Propeptide Cleavage Tend to Produce Features of EDS

The *N*-propeptide cleavage sites are located within exon 6, and mutations that result in skipping of this exon cause Ehlers–Danlos syndrome (EDS) VII, a condition associated with joint hypermobility and extremely stretchable skin (Weil et al. 1988, 1989a, 1990; D'Alessio et al. 1991; Nicholls et al. 1991; Vasan et al. 1991; Chiodo et al. 1992; Nicholls et al. 1992; Watson et al. 1992; Ho et al. 1994; Lehmann et al. 1994; Byers et al. 1997; Giunta et al. 1999; Nicholls et al. 2000; Malfait et al. 2006). Moreover, mutations near the N-terminal end of the helical domain can also interfere with propeptide cleavage, giving rise to overlap syndromes in which both skeletal fragility and ligamentous laxity are prominent features (Cabral et al. 2005). Disruption of C-propeptide processing can also lead to EDS features (Nicholls et al. 2001; Schwarze et al. 2004; Symoens et al. 2004; Cabral et al. 2005; Malfait et al. 2007).

16.2.2.6 Missense Mutations Affecting Amino Acids Other than Glycine Frequently Lead to Clinical Syndromes Other than OI

Most missense mutations in *COL1A1* and *COL1A2* substitute another amino acid for glycine. However, among those affecting other residues, 9 of 30 of the non-glycine mutations listed in the collagen mutation database lead to non-OI phenotypes (Phillips et al. 1990; Spotila et al. 1994; Symoens et al. 2004; Cabral et al. 2005; Gensure et al. 2005; Malfait et al. 2007; Suphapeetiporn et al. 2007).

16.2.2.7 Humans

There is a vast array of known pathogenic mutations in the *COL1A1* and *COL1A2* genes (Marini et al. 2007b). While some mutations have been observed in multiple kindreds, many others are private variants limited to a single family. Mutation analysis of the type I collagen genes and clinical correlation of these molecular data with patient characteristics have made it clear that mutations in these two genes encompass a broad range of phenotypes. These are catalogued online at the excellent site curated by Raymond Dalgleish (http://www.le.ac.uk/ge/collagen/) (Dalgleish 1997, 1998). A statistical summary of the mutations is given in Table 16.2. Not surprisingly, different type I collagen mutations each give rise to differing degrees of skeletal abnormality, including all clinical subtypes of OI, Ehlers–Danlos syndrome (EDS), most commonly type VII but also including other EDS subtypes (Steinmann et al. 1980; Eyre et al. 1985; Cole et al. 1986; Hata et al. 1988; Weil et al. 1988, 1989a,b, 1990; D'Alessio et al. 1991; Nicholls et al. 1991; Vasan et al. 1991; Chiodo et al. 1992; Nicholls et al. 1992; Watson et al. 1992; Wirtz et al. 1993; Ho et al. 1994; Lehmann et al. 1994; Nicholls et al. 1996; Byers et al. 1997; Giunta et al. 1999; Nicholls et al. 2000; Nuytinck et al. 2000; Schwarze et al. 2004; Malfait et al. 2006, 2007), and osteoporosis or osteopenia without clear-cut evidence of OI (Shapiro et al. 1989; Spotila et al. 1991; Shapiro et al. 1992; Spotila et al. 1994; Dawson et al. 1999). Conversely, individuals with type I collagen mutations may have normal bone mineral content even though this bone may have impaired structural performance (Paterson and Mole 1994). In addition, overlap syndromes in which features of both OI and EDS are present are well known (Feshchenko et al. 1998; Raff et al. 2000; Nicholls et al. 2001; Symoens et al. 2004; Cabral et al. 2005). Other reported clinical presentations of mutations in *COL1A1* and *COL1A2* include Marfan syndrome (Phillips et al. 1990), cervical artery dissection (Mayer et al. 1996), and Caffey disease, a pediatric cortical hyperostotic disorder (Gensure et al. 2005).

The challenge of understanding genotype–phenotype relationships is increased by the existence of significant intrafamilial variation within kindreds harboring type I collagen mutations. Two simple mechanisms have been documented to contribute to phenotypic variability within kindreds: heterozygosity or homozygosity for a mutant collagen allele (De Paepe et al. 1997) and mosaicism of mildly affected individuals (Constantinou et al. 1990; Wallis et al. 1990; Edwards et al. 1992; Constantinou-Deltas et al. 1993; Mottes et al. 1993; Cohen-Solal et al. 1996). These mechanisms account for a subset of the variability within kindreds, but most of the differences in OI severity among kindreds harboring the same mutation [e.g., (Superti-Furga et al. 1989; Zhuang et al. 1996)] and among individuals in a single kindred [e.g., (Tenni et al. 1991; Mottes et al. 1992; Zolezzi et al. 1997)] are not explained by these mechanisms. The existence of so much unexplained intrafamilial variation suggests that segregation of loci other than the type I collagen genes contributes to the overall clinical presentation.

Despite these variations, it is of interest to note that the clinical severity of OI can be correlated with the ultrastructure of bone, both in terms of the gross morphology and the width and polydispersity of the collagen fibrils (Sarathchandra et al. 2000). Thus for example, in Sillence type I OI, the mildest form, there were the fewest alterations in bone ultrastructure, and collagen fibrils observed by electron microscopy were larger than those in age-matched controls (Cassella et al. 1994), while in the most severe (type II OI) the cortical bone was thin, the collagen fibrils were thin, and the mineral was not oriented on the collagen matrix, as it would be in the healthy individual.

16.2.2.8 Mice

There are both naturally occurring and genetically engineered mouse models of osteogenesis imperfecta (Kamoun-Goldrat and Le Merrer 2007). The genetically engineered mice generated by ablating (knockout) or inserting (knockin) genetic material into specific regions of the mouse genome in embryonic stem cells by homologous recombination are implanted in fertilized eggs into pseudo-pregnant females. The techniques for preparing mutant mice, and the methods for identifying which of the embryos produced have the desired gene expression is reviewed in detail elsewhere (Houdebine 2002; Ristevski 2005). Whether the mouse models are derived or occur spontaneously, in each case, the mice have brittle bones that fracture more readily than those of the wild type. But most of them do not have all the other phenotypic characteristics of human osteogenesis imperfecta (blue sclera, hearing loss, vascular complications, etc.). Table 16.3 compares selected properties of some different models of OI. Because these mutations have been introduced in mice of different backgrounds, and mechanical analyses (generally torsion tests) were done at different ages and in different sexes, the table always compares data to the wild-type animals in the same study.

Table 16.3 Comparative mechanical properties* of femurs from mouse models of osteogenesis imperfecta

Model	Age	Change relative to wild type	Ref
Mov-13	8 week	17× increased porosity; 1.1× increased torsional failure load; no change in cortical area	a, b
	15 week	1.2× increased torsional failure load; 1.3× increased cortical area; 1.6× increased bending moment of inertia	b
oim	3 month	0.62× decreased torsional rigidity; 0.8× decreased cortical thickness	c
	14 week	0.66× decreased bending stiffness; 0.21× decreased work to failure; 0.86× decreased ultimate stress, 0.41× decrease in ultimate strain; 1.8× increase brittleness; .85× decreased cortical thickness	d,e
	6 month	0.56× decreased torsional rigidity	c
	11–13 month	0.52× decreased torsional rigidity	c, e
BrtlIV	1 month	0.55× decreased maximum load; 0.53× decreased yield energy; no change in elastic modulus	f
	2 month	0.66× decreased maximum load; no change in yield energy; 1.3× increased elastic modulus	f
	6 month	0.84× decreased maximum load; 2.1× increased yield energy; 1.7× increased elastic modulus	f
	12 month	0.75× decreased maximum load; no change in yield energy; no change in elastic modulus	f

* Relative values (mutant/wild type) are presented as fold (×) change due to lack of consistency of test methods and parameters reported in different studies.
a Jepsen et al. (1996).
b Bonadio et al. (1993).
c McBride et al. (1998).
d Miller et al. (2007).
e Camacho et al. (1999).
f Kozloff et al. (2004).

16.2.2.9 *Mov-13*

Jaenisch and colleagues performed a series of insertional mutagenesis experiments in which Moloney murine leukemia virus (MoMuLV) integrated into various genomic sites. One of these, *Mov-13* was found to be located in the first intron of *Col1a1*, leading to failure of transcription in most tissues (Schnieke et al. 1983; Harbers et al. 1984). The insertion is in the antisense orientation relative to *Col1a1*. This change in transcriptional activity is accompanied by changes in chromatin structure (Breindl et al. 1984; Jahner and Jaenisch 1985; Hartung et al. 1986). $Col1a1^{Mov-13/Mov-13}$ homozygotes die between days 12 and 14 of gestation of exsanguination following rupture of major blood vessels (Jaenisch et al. 1983; Lohler et al. 1984). The fibroblasts of the homozygous *Mov-13* mice (the mice are non-viable) produce only type I α2 chains and no collagen trimers were detected in the embryos (Lohler et al. 1984), perhaps due to the degradation of the proα2(I) chains which do not form stable triple helices in the absence of the proα1(I) chains (Tkocz and Kuhn 1969).

The viable heterozygotes and the rudiments from the stillborn homozygous fetuses provided the first information on the importance of collagen in the development of brittle bone disease. Interestingly, culture of the rudiments from these embryonic lethal mice, and their wild-type and heterozygous litter mates demonstrated normal development of lung, kidney, salivary glands, pancreas, and skin (Kratochwil et al. 1986), as well as the cellular morphogenesis of the cornea (Bard and Kratochwil 1987).

Bonadio and colleagues compared the biomechanical performance of *Mov-13* heterozygotes and littermate controls. *Mov-13* heterozygotes have a quantitative defect in the production of $\alpha 1$ chains and synthesize only about half the normal quantity of type I collagen (Bonadio et al. 1990). The phenotype of these mice resembles that of mild human OI, with decreased Young's modulus of bone, reduced ductility, decreased energy to failure, and poorer resistance to fatigue damage (Bonadio et al. 1990, 1993; Jepsen et al. 1996, 1997). *Mov-13* heterozygotes undergo marked periosteal expansion of their long bones compared to their wild-type littermates (Bonadio et al. 1993). This results in greater cross-sectional area and cross-sectional moment of inertia than in the wild-type mice, with consequent improvement in structural strength. Thus, an increase in bone size attenuates the fragility arising from defective collagen synthesis.

Interestingly, transcription is normal in the odontoblasts of these animals, so that *Mov-13* heterozygotes have no dental phenotype (Kratochwil et al. 1989; Schwarz et al. 1990). These mice also fail to develop deafness (Stankovic et al. 2007). They therefore lack two commonly encountered extraskeletal features of human osteogenesis imperfecta. Another important phenotypic feature of *Mov-13* mice is development of leukemia due to the presence of the endogenous virus, leading to premature death, typically between 2 and 6 months of age.

16.2.2.10 Lethal Mutagenesis Model

Created in the late 1980s by Jaenisch's group, transgenic mice with a glycine substitution (Stacey et al. 1988; Bateman et al. 1989) died in utero, and therefore resembled the lethal phenotype (Sillence type II) in humans. In these mice, the $\alpha 1(I)$ chain of human collagen was mutated so that one glycine (#859) was replaced with either cysteine or arginine, bulky groups that would disrupt the folding of the triple helix. Although the transgenic mice died in utero, they could be retrieved prior to birth by Cesarean section. The retrieved fetuses with the transgene had poorly mineralized skeletons, and the skin and bones contained less collagen. This mouse model which used a Moloney leukemia virus among others as a promoter was based on the *Mov-13*, had a similar phenotype, and will not be discussed separately, other than to note that breeding the heterozygotes provided the basis for the generation of a less severe model of osteogenesis imperfecta.

16.2.2.11 *Oim/Oim*

In contrast to the transgenic mice where type I $\alpha 2$ trimers fail to form, the naturally occurring *oim/oim* mouse cannot make $\alpha 2$ chains, but do make stable $\alpha 1(I)$

trimers (Kuznetsova et al. 2003) independent of the amount of posttranslational modification. The *oim/oim* mutation which consists of a premature termination codon in the type I α2 chain was first described by Chipman et al. (1993). Tissues in these mice contain only [pro-α1(I)]3 homotrimers instead of the normal α1(I)$_2$α2(I) heterotrimers. The *oim/oim* mice have marked skeletal fragility with fractures present when they are born, and additional fractures as they continue to grow, characteristic of a moderate-to-severe form of osteogenesis imperfecta (Camacho et al. 1999). The mice have a dentin phenotype which includes changes in pulp chamber size, tooth shape, and dentin ultrastructure (Lopez Franco et al. 2005, 2006), and while blue sclera and hearing abnormalities have not been reported, they do have abnormal vasculature (Pfciffer et al. 2005) and accumulate collagen fibrils in their kidneys (Phillips et al. 2002).

The increased fragility (Grabner et al. 2001) of the bones of the *oim/oim* animals seems to be more dependent on the impaired mineralization of the collagen than on the collagen per se, although the bones have a significant reduction in collagen content. The mineral in the *oim/oim* bones is abnormal both in terms of quantity and composition (Camacho et al. 1999). The mechanical strength of demineralized *oim/oim* bones does not differ significantly from age- and sex-matched controls, whereas testing the intact bones there are significant differences, thus implying that the improper mineralization of the tissue contributes to its brittle nature (Miller et al. 2007). The collagen in the tendons, however, is weaker in tension than that in the wild-type controls (Misof et al. 1997).

Structure studies of the tendons of the *oim/oim* animals demonstrate a decrease in the order of axial packing and a loss of crystalline lateral packing of the collagen fibrils (McBride et al. 1997). While there is only one report of mutations in humans resulting in α1 trimers (Deak et al. 1983), the *oim/oim* mice has proven to be a useful model for studies of therapeutic agents such as bisphosphonates (Camacho et al. 2001; McCarthy et al. 2002; Evans et al. 2003), growth hormone (King et al. 2005a, b), and gene therapy (Balk et al. 1997; Oyama et al. 1999; Niyibizi et al. 2001) because they so closely mimic the phenotype and biochemistry of human disease and more significantly, because they have a normal mouse lifespan.

16.2.2.12 Collagen Mini-gene Model

In 1991, Prockop's group reported the development of a transgenic mouse that expressed a truncated portion of the human type I collagen gene (Khillan et al. 1991). The mini-gene contained 2.5 kb of the human collagen promoter and lacked exons 6–46. The progeny from matings of these animals showed variable penetrance, with the mice expressing the highest levels of the truncated collagen having a lethal phenotype with multiple fractures and poorly mineralized bone (Khillan et al. 1991). Breeding the transgenic mice resulted in pups that had fractures and decreased cortical thickness (Pereira et al. 1993). The mice that carried the mini-gene and survived grew at a slower rate than the control animals lacking the mini-gene. They also had loose teeth that fractured easily requiring them to be given a soft food diet. Bone length, mineral content, and collagen content were less than normal in the transgenic mice. Interestingly, as the mice aged, they lost some of the osteogenesis imperfecta phenotype, similar to what is seen in some humans with OI (Pereira et al. 1995). Because of the variable penetrance there have been fewer therapy and structural studies in these mice.

16.2.2.13 Brittle II Mice

Forlino et al. (1999) created a conditional knockin mouse resulting in a dominant genetic transmission of a glycine–cysteine substitution in all tissues. The homozygous mice died shortly after birth from what the authors described as respiratory distress, mimicking Sillence type II OI, but the heterozygotes provided a model for the less severe OI types (*BrtlIV*). The embryonic *BrtlII* mice had decreased mineralization of all their long and flat bones relative to the wild type and fractures in their long bones and ribs.

16.2.2.14 Brittle IV Mice

The viable knockin mouse that was created by Forlino et al. (1999) is heterozygous for a glycine substitution, Gly^{349} Cys in one collagen $\alpha1(I)$ allele. This construct was selected based on analyses of Sillence type IV (moderate to severe) OI in a child (Sarafova et al. 1998). The heterozygous mutant mice mimicked findings in human OI, with balanced and tissue-specific expression of normal and mutant alleles, dominant genetic transmission, and moderately severe yet variable skeletal phenotype.

Similar to growing children with OI and to the *Mov-13* heterozygotes, the *BrtlIV* mice showed improvement with age (Kozloff et al. 2004). The bones of the young *BrtlIV* mice had reduced density, reduced cross-sectional area, and reduced moments of inertia, but the differences in these parameters between wild-type and *BrtlIV* mice decreased with age from 1 to 6 months. By 6 months there were no significant differences in the stiffness and brittle behavior of the bones compared to wild type. Some material properties in the *BrtlIV* mice remained different from those in the wild type even at 6 months, with reduced mineral content (as determined by Raman spectroscopy).

Structural studies (Kuznetsova et al. 2004) of the collagen fibrils in the *BrtlIV* mice and in cultures derived from their cells showed that about one-quarter of the $\alpha1(I)$ chains were disulfide-linked mutant dimers, whereas there are no disulfide links in the normal collagen molecule. As in humans with OI the individual α chains were over glycosylated, but there was no apparent discrimination of incorporation of mutant and normal chains into the collagen fibrils. The collagen structure was not otherwise different from normal in most tissues except in the tendon where the quasi-crystalline lateral packing of molecules was sometimes disrupted, leading the authors to postulate that the tissue-specific variation in mechanical properties were associated with collagen interactions with other molecules.

The difference in interactions with other molecules has been demonstrated in a recent proteomics study comparing calvarial protein expression in 1-day-old mice with the lethal and non-lethal *BrtlIV* phenotypes to wild-type mice (Forlino et al. 2007). In this study the expression of some proteins involved in intracellular machinery was altered. In lethal *BrtlIV*, the increased expression of the cartilaginous proteins proline/arginine-rich end leucine-rich repeat protein (PRELP), a small leucine-rich proteoglycan that binds collagen to cartilage, and the morphogenetic proteins Bmp6 and Bmp7, and the lower expression of the bone matrix proteins matrilin 4, microfibril-associated glycoprotein 2, and thrombospondin 3 suggested that there was both a delay in skeletal development and an alteration in extracellular matrix composition. This suggests that proteins other than collagen may be important determinants of the properties of OI bones.

16.2.2.15 Non-rodent Models

There are numerous examples of type I collagen mutations in other animals. The first large animal model of OI that was reported was found in a group of cattle (Jensen et al. 1976). Since that time, dogs (Campbell et al. 1997) and cats (Cohn and Meuten 1990; Evason et al. 2007) with confirmed collagen mutations, and even zebrafish models have been described. Skeletons of zebra fish with laboratory-induced mutations in their collagen genes have been shown by atomic force microscopy to be more brittle and have greater elastic moduli than wild-type animals (Wang et al. 2002; Zhang et al. 2002) and to have thinner collagen fibrils with abnormal patterns of mineralization (Wang et al. 2004). The underlying mutations have not been identified in all the animal models, but the thin collagen fibrils, thinner bones, and frequent skeletal fractures in all these animals demonstrate that a variant of osteogenesis imperfecta is most likely. Not every form of osteogenesis imperfecta is due to a collagen mutation, there are "OI phenocopies" that present a similar phenotype without there being an abnormality in the collagen.

16.2.3 OI Phenocopies in Mice and Men

16.2.3.1 Introduction

While the term phenocopy usually refers to an environmentally induced, non-hereditary variation, closely resembling a genetically determined trait, we use that term to describe alterations in molecules other than type I collagen which result in the presentation of a condition that is not distinguishable from osteogenesis imperfecta. Thus, each of the humans or animal models described has an abnormality in a processing enzyme or non-collagenous protein results in skeletal fragility (brittle bones), often accompanied by vascular, skin, ocular, and dental abnormalities.

16.2.3.2 Types V, VI, VII, and VIII OI

In recent years cases of OI have been identified that do not fall into the original Sillence classification scheme. Thus an autosomal recessive form of class II OI has been named class IIB by some authors, and other models that were distinguished from Sillence class IV because of the absence of detectable collagen mutations were designated classes V–VIII. All these new types, like the other forms of OI, have skeletal deformities and increased fractures.

Class V osteogenesis imperfecta has an autosomal-dominant pattern of inheritance (Primorac et al. 2001) and is characterized by hyperplastic callus formation in the lower extremities with and without the presence of fractures (Glorieux et al. 2000, Cheung et al. 2007). The subjects have no ocular or tooth involvement, but do show decreased bone formation without abnormalities in type I collagen. The genetic basis for this type of OI is not yet known.

Class VI OI is characterized by more frequent fractures, and improper organization of collagen fibrils, associated with a mineralization defect, but no evidence of blue sclerae or dentinogenesis imperfecta. All patients in the initially described group (Glorieux et al. 2002) had vertebral compression fractures and elevated levels of alkaline phosphatase, an enzyme whose activity increases in osteomalacia, yet there was no evidence of osteomalacia in the OI type VI patients. Despite a moderate-to-severe phenotype, there was no evidence of a collagen abnormality. The nature of the mutation is as yet unknown.

Class VII OI is distinct from the other types in being autosomal recessive. The phenotype is moderate to severe, characterized by fractures at birth, blue sclerae, early deformity of the lower extremities, and osteopenia, resembling in many ways Sillence type III. The disease has been localized to chromosome 3p22–24.1, which is outside the loci for type I collagen genes (Ward et al. 2002; Morello et al. 2006). The recent description of autosomal recessive in OI in patients with mutations in cartilage-associated protein (CRTAP) suggests that type VII may be related to 3-prolyl hydroxylase or CRTAP mutations; however, defects in CRTAP are also assigned to type IIB OI (Barnes et al. 2006) to distinguish it from the autosomal-dominant type IIA OI.

Class VIII OI is also autosomal recessive and is associated with a mutation in leprecan (Cabral et al. 2007), part of the 3-prolyl hydroxylase complex. The phenotype in these children is severe, associated with severe growth retardation and decreased mineralization.

16.2.3.3 3-Prolyl Hydroxylase, Cartilage-Associated Protein

There is only one site of 3-prolyl hydroxylation in vertebrate fibrillar collagen, pro 986 (Kefalides 1973), but this step in collagen processing seems to be very important, perhaps because of the processing or because of the way the enzyme binds to the type I fibrils. The hydroxylation is carried out by a complex consisting of prolyl 3 hydroxylase 1 (P3H1) also known as leprecan (Vranka et al. 2004), a proteoglycan, cartilage-associated protein (CRTAP) and cyclophilin B (Marini et al. 2007a). CRTAP associates with P3H1 and mice lacking CRTAP have a form of osteogenesis imperfecta characterized by severe osteoporosis (Morello et al. 2006). The CRTAP null mice are smaller than the wild type; they have less dense bones, less total bone, and have spinal curvature (skeletal kyphosis). The mice bones have normal numbers of bone-forming and bone-resorbing cells, but reduced rates of bone formation and mineral apposition along with decreased bone matrix synthesis. Collagen fibrils in the mutant dermis and cartilage were reported to be significantly thicker than in wild type suggesting that loss of the single 3-Hyp residue may affect collagen fibril assembly. The failure to 3-hydroxylate a single proline in fibrillar collagens I and II is associated with posttranslational overmodification, apparent altered rate of collagen production and matrix mineralization, and increased fibril diameter (Morello et al. 2006). Based on the phenotype in the mouse, Marini's group investigated whether children with OI without a known collagen defect had a CRTAP deficiency, and found it in three of ten children (Barnes et al. 2006). They also found mutations in leprecan, another member of the complex, in five children with severe forms of OI (Cabral et al. 2007).

The reason for the effect of one amino acid defect is not known. There are suggestions that 3-hydroxyproline destabilizes the helix while 4-hydroxyproline stabilizes it (Jenkins et al. 2003; Mizuno et al. 2004), but the thickened collagen fibrils and the severe phenotype in the bones of these animals and people makes one suspect additional mechanisms, perhaps involving cellular as well as extracellular matrix metabolism.

16.2.3.4 Lysyl Oxidase Mutations

Lysyl oxidase is responsible for the oxidative deamination of lysine in the individual α chains before assembly. It is essential for the formation of collagen cross-links. Bruck syndrome in humans is associated with defective collagen fibril formation and abnormal mineralization, and is a variant of OI. The other name for this syndrome is osteogenesis imperfecta with contractures of the larger joints. First identified in 1993 in humans (Brenner et al. 1993), it is associated with mutations in lysyl oxidase 2 (*PLOD2*) (Ha-Vinh et al. 2004). While there are knockouts of lysyl oxidase 1 and 3 (Myllyla et al. 2007), mice lacking *PLOD2* have not yet been described. Overexpression of the enzyme in cell culture resulted in the formation of smaller collagen fibrils and delayed onset of mineral deposition demonstrating the enzyme has a role in regulating collagen fibrillogenesis (Pornprasertsuk et al. 2005) but the direct effect of the mutations on this process has not been validated in a living animal.

16.2.3.5 *Fro/Fro*

The *fragilitas ossium (fro/fro)* model was developed in a chemical mutagenesis experiment with tris (1-aziridinyl)phosphine sulphide (Muriel et al. 1991; Sillence et al. 1993). The mutant mice were recognized as having an OI phenotype with bone fragility, dentinogenesis imperfecta, decreased mineral content, and alterations in some of the matrix protein expression, but there were no detectable collagen abnormalities. Positional cloning identified a deletion in the gene encoding neutral sphingomyelin phosphodiesterase 3 (*Smpd3*) that led to complete loss of enzymatic activity in these mice (Aubin et al. 2005). This enzyme degrades spingomyelin to ceramide and phosphoryl choline. There are three Smpd enzymes, located in different cellular compartments, each of which has been ablated in mice. When the acidic Smpd1 is knocked out the mutant (Niemann–Pick mouse) tissues develop massive lysosomal storage of sphingomelin (Otterbach and Stoffel, 1995). Knock-outs of the neutral *Smpd3* and Smpd2, and the *smpd2$^{-/-}$ smpd3$^{-/-}$* (double knock-out) mice have no such sphingomyelin accumulation (Zumbansen and Stoffel 2002; Stoffel et al. 2005) but have growth retardation starting from embryonic day 14 to age 2 years. In the mutant mice serum IGF1 and growth hormone were markedly reduced relative to control sera and impaired secretion of peptide-releasing hormones from hypothalamic neurosecretory neurons and pituitary hormone deficiency were proposed to be the reason for the growth defects.

Like the *fro/fro* mice, the *smpd3* null animals histologically appeared immature with delayed ossification and impaired endochondral ossification, spinal curvature, and malformed and misproportioned limbs. But unlike in *fro/fro* the bones of the *smpd3* null mice did not show radiographic fractures, and the dentition was visually normal (Stoffel et al. 2007). Thus while the *fro/fro* mice have brittle bones, the *smpd3* mice have a chondrodysplasia (cartilage abnormality). The long bone phenotype and the dwarfing in the *smpd3$^{-/-}$* cartilage were rescued by driving expression of the gene with a type II collagen promoter, demonstrating again that a cartilage defect is involved. The distinction between the *smdp3$^{-/-}$* mouse and the *fro/fro* mice (cartilage defect vs bone defect) might be due to differences in the background of the mice in which these mutants occurred, but as suggested elsewhere (Stoffel et al. 2007) are more likely to be due to other effects of the chemical mutagenesis.

16.2.3.6 Osteoporosis–Pseudoglioma Syndrome

The genetic basis of the ocular form of osteogenesis imperfecta was first described by Warman's group (Gong et al. 1996) as being associated with mutations in the low-density lipoprotein receptor (LRP5), affecting the wnt signaling pathway. This pathway is essential for the formation of osteoblasts, bone-forming cells (Bodine and Komm 2006). Children with this condition have excessive bone fractures, like children with OI, but they are blind. The bone fragility and blindness are both due to the deficit of collagen formation in the ocular tissues and in bone (Balemans and Van Hul 2007). The mouse model in which the LRP5 protein was knocked out (Kato et al. 2002) recapitulates the bone phenotype in humans with this disease, with decreased bone density, low bone density, shortened stature, increased fractures, decreased numbers of osteoblasts, and delayed mineralization. Additionally the *Lrp5* knockout mouse had impaired eye development due to persistent embryonic eye vascularization, resulting in blindness.

16.3 Other Fibrillar Collagen Mutations that Affect Tissue Structure and Mechanical Behavior

There are a large number of other so-called "collagen diseases" that because of the broad distribution of the collagens in structural and protective tissues affect the properties of these tissues and the ability of these tissues to function. The better known

Table 16.4 Some collagen mutations associated with human diseases

Disease*	Collagen type	Phenotype	Ref
OI	I	Brittle bones, blue sclera, dentin defects, short stature	a
EDS	V, I, III, VII	Joint laxity, bruising, vascular complications	b
SEDS, MEDS	II, X, IX, XI	Growth plate and growth abnormalities; abnormal cartilage calcification; in Stickler syndrome ocular defects	c
EB	VII, IV	Skin blistering	d
Alport Syndrome	IV	Renal and basement membrane disease	e

* OI – osteogenesis imperfecta, EDS – Ehlers–Danlos Syndromes, SEDS, MEDS – spondyloepiphyseal dysplasia and metaphyseal spondylodysplasia, EB – epidemylosis bulosa.
a Marini et al. (2007b); Primorac et al. (2001).
b Malfait and De Paepe (2005).
c Czarny-Ratajczak et al. (2001); Van Camp et al. (2006); Warman et al. (1993); Walter et al. (2007).
d Ee et al. (2007).
e Thorner (2007).

of these are summarized in Table 16.4 and include the Ehlers–Danlos syndromes (EDS, characterized by joint laxity), epidermolysis bullosa (EB, Masunaga 2006), characterized by skin blistering, and the spondyloepiphyseal dysplasias, associated with abnormalities in longitudinal skeletal growth. These will only be discussed as they reveal how type I collagen mutations can alter collagen structure and the mechanical performance of the tissue.

The Ehlers–Danlos syndromes, like OI, consist of a heterogeneous set of diseases (Beighton et al. 1998), mainly associated with joint laxity and hyermobile skin, weakened blood vessels, ocular problems, and easy bruising. It has been associated with mutations in types I, III, and V collagens, with the majority of the recent cases, as reviewed elsewhere, associated with mutations in the introns of the α1 chain of type V collagen (Fichard et al. 2003), but there are also reported cases with mutated lysyl oxidase (Seidler et al. 2006), mutations in tenascin-X (Bristow et al. 2005), and with abnormal accumulation of extracellular matrix proteins. Collagen fibrils are generally disorganized in all these syndromes, sometimes having unusual shapes that can explain the tissue fragility (Hermanns-Le and Pierard 2006). For example, in the tenascin-X knockout mouse model of EDS, the disorganization of the collagen fibrils provided proof that tenascin-X controls collagen organization and cross-linking (Lethias et al. 2006), and tissue mechanical properties.

Ehlers–Danlos syndromes (VIIa and b) are due to mutations in the N-terminal domain of type I collagen, with those in VIIa associated with mutations in proα1(I) and those in VIIb in proα2 (I). Recently cases of both OI and EDS VII were reported (Cabral et al. 2005). The mutations cause the collagen to fail to form triple helices properly because the N-termini are not cleaved, and the resulting fibrils are significantly smaller than those in controls (Cabral et al. 2005).

The "classical" EDS is due to decreased expression of colVα1, part of the [α1(V)]2α2(V) heterotrimer that coassembles with type I collagen fibrils regulating their formation. The colVα1 knockout mice are embryonic lethal due to vascular insufficiency, but heterozygous mice mimic classic EDS (Wenstrup et al. 2006). The aortas of these mice have decreased stiffness and breaking strength. Their skin is hyper-extensible and their fiber diameters are reduced by as much as 42% at all ages studied. The heterozygotes with 50% of the normal type V collagen show both qualitative and quantitative effects, and Wenstrup et al. suggest that this is due to the key role type V plays as a collagen fibril nucleator. Even in the growth zones of the long bones there is a reduced total collagen content, and some of these collagen fibrils have an abnormal (cauliflower-like) shape (Wenstrup et al. 2006).

16.4 Discussion

Missense mutations that replace the mandatory glycine in the Gly-X-Y sequence with another amino acid are the most common causes of osteogenesis imperfecta (Byers 1993; Marini et al. 2007b). The location of the mutation in the individual α chains, and the size of the substituted amino acid have definite effects on the severity of the disease, not only in osteogenesis imperfecta but also in the other diseases associated with collagen mutations discussed above. There are several models that explain how the mutations can affect collagen stability and function.

Originally suggested by Prockop (1984), the concept of protein suicide explains how OI mutations can be quantitative. Prockop suggested that even when normal collagen α chains are produced, the presence of abnormal chains triggers cellular protein destruction, and less collagen is formed. A similar effect is found in a model of EDS in which one type V collagen chain is mutated (Wenstrup et al. 2006).

The qualitative changes in the collagen diseases are more difficult to explain. In one model, the "Renucleation Model" the substitution of an alternate amino acid for glycine would result in delayed folding of the triple helix, not only accounting for the excessive posttranslational modification of these chains, but also accounting for the severity of the defect. According to the renucleation model (Byers et al. 1991; Hyde et al. 2006), sequences with high imino acid content (i.e., hydroxyproline and proline) allow a new folding start ("renucleation") of triple helix formation. Mutations adjacent to these regions may have less of an effect if the chains can "renucleate" – or begin to refold. In another model, the so-called "domain model", the concept is that the stability of the domain around the mutation site (i.e., one with low imino acid content) allows "micro-unfolding" domains, thus if the mutation is in this domain the effects would be less severe than those in regions with high imino acid content (Bachinger et al. 1993). According to this model, a mutation in the N anchor domain that is important for triple helix formation would be severe, and it has been observed (Makareeva et al. 2006) that such a mutation causes unfolding of the entire domain. According to both these models, it is not only whether the mutation is close to the N- or C-termini that determines the impact, but whether the effect is at a location in the chain that is important for stability. It is also likely that mutations in domains that serve other important functions (e.g., interaction with other matrix molecules, with receptors, etc. (Di Lullo et al. 2002) would be important. Mutations that affect the ability of the chains to form cross-links with other collagen fibrils or to interact with other molecules (matrix proteins, cell surface receptors) could be included under the heading of "domain" models.

16.5 Conclusion

The size, complexity, and large mutation spectrum of type I collagen lead to a diverse spectrum of clinical syndromes. Abnormal or insufficient collagen formation results in bones that are more fragile (more brittle). Altered collagen fibril structure, impaired mineralization, and altered cellular activities contribute to the mechanical instability. Ultimately, understanding the diseases and the mechanical properties of tissues that result from altered *COL1A1* and *COL1A2* will depend on fully understanding the biology of collagen synthesis, secretion, and assembly. In particular, better understanding of the higher-order assembly of collagen fibrils in the extracellular matrix and interactions between collagen and other matrix components (Marini et al. 2007b) are areas in which present knowledge is lagging. This knowledge will have to include tissue-specific as well as mutation-specific information, since the effects of any mutation on tissue properties may differ between bone, skin, tendon, dentin, and aorta, although the same collagen is produced in each of these tissues. An important task for the future is to help clinicians incorporate recent advances in our understanding of the biochemistry and biophysics of collagen synthesis into their thinking.

References

Aubin I, Adams CP, Opsahl S, Septier D, Bishop CE, Auge N, Salvayre R, Negre-Salvayre A, Goldberg M, Guenet JL, Poirier C (2005) A deletion in the gene encoding sphingomyelin phosphodiesterase 3 (Smpd3) results in osteogenesis and dentinogenesis imperfecta in the mouse. Nat Genet. 37:803–805.

Bachinger HP, Morris NP, Davis JM (1993) Thermal stability and folding of the collagen triple helix and the effects of mutations in osteogenesis imperfecta on the triple helix of type I collagen. Am J Med Genet. 45:152–162.

Balemans W, Van Hul W (2007) The genetics of low-density lipoprotein receptor-related protein 5 in bone: a story of extremes. Endocrinology.148:2622–2629.

Balk ML, Bray J, Day C, Epperly M, Greenberger J, Evans CH, Niyibizi C (1997) Effect of rhBMP-2 on the osteogenic potential of bone marrow stromal cells from an osteogenesis imperfecta mouse (oim). Bone. 21:7–15.

Bard JB, Kratochwil K (1987) Corneal morphogenesis in the Mov13 mutant mouse is characterized by normal cellular organization but disordered and thin collagen. Development. 101:547–555.

Barnes AM, Chang W, Morello R, Cabral WA, Weis M, Eyre DR, Leikin S, Makareeva E, Kuznetsova N, Uveges TE, Ashok A, Flor AW, Mulvihill JJ, Wilson PL, Sundaram UT, Lee B, Marini JC (2006) Deficiency of cartilage-associated protein in recessive lethal osteogenesis imperfecta. N Engl J Med. 355:2757–2764.

Bateman JF, Mascara T, Cole WG, Stacey A, Jaenisch R (1989) Collagen protein abnormalities produced by site-directed mutagenesis of the pro alpha 1(I) gene. Connect Tissue Res. 20: 205–212.

Beighton P, De Paepe A, Steinmann B, Tsipouras P, Wenstrup RJ (1998) Ehlers-Danlos syndromes: revised nosology, Villefranche, 1997. Ehlers-Danlos National Foundation (USA) and Ehlers-Danlos Support Group (UK). Am J Med Genet.77:31–37.

Bodine PV, Komm BS (2006) Wnt signaling and osteoblastogenesis. Rev Endocr Metab Disord. 7:33–39.

Bonadio J, Saunders TL, Tsai E, Goldstein SA, Morris-Wiman J, Brinkley L, Dolan DF, Altschuler RA, Hawkins JE, Bateman JF, Mascara T, Jaenisch R (1990) Transgenic mouse model of the mild dominant form of osteogenesis imperfecta. Proc Natl Acad Sci USA.87: 7145–7149.

Bonadio J, Jepsen KJ, Mansoura MK, Jaenisch R, Kuhn JL, Goldstein SA (1993) A murine skeletal adaptation that significantly increases cortical bone mechanical properties. Implications for human skeletal fragility. J Clin Invest. 92:1697–1705.

Breindl M, Harbers K, Jaenisch R (1984) Retrovirus-induced lethal mutation in collagen I gene of mice is associated with an altered chromatin structure. Cell. 38:9–16.

Brenner RE, Vetter U, Stoss H, Muller PK, Teller WM (1993) Defective collagen fibril formation and mineralization in osteogenesis imperfecta with congenital joint contractures (Bruck syndrome). Eur J Pediatr. 152:505–508.

Bristow J, Carey W, Egging D, Schalkwijk J (2005) Tenascin-X, collagen, elastin, and the Ehlers-Danlos syndrome. Am J Med Genet C Semin Med Genet. 139:24–30.

Byers PH (1993) Osteogenesis Imperfecta in Royce PM and Steinmann B (eds) Connective Tissues and Its Heritable Disorders, Wiley Liss, New York, pp 317–350.

Byers PH, Wallis GA, Willing MC (1991) Osteogenesis imperfecta: translation of mutation to phenotype. J Med Genet. 28:433–442.

Byers PH, Duvic M, Atkinson M, Robinow M, Smith LT, Krane SM, Greally MT, Ludman M, Matalon R, Pauker S, Quanbeck D, Schwarze U (1997) Ehlers-Danlos syndrome type VIIA and VIIB result from splice-junction mutations or genomic deletions that involve exon 6 in the COL1A1 and COL1A2 genes of type I collagen. Am J Med Genet. 72:94–105.

Cabral WA, Makareeva E, Colige A, Letocha AD, Ty JM, Yeowell HN, Pals G, Leikin S, Marini JC (2005) Mutations near amino end of alpha1(I) collagen cause combined osteogenesis imperfecta/Ehlers-Danlos syndrome by interference with N-propeptide processing. J Biol Chem. 280:19259–19269.

Cabral WA, Chang W, Barnes AM, Weis M, Scott MA, Leikin S, Makareeva E, Kuznetsova NV, Rosenbaum KN, Tifft CJ, Bulas DI, Kozma C, Smith PA, Eyre DR, Marini JC (2007) Prolyl 3-hydroxylase 1 deficiency causes a recessive metabolic bone disorder resembling lethal/severe osteogenesis imperfecta. Nat Genet. 39:359–365.

Camacho NP, Hou L, Toledano TR, Ilg WA, Brayton CF, Raggio CL, Root L, Boskey AL (1999) The material basis for reduced mechanical properties in oim mice bones. J Bone Miner Res. 14:264–272.

Camacho NP, Raggio CL, Doty SB, Root L, Zraick V, Ilg WA, Toledano TR, Boskey AL (2001) A controlled study of the effects of alendronate in a growing mouse model of osteogenesis imperfecta. Calcif Tissue Int. 69:94–101.

Campbell BG, Wootton JA, Krook L, DeMarco J, Minor RR (1997) Clinical signs and diagnosis of osteogenesis imperfecta in three dogs. J Am Vet Med Assoc. 211:183–187.

Cassella JP, Barber P, Catterall AC, Ali SY (1994) A morphometric analysis of osteoid collagen fibril diameter in osteogenesis imperfecta. Bone.15:329–334.

Cheung MS, Glorieux FH, Rauch F (2007) Natural history of hyperplastic callus formation in osteogenesis imperfecta type V. J Bone Miner Res. 22:1181–1186.

Chiodo AA, Hockey A, Cole WG (1992) A base substitution at the splice acceptor site of intron 5 of the COL1A2 gene activates a cryptic splice site within exon 6 and generates abnormal type I procollagen in a patient with Ehlers-Danlos syndrome type VII. J Biol Chem. 267:6361–6369.

Chipman SD, Sweet HO, McBride DJ Jr, Davisson MT, Marks SC Jr, Shuldiner AR, Wenstrup RJ, Rowe DW, Shapiro JR (1993) Defective pro alpha 2(I) collagen synthesis in a recessive mutation in mice: a model of human osteogenesis imperfecta. Proc Natl Acad Sci USA. 90: 1701–1705.

Cohen-Solal L, Zolezzi F, Pignatti PF, Mottes M (1996) Intrafamilial variable expressivity of osteo-
    genesis imperfecta due to mosaicism for a lethal G382R substitution in the COL1A1 gene. Mol
    Cell Probes. 10:219–225.
Cohn LA, Meuten DJ (1990) Bone fragility in a kitten: an osteogenesis imperfecta-like syndrome.
    J Am Vet Med Assoc. 197:98–100.
Cole WG, Chan D, Chambers GW, Walker ID, Bateman JF (1986) Deletion of 24 amino acids from
    the pro-alpha 1(I) chain of type I procollagen in a patient with the Ehlers-Danlos syndrome type
    VII. J Biol Chem. 261:5496–5503.
Constantinou CD, Pack M, Young SB, Prockop DJ (1990) Phenotypic heterogeneity in osteoge-
    nesis imperfecta: the mildly affected mother of a proband with a lethal variant has the same
    mutation substituting cysteine for alpha 1-glycine 904 in a type I procollagen gene (COL1A1).
    Am J Hum Genet. 47:670–679.
Constantinou-Deltas CD, Ladda RL, Prockop DJ (1993) Somatic cell mosaicism: another source
    of phenotypic heterogeneity in nuclear families with osteogenesis imperfecta. Am J Med Genet
    45:246–251.
Czarny-Ratajczak M, Lohiniva J, Rogala P, Kozlowski K, Perala M, Carter L, Spector TD,
    Kolodziej L, Seppanen U, Glazar R, Krolewski J, Latos-Bielenska A, Ala-Kokko L (2001)
    A mutation in COL9A1 causes multiple epiphyseal dysplasia: further evidence for locus het-
    erogeneity. Am J Hum Genet. 69:969–980.
D'Alessio M, Ramirez F, Blumberg BD, Wirtz MK, Rao VH, Godfrey MD, Hollister DW (1991)
    Characterization of a COL1A1 splicing defect in a case of Ehlers-Danlos syndrome type VII:
    further evidence of molecular homogeneity. Am J Hum Genet 49:400–406.
Dalgleish R (1997) The human type I collagen mutation database. Nucleic Acids Res. 25:181–187.
Dalgleish R (1998) The human collagen mutation database 1998. Nucleic Acids Res. 26:253–255.
Dawson PA, Kelly TE, Marini JC (1999) Extension of phenotype associated with structural muta-
    tions in type I collagen: siblings with juvenile osteoporosis have an alpha2(I)Gly436 $\longrightarrow$ Arg
    substitution. J Bone Miner Res. 14:449–455.
Deak SB, Nicholls A, Pope FM, Prockop DJ (1983). The molecular defect in a nonlethal variant
    of osteogenesis imperfecta. Synthesis of pro-alpha 2(I) chains which are not incorporated into
    trimers of type I procollagen. J Biol Chem. 258:15192–15197.
De Paepe A, Nuytinck L, Raes M, Fryns JP (1997) Homozygosity by descent for a COL1A2
    mutation in two sibs with severe osteogenesis imperfecta and mild clinical expression in the
    heterozygotes. Hum Genet 99:478–483.
Di Lullo GA, Sweeney SM, Korkko J, Ala-Kokko L, San Antonio JD (2002) Mapping the ligand-
    binding sites and disease-associated mutations on the most abundant protein in the human, type
    I collagen. J Biol Chem. 277:4223–4231.
Edwards MJ, Wenstrup RJ, Byers PH, Cohn DH (1992) Recurrence of lethal osteogenesis imper-
    fecta due to parental mosaicism for a mutation in the COL1A2 gene of type I collagen. The
    mosaic parent exhibits phenotypic features of a mild form of the disease. Hum Mutat. 1:47–54.
Ee HL, Liu L, Goh CL, McGrath JA (2007) Clinical and molecular dilemmas in the diagnosis of
    familial epidermolysis bullosa pruriginosa. J Am Acad Dermatol. 56:S77–81.
Evans KD, Lau ST, Oberbauer AM, Martin RB (2003) Alendronate affects long bone length and
    growth plate morphology in the oim mouse model for Osteogenesis Imperfecta. Bone. 32:
    268–274.
Evason MD, Taylor SM, Bebchuk TN (2007) Suspect osteogenesis imperfecta in a male kitten.
    Can Vet J. 48:296–298.
Eyre DR, Shapiro FD, Aldridge JF (1985) A heterozygous collagen defect in a variant of the
    Ehlers-Danlos syndrome type VII. Evidence for a deleted amino-telopeptide domain in the
    pro-alpha 2(I) chain. J Biol Chem. 260:11322–11329.
Feshchenko S, Brinckmann J, Lehmann HW, Koch HG, Muller PK, Kugler S (1998) Identification
    of a new heterozygous point mutation in the COL1A2 gene leading to skipping of exon 9 in a
    patient with joint laxity, hyperextensibility of skin and blue sclerae. Mutations in brief no. 166.
    Online. Hum Mutat. 12:138.

Fichard A, Chanut-Delalande H, Ruggiero F (2003) [The Ehlers-Danlos syndrome: the extracellular matrix scaffold in question] Med Sci (Paris). 19:443–452.

Forlino A, Porter FD, Lee EJ, Westphal H, Marini JC (1999) Use of the Cre/lox recombination system to develop a non-lethal knock-in murine model for osteogenesis imperfecta with an alpha1(I) G349C substitution. Variability in phenotype in BrtlIV mice. J Biol Chem. 274:37923–37931.

Forlino A, Tani C, Rossi A, Lupi A, Campari E, Gualeni B, Bianchi L, Armini A, Cetta G, Bini L, Marini JC (2007) Differential expression of both extracellular and intracellular proteins is involved in the lethal or nonlethal phenotypic variation of BrtlIV, a murine model for osteogenesis imperfecta. Proteomics. 7:1877–1891.

Francis MJ, Williams KJ, Sykes BC, Smith R (1981) The relative amounts of the collagen chains alpha 1(I), alpha 2 and alpha 1(III) in the skin of 31 patients with osteogenesis imperfecta. Clin Sci (Lond). 60:617–623.

Gensure RC, Makitie O, Barclay C, Chan C, Depalma SR, Bastepe M, Abuzahra H, Couper R, Mundlos S, Sillence D, Ala Kokko L, Seidman JG, Cole WG, Juppner H (2005) A novel COL1A1 mutation in infantile cortical hyperostosis (Caffey disease) expands the spectrum of collagen-related disorders. J Clin Invest. 115:1250–1257.

Giunta C, Superti-Furga A, Spranger S, Cole WG, Steinmann B (1999) Ehlers-Danlos syndrome type VII: clinical features and molecular defects. J Bone Joint Surg Am. 81:225–238.

Glorieux FH, Rauch F, Plotkin H, Ward L, Travers R, Roughley P, Lalic L, Glorieux DF, Fassier F, Bishop NJ (2000) Type V osteogenesis imperfecta: a new form of brittle bone disease. J Bone Miner Res. 15:1650–1658.

Glorieux FH, Ward LM, Rauch F, Lalic L, Roughley PJ, Travers R (2002) Osteogenesis imperfecta type VI: a form of brittle bone disease with a mineralization defect. J Bone Miner Res. 17:30–38.

Gong Y, Vikkula M, Boon L, Liu J, Beighton P, Ramesar R, Peltonen L, Somer H, Hirose T, Dallapiccola B, De Paepe A, Swoboda W, Zabel B, Superti-Furga A, Steinmann B, Brunner HG, Jans A, Boles RG, Adkins W, van den Boogaard MJ, Olsen BR, Warman ML (1996). Osteoporosis-pseudoglioma syndrome, a disorder affecting skeletal strength and vision, is assigned to chromosome region 11q12–13. Am J Hum Genet. 59:146–151.

Grabner B, Landis WJ, Roschger P, Rinnerthaler S, Peterlik H, Klaushofer K, Fratzl P (2001) Age- and genotype-dependence of bone material properties in the osteogenesis imperfecta murine model (oim). Bone. 29:453–457.

Harbers K, Kuehn M, Delius H, Jaenisch R (1984) Insertion of retrovirus into the first intron of alpha 1(I) collagen gene to embryonic lethal mutation in mice. Proc Natl Acad Sci USA 81:1504–1508.

Hartung S, Jaenisch R, Breindl M (1986) Retrovirus insertion inactivates mouse alpha 1(I) collagen gene by blocking initiation of transcription. Nature. 320:365–367.

Hata R, Kurata S, Shinkai H (1988) Existence of malfunctioning pro alpha2(I) collagen genes in a patient with a pro alpha 2(I)-chain-defective variant of Ehlers-Danlos syndrome. Eur J Biochem. 174:231–237.

Ha-Vinh R, Alanay Y, Bank RA, Campos-Xavier AB, Zankl A, Superti-Furga A, Bonafe L (2004) Phenotypic and molecular characterization of Bruck syndrome (osteogenesis imperfecta with contractures of the large joints) caused by a recessive mutation in PLOD2. Am J Med Genet A.131:115–120.

Ho KK, Kong RY, Kuffner T, Hsu LH, Ma L, Cheah KS (1994) Further evidence that the failure to cleave the aminopropeptide of type I procollagen is the cause of Ehlers-Danlos syndrome type VII. Hum Mutat. 3:358–364.

Hermanns-Le T, Pierard GE (2006) Collagen fibril arabesques in connective tissue disorders. Am J Clin Dermatol. 7:323–326.

Houdebine LM (2002) The methods to generate transgenic animals and to control transgene expression. J Biotechnol. 98:145–160.

Hyde TJ, Bryan MA, Brodsky B, Baum J (2006) Sequence dependence of renucleation after a Gly mutation in model collagen peptides. J Biol Chem. 281:36937–36943.

Jaenisch R, Harbers K, Schnieke A, Lohler J, Chumakov I, Jahner D, Grotkopp D, Hoffmann E (1983) Germline integration of moloney murine leukemia virus at the Mov13 locus leads to recessive lethal mutation and early embryonic death. Cell. 32:209–216.

Jahner D, Jaenisch R (1985) Retrovirus-induced de novo methylation of flanking host sequences correlates with gene inactivity. Nature. 315:594–597.

Jenkins CL, Bretscher LE, Guzei IA, Raines RT (2003) Effect of 3-hydroxyproline residues on collagen stability. J Am Chem Soc. 125:6422–6427.

Jensen PT, Rasmussen PG, Basse A (1976) Congenital osteogenesis imperfecta in Charollais cattle. Nord Vet Med. 28:304–308.

Jepsen KJ, Goldstein SA, Kuhn JL, Schaffler MB, Bonadio J (1996). Type-I collagen mutation compromises the post-yield behavior of Mov13 long bone. J Orthop Res. 14:493–499.

Jepsen KJ, Schaffler MB, Kuhn JL, Goulet RW, Bonadio J, Goldstein SA (1997). Type I collagen mutation alters the strength and fatigue behavior of Mov13 cortical tissue. J Biomech. 30: 1141–1147.

Kamoun-Goldrat AS, Le Merrer MF (2007) Animal models of osteogenesis imperfecta and related syndromes. J Bone Miner Metab. 25:211–218.

Kato M, Patel MS, Levasseur R, Lobov I, Chang BH, Glass DA 2nd, Hartmann C, Li L, Hwang TH, Brayton CF, Lang RA, Karsenty G, Chan L (2002) Cbfa1-independent decrease in osteoblast proliferation, osteopenia, and persistent embryonic eye vascularization in mice deficient in Lrp5, a Wnt coreceptor. J Cell Biol. 157:303–314.

Kefalides NA (1973) Structure and biosynthesis of basement membranes. Int Rev Connect Tissue Res. 6:63–104.

Khillan JS, Olsen AS, Kontusaari S, Sokolov B, Prockop DJ (1991) Transgenic mice that express a mini-gene version of the human gene for type I procollagen (COL1A1) develop a phenotype resembling a lethal form of osteogenesis imperfecta. J Biol Chem. 266:23373–23379.

King D, Chase J, Havey RM, Voronov L, Sartori M, McEwen HA, Beamer WG, Patwardhan AG (2005a) Effects of growth hormone transgene expression on vertebrae in a mouse model of osteogenesis imperfecta. Spine. 30:1491–1495.

King D, Jarjoura D, McEwen HA, Askew MJ (2005b) Growth hormone injections improve bone quality in a mouse model of osteogenesis imperfecta. J Bone Miner Res. 20:987–993.

Kozloff KM, Carden A, Bergwitz C, Forlino A, Uveges TE, Morris MD, Marini JC, Goldstein SA (2004) Brittle IV mouse model for osteogenesis imperfecta IV demonstrates postpubertal adaptations to improve whole bone strength. J Bone Miner Res. 19:614–622.

Kratochwil K, Dziadek M, Lohler J, Harbers K, Jaenisch R (1986) Normal epithelial branching morphogenesis in the absence of collagen I. Dev Biol. 117:596–606.

Kratochwil K, von der Mark K, Kollar EJ, Jaenisch R, Mooslehner K, Schwarz M, Haase K, Gmachl I, Harbers K (1989) Retrovirus-induced insertional mutation in Mov13 mice affects collagen I expression in a tissue-specific manner. Cell. 57:807–816.

Kuivaniemi H, Tromp G, Prockop DJ (1997) Mutations in fibrillar collagens (types I, II, III, and XI), fibril-associated collagen (type IX), and network-forming collagen (type X) cause a spectrum of diseases of bone, cartilage, and blood vessels. Hum Mutat. 9:300–315.

Kuznetsova NV, McBride DJ, Leikin S (2003) Changes in thermal stability and microunfolding pattern of collagen helix resulting from the loss of alpha2(I) chain in osteogenesis imperfecta murine. J Mol Biol. 331:191–200.

Kuznetsova NV, Forlino A, Cabral WA, Marini JC, Leikin S (2004) Structure, stability and interactions of type I collagen with GLY349-CYS substitution in alpha 1(I) chain in a murine Osteogenesis Imperfecta model. Matrix Biol. 23:101–112.

Lehmann HW, Mundlos S, Winterpacht A, Brenner RE, Zabel B, Muller PK (1994) Ehlers-Danlos syndrome type VII: phenotype and genotype. Arch Dermatol Res. 286:425–428.

Lethias C, Carisey A, Comte J, Cluzel C, Exposito JY (2006) A model of tenascin-X integration within the collagenous network. FEBS Lett. 580:6281–6285.

Lohler J, Timpl R, Jaenisch R (1984) Embryonic lethal mutation in mouse collagen I gene causes rupture of blood vessels and is associated with erythropoietic and mesenchymal cell death. Cell. 38:597–607.

Lopez Franco GE, Huang A, Pleshko Camacho N, Blank RD (2005) Dental phenotype of the col1a2(oim) mutation: DI is present in both homozygotes and heterozygotes. Bone. 36: 1039–1046.

Lopez Franco GE, Huang A, Pleshko Camacho N, Stone DS, Blank RD (2006). Increased Young's modulus and hardness of Col1a2 oim dentin. J Dent Res. 85:1032–1036.

Makareeva E, Cabral WA, Marini JC, Leikin S (2006) Molecular mechanism of alpha 1(I)-osteogenesis imperfecta/Ehlers-Danlos syndrome: unfolding of an N-anchor domain at the N-terminal end of the type I collagen triple helix. J Biol Chem. 281:6463–6470.

Malfait F, De Paepe A (2005) Molecular genetics in classic Ehlers-Danlos syndrome. Am J Med Genet C Semin Med Genet. 139:17–23.

Malfait F, Symoens S, Coucke P, Nunes L, De Almeida S, De Paepe A (2006) Total absence of the alpha2(I) chain of collagen type I causes a rare form of Ehlers-Danlos syndrome with hypermobility and propensity to cardiac valvular problems. J Med Genet. 43:e36.

Malfait F, Symoens S, De Backer J, Hermanns-Le T, Sakalihasan N, Lapiere CM, Coucke P, De Paepe A (2007) Three arginine to cysteine substitutions in the pro-alpha (I)-collagen chain cause Ehlers-Danlos syndrome with a propensity to arterial rupture in early adulthood. Hum Mutat. 28:387–395.

Marini JC (2006) Osteogenesis Imperfecta. Primer on the Metabolic Bone Diseases and Disorders of Mineral Metabolism. M. J. Favus. Washington, DC., American Society for Bone and Mineral Research:418–420.

Marini JC, Cabral WA, Barnes AM, Chang W (2007a) Components of the collagen prolyl 3-hydroxylation complex are crucial for normal bone development. Cell Cycle. 6:1675–1681.

Marini JC, Forlino A, Cabral WA, Barnes AM, San Antonio JD, Milgrom S, Hyland JC, Korkko J, Prockop DJ, De Paepe A, Coucke P, Symoens S, Glorieux FH, Roughley PJ, Lund AM, Kuurila-Svahn K, Hartikka H, Cohn DH, Krakow D, Mottes M, Schwarze U, Chen D, Yang K, Kuslich C, Troendle J, Dalgleish R, Byers PH (2007b). Consortium for osteogenesis imperfecta mutations in the helical domain of type I collagen: regions rich in lethal mutations align with collagen binding sites for integrins and proteoglycans. Hum Mutat. 28:209–221.

Masunaga T (2006) Epidermal basement membrane: its molecular organization and blistering disorders. Connect Tissue Res. 47:55–66.

Mayer SA, Rubin BS, Starman BJ, Byers PH (1996) Spontaneous multivessel cervical artery dissection in a patient with a substitution of alanine for glycine (G13A) in the alpha 1 (I) chain of type I collagen. Neurology 47:552, 556.

McBride DJ Jr, Choe V, Shapiro JR, Brodsky B (1997) Altered collagen structure in mouse tail tendon lacking the alpha 2(I) chain. J Mol Biol. 270:275–284.

McCarthy EA, Raggio CL, Hossack MD, Miller EA, Jain S, Boskey AL, Camacho NP (2002) Alendronate treatment for infants with osteogenesis imperfecta: demonstration of efficacy in a mouse model. Pediatr Res. 52:660–670.

Miller E, Delos D, Baldini T, Wright TM, Camacho NP (2007) Abnormal mineral-matrix interactions are a significant contributor to fragility in oim/oim bone. Calcif Tissue Int. 81:206–214.

Misof K, Landis WJ, Klaushofer K, Fratzl P (1997) Collagen from the osteogenesis imperfecta mouse model (oim) shows reduced resistance against tensile stress. J Clin Invest. 100:40–45.

Mizuno K, Hayashi T, Peyton DH, Bachinger HP (2004) The peptides acetyl-(Gly-3(S)Hyp-4(R)Hyp)10-NH2 and acetyl-(Gly-Pro-3(S)Hyp)10-NH2 do not form a collagen triple helix. J Biol Chem. 279:282–287.

Morello R., Bertin TK, Chen Y, Hicks J, Tonachini L, Monticone M, Castagnola P, Rauch F, Glorieux FH, Vranka J, Bächinger HP, Pace JM, Schwarze U, Byersn PH, Weis M, Fernandes RJ, Eyre DR, Yao Z, Boyce BF, Lee B (2006) CRTAP is required for prolyl 3- hydroxylation and mutations cause recessive osteogenesis imperfecta. Cell. 127:291–304.

Morike M, Brenner RE, Bushart GB, Teller WM, Vetter U (1992) Collagen metabolism in cultured osteoblasts from osteogenesis imperfecta patients. Biochem J. 286(Pt 1):73–77.

Mottes M, Sangalli A, Valli M, Gomez Lira M, Tenni R, Buttitta P, Pignatti PF, Cetta G (1992) Mild dominant osteogenesis imperfecta with intrafamilial variability: the cause is a serine for glycine alpha 1(I) 901 substitution in a type- I collagen gene. Hum Genet. 89:480–484.

Mottes M, Gomez Lira MM, Valli M, Scarano G, Lonardo F, Forlino A, Cetta G, Pignatti PF (1993) Paternal mosaicism for a COL1A1 dominant mutation (alpha 1 Ser-415) causes recurrent osteogenesis imperfecta. Hum Mutat. 2:196–204.

Muriel MP, Bonaventure J, Stanescu R, Maroteaux P, Guenet JL, Stanescu V (1991) Morphological and biochemical studies of a mouse mutant (fro/fro) with bone fragility. Bone. 12:241–248.

Myllyla R, Wang C, Heikkinen J, Juffer A, Lampela O, Risteli M, Ruotsalainen H, Salo A, Sipila L (2007) Expanding the lysyl hydroxylase toolbox: new insights into the localization and activities of lysyl hydroxylase 3 (LH3). J Cell Physiol. 212:323–329.

Nicholls AC, Oliver J, Renouf DV, McPheat J, Palan A, Pope FM (1991) Ehlers-Danlos syndrome type VII: a single base change that causes exon skipping in the type I collagen alpha 2(I) chain. Hum Genet. 87:193–198.

Nicholls AC, Oliver J, Renouf DV, Heath DA, Pope FM (1992) The molecular defect in a family with mild atypical osteogenesis imperfecta and extreme joint hypermobility: exon skipping caused by an 11-bp deletion from an intron in one COL1A2 allele. Hum Genet. 88:627–633.

Nicholls AC, Oliver J, McCarron S, Winter GB, Pope FM (1996) Splice site mutation causing deletion of exon 21 sequences from the pro alpha 2(I) chain of type I collagen in a patient with severe dentinogenesis imperfecta but very mild osteogenesis imperfecta. Hum Mutat. 7: 219–227.

Nicholls AC, Sher JL, Wright MJ, Oley C, Mueller RF, Pope FM (2000) Clinical phenotypes and molecular characterisation of three patients with Ehlers-Danlos syndrome type VII. J Med Genet. 37:E33.

Nicholls AC, Valler D, Wallis S, Pope FM (2001) Homozygosity for a splice site mutation of the COL1A2 gene yields a non-functional pro(alpha)2(I) chain and an EDS/OI clinical phenotype. J Med Genet. 38:132–136.

Niyibizi C, Smith P, Mi Z, Phillips CL, Robbins P (2001) Transfer of proalpha2(I) cDNA into cells of a murine model of human Osteogenesis Imperfecta restores synthesis of type I collagen comprised of alpha1(I) and alpha2(I) heterotrimers in vitro and in vivo. J Cell Biochem. 83: 84–91.

Nuytinck L, Freund M, Lagae L, Pierard GE, Hermanns-Le T, De Paepe A (2000) Classical Ehlers-Danlos syndrome caused by a mutation in type I collagen. Am J Hum Genet. 66:1398–1402.

Otterbach B, Stoffel W (1995) Acid sphingomyelinase-deficient mice mimic the neurovisceral form of human lysosomal storage disease (Niemann-Pick disease).Cell. 81:1053–1061.

Oyama M, Tatlock A, Fukuta S, Kavalkovich K, Nishimura K, Johnstone B, Robbins PD, Evans CH, Niyibizi C (1999) Retrovirally transduced bone marrow stromal cells isolated from a mouse model of human osteogenesis imperfecta (oim) persist in bone and retain the ability to form cartilage and bone after extended passaging. Gene Ther. 6:321–329.

Paterson CR, Mole PA (1994) Bone density in osteogenesis imperfecta may well be normal [see comments]. Postgrad Med J. 70:104–107.

Peltonen L, Palotie A, Prockop DJ (1980) A defect in the structure of type I procollagen in a patient who had osteogenesis imperfecta: excess mannose in the COOH-terminal propeptide. Proc Natl Acad Sci USA. 77:6179–6183.

Pereira R, Khillan JS, Helminen HJ, Hume EL, Prockop DJ (1993) Transgenic mice expressing a partially deleted gene for type I procollagen (COL1A1). A breeding line with a phenotype of spontaneous fractures and decreased bone collagen and mineral. J Clin Invest. 91: 709–716.

Pereira RF, Hume EL, Halford KW, Prockop DJ (1995) Bone fragility in transgenic mice expressing a mutated gene for type I procollagen (COL1A1) parallels the age-dependent phenotype of human osteogenesis imperfecta. J Bone Miner Res. 10:1837–1843.

Pfeiffer BJ, Franklin CL, Hsieh FH, Bank RA, Phillips CL (2005) Alpha 2(I) collagen deficient oim mice have altered biomechanical integrity, collagen content, and collagen crosslinking of their thoracic aorta. Matrix Biol. 24:451–458.

Phillips CL, Shrago-Howe AW, Pinnell SR, Wenstrup RJ (1990) A substitution at a non-glycine position in the triple-helical domain of pro alpha 2(I) collagen chains present in an individual with a variant of the Marfan syndrome. J Clin Invest 86:1723–1728.

Phillips CL, Pfeiffer BJ, Luger AM, Franklin CL (2002) Novel collagen glomerulopathy in a homotrimeric type I collagen mouse (oim). Kidney Int. 62:383–391.

Pollitt R, McMahon R, Nunn J, Bamford R, Afifi A, Bishop N, Dalton A (2006) Mutation analysis of COL1A1 and COL1A2 in patients diagnosed with osteogenesis imperfecta type I–IV. Hum Mutat. 27:716.

Pornprasertsuk S, Duarte WR, Mochida Y, Yamauchi M (2005) Overexpression of lysyl hydroxylase-2b leads to defective collagen fibrillogenesis and matrix mineralization. J Bone Miner Res. 20:81–87.

Primorac D, Rowe DW, Mottes M, Barisic I, Anticevic D, Mirandola S, Gomez Lira M, Kalajzic I, Kusec V, Glorieux FH (2001) Osteogenesis imperfecta at the beginning of bone and joint decade. Croat Med J. 42:393–415.

Prockop DJ (1984) Osteogenesis imperfecta: phenotypic heterogeneity, protein suicide, short and long collagen. Am J Hum Genet. 36:499–505.

Raff ML, Craigen WJ, Smith LT, Keene DR, Byers PH (2000) Partial COL1A2 gene duplication produces features of osteogenesis imperfecta and Ehlers-Danlos syndrome type VII. Hum Genet. 106:19–28.

Ristevski S (2005) Making better transgenic models: conditional, temporal, and spatial approaches. Mol Biotechnol. 29:153–163.

Rowe DW, Shapiro JR, Poirier M, Schlesinger S (1985) Diminished type I collagen synthesis and reduced alpha 1(I) collagen messenger RNA in cultured fibroblasts from patients with dominantly inherited (type I) osteogenesis imperfecta. J Clin Invest. 76:604–611.

Sarafova, AP, Choi H, Forlino A, Gajko A, Cabral WA, Tosi L, Reing CM, Marini JC (1998) Three novel type I collagen mutations in osteogenesis imperfecta type IV probands are associated with discrepancies between electrophoretic migration of osteoblast and fibroblast collagen. Hum Mutat. 11:395–403.

Sarathchandra P, Pope FM, Kayser MV, Ali SY (2000) A light and electron microscopic study of osteogenesis imperfecta bone samples, with reference to collagen chemistry and clinical phenotype. J Pathol. 192:385–395.

Schnieke A, Harbers K, Jaenisch R (1983) Embryonic lethal mutation in mice induced by retrovirus insertion into the alpha 1(I) collagen gene. Nature 304:315–320.

Schwarz M, Harbers K, Kratochwil K (1990) Transcription of a mutant collagen I gene is a cell type and stage-specific marker for odontoblast and osteoblast differentiation. Development 108:717–726.

Schwarze U, Hata R, McKusick VA, Shinkai H, Hoyme HE, Pyeritz RE, Byers PH (2004) Rare autosomal recessive cardiac valvular form of Ehlers-Danlos syndrome results from mutations in the COL1A2 gene that activate the nonsense-mediated RNA decay pathway. Am J Hum Genet. 74:917–930.

Seidler DG, Faiyaz-Ul-Haque M, Hansen U, Yip GW, Zaidi SH, Teebi AS, Kiesel L, Gottte M (2006) Defective glycosylation of decorin and biglycan, altered collagen structure, and abnormal phenotype of the skin fibroblasts of an Ehlers-Danlos syndrome patient carrying the novel Arg270Cys substitution in galactosyltransferase I (beta4GalT-7). J Mol Med. 84: 583–594.

Shapiro JR, Burn VE, Chipman SD, Velis KP, Bansal M (1989) Osteoporosis and familial idiopathic scoliosis: association with an abnormal alpha 2(I) collagen. Connect Tissue Res. 21: 117–123.

Shapiro JR, Stover ML, Burn VE, McKinstry MB, Burshell AL, Chipman SD, Rowe DW (1992) An osteopenic nonfracture syndrome with features of mild osteogenesis imperfecta associated with the substitution of a cysteine for glycine at triple helix position 43 in the pro alpha 1(I) chain of type I collagen. J Clin Invest 89:567–573.

Sillence DO, Senn A, Danks DM (1979) Genetic heterogeneity in osteogenesis imperfecta. J Med Genet. 16:101–116.

Sillence DO, Ritchie HE, Dibbayawan T, Eteson D, Brown K (1993). Fragilitas ossium (fro/fro) in the mouse: a model for a recessively inherited type of osteogenesis imperfecta. Am J Med Genet. 45:276–283.

Slayton RL, Deschenes SP, Willing MC (2000) Nonsense mutations in the COL1A1 gene preferentially reduce nuclear levels of mRNA but not hnRNA in osteogenesis imperfecta type I cell strains. Matrix Biol. 19:1–9.

Spotila LD, Constantinou CD, Sereda L, Ganguly A, Riggs BL, Prockop DJ (1991) Mutation in a gene for type I procollagen (COL1A2) in a woman with postmenopausal osteoporosis: evidence for phenotypic and genotypic overlap with mild osteogenesis imperfecta. Proc Natl Acad Sci USA. 88:5423–5427.

Spotila LD, Colige A, Sereda L, Constantinou-Deltas CD, Whyte MP, Riggs BL, Shaker JL, Spector TD, Hume E, Olsen N, et al. (1994) Mutation analysis of coding sequences for type I procollagen in individuals with low bone density. J Bone Miner Res. 9:923–932.

Stacey A, Bateman J, Choi T, Mascara T, Cole W, Jaenisch R (1988) Perinatal lethal osteogenesis imperfecta in transgenic mice bearing an engineered mutant pro-alpha 1(I) collagen gene. Nature. 332:131–136.

Stankovic KM, Kristiansen AG, Bizaki A, Lister M, Adams JC, McKenna MJ (2007) Studies of otic capsule morphology and gene expression in the Mov13 mouse – an animal model of type I osteogenesis imperfecta. Audiol Neurootol. 12:334–343.

Steinmann B, Tuderman L, Peltonen L, Martin GR, McKusick VA, Prockop DJ (1980) Evidence for a structural mutation of procollagen type I in a patient with the Ehlers-Danlos syndrome type VII. J Biol Chem. 255:8887–8893.

Stoffel W, Jenke B, Block B, Zumbansen M, Koebke J (2005) Neutral sphingomyelinase 2 (smpd3) in the control of postnatal growth and development. Proc Natl Acad Sci USA. 102: 4554–4559.

Stoffel W, Jenke B, Holz B, Binczek E, Gunter RH, Knifka J, Koebke J, Niehoff A (2007) Neutral sphingomyelinase (SMPD3) deficiency causes a novel form of chondrodysplasia and dwarfism that is rescued by Col2A1-driven smpd3 transgene expression. Am J Pathol. 171: 153–161.

Superti-Furga A, Pistone F, Romano C, Steinmann B (1989) Clinical variability of osteogenesis imperfecta linked to COL1A2 and associated with a structural defect in the type I collagen molecule. J Med Genet. 26:358–362.

Suphapeetiporn K, Tongkobpetch S, Mahayosnond A, Shotelersuk V (2007) Expanding the phenotypic spectrum of Caffey disease. Clin Genet. 71:280–284.

Sykes B, Francis MJ, Smith R (1977) Altered relation of two collagen types in osteogenesis imperfecta. N Engl J Med. 296:1200–1203.

Symoens S, Nuytinck L, Legius E, Malfait F, Coucke PJ, De Paepe A (2004) Met>Val substitution in a highly conserved region of the pro-alpha1(I) collagen C-propeptide domain causes alternative splicing and a mild EDS/OI phenotype. J Med Genet. 41:e96.

Tenni R, Biglino P, Dyne K, Rossi A, Filocamo M, Pendola F, Brunelli P, Buttitta P, Borrone C, Cetta G (1991) Phenotypic variability and abnormal type I collagen unstable at body temperature in a family with mild dominant osteogenesis imperfecta. J Inherit Metab Dis. 14:189–201.

Thorner PS (2007) Alport syndrome and thin basement membrane nephropathy. Nephron Clin Pract. 106:c82–c88.

Tkocz C, Kuhn K (1969) The formation of triple-helical collagen molecules from alpha-1 or alpha-2 polypeptide chains. Eur J Biochem. 7:454–462.

Van Camp G, Snoeckx RL, Hilgert N, van den Ende J, Fukuoka H, Wagatsuma M, Suzuki H, Smets RM, Vanhoenacker F, Declau F, Van de Heyning P, Usami S. (2006) A new autosomal recessive form of Stickler syndrome is caused by a mutation in the COL9A1 gene. Am J Hum Genet. 79:449–457.

Vasan NS, Kuivaniemi H, Vogel BE, Minor RR, Wootton JA, Tromp G, Weksberg R, Prockop DJ (1991) A mutation in the pro alpha 2(I) gene (COL1A2) for type I procollagen in Ehlers-Danlos syndrome type VII: evidence suggesting that skipping of exon 6 in RNA splicing may be a common cause of the phenotype. Am J Hum Genet 48:305–317.

Vranka JA, Sakai LY, Bachinger HP (2004) Prolyl 3-hydroxylase 1, enzyme characterization and identification of a novel family of enzymes. J Biol Chem. 279:23615–23621.

Wallis GA, Starman BJ, Zinn AB, Byers PH (1990) Variable expression of osteogenesis imperfecta in a nuclear family is explained by somatic mosaicism for a lethal point mutation in the alpha 1(I) gene (COL1A1) of type I collagen in a parent. Am J Hum Genet. 46:1034–1040.

Walter K, Tansek M, Tobias ES, Ikegawa S, Coucke P, Hyland J, Mortier G, Iwaya T, Nishimura G, Superti-Furga A, Unger S (2007) COL2A1-related skeletal dysplasias with predominant metaphyseal involvement. Am J Med Genet A. 143:161–167.

Wang XM, Cui FZ, Ge J, Zhang Y, Ma C (2002) Variation of nanomechanical properties of bone by gene mutation in the zebrafish. Biomaterials. 23:4557–4563.

Wang XM, Cui FZ, Ge J, Wang Y (2004) Hierarchical structural comparisons of bones from wild-type and liliput (dtc232) gene-mutated Zebrafish. J Struct Biol. 14:236–245.

Ward LM, Rauch F, Travers R, Chabot G, Azouz EM, Lalic L, Roughley PJ, Glorieux FH (2002) Osteogenesis imperfecta type VII: an autosomal recessive form of brittle bone disease. Bone. 31:12–18.

Warman ML, Abbott M, Apte SS, Hefferon T, McIntosh I, Cohn DH, Hecht JT, Olsen BR, Francomano CA (1993) A type X collagen mutation causes Schmid metaphyseal chondrodysplasia. Nat Genet. 5:79–82.

Watson RB, Wallis GA, Holmes DF, Viljoen D, Byers PH, Kadler KE (1992) Ehlers Danlos syndrome type VIIB. Incomplete cleavage of abnormal type I procollagen by N-proteinase in vitro results in the formation of copolymers of collagen and partially cleaved pNcollagen that are near circular in cross-section. J Biol Chem. 267:9093–9100.

Weil D, Bernard M, Combates N, Wirtz MK, Hollister DW, Steinmann B, Ramirez F (1988) Identification of a mutation that causes exon skipping during collagen pre-mRNA splicing in an Ehlers-Danlos syndrome variant. J Biol Chem. 263:8561–8564.

Weil D, D'Alessio M, Ramirez F, de Wet W, Cole WG, Chan D, Bateman JF (1989a) A base substitution in the exon of a collagen gene causes alternative splicing and generates a structurally abnormal polypeptide in a patient with Ehlers-Danlos syndrome type VII. Embo J. 8: 1705–1710.

Weil D, D'Alessio M, Ramirez F, Steinmann B, Wirtz MK, Glanville RW, Hollister DW (1989b) Temperature-dependent expression of a collagen splicing defect in the fibroblasts of a patient with Ehlers-Danlos syndrome type VII. J Biol Chem. 264:16804–16809.

Weil D, D'Alessio M, Ramirez F, Eyre DR (1990) Structural and functional characterization of a splicing mutation in the pro-alpha 2(I) collagen gene of an Ehlers Danlos type VII patient. J Biol Chem. 265:16007–16011.

Wenstrup RJ, Florer JB, Davidson JM, Phillips CL, Pfeiffer BJ, Menezes DW, Chervoneva I, Birk DE (2006) Murine model of the Ehlers-Danlos syndrome. col5a1 haploinsufficiency disrupts collagen fibril assembly at multiple stages. J Biol Chem. 281:12888–12895.

Willing MC, Deschenes SP, Scott DA, Byers PH, Slayton RL, Pitts SH, Arikat H, Roberts EJ (1994) Osteogenesis imperfecta type I: molecular heterogeneity for COL1A1 null alleles of type I collagen. Am J Hum Genet. 55:638–647.

Willing MC, Pruchno CJ, Atkinson M, Byers PH (1992) Osteogenesis imperfecta type I is commonly due to a COL1A1 null allele of type I collagen. Am J Hum Genet. 51:508–515.

Willing MC, Deschenes SP, Slayton RL, Roberts EJ (1996). Premature chain termination is a unifying mechanism for COL1A1 null alleles in osteogenesis imperfecta type I cell strains. Am J Hum Genet. 59:799–809.

Wirtz MK, Rao VH, Glanville RW, Labhard ME, Pretorius PJ, de Vries WN, de Wet WJ, Hollister DW (1993) A cysteine for glycine substitution at position 175 in an alpha 1 (I) chain of type I collagen produces a clinically heterogeneous form of osteogenesis imperfecta. Connect Tissue Res. 29:1–11.

Zhang Y, Cui FZ, Wang XM, Feng QL, Zhu XD (2002) Mechanical properties of skeletal bone in gene-mutated stopsel(dtl28d) and wild-type zebrafish (Danio rerio) measured by atomic force microscopy-based nanoindentation. Bone. 30:541–546.

Zhuang J, Tromp G, Kuivaniemi H, Castells S, Prockop DJ (1996) Substitution of arginine for glycine at position 154 of the alpha 1 chain of type I collagen in a variant of osteogenesis imperfecta: comparison to previous cases with the same mutation. Am J Med Genet. 61: 111–116.

Zolezzi F, Valli M, Clementi M, Mammi I, Cetta G, Pignatti PF, Mottes M (1997) Mutation producing alternative splicing of exon 26 in the COL1A2 gene causes type IV osteogenesis imperfecta with intrafamilial clinical variability. Am J Med Genet. 71:366–370.

Zumbansen M, Stoffel W (2002) Neutral sphingomyelinase 1 deficiency in the mouse causes no lipid storage disease. Mol Cell Biol. 22:3633–3638.

Chapter 17
Biomimetic Collagen Tissues: Collagenous Tissue Engineering and Other Applications

E.A. Sander and V.H. Barocas

Abstract Collagen gels provide an in-vivo-like, 3D environment suitable for studying cell–matrix interactions during proto-tissue formation. Cell-seeded collagen gels, reconstituted under a variety of conditions, are remodeled by cell-driven compaction and consolidation. The remodeled gel, or tissue equivalent (TE), possesses properties dependent on the organization of collagen fibrils in the network, which, in turn, is controlled by several environmental factors, particularly mechanical constraints on the gel boundaries. Mechanical tests performed under a variety of conditions suggest that many different physical processes are involved in the gel's mechanical response. Network restructuring under nonuniform loading conditions leads to mechanical anisotropy and nonlinearity at large strain. Although similar in behavior, collagen-based TEs do not yet possess sufficient mechanical properties to replace native tissues. Efforts are underway to improve TE properties by controlling ECM composition and organization.

17.1 Introduction

The goal of tissue engineering is to develop constructs that are functionally equivalent to lost or damaged tissues. Because function is dependent on structure, it is necessary to recreate the appropriate tissue micro-architecture, a major component of which is the extracellular matrix (ECM). The ECM structure has implications for both the mechanical performance of the tissue and the maintenance of an appropriate local environment for cellular activity. Since collagen is the primary component of the ECM, and therefore the most abundant protein in the body, a significant amount of attention has been placed on the use of collagen gels as a starting point for creating tissue equivalents (TEs). In addition to their use as proto-tissues, cell-seeded collagen gels, because of their relative simplicity, have also been studied as models for cell–matrix interactions. The gel provides a 3D environment, which promotes in-vivo-like cellular activity that differs from that of 2D cultures in many ways (Pedersen and Swartz 2005). Of particular interest are the cell-driven remodeling events that result in the requisite structure and mechanical properties needed for proper function. Cells interact with the collagen network in a complex manner dependent on many factors and can produce TEs with a wide range of mechanical properties.

P. Fratzl (ed.), *Collagen: Structure and Mechanics*,
© Springer Science+Business Media, LLC 2008

The remodeling process that leads to these properties can best be understood in terms of the behavior of the collagen network. To that end, this chapter first reviews the process by which collagen gels form an interconnected fibril network, followed by network rearrangement that ensues once cells are added and begin to compact the gel. Next, gel mechanics are described under a variety of loading conditions and in terms of the bulk and network properties of the gel. Finally, some techniques that provide control over TE fibril alignment and that improve mechanical properties are discussed.

17.2 Synthesis and Culture of Collagen Gels

17.2.1 Collagen Self-Assembly in Solution

Collagen gels are attractive because they are simple to produce and present a highly biomimetic, 3D environment. They are often created from cold, acid-solubilized type I collagen monomers (tropocollagen) that are neutralized and warmed to induce gelation, but gels can also be formed from enzyme-digested monomers or at room temperature (Stenzel et al. 1974; Holmes et al. 1986; Kadler et al. 1996; Silver et al. 2003). Monomers assemble end-to-end and laterally, forming fibrils that can further aggregate into fibril bundles, or fibers, and higher order structures, through an entropy-driven self-assembly process that is strongly affected by the nature of the collagen monomer and is sensitive to temperature, ionic strength, and pH (Wood and Keech 1960; Kadler et al. 1996; Christiansen et al. 2000; Roeder et al. 2002; Raub et al. 2007). Monomer self-assembly, alignment, and packing order are also dependent on the retention of non-helical end regions (telopeptides), which can be clipped or completely removed through enzymatic digestion (Stenzel et al. 1974). Together, these factors affect the kinetics of fibril self-assembly by altering electrostatic, hydrophobic, and covalent interactions between monomers, resulting in a range of fibril sizes. Typical fibril sizes span between 10 and 200 nm in diameter, although it is sometimes difficult to discern large single fibrils from fibril bundles or fibers. Hydrated, random networks of entangled fibrils form (Fig. 17.1), typically between 0.1 and 0.5% protein by weight (Tranquillo 1999), a substantially lower density than exists in native tissues, e.g., skin > 5% (Nakagawa et al. 1989). Network pore sizes are dependent on the reconstitution conditions, but are generally on the order of 1 μm (Knapp et al. 1997).

17.2.2 FPCL/Tissue Equivalent (TE)

The seminal work of Bell et al. presented the fibroblast-populated collagen lattice (FPCL) as a quantitative assay for the study of fibroblast and collagen gel interactions (Bell et al. 1979). The assay has proven useful for understanding the process by which fibroblasts reorganize collagen gels, for providing insight into wound-healing

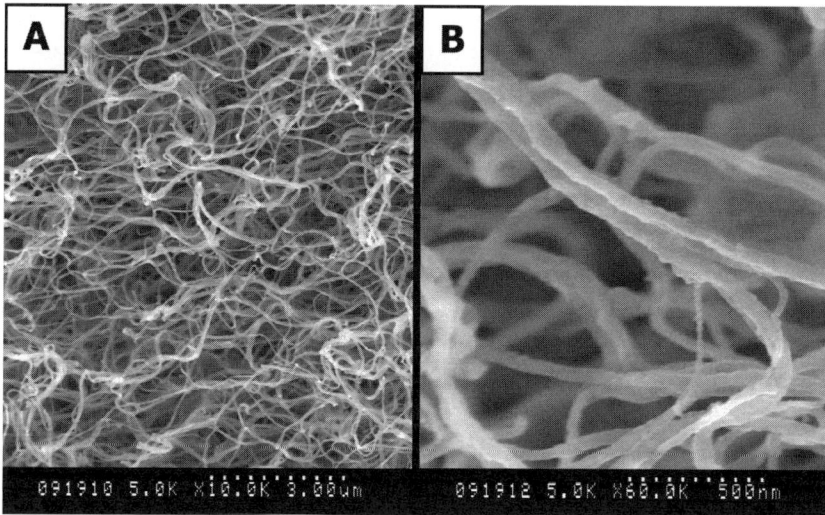

Fig. 17.1 Scanning electron micrographs of a type I collagen gel. Depicted are images of a 1.5 mg/mL collagen gel cast from acid-soluble bovine patellar tendon (Organogenesis, Inc.). (**A**) A random, entangled network of fibrils can be seen. (**B**) Higher magnification reveals the characteristic pattern of banding

events, and for serving as a basis for the first dermal "skin equivalent" (Bell et al. 1981b). Variants of the FPCL assay have assumed other names based on the particulars of the experiment, such as fibroblast-populated microspheres (Moon and Tranquillo 1993) or fibroblast-populated matrix (Huang et al. 1993), but a more generalized term consistent with the notion of producing an in vitro tissue analog has also emerged, that of tissue equivalent (TE) (Nusgens et al. 1984; Tranquillo 1999). Regardless of the nomenclature, the distinguishing feature of cell-populated collagen gels is compaction and consolidation of the fibrillar network through the action of the cells entrapped by the reconstituted matrix (Tranquillo 1999).

17.2.3 Entrapment of Cells/Compaction

When cells are added to gels, either during or after reconstitution, substantial remodeling occurs. Initially, cells migrate through the gel before exerting traction forces on collagen network fibers. Rounded cells extend processes into the matrix and, in a matter of hours, develop stellate or bipolar morphologies in a manner dependent on the initial collagen concentration (Pizzo et al. 2005) and mechanical constraints on the gel (Stopak and Harris 1982; Huang et al. 1993; Wakatsuki and Elson 2003). Cells entrapped in free-floating gels often assume a stress fiber-free stellate morphology, while cells in attached gels develop stress fibers and a bipolar morphology (Roy et al. 1997). Cells form adhesions to the ECM through integrins, a heterodimeric protein complex that mechanically couples the cytoskeleton to the ECM (Carver et al. 1995; Cooke et al. 2000; Wakatsuki and Elson 2003). Myosin activation generates cell contractile forces that result in compaction and significant gel stiffening that begins within hours and continues for days (Wakatsuki et al. 2000). Compaction can only proceed when cell tractions exceed the mechanical resistance of the matrix. When traction forces are high enough, network fibers are drawn inward. Pericellular fibril density increases through local fibril translations that transmit rapidly through the network and lead to large-scale fibrillar rearrangements (Grinnell and Lamke 1984; Tranquillo 1999; Sawhney and Howard 2002; Pizzo et al. 2005). These events cause a reduction in gel volume and an increase in solid volume fraction through matrix compaction and fluid exudation (Fig. 17.2). Although dependent on many factors, volume reduction can be substantial – 85–99% (Bell et al. 1979; Guidry and Grinnell 1985). Matrix densification also reduces pore size, particularly in the pericellular region.

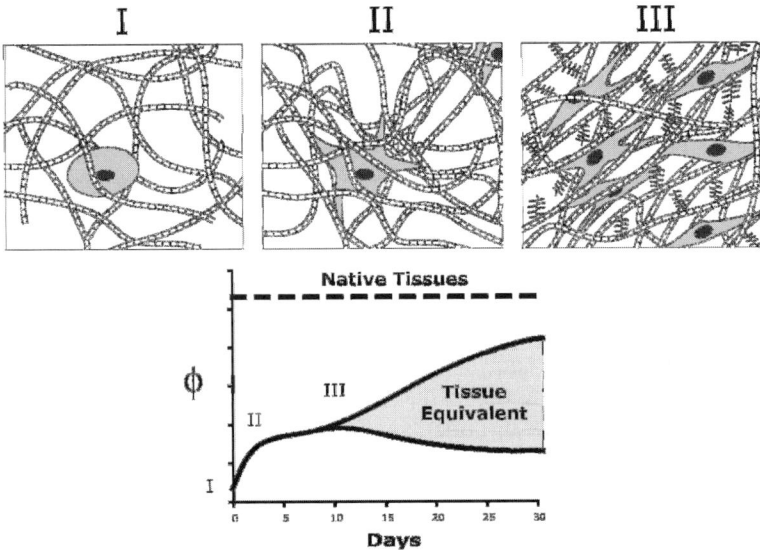

Fig. 17.2 Gel compaction and remodeling: change in solid volume fraction over time. (**I**) Initially, rounded cells are dispersed in a dilute gel with low volume fraction (ϕ). (**II**) In a matter of hours, cells migrate and extend processes into the collagen network. Adhesions form and cell traction develops, leading to increased pericellular fibril density. Fibrils distal to the cells translate and rotate in response to tensile forces propagated throughout the network. The gel shrinks along unconstrained dimensions, and the volume fraction increases rapidly. (**III**) Over the next few weeks, cell number increases, some of the initial collagen degrades, and de novo collagen and additional matrix components, such as proteoglycans, are synthesized. Under proper mechanical constraints on the gel, network alignment develops, and ϕ increases, approaching the levels in native tissues. Of course, several factors affect the gel composition and organization resulting in a range of outcomes

17.2.4 Generation of Fiber Alignment

Cell-driven compaction of the matrix is a necessary step for increasing gel density and producing fibril alignment in a TE. Many interdependent factors contribute to the formation of the microstructure during gel compaction, including cell density, collagen density, and culture medium. While all of these are important, their effect on creating alignment in cell-compacted gels is ultimately dependent on the gel's boundary conditions.

17.2.5 Free Floating or Constrained

Constraints to the gel, most often in the form of adherent surfaces or embedded obstructions, provide opposition to cell-induced tractions propagated through the collagen network. As a result, the gel compacts along directions perpendicular to free surfaces. The differences observed between free-floating and attached gels clearly demonstrate the effect of constraints. Unconstrained gels, freely floating in medium, compact the matrix into a random, isotropic fibrillar network, significantly reduced in both diameter and thickness (Bell et al. 1979). Gels that remain attached to the walls and base of a tissue culture well compact through the thickness but maintain the diameter (Guidry and Grinnell 1985). In this process, both cells and fibrils align perpendicular to the direction of compaction.

Harris and colleagues conducted the first studies in which fixed boundaries were used to create specific alignment patterns in collagen gels (Harris et al. 1981; Stopak and Harris 1982). In these studies, cell tractions generated strong fibril alignment along the axis between tissue explants embedded in collagen gels, a tissue explant and a fixed cylindrical post, and in dispersed fibroblast gels containing fixed boundaries at different locations. In each case, the resulting alignment arose from the intrinsic properties of the fibril network. Cell tractions that pull fibrils inward create forces that are transmitted between fibrils through interconnections that behave like cross-links rather than entanglements (Sawhney and Howard 2002; Tower et al. 2002; Chandran and Barocas 2006). Intact networks without mechanical constraints cannot oppose contraction forces, and the fibrils translate freely in the direction of cell traction. Adherent surfaces prevent fibril translation. As a result, constrained fibrils rotate into alignment with the tension, provided that cross-links and steric interactions permit it. Network disruptions limit force propagation and network reorganization to regions proximal to contracting cells (Guidry and Grinnell 1987b).

Fibrils can also influence compaction through contact guidance, the tendency of cells to orient codirectionally with an aligned matrix. The aligned cells migrate and exert tractions along their polarity axes, which generally align with the matrix fibers (Tranquillo 1999). The contact guidance response is particularly evident when cells are added to pre-aligned fibril networks, which align preferentially along the fibrils (Guido and Tranquillo 1993; Dickinson et al. 1994; Tranquillo et al. 1996). In the absence of constraints, however, a negative feedback effect arises, with cells aligning and inducing compaction primarily along the network alignment direction, driving the system to isotropy (Barocas et al. 1998). In contrast, conditions that favor inhomogeneous network deformations, such as mechanical constraints that prevent compaction or nonuniform cell density, can generate network alignment (Barocas and Tranquillo 1997a). The underlying cellular basis for contact guidance is not well understood. It may result from steric or mechanical effects of the network, from focal adhesions located along the fibril, or another source entirely (Tranquillo 1999).

17.2.6 Collagen and Cell Concentration

The gel's initial cell and collagen concentrations affect the degree and rate of gel contraction (Guidry and Grinnell 1987a; Moon and Tranquillo 1993; Barocas and Tranquillo 1997a; Newman et al. 1997; Zhu et al. 2001; Nirmalanandhan et al. 2006). Early studies demonstrated that higher cell number in free-floating collagen gels increased both the total amount of compaction and the rate at which it occurred (Fig. 17.3). Similarly, for a given cell density, the lower the initial collagen concentration, the greater the amount of total compaction, the faster the rate of compaction, and the higher the final collagen concentration (Zhu et al. 2001). In addition, the extent of rearrangement is dependent on the collagen density;

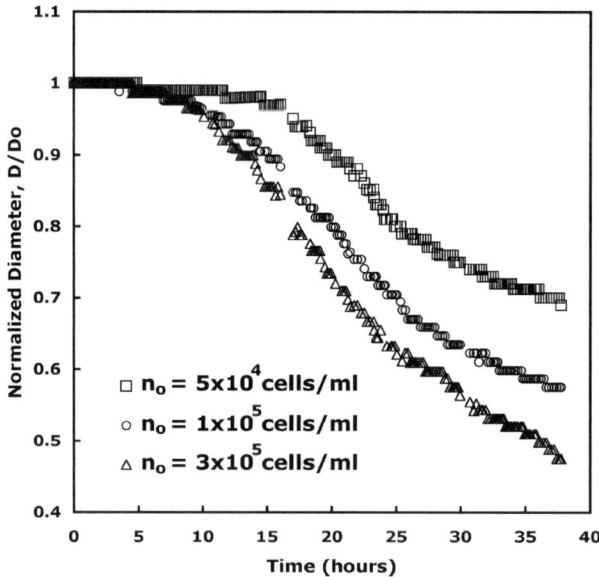

Fig. 17.3 Transient compaction of free-floating spherical gels with different initial fibroblast concentrations. Unconstrained gels shrink rapidly and in a manner dependent on the initial cell number (n_o). (Data from Moon 1993)

higher-density gels offer more resistance and limit reorganization more proximal to the cell than their low-density counterparts, e.g., 0.1 versus 0.3% (Pizzo et al. 2005). However, cell and collagen concentrations are coupled. As one might expect, total gel contraction is greatest in low-collagen, high-cell-density constructs and lowest in high-collagen, low-cell-density constructs. Although collagen concentration is more important than cell density, a threshold cell number exists, above which contraction does not change, potentially due to a lack of binding sites (Nirmalanand-han et al. 2006).

As the compaction process begins, cell viability appears dependent on whether the gel is constrained and whether the network is capable of sustaining tension. Unconstrained compaction in floating gels has been shown to induce quiescence (Rosenfeldt and Grinnell 2000) and apoptotic cell death (Nakagawa et al. 1989; Grinnell et al. 1999; Zhu et al. 2001), although a modest increase in cell number over the first few days has also been reported (Barocas et al. 1995). In contrast, axially constrained gels compacted to less than 1% of the original volume increased in cell number (Nakagawa et al. 1989; Wille et al. 2006).

Initial cell density can also modulate the effect of exogenous chemical factors and medium additives, such as platelet-derived growth factor (PDGF) and serum (Ehrlich and Rittenberg 2000). Serum is essential for gel compaction (Guidry and Grinnell 1985; Newman et al. 1997; Wakatsuki and Elson 2003) – serum-free gels do not generate force (Wakatsuki and Elson 2003) – but the increased compaction witnessed in response to higher serum levels is more pronounced in moderate density (0.5×10^5 cells/mL) than in high density (5.0×10^5 cells/mL) floating gels (Ehrlich and Rittenberg 2000). Many other chemical factors have also been investigated and found to affect cell-driven gel compaction, such as PDGF (Gullberg et al. 1990; Shreiber et al. 2001; Parekh and Velegol 2007), transforming growth factor beta (TGF-β) (Grinnell and Ho 2002; Parekh and Velegol 2007), angiotensin II (Burgess et al. 1994), and thrombin (Kolodney and Wysolmerski 1992).

17.2.7 Matrix Synthesis and Cross-Link Formation

The remodeling process that occurs during gel compaction is largely due to rearrangement of existing collagen fibrils. Over the short term (e.g., for 3 days or less), only a fraction of the initial collagen is degraded, typically less than 5% (Guidry and Grinnell 1985; Zhu et al. 2001). Over longer culture periods, original collagen is progressively degraded, as much as 80% in 4 weeks (Nakagawa et al. 1989). Collagen synthesis increases in attached gels but decreases in floating gels (Nakagawa et al. 1989).

Additional matrix components are synthesized and incorporated into the matrix, but in a manner dependent on the cell type and culture conditions. The amount of proteoglycans and glycosaminoglycans – the most commonly assayed ECM components – increases rapidly over the first few weeks before slowing (Huang et al. 1993; Ahlfors and Billiar 2007). Other ECM components, such as elastin (Isenberg and Tranquillo 2003), are conspicuously absent or present at very low concentrations.

Over time, fibrils brought together by cellular contraction stabilize the network through non-covalent chemical interactions (Guidry and Grinnell 1987a). These associations can be replicated in acellular collagen gels subjected to centrifugation (Guidry and Grinnell 1985), suggesting that interfibrillar associations may result from electrostatic and hydrophobic interactions made permissible by contact. Although gels partially expand when cells are removed or inactivated through the addition of detergent, trypsin/EDTA, or cytochalasin D, pericellular network rearrangements are not observed (Guidry and Grinnell 1985, 1987a). Covalent modifications are not formed, at least not over the short term (Guidry and Grinnell 1985), but rather require from several days to weeks to form, through lysyl oxidase activity (Huang et al. 1993; Redden and Doolin 2003).

17.2.8 Conclusions

Although the number of factors involved in cell-driven collagen gel remodeling is obviously large, certain general trends emerge from the literature. Cells will compact the gel and rearrange the fibril network in a reciprocal manner based on physical constraints, cell number, collagen concentration, and the chemical composition of the medium. The remodeling process not only changes the gel's mechanical behavior (Section 17.3), but also the local signaling environment (Section 17.4), which influences whether cells adopt an active or quiescent phenotype. Given the importance of the compaction process, theoretical and computational approaches to understanding it are being explored by a number of groups (Barocas and Tranquillo 1997a, 1997b; Ferrenq et al. 1997; Zahalak et al. 2000; Breuls et al. 2002; Marquez et al. 2005; Ohsumi et al. 2007).

17.3 Mechanical Properties of Collagen Gels and TEs

Because of the demonstrated importance of gel mechanics in mediating compaction and the need to match native properties of mechanical tissues (e.g., artery), mechanical tests have been performed on collagen gels, with and without cells present, under a wide range of loading conditions. Data from these tests can be used to assess the extent and success of the remodeling process in replicating the organization that exists in native tissues. In addition, collagen gels and TEs are relatively simple constructs that share some of the properties of native tissues. Although the mechanical response of native tissues differs, the study of these constructs under different testing modes is important for elucidating the interactions that initiate at the microstructural level and determine the overall mechanical behavior of the construct.

Collagen gels have been tested under a wide range of testing configurations, including shear, compression, and extension in one or two directions. The mechanical behavior is determined not only by the test geometry, but also by the cellularity and chemical composition, the culture conditions (e.g., temperature or medium ionic strength), and loading conditions (e.g., frequency of a dynamic test). The following section reviews the mechanical responses observed in collagen gels subjected to shear, extension, and compression.

17.3.1 Shear

In shear, the fluid and solid phases deform together, such that the material changes shape but preserves macroscopic volume. Most of the material response is attributed to elastic and viscoelastic resistance by the network solid phase with little contribution from the fluid phase. Oscillatory shear tests have been conducted using concentric cylinder (Barocas et al. 1995; Knapp et al. 1997), cone and plate (Hsu et al. 1994; Kuntz and Saltzman 1997; Wu et al. 2005), and parallel plate (Velegol and Lanni 2001; Raub et al. 2007) geometry. The amount of phase shift between the gel stress response and the imposed oscillatory shear strain within the linear viscoelastic limit is a reflection of energy dissipated during a loading cycle. The in-phase stress response, termed the storage modulus (G'), and the out-of-phase stress response, termed the loss modulus (G''), correspond to the elastic and viscous components of the material or the amount of energy recovered and lost, respectively (Findley et al. 1976). The loss tangent (tan $\delta = G'/G''$) describes the viscoelastic character of a material: for an elastic material tan $\delta = 0$ (response in phase with displacement) and for a completely viscous material tan $\delta = \infty$ (response out of phase with displacement, thus in phase with velocity).

Significant differences in storage and loss modulus values have been observed for gels formed under a variety of conditions (Table 17.1). For shear strains up to 10% and frequencies 0.1–100 rad/s, modulus values decrease only gradually with decreasing frequency (Fig. 17.4), indicating that although some viscoelasticity is present, the majority of the response is elastic. This is also borne out in the loss tangent, which ranges from 0.1 to 0.27. The elastic character likely stems from cross-links in the network that reduce fibril mobility. Non-covalent cross-links, rather than fibril entanglements, appear to be largely responsible for physical linkages between

Table 17.1 Acellular collagen gel mechanical properties

Test	Range	Study
Shear		
Shear storage modulus	0.3–150 Pa	Barocas (1995), Hsu (1994), Knapp (1997), Kuntz (1997), Parsons (2002), Raub (2007), Velegol (2001), and Wu (2005)
Shear loss modulus	0.02–29 Pa	Barocas (1995), Hsu (1994), Knapp (1997), Kuntz (1997), Parsons (2002), Raub (2007), Velegol (2001), and Wu (2005)
Extension		
Linear modulus	1.5–60 kPa	Osborne (1998) and Roeder (2002)
Instantaneous modulus	2.0–57 kPa	Ozerdem (1995) and Pryse (2003)
Equilibrium modulus	0.4–5.5 kPa	Krishnan (2004), Ozerdem (1995), and Pryse (2003)

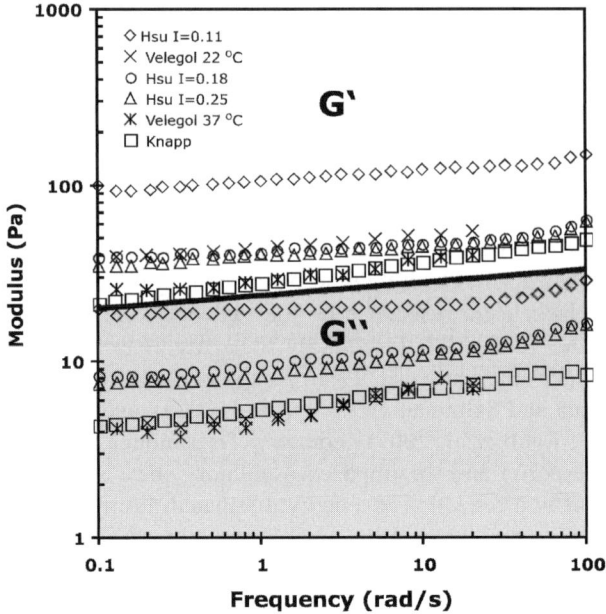

Fig. 17.4 Dynamic shear moduli for collagen gels. The frequency dependence of the storage (G') and loss modulus (G'') for type I collagen gels under oscillatory shear and constant strain amplitude. Gels were reconstituted and tested under different conditions, including temperature, ionic strength (I), collagen concentration, and strain amplitude. (Data from Knapp et al. 1997, Velegol and Lanni 2001, and Hsu et al. 1994)

fibrils because shear modulus rapidly increases at large strain ($> 10\%$) before abrupt failure, a phenomenon associated with cross-links rather than fibril entanglements (Barocas et al. 1995; Chandran and Barocas 2004).

Long-time creep experiments show immediate and steep gel compliance before developing slow and steady creep (Barocas et al. 1995). When the load is removed, much of the recovery is immediate, though additional recovery proceeds slowly, indicating viscoelastic behavior with a broad relaxation spectrum. The incomplete recovery observed in that study is characteristic of fluid-like behavior. In addition, over short time scales (0–100 s), the creep compliance and recovery of collagen gels are independent of shear stress for applied shears up to 0.7 Pa. Taken together, these results indicate that acellular collagen gels can be treated as linear viscoelastic Maxwell fluids up to 10% shear strain (Barocas et al. 1995).

The shear properties of a gel are directly related to the conditions that lead to the formation of the underlying collagen network. For example, the polymerization temperature during gelation influences the storage and loss modulus of the gel by affecting fibril size and network connectivity (Raub et al. 2007). In the study by Raub et al., both G' and G'' increased with polymerization temperature even though higher temperatures produced thinner fibers (Raub et al. 2007). The increase in modulus presumably reflects more interactions between many thin fibers compared to a fewer associations between a smaller number of larger diameter fibers. However, the nature of these interactions is unclear. Although G' and G'' were relatively insensitive to a frequency sweep, G'' was higher that G'. This difference might be attributable to the polymerization time, which was significantly reduced in comparison to other rheology studies (Hsu et al. 1994; Barocas et al. 1995; Knapp et al. 1997; Velegol and Lanni 2001; Wu et al. 2005) and may reflect a reduction in the number of non-covalent cross-links and fiber–fiber associations that take time to develop (Roeder et al. 2002). The addition of other matrix components during gel formation has also been shown to influence modulus values (Hsu et al. 1994; Kuntz and Saltzman 1997). The gel's shear properties are altered by changes in fibril diameter, pore size, and interfibrillar associations that occur during the assembly process, all of which affect the network structure.

The previous set of studies measured bulk shear properties of the gel and did not account for local heterogeneity throughout the sample. Velegol and Lanni used a laser-trap microrheometry technique to probe the gel locally at the micron level (Velegol and Lanni 2001). The displacement of embedded 2.1 μm microspheres was in phase with the trap indicating elastic behavior locally. Shear modulus varied substantially with location, ranging from 2 to 90 Pa within one sample, presumably reflecting differences in microstructure. Average shear modulus increased with collagen concentration (e.g., 7 and 55 Pa for 0.5 and 2.3 mg/mL gel, respectively) and proved slightly higher than macroscopic measurements made with a parallel plate rheometer. Parsons and Coger (2002) measured the viscoelastic properties of collagen gels with a custom-built piezoelectrically actuated linear rheometer that uses a cantilever configuration. They found the local storage modulus, but not the loss modulus, much lower than bulk measurements.

17.3.2 Extension

Gel extension is principally resisted by tension developed in the collagen network. Uniaxial tensile tests on gels generate nonlinear stress–strain curves similar to those of native soft tissues (Roeder et al. 2002, 2004; Tower et al. 2002; Voytik-Harbin et al. 2003; Krishnan et al. 2004). Curves begin with a compliant, nonlinear "toe" region that is followed by a stiff "linear" region and an exponential failure region (Fig. 17.5). Polarized light microscopy has shown that this behavior can be explained in part by the mechanics of the microstructure (Tower et al. 2002). Under low load, network fibrils rotate and realign with the stretch direction before rapidly stiffening in response to higher loads necessary to axially deform the fibrils (Tower et al. 2002; Chandran and Barocas 2006, 2007). When gels are loaded repeatedly at the same rate and the same extent, the peak force reached and the amount of hysteresis between the loading and unloading curves decrease to a stable value. This preconditioning has long been associated with changes in the microstructure of native tissues during testing (Fung 1993). Changes in fiber alignment as a result of preconditioning have been observed, but other processes including a reduction

Fig. 17.5 Typical isotropic
collagen gel stress–strain
response to uniaxial tension
by an initially isotropic
collagen gel. (**I**) In the
unloaded configuration, the
collagen network is isotropic.
Two enlarged fibers are
depicted to illustrate their
response to the stretch at the
indicated points on the
stress–strain curve. (**II**)
Under low load, fibrils rotate
and align in the direction of
stretch, possibly restricted by
entanglements or cross-links.
(**III**) The gel stiffens and
higher loads are required to
axially deform the aligned
fibrils

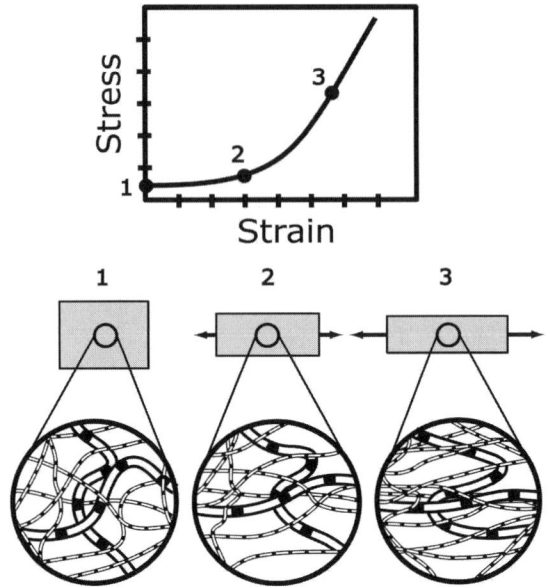

of interfibrillar associations and the extrusion of water may be involved as well (Tower et al. 2002). Local 3D strain measurements using confocal microscopy and digital image correlation methods show individual fibrils recruited in the direction of stretch and the largest network deformations transverse to the loading direction (Voytik-Harbin et al. 2003; Roeder et al. 2004).

Despite differences in gel fabrication and testing procedures, some general observations can be made. Uniaxial extension tests show that increased collagen concentration generally translates to higher mechanical properties, particularly the modulus (Ozerdem and Tozeren 1995; Roeder et al. 2002; Krishnan et al. 2004). Krishnan (2004) suggested that more interfibrillar cross-links would likely form when collagen fibril density increases, which would result in greater fibril recruitment and a stiffer response. Cross-linking agents (Osborne et al. 1998; Sheu et al. 2001) and glycosaminoglycans (Osborne et al. 1998) have also been shown to improve the stress response. Large contractions transverse to the loading direction are typical (Roeder et al. 2002, 2004), as would be expected from network mechanics theory (Stylianopoulos and Barocas 2007). The gel response is also strain- (Ozerdem and Tozeren 1995; Pryse et al. 2003; Krishnan et al. 2004) and strain-rate-dependent (Roeder et al. 2002; Krishnan et al. 2004), indicating that viscoelasticity is important. More of the load is resisted by axial fiber deformations rather than matrix reorganization through rotation and translation.

Krishnan et al. performed uniaxial extension tests without the difficulties that arise from grips by casting collagen gels as rectangular slabs with embedded fixed and actuating posts (Krishnan et al. 2004). In addition, this configuration provided a uniform strain field through much of the gel's central region. The collagen concentration, frequency, and equilibrium strain levels of the test significantly influenced the dynamic and equilibrium response. The equilibrium stress–strain response was nearly linear at lower collagen concentrations (1.5 and 3.0 mg/mL) but then became nonlinear (concave up) at higher concentrations (4.5 mg/mL). A greater amount of cross-linking at the higher collagen density could be responsible by increasing fiber recruitment in the direction of stretch (Krishnan et al. 2004). For the conditions under which the gels were tested, a considerable proportion of the material response was viscoelastic, as demonstrated by the difference between the dynamic stiffness and equilibrium modulus (21.84 ± 0.21 kPa versus 1.53 ± 0.01 kPa, respectively, at 6% strain and 5 Hz). Here, dynamic stiffness is defined as the ratio of the amplitudes of the cyclic stress-time to strain-time data. The mechanical response did not significantly change over the 7 days of culture, a finding contrary to that of Roeder et al. (2002).

The mechanical behavior of TEs has also been the subject of much investigation. In many respects, TE behavior is similar to that of acellular gels. However, the TE fibril network is denser, often aligned, and complicated by cell modifications that include synthesis, degradation, and cross-linking. In addition, the cells themselves contribute to the TE mechanical properties (Wakatsuki et al. 2000) both actively, in terms of cell-generated tension, and passively, as a component of a composite material. Furthermore, cells can influence the mechanical properties in response to chemical signals in the medium. Laser-trap microrheometry measurements showed that the addition of PDGF and TGF induced fibroblast contractions that increased the heterogeneity and anisotropy of the TE's local mechanical properties (Parekh and Velegol 2007).

The mechanical properties of unconstrained, free-floating TEs improve over the first few weeks of culture, presumably due to matrix densification, but eventually decline as the culture period is extended up to 10 weeks (Feng et al. 2003). In contrast, mechanically constrained TEs generally improve over time. Network density increases, fibrils align, and the provisional fibril network is gradually remodeled into a cell-synthesized ECM – although this ECM still differs from that in native tissues.

As with acellular gels, the mechanical behavior observed in a TE is dependent on the stretch magnitude, stretch rate, rest period, and preconditioning protocol (Wagenseil et al. 2003). TE preconditioning is typically employed in an extension test because successive uniaxial stretches do not generate as much force to reach the same strain as the first stretch (Wakatsuki et al. 2000). To confound matters even more, it has been suggested that preconditioning cannot be done on TEs because they never reach a repeatable steady-state force–stretch response (Wagenseil et al. 2003). This observation was made for relatively young TEs (2 days compaction) and may not be widely applicable for older TEs, particularly those that have undergone extensive remodeling over several weeks. The largest drop-off in TE force occurs between the first and second loading cycles (Wagenseil et al. 2003). Interestingly, the shape of the first force curve differs from that of all subsequent ones; the first curve is slightly concave down, whereas all subsequent curves are concave up (Wagenseil et al. 2003; Wille et al. 2006). Cyclic stretch applied for 12 h saw the force reduced substantially further (Wille et al. 2006). The viscoelastic behavior of TEs is generally associated with internal changes in structure, the nature of which remains unclear. Potential sources include network rearrangements and changes in interfibrillar associations.

Multiaxial extension tests provide a better means of characterizing TE mechanical properties. These tests are more difficult to perform, which explains in part the scarcity of this type of data in the literature. In one study, pressure–diameter, force–length testing was conducted on tubular TEs with either axial or circumferential alignment (Wagenseil et al. 2004). TE extensibility was significantly greater in the direction perpendicular to cellular alignment. In another study, an inflation device was used to quantify the biaxial properties of nominally isotropic TEs (Billiar et al. 2005). Modulus values ranged from 11 to 68 kPa, typically lower than those obtained from uniaxial tests.

17.3.3 Compression

The viscoelastic behavior observed in shear and extension is attributed almost entirely to the collagen network. Damping and stress relaxation as a result of fluid flow in a multiphase medium, such as arises in confined compression, is a different phenomenon. Recognizing that poroelasticity theory is but one of many possible approaches to the interstitial flow problem (mixture theory being a highly successful example for articular cartilage (Mow et al. 1980; Soltz and Ateshian 2000)), we use the term poroelasticity to describe dissipation via interstitial flow because of its broad use across disciplines and its natural contrast to viscoelasticity, dissipation due to molecular interactions within the fiber network. In confined compression, a porous piston that mates with a sample chamber is depressed against the sample. In order to conserve volume, fluid flows out of the laterally constrained solid network and through the solid-impenetrable piston pores. Relative motion between the phases can generate interstitial drag that pressurizes the fluid. Interstitial drag dominates the short-term response. Over time, stress dissipates as fluid flows out of the gel, and the pressure equilibrates. As a result, short-term creep behavior is different from that in shear, but becomes similar over time as the viscous behavior of the network dominates (Knapp et al. 1997). Consequently, the material properties are strongly dependent on the time scale of the deformation, with tests performed at a constant strain rate demonstrating much higher modulus values compared to long-time creep (Knapp et al. 1997).

Both viscoelastic and poroelastic mechanisms operate to dissipate stress. Distinguishing between the two during a confined compression test is difficult and requires knowledge of the micromechanical response at the level of the gel network. A series of experiments conducted in concert with real-time birefringence mapping was able to show local changes in network organization in response to step- (Girton et al. 2002; Chandran and Barocas 2004) and ramp-confined compression (Chandran and Barocas 2004). Alignment, orientation, and concentration profiles indicate network restructuring is dependent on which dissipative mechanism dominates, and hence the loading protocol (Chandran and Barocas 2004).

17.3.4 Effect of Gelation Conditions on Properties of Gels

As we have seen, the gel's mechanical properties are derived from the underlying gel microstructure, which is dependent on assembly conditions. In one study, increased polymerization time up to 10 h improved linear modulus and failure stress but not failure strain (Roeder et al. 2002). In addition, network formation under acidic conditions reduced mechanical properties despite increasing fibril diameter. Basic conditions produced stronger gels composed of longer small-diameter fibrils. In contrast, Christiansen et al. found acid-formed gels had higher ultimate tensile strength than neutral and alkaline-formed gels (Christiansen et al. 2000). While thicker fibers should prove stiffer, network connectivity is more important. Consequently, for a finite amount of protein, gelation conditions that produce many interconnected, small-diameter fibrils will prove stronger than conditions that produce fewer, less-connected, larger-diameter fibrils.

17.3.5 Conclusions

In spite of the broad range of tests and conditions, certain themes recur in the study of collagen gel and TE mechanics. In general, samples with a higher collagen density (e.g., because of a greater degree of compaction) are stiffer than those with lower collagen densities, and the same is true of cellularity. Gels and TEs are stiffer in tension than in shear, and stiffer in shear than in compression, indicative of the relative inextensibility of the collagen fiber. The fibers in the collagen network restructure dramatically under nonuniform loading conditions, leading to anisotropy in the gel's optical and mechanical properties and creating a considerable nonlinearity at large strain. The broad relaxation spectrum and the extreme sensitivity to loading conditions suggest that many different physical processes are involved in the mechanical response of the gel – fibrils may stretch, bend, rotate, buckle, and slide past each other over varying time scales, while relative motion between the interstitial fluid and the network may generate high internal pressures that resist compression and prevent network collapse. Finally, recent experimental observations (Evans 2007; Roeder et al. 2004) suggest that small-scale gel inhomogeneity may affect compacted TE properties, which will require considerable theoretical advances over existing work before a model is available to predict TE properties based on composition and history.

17.4 Applications

Collagen has been used in a variety of biomedical applications (reviewed in (Lee et al. 2001)), including as a scaffolding material for tissue-engineered tissues. Scaffolds serve as a template that provides support and guidance while cell remodeling restores functionality. Collagen's biocompatibility, high cellularity, and high-tensile

Table 17.2 Some collagen gel based tissue equivalent studies

Tissue	Study
Dermal	Bell (1981a, 1981b) and Lopez Valle (1992)
Cardiovascular	
Blood vessels	Barocas (1998), Berglund et al. (2005), Cummings et al. (2004), Isenberg (2003), L'Heureux (1993), Seliktar (2000, 2001, 2003), Stegemann (2003), Wagenseil (2004), Weinberg (1986), Yamamura et al. (2007), and Ziegler et al. (1995)
Heart muscle	Eschenhagen (1997), Fink (2000), and Zimmermann et al. (2000)
Heart valves	Neidert (2006) and Shi et al. (2005)
Chordae	Shi (2003, 2006)
Musculoskeletal	
Cartilage	Galois (2006), Iwasa et al. (2003), Noguchi et al. (1994), and Wakitani et al. (1998)
Ligament	Bellows et al. (1981, 1982), Gentleman et al. (2006a), Huang (1993), and Langelier (1999)
Tendon	Awad et al. (2000), Garvin (2003), Juncosa-Melvin (2006), Nirmalanandhan (2006), and Shi (2006)
Intervertebral disc	Gruber et al. (2003, 2004)
Muscle	Okano et al. (1997)
Other tissues	
Cornea	Germain et al. (1999) and Minami et al. (1993)
Bladder	Fujiyama et al. (1995)
Adipose	Gentleman et al. (2006b)

strength make it an obvious choice for such a purpose. Collagen-based scaffolds include gels (Table 17.2), sponges (Roche et al. 2001; Juncosa-Melvin et al. 2007), and hybrid-polymer materials (Yannas et al. 1989; Alini et al. 2003; Prajapati et al. 2000; Chen et al. 2003; Engelmayr et al. 2006). They are seeded with cells, often tissue specific (e.g., rat aortic smooth muscle cells are used to produce bioartificial arteries (Grassl et al. 2002)), and cultured under a variety of conditions to produce functional TEs.

17.4.1 Control of TE Properties

Without sufficient mechanical properties, both for the overall integrity and the appropriate cellular microenvironment, a TE will fail in vivo (Guilak et al. 2003). Unfortunately, cell-seeded gel compaction per se is insufficient to produce a viable replacement tissue. TEs possess some of the features of native tissues, but the matrix is not compositionally or structurally the same, explaining in part the fact that TE mechanical properties are insufficient to withstand in vivo loads. As a result, researchers seek to improve TE mechanical properties by controlling the ECM composition and organization. Many approaches attempt to reproduce mechanical and chemical cues that occur in vivo during development, in the hope of encouraging a functional bioartificial tissue; other strategies involve the introduction of exogenous cross-linkers to stiffen and strengthen the TE.

17.4.2 Boundary Conditions – Free Surfaces and Mechanical Constraints

When designing a TE, the combination of mold geometry and gel boundary conditions can be used to control cell and matrix alignment in the construct and improve mechanical properties. The network alignment that develops strongly influences the mechanical properties (Agoram and Barocas 2001). Boundary conditions include free surfaces and mechanical constraints. Free surfaces are gel boundaries exposed to the culture medium, leaving them free to translate in response to tension developed in the network. Mechanical constraints can be classified as either adherent or free-slip surfaces. Adherent surfaces (e.g., Velcro, porous polyethylene) oppose compaction by restricting translation normal and parallel to the surface. Most adherent surfaces are stationary, but constant loads can also be applied. Free-slip surfaces (e.g., Teflon) are not attached to the gel and provide a barrier to gel movement through the surface but not parallel to it.

Mechanical constraints function by placing local limits on network rearrangement, which combine globally with free surfaces to give TE anisotropic mechanical properties similar to native tissues. In the absence of mechanical constraints, the TE remains isotropic due to equal compaction in all directions (Fig. 17.6A). Perhaps the most basic example of using boundary conditions to control network alignment occurs in the configuration used to generate transversely isotropic TEs, such as model ligaments (Huang et al. 1993), tendons (Juncosa-Melvin et al. 2006; Nirmalanandhan et al. 2006), and mitral valve chordae (Shi and Vesely 2003). Here, a slab is cast between two opposite and adherent surfaces. The remaining four surfaces are free. As the cells compact the matrix, an anisotropic strain field develops, and the slab compacts through its width and thickness, and collagen and cell alignment develops along the long axis of the construct (Fig. 17.6C). More complex configurations that incorporate multiple constraints to produce net-like geometries (Stopak and Harris 1982) and bifurcating patterns (Shi et al. 2006) have also been explored.

Free-slip surfaces are frequently used to produce strong circumferential alignment in bioartificial arteries, similar to the alignment present in the medial layer of the native artery (L'Heureux et al. 1993; Barocas et al. 1998). In these studies, collagen is reconstituted in the annular region between a non-adherent inner mandrel and an outer tubular mold, which is removed soon after gelation occurs. The free-slip surface of the inner mandrel allows axial and radial (but not circumferential) compactions, which combine to produce circumferentially aligned cells and fibers (Fig. 17.6D). In contrast, an adherent mandrel surface prevents axial compaction, and predominantly axial alignment develops instead.

More complex TEs with spatially varying alignment can also be created with carefully designed molds and appropriate boundary conditions (Fig. 17.7). Neidert

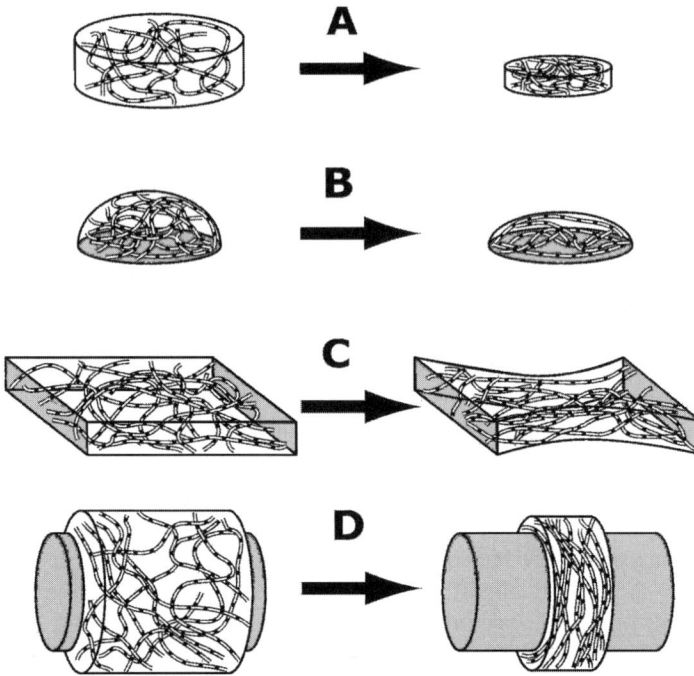

Fig. 17.6 The use of mechanical constraints on the compaction process to induce fibril alignment. The boundary conditions and initial gel geometry can be used to control the fibril alignment in a cell-compacted collagen gel. Mechanically constrained gel surfaces are shaded, free surfaces are not. Cells are not depicted for clarity. (**A**) A free-floating TE disk geometry has no mechanical constraints on it. As it compacts, each dimension shrinks, and the fibril network remains macroscopically isotropic. (**B**) A hemisphere is attached at its base to a fixed surface and compacts through its thickness to produce fibril alignment parallel to the base. (**C**) A rectangular slab whose ends are constrained to an adherent surface. The gel compacts perpendicular to its free surfaces, and the fibrils align along the axis. (**D**) A gel is cast around a nonadhesive mandrel. The gel compacts axially and radially to produce circumferentially aligned fibrils

and Tranquillo used these principles to produce heart valve equivalents that possessed circumferential alignment in the root and commissure-to-commissure alignment in the leaflets, similar to native valves (Neidert and Tranquillo 2006). In this configuration, the root of the valve has a free-slip surface that results in circumferential alignment. The leaflets develop commissural alignment due to a radial constraint that arises from attachment to the root.

Local variations in fibril alignment arise from spatial differences in cell concentration and imperfect boundary conditions. In addition, the influence of the boundary conditions decreases with distance from the boundary, so that interior regions may not be as aligned as edge regions (Costa et al. 2003; Neidert and Tranquillo 2006).

TEs produced in this manner possess improved mechanical properties, but are still weak compared to native tissues. TE microstructure is morphologically similar, primarily in gross ECM alignment, but the ECM is not mechanically coupled to the